Second Edition

Electric Power System Applications of Optimization

Second Edition

Electric Power System Applications of Optimization

James A. Momoh

CRC Press
Taylor & Francis Group
Boca Raton London New York

CRC Press is an imprint of the
Taylor & Francis Group, an **informa** business

CRC Press
Taylor & Francis Group
6000 Broken Sound Parkway NW, Suite 300
Boca Raton, FL 33487-2742

First issued in paperback 2019

© 2009 by Taylor & Francis Group, LLC
CRC Press is an imprint of Taylor & Francis Group, an Informa business

No claim to original U.S. Government works

ISBN-13: 978-1-4200-6586-2 (hbk)
ISBN-13: 978-0-367-38616-0 (pbk)

Library of Congress Cataloging-in-Publication Data

Momoh, James A., 1950-
    Electric power system applications of optimization / James A. Momoh. -- 2nd ed.
        p. cm.
    "A CRC title."
    Includes bibliographical references and index.
    ISBN 978-1-4200-6586-2 (alk. paper)
    1. Electric power systems--Mathematical models. 2. Mathematical optimization. 3. Electric power production--Mathematics. I. Title.

TK1001.M5987 2008
621.31--dc22   2008049152

Visit the Taylor & Francis Web site at
http://www.taylorandfrancis.com

and the CRC Press Web site at
http://www.crcpress.com

Contents

Preface

Electric Power System Applications of Optimization is intended to introduce optimization, system theory, foundations of different mathematical programming techniques, and application to selected areas of electrical power engineering. The idea is to present theoretical background material from a practical power system point of view and then proceed to explore applications of optimization techniques, new directions, and continuous application problems. The need for such a book stems from the extensive and diverse literature on topics of optimization methods in solving different classes of utility operations and planning problems.

Optimization concepts and algorithms were first introduced to power system dispatching, resource allocation, and planning in the mid-1960s in order to mathematically formalize decision making with regard to myriad objectives subject to technical and nontechnical constraints. There has been a phenomenal increase in research activities aimed at implementing dispatched, resource allocation problems and at planning optimally. This increase has been facilitated by several research projects (theoretical papers usually aimed at operation research communities) that promote usage of commercial programs for power system problems but do not provide any relevant information for power engineers working on the development of power system optimization algorithms. Most recently, there has been a tremendous surge in publications on research applications, especially on optimization in electric power engineering.

However, currently no book serves as a practical guide to the fundamental and application aspects of optimization for power system work. This book is intended to meet the needs of a diverse range of groups interested in optimization application. They include university faculty, researchers, students, and developers of power systems who are interested in or who plan to use optimization as a tool for planning and operation. The focus of this book is exclusively on the development of optimization methods, foundations, and algorithms and their application to power systems.

The focus was based on the following factors. First, good references that survey optimization techniques for planning and operation are currently available but they do not detail theoretical formulation in one complete environment. Second, optimization analysis has become so complex that examples that deal with nonpower system problems are only studied and many issues are covered by only a few references for the utility industry. Finally, in the last decade, new optimization technologies such as interior point methods and genetic algorithms (GAs) have been successfully introduced to deal with issues of computations and have been applied to new areas in power system planning and operation.

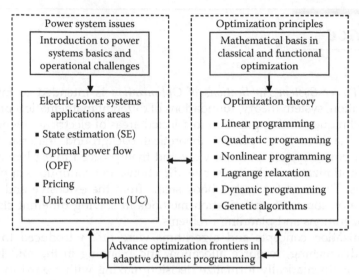

FIGURE P.1
Summary of the main topics in text.

This book provides both the analytical formulation of optimization and the various algorithmic issues that arise in the application of various methods in power system planning and operation. In addition, it also provides a blend of theoretical approach and application based on simulation. Figure P.1 shows a summary of the main areas and topics covered for the benefit of power/non-power engineers as well as other optimization experts. The readers here are exposed to the foundations of classical optimization theories, which are extended to linear and nonlinear programming, integer programming, and dynamic programming (DP).

This book then provides direct applications of these technologies from the operation research domain to electric power systems. It also provides foundation knowledge and references in power systems operation, optimization, and control as background to the new readers. State estimation (SE), optimal power flow (OPF), pricing, and unit commitment (UC) are presented as applications in this book. In addition, new advances in the field of adaptive critics design have spurred interest in research and practical applications of approximate DP. The final chapter of the book combines fundamental theories and theorems from functional optimization, optimal control, and DP to explain new adaptive or approximate dynamic programming (ADP) concepts and its variants.

A summary of the chapters is as follows: in Chapter 2, we revise electric power system models, power system component modeling, reactive capabilities, ATC, and AGC. It concludes with illustrated examples. In Chapter 3, we introduce the theoretical concepts and algorithms for power-flow computation using different numerical methods with illustrative examples and applications for practical simulation studies.

To treat the problem of optimization in a concise form, Chapter 4 deals with classical unconstrained and constrained techniques with simple applications to power systems. This chapter concludes with illustrative examples. Chapter 5 is dedicated to linear programming theory, methods, and its extension to integer programming, postoptimization (often referred to as sensitivity analysis), and its application to power systems, with illustrative examples at the end. Chapter 6 deals with new trends in optimization theory such as interior point optimization methods for both linear and quadratic formulation. It includes examples and applications to power systems. In Chapter 7, we discuss the nonlinear programming technique and its extension to recent interior point methods such as barrier methods. The computational algorithm for each of the nonlinear programming variants is also presented.

Chapter 8 presents the DP optimization algorithm with illustrative examples. In Chapter 9, the Lagrangian relaxation concept and algorithm are discussed. Their applicability to UC and resource allocation is also described. In Chapter 10, the decomposition method for solving large-scale optimization problems is presented with illustrative examples following the procedure.

In Chapter 11, for operation and computation of system variables and constraints used in OPF, power flow, and power system security assessment, accurate representation of the system data, information, and parameters is mandatory. Several techniques are used, such as SE algorithms, filtering algorithms, and predictive control. We summarize herein the SE with applications for power system data processing.

In Chapter 12, OPF modeling and selected programming techniques derived from earlier chapters are used for solving difficult objective functions with constraints in power system operation and planning. Illustrative examples are also included.

In Chapter 13, as a result of power system restructuring, market strategies and pricing have recently become important issues. They are designed for the principle of OPF and hence from natural extension of well-known economic description with constraints. We present here a summary of economic principles and price theory necessary for optimal zonal control and marginal prices. These formulations are given in a general form and are amenable for the use of several of the classical static or dynamic optimization concepts provided in earlier chapters.

Chapter 14 addresses UC concepts, formulation, and algorithms. Examples and applications to power systems dispatching are also presented here. Chapter 15 presents GAs as tools for optimization and discusses the definition of GA computation, approach, and algorithm. Application areas of GAs as a computational tool in power system operation and planning are also described.

Chapter 16 introduces a new topic that is of current interest to engineers and scientists with advantages such as overcoming several shortfalls of static optimization with a deterministic variable. Static optimization is used in most power system operations and planning. These limit the problems that are to be solved. In cases where a system model has dynamics or predictability, some level of stochastic knowledge of optimal control is needed. However, the use of DP and optimal control has been discussed in nonlinear programming books. Recent progress made to accelerate both predictive and stochastic nature is handled by adaptive dynamic critics (reinforcing learning algorithm). As we seek solutions to power system performance in real time, new advances in optimization technologies are proposed. This forms the basis of the revised Chapter 16. Here, we summarize the Euler–Lagrange/Pontryagin principle with the unified theory of Hamilton–Jacobian–Bellman (HJB) from their derivatives, leading to the basis for adaptive dynamic critics. The capability of these new advances is used in solving problems of optimization, placement of flexible AC transmission devices, UC, real-time pricing, and several others.

Notably, in order to accommodate the determination of optimal power system performances for practical, real-time systems with uncertainty and stochasticity in data and system modeling, this book introduces new advancement in adaptive dynamic critics for optimal control. Several case studies in support of the new types of advanced dynamic optimization are presented as new material in this book. They involve general optimizations for the example applications of ADP to stochastic OPF, UC, etc.

It is hoped that the application areas discussed here will offer the reader an overview of classical optimization methods without sacrificing the rudiments of the theory. Those working in the field or willing to engage in OPF will find the material useful and interesting as a reference or as a good starting point to engage in power system optimization studies.

A significant portion of the material presented here is derived from new ideas generated through sponsored projects, professional society meetings, panel sessions, and popular texts in operation research in which I have had personal involvement. These include research and development efforts, which were generally supported by funding agencies such as the Electric Power Research Institute, the National Science Foundation, and Howard University. I wish to acknowledge the significant contribution made by the engineers of Bonneville Power Authority, Commonwealth Edison, and the Department of Energy in the development and testing of OPF using variants of optimization techniques such as GAs and interior point methods.

This book would not have been possible without the help of the students in the optimization and power system group at the Center for Energy Systems and Control (CESaC) at Howard University and CESaC's research staff who provided dedicated support in OPF algorithm testing, problem solving, and the tasks of preparing this book for publication. I am indebted to my colleagues for their keen interest in the development of the first and this subsequent edition of the book. These include Professor Kenneth Fegly of the University of Pennsylvania, Professor Bruce Wollenberg of the University of Minnesota, Professor Emeritus Hua Ting Chieh of Howard University, and

Professor Mohammed El Hawary of Dalhousie University who offered valuable criticism during the preparation of this book.

Finally, I wish to thank my admirable students in the CESaC family for their help in typing and editing. Finally, I offer my deepest personal thanks to those closest to me who have provided support during the time-consuming process of writing this book.

James A. Momoh

Author

James A. Momoh is a professor at the Department of Electrical and Computer Engineering and the director of the Center for Energy Systems and Control at Howard University, Washington, D.C. At present, he develops interdisciplinary research and education programs in systems engineering, energy systems, and power economics; and new teaching pedagogy and curricula in power systems engineering at Howard University. Dr. Momoh additionally serves as principal consultant at Bonneville Power Administration, Portland, Oregon (and several other power companies and agencies) and was an affiliate staff scientist at Pacific Northwest Laboratory, Seattle.

Dr. Momoh has contributed to several book developments in his field and authored over 225 technical publications and reports in the field of power engineering. He is an associate editor of the *Journal of Power Letters* and *Journal of Electric Machines and Power Systems*. He has authored several books including *Electric Power Distribution, Automation, Protection, and Control* (Taylor & Francis Group LLC) and *Electric Power Systems Applications of Optimization* (Marcel Dekker). He has also coauthored other textbooks including *Electric Systems, Dynamics, and Stability with Artificial Intelligence Applications* (Marcel Dekker) and has written on subjects in the areas of adaptive dynamic programming applications and applied mathematics for power systems.

Dr. Momoh received his BSEE (1975) with honors from Howard University, his MSEE (1976) from Carnegie Mellon University, his MS (1980) in systems engineering from the University of Pennsylvania, and his PhD (1983) in electrical engineering from Howard University. In addition, he received his MA (1991) in theology from the School of Divinity at Howard University. Professor Momoh was the chair of the Electrical Engineering Department at Howard University (1990–2000). In 1987, he received a National Science Foundation Presidential Young Investigator Award. He was the program director of the power program within the Electrical and Communications Systems Division at the National Science Foundation from 2001 to 2004.

Dr. Momoh is a fellow of the Institute of Electronics and Electrical Engineering and a distinguished fellow of the Nigerian Society of Engineers. He was inducted as a fellow member of the Nigerian Academy of Engineering in 2004. He holds membership in numerous other awards in professional and honor societies.

1

Introduction

1.1 Structure of a Generic Electric Power System

Although no two electric power systems are alike, all share some common fundamental characteristics including the following:

1. Electric power is generated using synchronous machines that are driven by turbines (steam, hydraulic, diesel, or internal combustion).
2. Generated power is transmitted from the generating sites over long distances to load centers that are spread over wide areas.
3. Three-phase AC systems comprise the main means of generation, transmission, and distribution of electric power.
4. Voltage and frequency levels are required to remain within tight tolerance levels to ensure a high-quality product.

The basic elements of a generic electric power system [1,3] are displayed in Figure 1.1. Electric power is produced at generating stations (GS) and transmitted to consumers through an intricate network of apparatus including transmission lines, transformers, and switching devices.

A transmission network is classified as

1. Transmission system
2. Subtransmission system
3. Distribution system

The transmission system interconnects all major GS and main load centers in the system. It forms the backbone of the integrated power system and operates at the highest voltage levels (typically, 230 kV and above). The generator voltages are usually in the range of 11–35 kV. These are stepped up to the transmission voltage level, and power is transmitted to transmission substations where the voltages are stepped down to the subtransmission

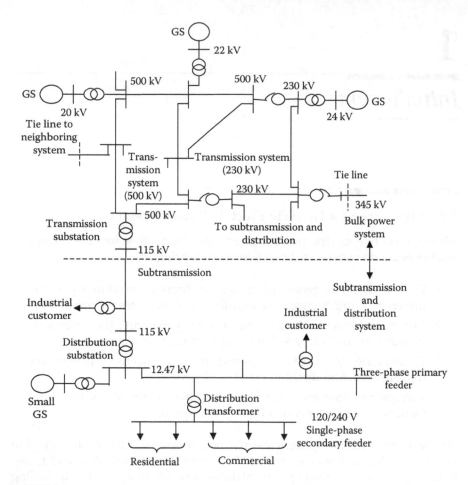

FIGURE 1.1
Basic elements of a power system.

level (typically, 69–138 kV). The generation and transmission subsystems are often referred to as the bulk power system.

The subtransmission system transmits power at a lower voltage and in smaller quantities from the transmission substation to the distribution substations. Large industrial customers are commonly supplied directly from the subtransmission system. In some systems, as expansion and higher voltage levels become necessary for transmission, the older transmission lines are often relegated to subtransmission function.

The distribution system is the final stage in the transfer of power to the individual customers. The primary distribution voltage is typically between 4.0 and 34.5 kV. Primary feeders at this voltage level supply small industrial customers. The secondary distribution feeders supply residential and commercial customers at 120–240 V.

1.2 Power System Models

In order to be able to control the power system from the point of view of security, one is required to know two basic things: the mathematical models of the system and the variety of control functions and associated objectives that are used. In this section some general remarks about power system models are given.

In Figure 1.2, we show the basic decomposition of the system into a set of generators that deliver electrical power to the distributed load by means of a network. In our subsequent discussion we start by describing the load, then the network, and finally the generators.

In assessing load behavior as seen from a substation, one is interested typically in items such as the following:

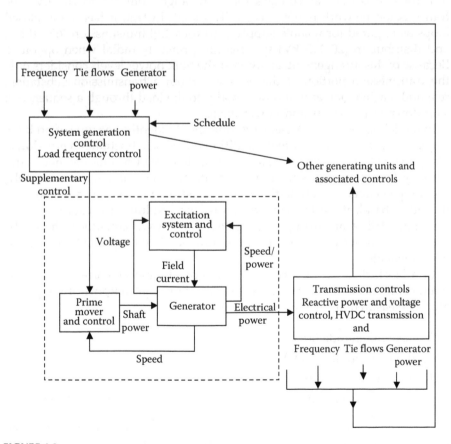

FIGURE 1.2
Subsystems of a power system and associated controls.

- Present value of real power consumed in megawatt (MW) and the associated power factor (or reactive power)
- Forecast values of real and reactive power over a range of future times—next few minutes, to days and years
- Load response characteristics (e.g., lumped circuit representation or transfer function) for fluctuations in substation voltage and frequency

By knowing the real and reactive components of substation loads (present or forecast), one can establish a complete picture for a steady-state bulk power demand in the system. Furthermore, by identifying load response characteristics one can adequately evaluate the dynamic behavior of the demand in the presence of disturbances and feedback controllers.

The network portion of the system consists of the transmission, subtransmission, and distribution networks. This division is based on voltage levels, and, consequently, on the ratings of various circuits. Typically, power transmission is done at voltages that can range from 115 to 765 kV. The transmission network is not necessarily radial. In fact, it has many closed loops as required for reliable supply purposes. Subtransmission (65–40 kV) and distribution (20–115 kV) systems are primarily radial when operated. Because of this arrangement, analysis at the bulk power levels considers only the transmission portion of the network. From a transmission substation, real and reactive power will flow radially to the load through a sequence of step-down transformers and power lines.

In modeling the network elements, one should identify the type of problem being analyzed. Under normal conditions, the load fluctuates very slowly compared to transient time constants associated with transmission lines. And since system frequency is maintained at its nominal value quite accurately, the lumped circuit representation of transmission lines is quite adequate. On the other hand, if electromagnetic transients resulting, for example, from lightning strikes are being investigated, then wave equations should be considered. In our present context, the lumped circuit representation can be very adequate.

In a block diagram of a typical generator, the blocks correspond to the main components of the power plant. The significant outputs of the generator as measured at the terminals are its

Voltage magnitude (kV)

Real power produced (MW)

Reactive power produced (MVAr)

Speed (in radians and denoted by $w = 2\pi f$, where f is the frequency)

Under normal conditions, three of those quantities are continuously controlled by the power plant. These are the terminal voltage, the frequency (speed), and real power output. The output voltage V is subtracted from the specified voltage V_0 and the difference is an error signal that drives the exciter system. The exciter, in turn, modifies the field voltage of the turbo-alternator in such a way that V becomes closer in value to V_0. The same feedback concept applies to the control of frequency and real power. In this case, however, the corresponding error signal drives the governor system. The governor controls valve openings of the turbine, which in turn control its speed and mechanical power output. The power error signal can also go back to the prime mover (boiler, in the case of steam generation), so that more, or less, steam is generated.

The exciter normally has a fast response (10^{-2}–10^{-1} s). The governor-turbine system is slower (0.1–1 s) in its response. However, since the load is much slower in its changes than the response times, it is safe to assume that perfect control is always present; that is, normal operation is, to a high degree, sinusoidal steady-state operation. Only when events that are fast relative to governor-turbine or exciter response times, one would worry about the steady-state operation. Thus, in the presence of network faults, or immediately following switching operations, one needs to consider transient or dynamic representation of the system.

1.3 Power System Control

Before discussing how the system is controlled, one needs to briefly summarize the means by which control action is obtained.

First, let us understand the meaning of control as it applies to the power system case. The system is normally designed so that certain quantities can be manipulated by means of devices. Some of these are the so-called status quantities. By means of power circuit breakers, a transmission line is open (status = OFF) or closed (status = ON). Some of them are integer variables (tap-settings on power transformers). And the rest are continuous variables such as the real power output of a generator. The control devices can be simple, like fuses, or highly complex dynamic systems, like exciters and governors.

Control action is attained by the manipulation of all control devices that exist on the system. This is achieved in order to meet different, but consistent objectives, and through a variety of means. The general objectives of system control are listed in order as follows:

1. Protection of major pieces of equipment and of system integrity

2. Continuity of high-quality service

3. System secure operation

4. System economic and environmentally acceptable operation

5. Emergency state control

6. Restorative control in minimum time

As a rule, control action is based on information derived from direct measurements and inferred data. Each control device will require certain kinds of information based on the following considerations:

Speed-of-response requirements

Impact of control action (i.e., global vs. local)

Relative importance of different pieces of information (e.g., local vs. distant information)

Some examples of this are now in order. For a short-circuit fault on a transmission line, the main objective is to protect the system from going unstable (losing synchronism) and to protect the line from serious damage. This is achieved by correct breaker action, which will open that line and isolate it from the system. Normally, however, other neighboring lines and transformers feel the effect of the short circuit. Hence, it is important to open the faulted line first. By means of offline short-circuit analysis, relay settings are established so that the faulted line will open first. Hence, the only needed online information for that purpose is line current. This is strictly local information.

In a more complicated situation, we can look at the problem of maintaining a satisfactory voltage profile in the system. Scheduled generator terminal voltage is attained by means of local feedback control to the exciter. The values of scheduled voltages, which are the set points in the feedback loop, are established from an analysis of the entire system's operating conditions. In most cases, offline analysis of the system yields values of scheduled voltages. Modern, energy control centers (ECC) have the capability of processing global online information and updating voltage profile set points [5].

The function of an electric power system is to convert energy from one of the naturally available forms to the electrical form and to transport it to the points of consumption. Energy is seldom consumed in the electrical form but is rather converted to other forms such as heat, light, and mechanical energy. The advantage of the electrical form of energy is that it can be transported and controlled with relative ease and with a high degree of efficiency and reliability. A properly designed and operated power system should, therefore, meet the following fundamental requirements:

1. It must be able to meet the continually changing load demand for active and reactive power. Unlike other types of energy, electricity cannot be conveniently stored in sufficient quantities. Therefore, adequate spinning reserve of active and reactive power should be maintained and appropriately controlled at all times.

2. It should supply energy at minimum cost and with minimum ecological impact.

3. The "quality" of the power supply must meet certain minimum standards with regard to the factors:

   a. Constancy of frequency

   b. Constancy of voltage

   c. Level of reliability

Several levels of controls involving a complex array of devices are used to meet the above requirements. These are depicted in Figure 1.2, which identifies the various subsystems of a power system and the associated controls. In this overall structure, there are controllers operating directly on individual system elements. In a generating unit these consist of prime mover controls and excitation controls. The prime mover controls are concerned with speed regulation and control of energy supply system variables such as boiler pressures, temperatures, and flows. The function of the excitation control is to regulate generator voltage and reactive power output. The desired MW outputs of the individual generating units are determined by the system generation control.

The primary purpose of the system generation control is to balance the total system generation against system load and losses so that the desired frequency and power interchange with neighboring systems (tie flows) is maintained.

The transmission controls include power and voltage control devices, such as static VAr compensators, synchronous condensers, switched capacitors and reactors, tap-changing transformers, phase-shifting transformers, and high voltage direct current (HVDC) transmission controls.

These controls described above contribute to the satisfactory operation of the power system by maintaining system voltages and frequency and other system variables within their acceptable limits. They also have a profound effect on the dynamic performance of the power system and its ability to cope with disturbances.

The control objectives are dependent on the operating state of the power system. Under normal conditions, the control objective is to operate as efficiently as possible with voltages and frequency close to nominal values. When an abnormal condition develops, new objectives must be met to restore the system to normal operation.

Major system failures are rarely the result of a single catastrophic disturbance causing collapse of an apparently secure system. Such failures are usually brought about by a combination of circumstances that stress the network beyond its capability. Severe natural disturbances (such as a tornado, severe storm, or freezing rain), equipment malfunction, human error, and inadequate design combine to weaken the power system and eventually lead to its breakdown. This may result in cascading outages that must be contained within a small part of the system if a major blackout is to be prevented.

Protecting isolated systems has been a relatively simple task, which is carried out using overcurrent directional relays with selectivity being obtained by time grading. High-speed relays have been developed to meet the increased short-circuit currents due to the larger size units and the complex interconnections.

For reliable service, an electric power system must remain intact and be capable of withstanding a wide variety of disturbances. It is essential that the system be operated so that the more probable contingencies can be sustained without loss of load (except that connected to the faulted element) and so that the most adverse possible contingencies do not result in widespread and cascading power interruptions.

The November 1965 blackout in the northeastern part of the United States and Ontario had a profound impact on the electric utility industry. Many questions were raised and led to the formation of the National Electric Reliability Council in 1968. The name was later changed to the North American Electric Reliability Council (NERC). Its purpose is to augment the reliability and adequacy of bulk power supply in the electricity systems of North America. The NERC is composed of nine regional reliability councils and encompasses virtually all the power systems in the United States and Canada. Each regional council has established reliability criteria for system design and operation. Since differences exist in geography, load pattern, and power sources, criteria for the various regions differ to some extent.

Design and operating criteria play an essential role in preventing major system disturbances following severe contingencies. The use of criteria ensures that, for all frequently occurring contingencies, the system will, at worst, transit from the normal state to the alert state, rather than to a more severe state such as the emergency state or the in extremis state. When the alert state is entered following a contingency, operators can take actions to return the system to the normal state.

1.4 Power System Security Assessment

Power system security is the ability of the bulk power electric power system to withstand sudden disturbances such as electric short circuits or unanticipated loss of system components [5]. In terms of the requirements for the proper planning and operation of the power system, it means that following the occurrence of a sudden disturbance, the power system will

1. Survive the ensuing transient and move into an acceptable steady-state condition.
2. In this new steady-state condition, all power system components operate within established limits.

Electric utilities require security analysis to ensure that, for a defined set of contingencies, the above requirements are met. The analysis required to survive a transient is complex, because of increased system size, greater dependence on controls, and more interconnections. Additional complicating factors include the operation of the interconnected system with greater interdependence among its member systems, heavier transmission loading, and concentration of the generation among few large units at light loads.

After the 1965 blackout, various efforts went into improving reliable system operation. Several reliability criteria and emergency guidelines were introduced by the Federal Power Commission (FPC) and North American Power System Interconnection Committee (NAPSIC). Summaries of these guidelines are given in Ref. [2, Appendix]. These guidelines and criteria represent efforts by the government and the utilities to improve control and operational practices.

More important, however, were the efforts by various researchers and specialists in what has come to be known as the secure control of the power system. In DyLiacco's pioneering work [6], the power system is judged to reside at any instant of time in any of three operating states: normal, emergency, and restorative.

Under normal steady-state operating conditions all customer demands are met and all equipments operate below its rated capacity. Theoretically speaking, the requirement of meeting customer demands is expressed mathematically by means of a set of equations (equality constraints) of the type

$$h_1(x_1, \ldots, x_n; u_1, \ldots, u_m) = 0$$
$$h_2(x_1, \ldots, x_n; u_1, \ldots, u_m) = 0$$
$$\vdots$$
$$h_n(x_1, \ldots, x_n; u_1, \ldots, u_m) = 0,$$

(1.1)

where

$x_1, \ldots, x_n$ is a set of dependent (state) variables

$u_1, \ldots, u_m$ is a set of independent (input, demand, or control) variables

Typically these equality constraints correspond to the so-called load-flow equations. The constraints relative to equipment can be written, in general, in the following form:

$$g_1(x_1, \ldots, x_n; u_1, \ldots, u_m) \leq 0$$
$$g_2(x_1, \ldots, x_n; u_1, \ldots, u_m) \leq 0$$
$$\vdots$$
$$g_l(x_1, \ldots, x_n; u_1, \ldots, u_m) \leq 0.$$

(1.2)

They correspond to items such as upper and lower limits on power generation by a given unit, current limits on transmission lines and transformers, and so on.

Mathematically, the normal operating state is defined whenever the utility system considered satisfies Equations 1.1 and 1.2.

Following certain disturbance events (short circuits due to faults, loss of generation, loss of load, and others) some of the inequality constraints may be violated. For example, a line may become overloaded, or system frequency may drop below a certain limit. In these cases the system is in the emergency operating state.

Finally, the system may exist in a situation where only a fraction of the customers are satisfied without overloading any equipment. In this case only a portion of the system is in the normal state. As a result, not all the equality constraints are satisfied, but the inequality constraints are. Such a state is called the restorative operating state.

Symbolically, we can rewrite Equations 1.1 and 1.2 in the following form.

$$\left.\begin{array}{l} h(x,u) = 0 \\ g(x,u) \leq 0 \end{array}\right\}. \tag{1.3}$$

With this notation, we summarize our definition of the three operating states as follows:

Normal state

$$h(x,u) = 0$$

$$g(x,u) \leq 0.$$

Emergency state (need to change $g(x,u)$ to not less than or equal to)

$$h(x,u) = 0$$

$$g(x,u) \leq 0.$$

Restorative state

$$h(x,u) \neq 0$$

$$g(x,u) \leq 0.$$

The security of the system is defined relative to its ability to withstand a number of postulated disturbances. A normal state is said to be secure if, following any one of the postulated disturbances, the system remains in the normal state. Otherwise, it is insecure.

In the online operation of the system, one monitors the different variables that correspond to its operating conditions. This monitoring process is called security monitoring. The process of determining whether the system is in the secure normal state is called security assessment.

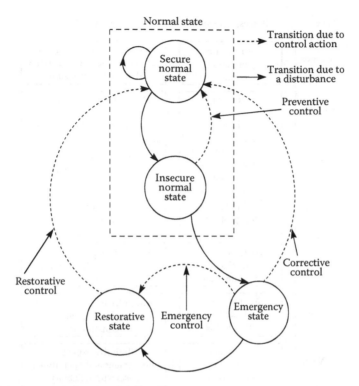

FIGURE 1.3
Operating states of a power system.

In the process of security assessment it may be concluded that the system is in the insecure normal state. In that case, the system operator will attempt to manipulate system variables so that the system is secure. This action is called preventive control. If, on the other hand, the system is in an emergency state, then two types of control action are possible. In the first type, called corrective control, action is possible whereby the system is sent back to the normal state. If corrective control is not possible, then emergency control is applied [6]. This control can be due to relay-initiated action, automatic control, or operator control. In any case, the system will drift to the restorative state as a result of emergency control. Finally, in the restorative state, control action is initiated to restore all services by means of restorative control. This should put the system back to the normal state. Figure 1.3 illustrates the various transitions due to disturbances as well as various control actions.

1.5 Power System Optimization as a Function of Time

The hourly commitment of units, the decision whether a unit is on or off at a given hour, is referred to as unit commitment. Hourly production

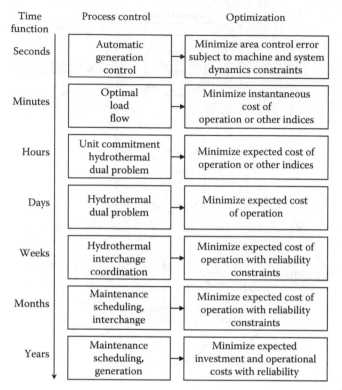

Time function	Process control	Optimization
Seconds	Automatic generation control	Minimize area control error subject to machine and system dynamics constraints
Minutes	Optimal load flow	Minimize instantaneous cost of operation or other indices
Hours	Unit commitment hydrothermal dual problem	Minimize expected cost of operation or other indices
Days	Hydrothermal dual problem	Minimize expected cost of operation
Weeks	Hydrothermal interchange coordination	Minimize expected cost of operation with reliability constraints
Months	Maintenance scheduling, interchange	Minimize expected cost of operation with reliability constraints
Years	Maintenance scheduling, generation	Minimize expected investment and operational costs with reliability

FIGURE 1.4
Time horizon of the power system optimization problem.

of hydroelectric plants based on the flexibility of being able to manage water reserve levels to improve system performance is referred to as the hydrothermal problem and hourly production of coal generation or a dual purpose plant is called the dual purpose problem. Scheduling of unit maintenance without violating reserve capacity while minimizing the production cost is referred to as a maintenance scheduling problem. The interdependence among the various control optimization problems as the time horizon expands from seconds to years is shown in Figure 1.4.

In power system operation and planning, there are many optimization problems that require real-time solutions such that one can determine the optimal resources required at minimum cost within a given set of constraints. This scheduling is done over time (minutes, hours, days, etc.). In this regard, we classify the problem as either operational or planning. Notably, in the operations scheduling problem, we usually extend the studies up to 24 h. On the other hand, planning problems are solved in the time frame of years.

In analyzing the optimization problem, there are many controllable parameters of interests. There are many objective functions and constraints that must be satisfied for economic operation. (These objectives and constraints are quantified later.) Methods exist for solving the resulting economic dispatch problem as a function of time when we incorporate the constraints of the system and typically the economic dispatch problem evolves. It uses mathematical techniques such as linear programming (LP), unconstrained optimization techniques (using Lagrange multipliers), and nonlinear programming (NLP) to accommodate the constraints [4]. The availability of these techniques in addressing this problem has been noted. Other variations on the economic dispatch problem are hydrothermal and unit commitment problems.

Dynamic programming (DP), Lagrange relaxation technique, and Bender's decomposition algorithm are used to solve this class of optimization problem. Another method in power system operation and control is the optimal maintenance of units and generators.

Finally, in the same realm is the optimal power flow (OPF), which holds the promise of extending economic dispatch to include the optimal setting of under load tap-changers, generator real and reactive powers, phase-shifter taps, and the like. OPF has been expanded as new problems arise to include new objective functions and constraints. And OPF has attracted researchers to the development of new optimization algorithms and tests as a routine base. Other applications extending the work to optimization of the network include VAr planning, network expansion, and availability transfer capability.

At the distribution end, loss minimization, data estimation, and network reconfiguration have demanded optimum decision making as a planning problem as well as an operations problem. There are mathematical optimization techniques ranging from LP to evolutionary search techniques that can be employed to obtain optimum distribution networks.

There is a need to summarize the essential mathematical methods that have been fully developed, tested, and utilized on a routine basis for security analysis of the power system. The selection of the appropriate optimization technique depends on the system as defined by the objective functions and the constraints. The constraints are divided into two classes, namely, technical and nontechnical. The class of technical constraints includes network, equipment, and device constraints. The class of nontechnical constraints includes social, environmental, and economic limitations.

1.6 Review of Optimization Techniques Applicable to Power Systems

In the early days of power system operation, the OPF tool was defined in terms of the conventional economic dispatch problem aimed at determining the optimal settings for control variables in a power system with respect to various constraints. However, the capability of power system optimization tools has been broadened to provide solutions for a wide range of utility-dependent problems. Today, the OPF tool is used to solve a static constrained nonlinear optimization problem whose development has followed closely the advances in numerical optimization techniques and computer technology. Commonly available OPF packages can solve very large and complex power system formulations in a relatively short time.

Generally, OPF requires solving a set of nonlinear equations, describing optimal and secure operation of a power expressed as

> *Minimize* $\quad F(\mathbf{x},\mathbf{u})$
> while satisfying $\quad \mathbf{g}(\mathbf{x},\mathbf{u}) = 0$
> $\quad\quad\quad\quad\quad \mathbf{h}(\mathbf{x},\mathbf{u}) \leq 0,$

where
   $\mathbf{g}(\mathbf{x},\mathbf{u})$ is the set of nonlinear equality constraints (power flow equations)
   $\mathbf{h}(\mathbf{x},\mathbf{u})$ is the set of inequality constraints of vector arguments $\mathbf{x}$ and $\mathbf{u}$
   $\mathbf{x}$ is the vector of dependent variables consisting of bus voltage magnitudes and phase angles, as well as MVAr loads, fixed bus voltages, line parameters, and so on
   $\mathbf{u}$ is the vector of control variables

Vector $\mathbf{u}$ includes the following:

   Real and reactive power generation
   Phase-shifter angles
   Net interchange
   Load MW and MVAr (load shedding)
   DC transmission line flows
   Control voltage settings
   Load tap changer (LTC) transformer tap settings

Common objectives in a power system include

   Active power cost minimization
   Active power loss minimization
   Minimum control shift
   Minimum number of controls scheduled

Examples of the associated equality and inequality constraints are

   Limits on all control variables
   Power flow equations

Generation/load balance

Branch flow limits

Bus voltage limits

Active and reactive reserve limits

Generator MVAr limits

Corridor (transmission interface) limits

The optimization methods that are incorporated in the OPF tools can be classified based on optimization techniques such as

LP based methods

NLP based methods

Integer programming (IP) based methods

Separable programming (SP) methods

Notably, LP is recognized as a reliable and robust technique for solving a wide range of specialized optimization problems characterized by linear objectives and linear constraints. Many commercially available power system optimization packages contain powerful LP algorithms for solving power system problems for both planning and operator engineers. LP has extensions in the simplex method, revised simplex method, and interior point techniques.

Interior point techniques are based on the Karmarkar algorithm and encompass variants such as the projection scaling method, dual affine method, primal affine method, and barrier algorithm.

In the case of the NLP optimization methods, the following techniques are introduced:

Sequential quadratic programming (SEQ)

Augmented Lagrangian methods

Generalized reduced gradient method

Projected augmented Lagrangian

Successive LP

Interior point methods

The basic formulation is then extended to include security and environmental constraints, which have become very important factors in power system operation in the past few decades. Special decomposition strategies are also applied in solving large-scale system problems. These include Benders decomposition, Lagrangian relaxation, and Talukdar–Giras optimization techniques.

In recent years, the advancement of computer engineering and the increased complexity of the power system optimization problem have led to greater need and application of specialized programming techniques for large-scale problems. These include DP, Lagrange multiplier methods, and evolutionary computation methods such as genetic algorithms. These techniques are often hybridized with many other techniques of intelligent systems, including artificial neural networks, expert systems, Tabu-search algorithms, and fuzzy logic.

References

1. Bergen, A. R. and Vittal, V., *Power Systems Analysis*, 2nd edn., Prentice-Hall, Upper Saddle River, NJ, 1999.
2. Elgerd, O. I., *Electrical Energy Systems Theory—An Introduction*, McGraw-Hill, New York, 1982.
3. Gonen, T., *Electric Power Distribution System Engineering*, McGraw-Hill, New York, 1986.
4. Luenberger, D. G., *Introduction to Linear and Nonlinear Programming*, Addison-Wesley, Reading, MA, 1975.
5. Wood, A. J. and Wollenberg, B. F., Ed., *Power Generation, Operation, and Control*, Wiley, New York, 1996.
6. DyLiaco, T. E., *An Overview of Power System Control Centres*, IEEE Tutorial Course on Energy Control Center Design, 1977.

2

Electric Power System Models

2.1 Introduction

The power industry in the United States has been engaged in a changing business environment for quite some time now, moving away from a centrally planned system to one in which players operate in a decentralized fashion with little knowledge of the full state of the network, and where decision making is likely to be market driven rather than based on technical considerations.

The new environment differs markedly from the one in which the system previously has been operated. This leads to the requirement of some new techniques and analysis methods for system operation, operational and long-term planning, and the like.

Electrical power systems vary in size, topography, and structural components. However, the overall system can be divided into three subsystems: generation, transmission, and distribution. System behavior is affected by the characteristics of every major element of the system. The representation of these elements by means of appropriate mathematical models is critical to the successful analysis of system behavior. For each different problem, the system is modeled in a different way. Here, power system under steady constraints has been extended to include load conditions, thermal limits, stability constraints, congestion limits, and other constraints in pricing mechanism, which are introduced in numerous real-life applications discussed in subsequent chapters of this book.

This chapter describes some system models for analysis purposes and introduces concepts of power expressed as active, reactive, and apparent, followed by a brief review of three-phase systems and the per unit system representations (Sections 2.3 and 2.4). Section 2.5 discusses modeling the synchronous machine from an electric network standpoint. Reactive capability curves are examined in Section 2.6 followed by discussion of static and dynamic load models later in Section 2.11. We now introduce some fundamental concepts and background knowledge required in understanding power systems analysis.

2.2   Complex Power Concepts

The electrical power systems specialist is, in many instances, more concerned with electrical power in the circuit rather than current. As the power into an element is basically the product of the voltage across and current through it, it seems reasonable to swap the current for power without losing any information.

In treating sinusoidal steady-state behavior of an electric circuit, some further definitions are necessary. To illustrate, a cosine representation of the sinusoidal waveforms involved is used.

Consider an impedance element $\overline{Z} = Z/\phi$ connected to a sinusoidal voltage source $v(t)$ that is given by $v(t) = V_m \cos \omega t$. Figure 2.1 shows the typical load circuit. The instantaneous current in the circuit shown in Figure 2.1 is

$$i(t) = I_m \cos(\omega t - \phi),$$

where the current magnitude is

$$I_m = V_m/|Z|.$$

The instantaneous power is given by

$$p(t) = v(t)i(t) = V_m I_m [\cos(\omega t) \cos(\omega t - \phi)].$$

Using the trigonometric identity

$$\cos \alpha \cos \beta = \frac{1}{2} [\cos(\alpha - \beta) + \cos(\alpha + \beta)],$$

we can write the instantaneous power as

$$p(t) = \frac{V_m I_m}{2} [\cos \phi + \cos(2\omega t - \phi)].$$

The average power p_{av} is seen to be

$$p_{av} = \frac{V_m I_m}{2} \cos \phi. \qquad (2.1)$$

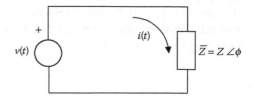

FIGURE 2.1
Load circuit.

Since through one cycle, the average of $\cos(2\omega t - \phi)$ is zero, this term therefore contributes nothing to the average of p.

It is more convenient to use the effective (or root mean square [rms]) values of voltage and current than the maximum values. Substituting $V_m = \sqrt{2}(V_{rms})$, and $I_m = \sqrt{2}(I_{rms})$, we get

$$p_{av} = V_{rms}I_{rms}\cos\phi. \tag{2.2}$$

Thus, the average power entering any network is the product of the effective values of terminal voltage and current and the cosine of the phase angle, which is called the power factor (PF). This applies to sinusoidal voltages and currents only. For a purely resistive load $\cos\phi = 1$, the current in the circuit is fully engaged in conveying power from the source to the load resistance. When reactive (inductive or capacitive) as well as resistive elements are present in the network, a component of the current in the circuit is engaged in conveying energy that is periodically stored in and discharged from the reactance. This stored energy, being shuttled to and from the magnetic field of an inductance or the electrostatic field of a capacitance, adds to the magnitude of the current in the circuit but does not add to the average power.

The average power in a circuit is called active power and the power that supplies the stored energy in reactive elements is informally called reactive power. Active power is denoted P, and the reactive power is designated Q. They are expressed as

$$P = VI\cos\phi \tag{2.3}$$

$$Q = VI\sin\phi, \tag{2.4}$$

where
   V and I are the rms values of terminal voltage and current
   ϕ is the phase angle by which the current lags the voltage
   P and Q are of the same dimension (J/s)

We define a quantity called the complex or apparent power, designated S, of which P and Q are orthogonal components. By definition

$$S = P + jQ = \overline{V}I^*$$
$$= VI\cos\phi + jVI\sin\phi$$
$$= VI(\cos\phi + j\sin\phi).$$

Using Euler's identity, we thus have

$$S = VIe^{j\phi}$$
$$= VI\angle\phi.$$

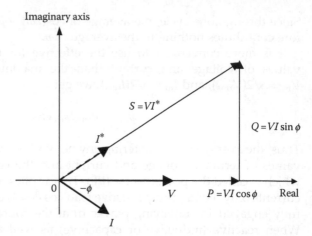

FIGURE 2.2
Phasor diagram for complex
power relationships.

If we introduce the conjugate current defined by the symbol asterisk (*),

$$I^* = |I| \angle \phi,$$

it becomes obvious that an equivalent definition of complex or apparent power is

$$S = VI^*. \tag{2.5}$$

We can write the complex power in two alternative forms by using the relationships $\overline{V} = Z\overline{I}$ and $\overline{I} = Y\overline{V}$.

Multiplying the phasors by V, we obtain the complex power diagram in Figure 2.2. Inspection of the diagram as well as the previous development leads to a relation for the PF of the circuit:

$$\cos \phi = \frac{P}{|S|}.$$

2.3 Three-Phase Systems

The major portion of all the electric power presently used is generated, transmitted, and distributed using balanced three-phase voltage systems. The single-phase voltage sources referred to in Section 2.2 originate in many instances as part of a three-phase system [1]. Three-phase operation is preferable to single-phase because a three-phase winding makes more efficient use of generator copper and iron. Power flow in single-phase circuits was shown in the previous section to be pulsating. This drawback is not

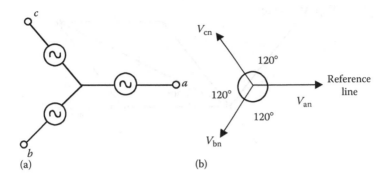

FIGURE 2.3
(a) Y-connected three-phase system and (b) corresponding phasor diagram.

present in a three-phase system as shown later. Also, three-phase motors start more easily and, having constant torque, run more satisfactorily than single-phase motors. However, the complications of additional phases are not compensated by the slight increase of operating efficiency when systems higher than three-phase systems are used.

A balanced three-phase voltage system consists of three single-phase voltages having the same magnitude and frequency but time-displaced from one another by 120°. Figure 2.3a shows a schematic representation where the single-phase voltage sources appear in a Y connection; a Δ configuration is also possible as discussed later. A phasor diagram showing each of the phase voltages is given in Figure 2.3b. As the phasors revolve at the angular frequency ω with respect to the reference line in the counterclockwise (designated as positive) direction, the positive maximum value first occurs for phase a and then in succession for phases b and c. Stated in a different way, to an observer in the phasor space, the voltage of phase a arrives first followed by that of b and then that of c. For this reason the three-phase voltage of Figure 2.3 is said to have the phase sequence abc (order, phase sequence, and rotation all mean the same thing). This is important for certain applications. For example, in three-phase induction motors, the phase sequence determines whether the motor rotates clockwise or counterclockwise.

2.3.1 Y-Connected Systems

With reference to Figure 2.4, the common terminal n is called the neutral or star (Y) point. The voltages appearing between any two of the line terminals a, b, and c have different relationships in magnitude and phase to the voltages appearing between any one line terminal and the neutral point n. The set of voltages V_{ab}, V_{bc}, and V_{ca} are the line voltages, and the set of voltages V_{an}, V_{bn}, and V_{cn} are referred to as the phase voltages. Consideration of the phasor diagrams provides the required relationships [1,3].

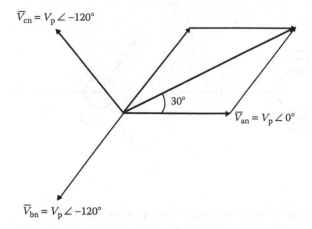

FIGURE 2.4
Phase and magnitude relations between the phase and line voltage of a Y-connection.

The effective values of the phase voltages are shown in Figure 2.5 as V_{an}, V_{bn}, and V_{cn}. Each has the same magnitude, and each is displaced 120° from the other two phasors. To obtain the magnitude and phase angle of the line voltage from a to b (i.e., V_{ab}), we apply Kirchhoff's voltage law:

$$V_{ab} = V_{an} + V_{nb}. \tag{2.6}$$

Equation 2.6 states that the voltage existing from a to b is equal to the voltage from a to n (i.e., V_{an}) plus the voltage from n to b. Thus, Equation 2.6 can be rewritten as

$$V_{ab} = V_{an} - V_{bn}. \tag{2.7}$$

Since for a balanced system, each phase voltage has the same magnitude, let us set

$$|V_{an}| = |V_{bn}| = |V_{cn}| = V_p, \tag{2.8}$$

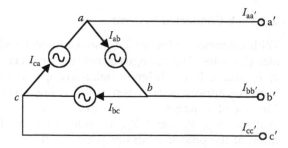

FIGURE 2.5
Relation between phase and line currents in a Δ connection.

where V_p denotes the effective magnitude of the phase voltage. Accordingly, we may write

$$V_{an} = V_p \angle 0° \tag{2.9}$$

$$V_{bn} = V_p \angle -120° \tag{2.10}$$

$$V_{cn} = V_p \angle -240° = V_p \angle 120°. \tag{2.11}$$

Substituting Equations 2.9 and 2.10 in Equation 2.7 yields

$$V_{ab} = V_p(1 - 1\angle -120°)$$
$$= \sqrt{3}V_p \angle 30°. \tag{2.12}$$

Similarly we obtain

$$V_{bc} = \sqrt{3}V_p \angle -90° \tag{2.13}$$

$$V_{ca} = \sqrt{3}V_p \angle 150°. \tag{2.14}$$

The expressions obtained above for the line voltages, V_L, show that they constitute a balanced three-phase voltage system whose magnitudes are $\sqrt{3}$ times as those of the phase voltages. Thus, we write

$$V_L = \sqrt{3}V_p. \tag{2.15}$$

A current flowing out of a line terminal a (or b or c) is the same as that flowing through the phase source voltage appearing between terminals n and a (or n and b or n and c). We can thus conclude that for a Y-connected three-phase source, the line current equals the phase current. Thus

$$I_L = I_p, \tag{2.16}$$

where
I_L denotes the effective value of the line current
I_p denotes the effective value for the phase current

2.3.2 Delta-Connected Systems

We now consider the case when the three single-phase sources are rearranged to form a three-phase Δ connection as shown in Figure 2.6. It is clear from inspection of the circuit shown that the line and phase voltages have the same magnitude:

$$|V_L| = |V_p|. \tag{2.17}$$

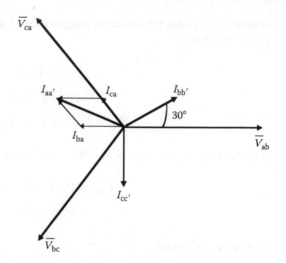

FIGURE 2.6
A Δ-connected three-phase source.

The phase and line currents, however, are not identical, and the relationship between them can be obtained using Kirchhoff's current law at one of the line terminals.

In a manner similar to that adopted for the Y-connected source, let us consider the phasor diagram shown in Figure 2.6. Assume the phase currents to be

$$I_{ab} = I_p\angle 0$$
$$I_{bc} = I_p\angle -120°$$
$$I_{ca} = I_p\angle 120°.$$

The current that flows in the line joining a to a' is denoted $I_{aa'}$ and is given by

$$I_{aa'} = I_{ca} - I_{ab}.$$

As a result, we have

$$I_{aa'} = I_p[1\angle 120° - 1\angle 0],$$

which simplifies to

$$I_{aa'} = \sqrt{3}I_p\angle 150°.$$

Similarly

$$I_{bb'} = \sqrt{3}I_p\angle 30°$$
$$I_{cc'} = \sqrt{3}I_p\angle -90°.$$

2.3.3 Power Relationships

Note that a set of balanced three-phase currents yields a corresponding set of balanced line currents whose magnitudes are $\sqrt{3}$ times the magnitudes of the phase values:

$$I_L = \sqrt{3}I_p,$$

where I_L denotes the magnitude of any of the three line currents.

Assume that a three-phase generator is supplying a balanced load with the three sinusoidal phase voltages:

$$v_a(t) = \sqrt{2}V_p \sin \omega t$$
$$v_b(t) = \sqrt{2}V_p \sin(\omega t - 120°)$$
$$v_c(t) = \sqrt{2}V_p \sin(\omega t + 120°),$$

with the currents given by

$$i_a(t) = \sqrt{2}I_p \sin(\omega t - \phi)$$
$$i_b(t) = \sqrt{2}I_p \sin(\omega t - 120° - \phi)$$
$$i_c(t) = \sqrt{2}I_p \sin(\omega t + 120° - \phi),$$

where ϕ is the phase angle between the current and voltage in each phase. The total power in the load is

$$p_{3\phi}(t) = v_a(t)i_a(t) + v_b(t)i_b(t) + v_c(t)i_c(t).$$

This can be expanded as

$$p_{3\phi}(t) = 2V_pI_p[\sin(\omega t)\sin(\omega t - \phi)$$
$$+ \sin(\omega t - 120°)\sin(\omega t - 120° - \phi)$$
$$+ \sin(\omega t + 120°)\sin(\omega t + 120° - \phi)].$$

Using a trigonometric identity, we get

$$p_{3\phi}(t) = V_pI_p\{3\cos\phi - [\cos(2\omega t - \phi) + \cos(2\omega t - 240° - \phi) + \cos(2\omega t + 240° - \phi)]\}.$$

Note that the last three terms in the above equation add up to zero. Thus, we obtain

$$p_{3\phi}(t) = 3V_pI_p \cos \phi.$$

When referring to the voltage level of a three-phase system, by convention, one invariably understands the line voltages. From the above discussion, the relationship between the line and phase voltages in a Y-connected system is

$$|V_L| = \sqrt{3}|V|.$$

The power equation thus reads in terms of line quantities:

$$p_{3\phi} = \sqrt{3}|V_L||I_L|\cos\phi.$$

We note that the total instantaneous power is constant, having a magnitude of three times the real power per phase.

We may be tempted to assume that the reactive power is of no importance in a three-phase system since the Q terms cancel out. However, this situation is analogous to the summation of balanced three-phase currents and voltages that also cancel out. Although the sum cancels out, these quantities are still very much in evidence within each phase.

We extend the concept of complex or apparent power (S) to three-phase systems by defining

$$S_{3\phi} = 3V_p I_p^*,$$

where the active and reactive powers are obtained from

$$S_{3\phi} = P_{3\phi} + jQ_{3\phi}$$

as

$$P_{3\phi} = 3|V_p||I_p|\cos\phi$$
$$Q_{3\phi} = 3|V_p||I_p|\sin\phi.$$

In terms of line values, we can assert that

$$S_{3\phi} = \sqrt{3}V_L I_L^*$$

and

$$P_{3\phi} = \sqrt{3}|V_L||I_L|\cos\phi$$
$$Q_{3\phi} = \sqrt{3}|V_L||I_L|\sin\phi.$$

2.4   Per Unit Representation

In power system analysis, it is usually convenient to use a per unit system to normalize system variables. The per unit system offers computational simplicity by eliminating units and expressing system quantities as a dimensionless ratio. Thus

$$\overline{X}_{\text{(in per unit)}} = \frac{X_{\text{(actual quantity)}}}{X_{\text{(base value of equation)}}}.$$

A well-chosen per unit system can minimize computational effort, simplify evaluation, and facilitate understanding of system characteristics. Some base quantities may be chosen independently and quite arbitrarily, while others follow automatically depending on fundamental relationships between system variables. Normally, the base values are chosen so that the principal variables will be equal to one per unit under rated operating conditions.

In the case of a synchronous machine, the per unit system may be used to remove arbitrary constants and simplify mathematical equations so that they may be expressed in terms of equivalent circuits. The basis for selecting the per unit system for the stator is straightforward, but it requires careful consideration for the rotor. The L_{ad}-base reciprocal per unit system is discussed here.

The following base quantities for the stator are chosen (denoted by subscripts).

$e_{s \text{ base}}$, peak value of rated line-to-line voltage (V)
$i_{s \text{ base}}$, peak value of rated line-to-line current (A)
f_{base}, rated frequency (Hz)

The base value of each of the remaining quantities is automatically set and depends on the above as follows:

$$\omega_{base} = 2\pi f_{base} \text{ (electrical rad/s)}$$

$$\omega_{m \text{ base}} = \omega_{base} \frac{2}{\pi f} \text{ (mechanical rad/s)}$$

$$Z_{s \text{ base}} = \frac{e_{s \text{ base}}}{i_{s \text{ base}}} \text{ (}\Omega\text{)}$$

$$L_{s \text{ base}} = \frac{e_{s \text{ base}}}{\omega_{base}} \text{ (H)}$$

$$\psi_{s \text{ base}} = L_{s \text{ base}} \times i_{s \text{ base}} = \frac{e_{s \text{ base}}}{\omega_{m \text{ base}}} \text{ (Weber turns)}$$

$$VA_{3\phi,base} = 3E_{\text{rms base}} I_{\text{rms base}}$$

$$= \frac{3}{2} e_{s \text{ base}} \times i_{s \text{ base}} \text{ (VA)}.$$

2.5 Synchronous Machine Modeling

In power system stability analysis, there are several types of models used for representing the dynamic behavior of the synchronous machine. These models are deduced by using some approximations to the basic machine equations. Section 2.5 begins with a brief introduction to machine equations [2,5].

The stator voltage equations expressed in per unit notations are given by

$$\bar{e}_d = \frac{d\bar{\psi}_d}{d\bar{t}} - \bar{\psi}_q \bar{\omega} - \bar{R}_a \bar{i}_d$$

$$\bar{e}_q = \frac{d\bar{\psi}_q}{d\bar{t}} + \bar{\psi}_d \bar{\omega} - \bar{R}_a \bar{i}_q \qquad (2.18)$$

$$\bar{e}_0 = \frac{d\bar{\psi}_0}{d\bar{t}} - \bar{R}_a \bar{i}_0.$$

The corresponding flux linkage equations may be written as

$$\bar{\psi}_d = -\bar{L}_d \bar{i}_d + \bar{L}_{afd} \bar{i}_{fd} + \bar{L}_{akd} \bar{i}_{kd}$$

$$\bar{\psi}_q = -\bar{L}_q \bar{i}_q + \bar{L}_{akq} \bar{i}_q \qquad (2.19)$$

$$\bar{\psi}_0 = -\bar{L}_0 \bar{i}_0.$$

The rotor circuit base quantities should be chosen so as to make the flux linkage equations simple by satisfying the following:

1. Per unit mutual inductances between different windings are to be reciprocal. This will allow the synchronous machine model to be represented by simple equivalent circuits.
2. All per unit mutual inductances between stator and rotor circuits in each axis are to be equal.

The following base quantities for the rotor are chosen, in view of the L_{ad}-base per unit system choice.

$$\bar{e}_d = \frac{d\bar{\psi}_d}{d\bar{t}} - \bar{\psi}_q \bar{\omega} - \bar{R}_a \bar{i}_d$$

$$\bar{e}_q = \frac{d\bar{\psi}_q}{d\bar{t}} + \bar{\psi}_d \bar{\omega} - \bar{R}_a \bar{i}_q \qquad (2.20)$$

$$\bar{e}_0 = \frac{d\bar{\psi}_0}{d\bar{t}} - \bar{R}_a \bar{i}_0.$$

Per unit rotor flux linkage equations are given by

$$\bar{\psi}_d = -\bar{L}_d \bar{i}_d + \bar{L}_{afd} \bar{i}_{fd} + \bar{L}_{akd} \bar{i}_{kd}$$
$$\bar{\psi}_q = -\bar{L}_q \bar{i}_q + \bar{L}_{akq} \bar{i}_q \qquad\qquad (2.21)$$
$$\bar{\psi}_0 = -\bar{L}_0 \bar{i}_0.$$

Since all quantities in Equations 2.18 through 2.21 are in per unit, the super bar notation is dropped in subsequent discussions.

If the frequency of the stator quantities is equal to the base frequency, the per unit reactance of a winding is numerically equal to the per unit inductance. For example:

$$X_d = 2\pi f L_d \ (\Omega).$$

Dividing by $Z_{s\ base} = 2\pi f_{base} L_{s\ base}$, if $f = f_{base}$, the per unit values of X_d and L_d are equal.

2.5.1 Classical Representation of the Synchronous Machine

The per unit equations completely describe the electrical and dynamic performance of a synchronous machine. However, except for the analysis of very small systems, these equations cannot be used directly for system stability studies. Some simplifications and approximations are required for the representation of synchronous machines in stability studies. For stability analysis of large systems, it is necessary to neglect the transformer voltage terms $p\psi_d$ and $p\psi_q$ and the effect of speed variations. Therefore, the machine equation [3] described by Equations 2.20 and 2.21 becomes

$$\begin{cases} e_d = -\psi_q - R_a i_a \\ e_q = \psi_d - R_a i_a \\ e_{fd} = p\psi_{fd} + R_{fd} i_{fd}. \end{cases} \qquad (2.22)$$

For studies in which the period of analysis is small in comparison to T'_{d0} the machine model is often simplified by assuming E'_q constant throughout the study period. This assumption eliminates the only differential equation associated with the electrical characteristics of the machine. A further approximation, which simplifies the machine representation significantly, is to ignore transient saliency by assuming that $x'_d = x'_q$ and to assume that the flux linkage also remains constant. With these assumptions, the voltage behind the transient impedance $R_a + jx'_d$ has a constant magnitude. The equivalent circuit is shown in Figure 2.7. The machine terminal voltage phasor is represented by

$$\tilde{V}_t = E' \angle \delta - (R_a + jx'_d)\tilde{I}_t.$$

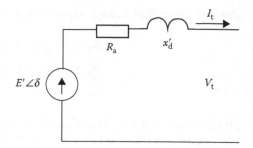

FIGURE 2.7
Equivalent circuit of machine.

2.6 Reactive Capability Limits

Synchronous generators are usually rated in terms of the maximum kVA load at a specific voltage and PF (often 80%, 85%, or 90% lagging) that they can carry continuously without overheating. The active power output of the generator is usually limited to a value within the kVA rating by the capability [4] of its prime mover. By virtue of its voltage regulating system, the machine normally operates at a constant voltage whose value is within +5% of rated voltage. When the active-power loading and voltage are fixed, either armature or field heating limits the allowable reactive-power loading. A typical set of reactive-power capability curves for a large turbine generator are shown in Figure 2.8. They give the maximum reactive-power loadings corresponding

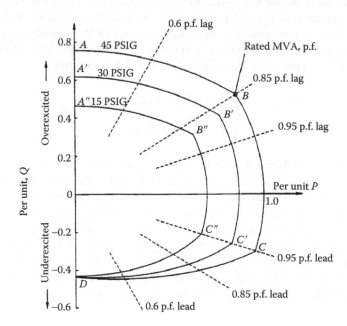

FIGURE 2.8
Reactive capability curves of a hydrogen-cooled generator at rated voltage.

to various power loadings with operation at rated voltage. Armature heating is the limiting factor in the region from unity to the rated PF (0.85 in Figure 2.8). For lower PFs, field heating is limiting. Such a set of curves forms a valuable guide in planning and operating the system of which the generator is a part.

Also shown in Figure 2.8 is the effect of increased hydrogen pressure on allowable machine loadings. The PF at which a synchronous motor operates, and hence its armature current, can be controlled by adjusting its field excitation. The curve showing the relation between armature current and field current at a constant terminal voltage and with a constant shaft load is known as a V-curve because of its characteristic shape.

2.7   Prime Movers and Governing Systems

The prime sources of electrical energy supplied by utilities are the kinetic energy of water and the thermal energy derived from fossil fuels and nuclear fission. The prime movers convert these sources of energy into mechanical energy that, in turn, is converted to electrical form by the synchronous generator. The prime mover governing system provides a means of controlling power and frequency. The functional relationships among the basic elements associated with power generation and control are shown in Figure 2.9. This section introduces the models for hydraulic and steam turbines and their respective governing systems.

2.7.1   Hydraulic Turbines and Governing Models

The hydraulic turbine model describes the characteristics of gate opening *m* and output mechanical power. In power system dynamic analysis, the

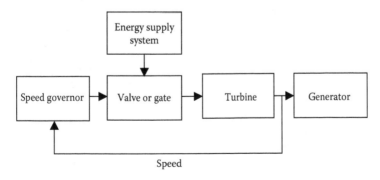

FIGURE 2.9
Power generation and control.

hydraulic turbine is usually modeled by an ideal lossless turbine along with consideration of the "water hammer" effect caused by the water inertia, given by

$$\frac{P_m}{\mu} = \frac{1 - T_w s}{1 + T_w s},\tag{2.23}$$

where T_w is water-starting time.

Because of the water hammer effect, a change in gate position produces an initial turbine power change that is opposite to that which is sought. For stable control performance, a large transient (temporary) droop with a long resetting time is required. This is accomplished by introducing a transient gain-reduction compensation in the governing system. The compensation retards or limits the gate movement until the water flow and power output have time to catch up. The governing system model is shown in Figure 2.10.

The governing system model is given by

$$\frac{dX_1}{dt} = \eta$$

$$T_G \frac{d\mu}{dt} = -\mu + X_1$$

$$T_R \frac{dX_2}{dt} = -X_2 + R_T T_R \eta \tag{2.24}$$

$$T_P \frac{d\eta}{dt} = -\eta + K_s(\omega_{\text{ref}} - \omega_r - R_p X_1 - X_2),$$

where

T_p is the pilot valve and servomotor time constant
K_s is the servo gain
T_a is the main servo time constant

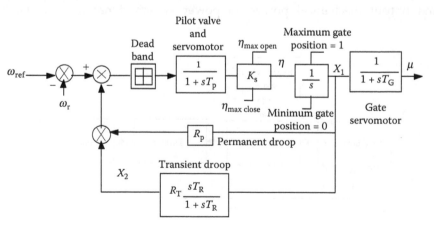

FIGURE 2.10
Governing system model.

R_p is the permanent droop
R_T is the temporary droop
T_R is the reset time
$\eta_{max\ open}$ is the maximum gate opening rate
$\eta_{max\ close}$ is the maximum gate closing rate
μ is the gate position

2.7.2 Steam Turbines and Governing System Models [2,5]

A steam turbine converts stored energy of high pressure and high temperature steam into rotating energy. The input of the steam turbine is the control valve position (ΔV_{cv}), while its output is the torque (ΔT_m). In power stability analysis, a first-order model is used for a steam turbine; that is

$$\frac{\Delta T_m}{\Delta V_{cv}} = \frac{1}{1 + sT_{CH}}, \tag{2.25}$$

where T_{CH} is the time constant.

Comparing the turbine models for hydraulic and steam turbines, it is clear that the response of a steam turbine has no peculiarity such as that exhibited by a hydraulic turbine due to water inertia. The governing requirements of steam turbines, in this respect, are more straightforward. There is no need for transient droop compensation.

The governing system model is given by

$$T_G \frac{d\Delta V_{cv}}{dt} = -\Delta V_{cv} + X_1$$
$$\frac{dX_1}{dt} = K_s(\omega_{ref} - \omega_r - R_p X_1). \tag{2.26}$$

A typical governing model for a steam turbine is shown in Figure 2.11.

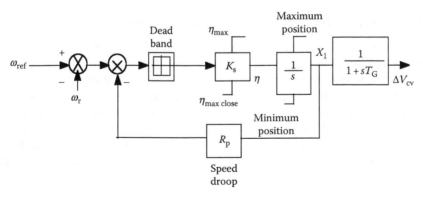

FIGURE 2.11
Governing model for steam turbine.

2.8 Automatic Gain Control

Within an energy management system, the generation scheduling and control function control the electrical power output of generators so as to supply the continuously changing customer power demand in an economical manner. This function is largely provided by an automatic generation control (AGC) program operating within the control center computer. The power system dispatcher also plays an important role, interacting with the program to incorporate current operating conditions. The basic objectives of power system operation during normal operating conditions associated with AGC are

1. Matching total system generation to total system load
2. Regulating system electrical frequency error to zero
3. Distributing system generation among control areas so that net area tie flows match net area tie flow schedules
4. Distributing area generation among area generation sources so that area operating costs are minimized

The first objective is conventionally associated with the terms primary or governor speed control; turbine speed governors respond proportionally to local frequency deviations and normally bring the frequency rate of change to zero within a time frame of several seconds. The latter three objectives are accomplished by supplementary controls directed from area control centers. The second and third AGC objectives are classically associated with the regulation function, or load-frequency control, while the fourth is associated with the economic dispatch function of AGC. The regulation and economic dispatch functions typically operate in time frames of several seconds and several minutes, respectively.

2.8.1 Power Control in a Multigenerator Environment

Consider the case of a number of generating units connected by a transmission system, with local loads at each generator bus as shown in Figure 2.12. We are interested in active power flows and assume that we can use the active power model with all voltages set equal to their nominal values. The objective of the analysis is to obtain an incremental model relating power commands, electrical and mechanical power outputs, and power angles for the different generators. We note that the analysis may be easily extended to include pure load buses in the model; however, here we assume that every bus is a generator bus.

At each bus, $P_{gi} = P_{Di} + P_i$, where P_i is the total injected active power from the ith bus into the transmission system. Assume an operating point

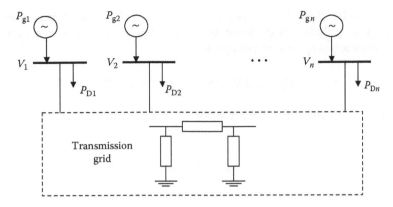

FIGURE 2.12
Simplified network model for an *n*-generator power system.

given by $P_{gi}^o = P_{Di}^o + P_i^o$ and then $P_{gi}^o = P_{gi}^o + \Delta P_{gi}$, $P_{Di}^o = P_{Di}^o + \Delta P_{Di}$, and $P_i^o = P_i^o + \Delta P_i$; we get the relationship between increments:

$$\Delta P_{gi} = \Delta P_{Di}^o + \Delta P_i.$$

For each generator we use the linearized version of the swing equation. Thus, we consider

$$M_i \Delta \ddot{\delta}_i + D_i \Delta \dot{\delta}_i + \Delta P_{Gi} = \Delta P_{Mi}, \qquad (2.27)$$

where δ_i is the angle of the internal voltage of the *i*th generator. We number the generators so that the *i*th generator is connected to the *i*th bus.

We next consider the behavior of P_{Di}. In general, P_{Di} depends on both the voltage magnitude and frequency. In our active power model we assume that the voltage magnitudes are fixed at their nominal values, and we therefore need not consider the voltage magnitude dependence. Concerning the frequency dependence, we assume that P_{Di} increases with frequency. This assumption is in accordance with observed behavior. We would also like to be able to introduce changes in load as external (disturbance) inputs. Thus, we can imagine the sudden increments in load due to the switching of an electrical device that draws a specified power. In accordance with this description, for small changes in frequency, we assume that

$$P_{Deei} = P_{Di}^o + \frac{\partial P_{Di}(\omega^o)}{\partial \omega_i} \Delta \omega_i + \Delta P_{Li}, \qquad (2.28)$$

where the first two terms describe the linearization of the frequency dependence of the load and ΔP_{Li} is the change of load input, by switching, referred to previously. Note that the operating-point frequency is ω^o, which is expected to be close to but not necessarily equal to the nominal system frequency $\omega_o = 2\pi 60$. In defining the frequency at each bus, we note that the instantaneous (phase a) voltage is

$$v_i(t) = \sqrt{2}|V_i|\cos\left[\omega^o t + \theta_i^o + \Delta\theta_i(t)\right]. \tag{2.29}$$

Thus, by taking the time derivative of the argument of the cosine function, we get

$$\omega_i = \omega^o + \Delta\dot{\theta}_i \tag{2.30}$$

and thus

$$\Delta\omega_i = \omega_i - \omega^o = \Delta\dot{\theta}_i. \tag{2.31}$$

Defining $D_{Li} = \partial P_{Di}(\omega_o)/\partial\omega_i$, and using Equation 2.31, we can replace Equation 2.28 with

$$\Delta P_{Di} = D_{Li}\Delta\dot{\theta}_i + \Delta P_{Li}. \tag{2.32}$$

Also, we can get

$$M_i\Delta\ddot{\delta}_i + D_i\Delta\dot{\delta}_i + D_{Li}\Delta\dot{\theta}_i + \Delta P_{Li} + \Delta P_i = \Delta P_{Mi}. \tag{2.33}$$

We next calculate ΔP_i, assuming that the line admittance parameters are purely imaginary,

$$P_i = \sum_{k=1}^{n}|V_i||V_k|B_{ik}\sin(\theta_i - \theta_j), \tag{2.34}$$

where θ_i is the phase angle of the bus voltage V_i. Assuming that voltage magnitudes $|V_i|$ are constants, then the terms $|V_i||V_k|B_{ik}$ are constants by linearizing Equation 2.34 around the operating point. For variety, instead of using the Taylor series, we can proceed as follows, using trigonometric identities.

$$P_i = P_i^o + \Delta P_i = \sum_{k=1}^{n}|V_i||V_j|B_{ik}\sin\left(\theta_i^o + \Delta\theta_i - \theta_k^o - \Delta\theta_k\right)$$

$$= \sum_{k=1}^{n}|V_i||V_j|B_{ik}\sin\left(\theta_i^o - \theta_k^o\right)\cos(\Delta\theta_i - \Delta\theta_k)$$

$$+ \cos\left(\theta_i^o - \theta_k^o\right)\sin(\Delta\theta_i - \Delta\theta_k). \tag{2.35}$$

This result is exact. As we let the increments go to zero we get

$$\Delta P_i = \sum_{k=1}^{n} |V_i||V_j|B_{ik} \cos\left(\theta_i^o - \theta_k^o\right)(\Delta\theta_i - \Delta\theta_k), \qquad (2.36)$$

which is the same as the linearization result found by using the Taylor series. Defining $T_{ik} = |V_i||V_k|B_{ik} \cos\left(\theta_i^o - \theta_k^o\right)$, we get

$$\Delta P_i = \sum_{k=1}^{n} T_{ik}(\Delta\theta_i - \Delta\theta_k). \qquad (2.37)$$

The constants T_{ik} are called stiffness or synchronizing power coefficients. The larger the T_{ik}, the greater the exchange of power for a given change in bus phase angle. We are now in a position to substitute Equation 2.37 into Equation 2.33. First, however, we make a further approximation. We assume coherency between the internal and the terminal voltage phase angles of each generator so that these angles tend to "swing together." Stated differently, we assume that the increments $\Delta\delta_i$ and $\Delta\theta_i$ are equal. The assumption that $\Delta\delta_i = \Delta\theta_i$ significantly simplifies the analysis and gives results that are qualitatively correct.

Making the assumption, and substituting Equation 2.37 into Equation 2.33, we get for i 2,..., n,

$$M_i\Delta\ddot{\delta}_i + \tilde{D}_i\Delta\dot{\delta}_i + \Delta P_i = \Delta P_{Mi} - \Delta P_{Li}, \qquad (2.38)$$

where
$\tilde{D}_i = D_i + D_{Li}$
$\Delta P_i = \sum_{k=1}^{n} T_{ik}(\Delta\delta_i - \Delta\delta_k)$

From the way D_{Li} adds to D_i, we expect the (positive) load dependence on frequency to contribute to system damping.

The preceding relations are shown in the form of block diagram in Figure 11.10. The reader is invited to check that with $K_{Pi} = 1/\tilde{D}_i$ and $T_{Pi} = M_i/\tilde{D}_i$, Figure 11.10 represents Equation 2.38 in form of block diagram. In Figure 2.13

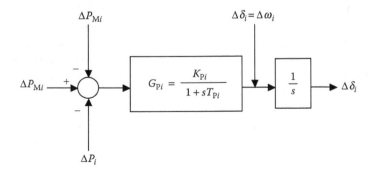

FIGURE 2.13
Generator block diagram.

we have $\Delta\omega_i$ as an output and ΔP_{Mi} as an input and can close the power control loop by introducing the turbine governor block diagram.

2.8.2 AGC System Models

2.8.2.1 Case A: Two Generating Units

In this case, we have two synchronizing coefficients to consider, T_{12} and T_{21}.

$$T_{12} = |V_1||V_2|B_{12}\cos\left(\theta_1^o - \theta_2^o\right)$$
$$T_{21} = |V_1||V_2|B_{21}\cos\left(\theta_2^o - \theta_1^o\right).$$

Since $T_{12} = T_{21}$, the block diagram for the two units may be combined and we can describe the operation of the system in qualitative terms as follows. Starting in the steady-state $\Delta\omega_1 = \Delta\omega_2 = 0$, suppose that additional load is switched onto bus 2 (i.e., ΔP_{L2} is positive). Initially, the mechanical power input to generator 2 does not change; the power imbalance is supplied from the rotating kinetic energy. Thus ω_2 begins to drop (i.e., $\Delta\omega_2$ goes negative). As a consequence, $\Delta\delta_2$ decreases and the phase angle $\delta_1 - \delta_2$ across the connecting transmission line increases. Thus the transferred power P_{12} increases. The sending-end power P_{12} and the receiving-end power $-P_{12}$ both increase. Unit 1 sees an increased power demand and ω_1 begins to drop. Each governor senses a frequency (speed) drop and acts to increase the mechanical power output from its turbine. In the steady-state there is a new (lower) system frequency and an increased P_{12}. The frequency may then be restored and P_{12} adjusted, if desired, by operator action (i.e., by adjustment of P_{c1} and P_{c2}).

We next consider an example in which we calculate the steady-state change frequency by automatic (governor) control and see the benefit of the interconnection in quantitative terms. We consider the more general case when both P_{L1} and P_{L2} change.

1. *Controlled two-area systems*: The response curves indicate that, as in the single-area case, we must add integral control to our system. Let us state, first, the minimum requirements the system should meet.

2. *Suggested control-system specifications*: We require that our system meet the four-point specifications that we stipulated for the single-area system. In addition, we require that the steady-state tie line power variation, following a step load change, must be zero. This requirement guarantees that each area, in steady state, absorbs its own load, the guiding principle in pool operation.

3. *Tie line bias control strategy*: Since we must now use a strategy that will cause both the frequency and the tie line deviations to vanish, we adopt integral control, as in the single-area case, but with the tie line deviation added to our area control error; that is, we attempt

$$ACE_1 = \Delta P_{tie1} + B_1 \Delta f_1$$
$$ACE_2 = \Delta P_{tie2} + B_2 \Delta f_2. \qquad (2.39)$$

The speed changer commands are thus of the form

$$\Delta P_{c1} \underline{\Delta} - K_{I1} \int (\Delta P_{tie1} + B_1 \Delta f_1)\, dt$$

$$\Delta P_{c2} \underline{\Delta} - K_{I2} \int (\Delta P_{tie2} + B_2 \Delta f_2)\, dt. \qquad (2.40)$$

The constants K_{I1} and K_{I2} are integrator gains, and the constants B_1 and B_2 are the frequency bias parameters. The minus signs must be included since each area should increase its generation if either its frequency error Δf_i or its tie line power increment $\Delta P_{tie.i}$ is negative.

4. *Static system response*: The chosen strategy will eliminate the steady-state frequency and tie line deviations for this reason: following a step load change in either area, a new static equilibrium, if it exists, can be achieved only after the speed changer commands ΔP_{c1} and ΔP_{c2} have reached constant values. But this evidently requires that both integrands in Equation 2.40 be zero; that is

$$\Delta P_{tie1,stat} + B_1 \Delta f_{stat} = 0$$
$$\Delta P_{tie2,stat} + B_2 \Delta f_{stat} = 0. \qquad (2.41)$$

In view of Equation 2.41, these conditions can be met only if

$$\Delta f_{stat} = \Delta P_{tie1,stat} = \Delta P_{tie2,stat} = 0.$$

Note that this result is independent of the B_1 and B_2 values. In fact, one of the bias parameters (but not both) can be zero, and we still have a guarantee that the previous equation is satisfied.

5. *Prime mover response*: At this point, we can construct a block diagram of a governor-prime mover obtaining mass/load model. Suppose that this governor experiences a step increase in load

$$\Delta P_L(s) = \frac{\Delta P_L}{s}. \qquad (2.42)$$

The transfer function relating the load change ΔP_L to the frequency change $\Delta \omega$ is

$$\Delta \omega(s) = \Delta P_L(s) \left[\frac{(-1/Ms + D)}{1 + (1/R)(1/1 + sT_G)(1/1 + sT_{CH})(1/Ms + D)} \right]. \qquad (2.43)$$

The steady-state value of $\Delta\omega(s)$ may be found by

$$\Delta\omega_{\text{steady state}} = \lim_{s\to 0}[s\Delta\omega(s)] = \frac{-\Delta P_L(1/D)}{1+(1/R)(1/D)} = \frac{-\Delta P_L}{(1/R)+D}. \qquad (2.44)$$

Note that if D were zero, the change in speed would simply be $\Delta\omega = -R\Delta P_L$. If several generators (each having its own governor and prime mover) were connected to the system, the frequency change would be

$$\Delta\omega = \frac{-\Delta P_L}{(1/R_1)+(1/R_2)+\cdots+(1/R_n)+D}. \qquad (2.45)$$

2.9 Transmission Subsystems [3,6]

The transmission network subsystem involves complexities in busbar inter-ties, transformers, and in the transmission links themselves.

1. Busbars are often interconnected by circuit breakers in such a manner to allow switching flexibility (i.e., single or groups of circuits may be outaged without undesirable outaging of key buses or other circuits). A bus tie circuit breaker presents somewhat of a mathematical diffi-culty in that opening of the breaker in the computer study requires "outaging" of a zero impedance tie. The problem is overcome by considering adjacent buses tied by a bus tie-breaker as being tied by a very low impedance "line" (e.g., $j0.00001$ per unit impedance).

2. Transformer-magnetizing branch is occasionally required in EHV studies at off-peak hours when the magnetizing current may be a considerable fraction of the total transformer current. If the trans-former is located at bus i, inclusion of the magnetizing reactance as a lumped ground tie at bus i is usually adequate.

3. Models for tap-changing transformers and phase-shifters should also include tap and phase-shift limits. These limits may not be symmet-rical about the nominal setting. If a tap-changing under load (TCUL) voltage regulating the transformer or phase-shifter hits a limit, that limit should be retained regardless of the results of calculating $\Delta\phi$ and Δt.

4. Transmission line itself is usually modeled as a lumped series imped-ance and two lumped-shunt capacitive susceptances. The latter occur on each line terminal and represent line-charging capacitance. If the

total line-charging susceptance is jB, a tie of $jB/2$ siemens per unit occurs at each line terminal. The value of B depends on the line configuration and in the case of cables is usually obtained from a computer program.

5. For very long transmission circuits, occasionally the long-line model is used. At 60 Hz, this factor is significant only for the longest lines (e.g., 625 km, which is approximately 1/8 the wavelength). In some, higher frequency signals are involved, and the corresponding shorter 1/8 wavelength point is approximately $37{,}500/f$ km (where f is in hertz). The long-line equations for voltage $V(x)$ and current $I(x)$ at a point x measured from bus 1 are

$$V(x) = V_1 \cosh \gamma x - I_1 Z_c \sinh \gamma x \qquad (2.46)$$

$$I(x) = -I_1 \cosh \gamma x + \frac{V_1}{Z_c} \sinh \gamma x, \qquad (2.47)$$

where
   subscript 1 denotes bus 1
   γ and Z_c are the propagation constant and characteristic impedance of the line, respectively

The sign convention in Equations 2.46 and 2.47 is such that $+I_1$ flows in the direction of $+x$. Let bus 2 be the other line terminal located at $x = l$, and let the voltage at bus 2 be V_2 and the current be $+I_2$,

$$V_2 = V_1 \cosh \gamma l - I_1 Z_c \sinh \gamma \ell$$
$$I_2 = -I_1 \cosh \gamma l - \frac{V_1}{Z_c} \sinh \gamma \ell. \qquad (2.48)$$

Solving for the currents in terms of voltages,

$$\begin{bmatrix} I_1 \\ I_2 \end{bmatrix} = \begin{bmatrix} \dfrac{1}{Z_c \tanh \gamma l} & \dfrac{-1}{Z_c \sinh \gamma l} \\ \dfrac{-1}{Z_c \sinh \gamma l} & \dfrac{1}{Z_c \tanh \gamma l} \end{bmatrix} \begin{bmatrix} V_1 \\ V_2 \end{bmatrix} \qquad (2.49)$$

is obtained. Equation 2.48 involves a bus admittance matrix, which can be realized by the two-port *pi* section shown in Figure 2.14. The bus 1 to ground tie is the row sum 1, the bus 2 ground tie is identical, and the off-diagonal entry in Equation 2.48 is the negative of the admittance between buses 1 and 2.

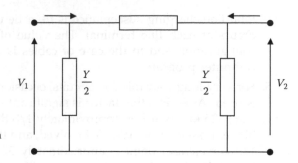

FIGURE 2.14
Simplified network model for an
n-generator power system.

2.10 Y-Bus Incorporating the Transformer Effect

The following types of transforms are discussed:

1. Fixed tap transformer
2. Tap-changing under load (TCUL) transformer
3. Phase-shifting transformer

2.10.1 Fixed Tap-Setting Transformer

A transformer with a fixed-tap setting is represented by an impedance or admittance Y_{pq} in series with an ideal autotransformer. For such a transformer at bus p in the line $L_{p\text{-}q}$, we have from Figure 2.12 the transformer ratio:

$$\frac{V_p}{V_t} = \frac{i_{tq}}{I_p} = a. \tag{2.50}$$

Hence

$$I_p = \frac{i_{tq}}{a} = (V_t - V_q)\frac{y_{pq}}{a}. \tag{2.51}$$

But from Equation 2.49 we have

$$V_t = \frac{V_p}{a}. \tag{2.52}$$

Substituting Equation 2.52 into Equation 2.51, we have

$$I_p = (V_p - aV_q)\frac{y_{pq}}{a^2} \tag{2.53}$$

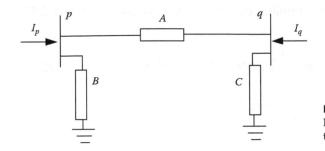

FIGURE 2.15
Equivalent π-circuit of fixed tap transformer.

and also

$$I_q = (V_q - V_t)y_{pq}$$
$$= (aV_q - V_p)\frac{y_{pq}}{a}. \tag{2.54}$$

Such a transformer at bus p in the line p–q can be represented by the equivalent π-circuit shown in Figure 2.15.

The corresponding currents of the equivalent circuit (Figure 2.15) are

$$I_p = (V_p - V_q)A + V_pB \tag{2.55}$$
$$I_q = (V_q - V_p)A + V_qC. \tag{2.56}$$

Now equating the terminal currents I_p and I_q of Figure 2.15, and letting $V_p = 0$, $V_q = 1$, we have from Equations 2.53 and 2.55,

$$I_p = -\frac{y_{pq}}{a} = -A;$$

that is

$$A = \frac{y_{pq}}{a}. \tag{2.57}$$

Similarly, we get

$$I_q = y_{pq} = A + C.$$

Hence

$$C = y_{pq} - A = y_{pq} - \frac{y_{pq}}{a}$$
$$= \left(1 - \frac{1}{a}\right)y_{pq}. \tag{2.58}$$

Similarly for B, we equate terminal current I_p for Equations 2.53 and 2.55. Thus, we get

$$(V_p - aV_q)\frac{Y_{pq}}{a^2} = (V_p - V_q)A + V_pB. \tag{2.59}$$

Substituting for $A = Y_{pq}/a$, and letting $V_p = 1$, $V_q = 0$, we get

$$B = \frac{Y_{pq}}{a^2} - \frac{y_{pq}}{a} = \left[\frac{1}{a^2} - \frac{1}{a}\right]y_{pq}$$

$$= \frac{1}{a}\left[\frac{1}{a} - 1\right]y_{pq}. \tag{2.60}$$

With the substitutions of A, B, and C from Equations 2.57, 2.58, and 2.60, Figure 2.15 takes the form of Figure 2.16.

Thus to represent this transformer in the elements of a Y_{BUS} matrix:

$$Y_{pp} = y_{p1} + \cdots + \frac{y_{pq}}{a} + \cdots + y_{pn} + \frac{1}{a}\left[\frac{1}{a} - 1\right]y_{pq}$$

$$= y_{p1} + \cdots + \frac{y_{pq}}{a^2} + \cdots + y_{pn}; \tag{2.61}$$

that is, this element is changed. Moreover,

$$Y_{pq} = Y_{qp} = -\frac{y_{pq}}{a} \tag{2.62}$$

and also,

$$Y_{qp} = y_{q1} + \cdots + \frac{y_{pq}}{a} + \cdots + y_{qn}, \cdots \left[1 - \frac{1}{a}\right]y_{pq}$$

$$= y_{q1} + \cdots + y_{qp} + \cdots + y_{qn}; \tag{2.63}$$

that is, there is no change in this element.

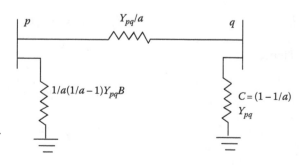

FIGURE 2.16
Equivalent π-circuit of the transformer.

2.10.2 TCUL Transformer

In this case, the taps of the transformer are changed to maintain voltage magnitude within a specified tolerance. Normally, taps are changed once in two iterations and the corresponding values of Y_{pp}, Y_{pq}, and Y_{qq} are calculated such that for any bus p, we have

$$\left| V_p^k - V_p^{\text{scheduled}} \right| < \rho. \tag{2.64}$$

However, to avoid excessive calculations, the series impedance of the equivalent π-circuit is set equal to the series impedance of the transformer and the shunt parameters B and C are changed to reflect the changes in taps.

$$(V_p - V_q)A + V_p B = (V_p - aV_q)\frac{y_{pq}}{a^2} \tag{2.65}$$

with $A = y_{pq}$,

$$(V_p - V_q)y_{pq} + V_p B = (V_p - aV_q)\frac{y_{pq}}{a^2};$$

that is

$$B = \left(\frac{1}{a} - 1\right)\left[\left(\frac{1}{a} + 1\right) - \frac{V_q}{V_p}\right] y_{pq}. \tag{2.66}$$

Similarly taking Equations 2.53 and 2.55 for I_q and substituting $A = y_{pq}$, we have

$$[V_q - V_p]y_{pq} + V_q C = (aV_q - V_p)\frac{y_{pq}}{a}. \tag{2.67}$$

From Equation 2.67, we obtain

$$C = \left(1 - \frac{1}{a}\right)y_{pq}\frac{V_p}{V_q}. \tag{2.68}$$

Subsequently, only the elements Y_{pp} and Y_{qq} are calculated with the change in tappings.

2.10.3 Phase-Shifting Transformer

Tap-changing is a means of voltage control and accompanying reactive-power control. Transformers may also be used to control phase angle and, therefore, active-power flow. Such special transformers are termed phase-shifting transformers or simply phase-shifters. Beyond the limits of rating,

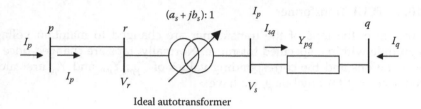

Ideal autotransformer

FIGURE 2.17
Phase-shifting transformer model.

tap-setting range and phase-shift range, there are economic considerations. Phase-shifters are costly and there are numerous economically more advantageous alternative methods of active-power control. The principal use of phase-shifters is at major intertie buses where the control of active-power exchange is especially important.

A phase-shifter is a three-phase device, which inserts a voltage V_{psr} into each line. If V_{ser} is out of phase with the line-to-neutral voltage of the source V_{an}, the phase-shifter output voltage V'_{an} will be controllable. Thus the phase of V'_{an} depends on V_{psr}. The magnitude of V'_{an} also varies with V_{psr}. It is possible to construct a controller that will cause only the phase of V'_{an} to vary while keeping the magnitude unchanged. The details of the dependence of $|V'_{an}|$ with the amount of phase shift depends on the phase-shifter design.

A phase-shifting transformer is represented by an ideal transformer with a complex turn ratio $(a + jb)$ in series with impedance or admittance y_{pq} as shown in Figure 2.17. We have $V_p = V_r$ and

$$\frac{V_p}{V_s} = \frac{V_r}{V_s} = a_s + jb_s. \tag{2.69}$$

Since the power loss in the ideal autotransformer is negligible, we have from Figure 2.17

$$V_p^* i_{pr} = V_s^* i_{sq};$$

that is

$$i_{sq}/i_{pr} = V_p^*/V_s^* = a_s - jb_s. \tag{2.70}$$

Now

$$i_{sq} = (V_s - V_q)y_{pq};$$

then

$$i_{pr} = \frac{i_{sq}}{a_s - jb_s} = [(V_s - V_q)]y_{pq}/(a_s - jb_s). \tag{2.71}$$

But since from Equation 3.44,

$$V_s = \frac{V_p}{a_s + jb_s},$$
(2.72)

by substituting Equation 2.73 in Equation 2.72 we obtain

$$i_{pr} = \left[(V_p - (a_s + jb_s)V_q)\right]\left(y_{pq}/a_s^2 + b_s^2\right).$$
(2.73)

Similarly

$$\begin{aligned}
i_{qs} &= (V_q - V_s)y_{pq} \\
&= \left[(a_s + jb_s)V_q - V_p\right]\frac{y_{pq}}{a_s + jb_s}.
\end{aligned}$$
(2.74)

Now we calculate the elements of the Y_{BUS} matrix by conducting a short-circuit test. The diagonal element Y_{pp} is found by connecting a unit voltage source at the pth bus (i.e., $V_p = 1$ pu) and short circuiting the remaining buses (i.e., substituting the voltages at other buses as zero); then we get

$$\begin{aligned}
I_p = Y_{pp} &= I_{p1} + I_{p2} + \cdots + i_{pr} \\
&= [V_p - V_1]y_{p1} + [V_p - V_{p2}]y_{p2} + \cdots + i_{pr} \\
&= y_{p1} + y_{p2} + \cdots + \left[V_p - (a_s + jb_s)V_q\right]y_{pq}/a_s^2 + b_s^2.
\end{aligned}$$
(2.75)

As $V_p = 1$ pu and $V_i = 0$ for $i = 1, \ldots, n$, $i \neq p$, we get

$$I_p = Y_{pp} = y_{p1} + y_{p2} + \cdots + \left(y_{pq}/a_s^2 + b_s^2\right) + \cdots;$$
(2.76)

that is, this element is changed.

Similarly for Y_{qq}, we let $V_q = 1$ pu and the rest of the bus voltages be set to zero; we get

$$\begin{aligned}
I_q = Y_{qq} &= I_{q1} + I_{q2} + \cdots + I_{qs} \\
&= [V_q - V_1]y_{q1} + [V_q - V_2]y_{q2} + \left((a_s + jb_s)V_q - V_p\right)y_{qq}/a_s + jb_s.
\end{aligned}$$
(2.77)

Since $V_q = 1$ pu and the rest of the bus voltages viz. $V_1 = V_2 \cdots = 0$, we get

$$I_q = Y_{qq} = y_{qi} + y_{q2} + \cdots + y_{pq},$$
(2.78)

that is no change.

For mutual admittances, say Y_{qp}, we apply $V_p = 1$ pu and measure I_q keeping other bus voltages equal to zero (i.e., $I_q = Y_{qp}V_p$) but since $V_p = 1$ pu we get from Equation 2.75,

$$I_q = Y_{qp} = -\frac{y_{pq}}{a_s + jb_s}. \tag{2.79}$$

For Y_{pq} we keep $V_q = 1$ and the rest of the buses are short circuited; then from Equation 2.75 we get

$$I_p = Y_{pq} = -\frac{y_{pq}}{a_s - jb_s}. \tag{2.80}$$

Moreover, $Y_{pq} \neq Y_{qp}$. Now the complex turn ratio $a_s + jb_s = a[\cos\theta + j\sin\theta]$ where

$$|V_p| = a|V_s|. \tag{2.81}$$

However, if θ is positive, the phase of $|V_p|$ is advanced, that is, leading with respect to that of $|V_s|$ or V_q.

Figure 2.18 shows a generalized model of a phase-shifter in which both voltage magnitude and phase angle vary. The equivalent tap position is $t(\phi)$ which is only a function of the phase setting ϕ. If no voltage variation occurs, $t = 1$. The phase-shifter admittance is Y_{ij}. Figure 2.19 shows a further equivalent. Note that in Figure 2.19

$$|V_i| = |V_i'|$$
$$\delta_{i'} = \delta_i + \phi$$

and the complex volt–amperes entering the phase-shifter at bus i are delivered to bus i'. Let bus i' be introduced into the $[\delta, V]^T$ vector, but only in the δ

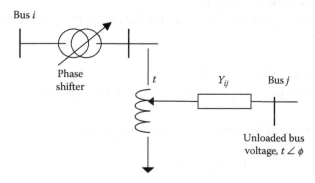

FIGURE 2.18
Generalized model of a phase-shifter.

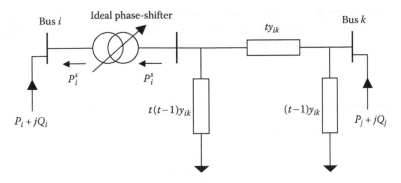

FIGURE 2.19
Equivalent circuit of a phase-shifter.

subvector. Also, rather than writing $\delta_{i'}$ in this subvector, we introduce the unknown phase shift ϕ:

$$[\delta_2, \ldots, \delta_N, \phi_1, \ldots, \phi_N, V_2, \ldots, V_i, V_{i+1}, \ldots, V_N]^T.$$

Note that V_i' is missing from the $|V_{\text{bus}}|$ subvector. Also note that the partial derivatives $\partial P/\partial \delta_{i'}$ and $\partial Q/\partial \delta_{i'}$ are equal to $\partial P/\partial \phi$ and $\partial Q/\partial \phi$, respectively, since $\delta_i' = \delta_i + \phi$. The missing voltage magnitude element is counterbalanced by a missing $\Delta Q_i'$ expression. Thus the size of J is $2(N-1) - n_\phi$ by $2(N-1) - n_\phi$, where n_ϕ is the number of phase-shifters in the network.

The procedure for handling a phase-shifter bus where ϕ adjusts itself to give P_i^ϕ (see Figure 2.18) injected into bus i is as follows:

1. Construct the Y_{bus} using the conventional algorithm without connecting buses i' and i, or buses i' and k.
2. Use the equivalent circuit in Figure 2.19 to connect buses i' and k. Use the initialized value of ϕ to find t. The equivalent tap-setting is a function of ϕ in general (this depends on the phase-shifter design).
3. Construct the Jacobian matrix. Eliminate the ΔQ expression at bus i and eliminate $|V|$ at bus i.
4. Consider P_i^ϕ the specified power through the phase-shifter, as injected into i and as the only load at i'.
5. Perform one iteration in the power flow study and repeat as required.

To determine the effect of changes in real and reactive generation sources on the overall power flow, the simplest approach is by trial and error. The generators, power, and voltage levels are maintained to match the desired load and voltage profile. A power-flow analysis is made and the corresponding flow picture is examined. If equipment loadings are outside their desired limits, adjustments are made to the set points and a new power-flow analysis is performed. It is hoped that the procedure would converge upon a suitable power-flow picture iteratively.

2.11   Load Models [3]

The prespecified S load (i.e., independent of bus voltage) is a reasonable representation for buses with a large percentage of rotating machines, loads, and voltage-regulated loads. Other alternative load models include the following:

1. Loads whose $|S|$ demand is proportional to $|V|^2$. This type of load is a prespecified impedance load and is common when the load contains significant incandescent lighting and resistive heating.
2. Loads whose $|S|$ demand is proportional to $|V|^1$. This type of load is a prespecified current load and is common when the load contains significant rectifier loads and certain types of synchronous machine loads.

It has been suggested that the actual load be modeled as a linear combination of each of the load types mentioned (constant S load in which $|S| \; \alpha \; |V|^0$, constant current load in which $|S| \; \alpha \; |V|^1$, and constant impedance load in which $|S| \; \alpha \; |V|^2$). Such an approach would require considerable knowledge of the load composition or a knowledge of the active and reactive-power variation with $|V|$ (from a historical perspective). Load models are traditionally classified into two broad categories: static models and dynamic models.

2.11.1   Static Load Models

A static load model expresses the characteristic of the load at any instant of time as algebraic functions of the bus voltage magnitude and frequency at that instant. The active power component P and the reactive-power component Q are considered separately.

The load characteristics in terms of voltage are represented by the exponential model:

$$P = P_0 \left(\frac{V}{V_0} \right)^a$$

$$Q = Q_0 \left(\frac{V}{V_0} \right)^a,$$

(2.82)

where
  P and Q are active and reactive components of the load when the bus
    voltage magnitude is V
  Subscript "0" identifies the values of the respective variables at the
    initial operating condition

The parameters of this model are the exponents a and b. With these exponents
equal to 0, 1, or 2, the model represents constant power, constant current, or
constant impedance characteristics, respectively. For composite loads, their
values depend on the aggregate characteristics of load components.

An alternative model that has been widely used to represent the voltage
dependence of loads is the polynomial model:

$$P = P_0\left[p_1\left(\frac{V}{V_0}\right)^a + p_2\left(\frac{V}{V_0}\right) + p_3\right]$$

$$Q = Q_0\left[q_1\left(\frac{V}{V_0}\right)^a + q_2\left(\frac{V}{V_0}\right) + q_3\right].$$

(2.83)

This model is commonly referred to as the ZIP model, since it is composed of
constant impedance (Z), constant current (I), and constant power (P) components. The parameters of the model are the coefficients P_1–P_3 and Q_1–Q_3
which define the proportion of each component.

Multiplying the exponential model or the polynomial model as follows
represents the frequency dependence of load characteristics by a factor.

$$P = P_0\left(\frac{V}{V_0}\right)^a(1 + K_{pf}\Delta f)$$

$$Q = Q_0\left(\frac{V}{V_0}\right)^a(1 + K_{qf}\Delta f)$$

(2.84)

or

$$P = P_0\left[p_1\left(\frac{V}{V_0}\right)^a + p_2\left(\frac{V}{V_0}\right) + p_3\right](1 + K_{pf}\Delta f)$$

$$Q = Q_0\left[q_1\left(\frac{V}{V_0}\right)^a + q_2\left(\frac{V}{V_0}\right) + q_3\right](1 + K_{qf}\Delta f),$$

(2.85)

where Δf is the frequency deviation, given as $(f - f_0)$.

2.12   Available Transfer Capability

The introduction of competition in the power industry will increase the search for better utilization of transmission facilities. A necessary condition to support competition is open access by market participants to the transmission network. Several open access schemes are currently being discussed. Transmission companies are facing several challenges in the new economic environment including the assessment of adequate transmission charges and the choice of the best transmission investment options. These aspects have motivated the development of methodologies to evaluate existing power transfer capabilities and transmission margins. Under FERC Orders 888 and 889, all transmission providers must determine and offer for sale ATC taking into account existing obligations and allowing appropriate margins to maintain reliability. There is a growing body of work dealing with determination of ATC. The main intention is to evaluate the ability of a network to allow for the reliable movement of electric power from areas of supply to areas of demand. The framework of analysis may or may not take into account the stochastic nature of loads and equipment availability. Also, either DC or AC models can assess the performance of the electric network.

2.12.1   ATC Definition and Formulation

Total transfer capability (TTC) determines the amount of electric power that can be transferred over the interconnected transmission network in a reliable manner based on all the following conditions.

For the existing or planned system configuration, and with normal (precontingency) operating procedures in effect, all facility loadings must be within normal ratings and all voltages must be within normal limits. The electric systems must be capable of absorbing the dynamic over swings, and remain stable, following a disturbance that results in the loss of any single electric system element, such as a transmission line, transformer, or generating unit. Dynamic power savings subside following a disturbance that results in the loss of any single electric system element as described and after operation of any automatic operating systems. But before any postcontingency operator-initiated system adjustments are implemented, all transmission facility loadings should be within emergency ratings and all voltages within emergency limits. When precontingency facility loadings reach normal thermal ratings at a transfer level below that at which any first contingency transfer limits are reached, the transfer capability is defined as the transfer level at which such normal ratings are reached. In some cases depending on the geographical area, specified multiple contingencies may be required to determine the transfer capability limits. If these limits are more restrictive than the single contingency limits, they should be used. With this very general definition, we note that the TTC is a function of system thermal, voltage, and transient stability limits and is given by

$$TTC = \min[\text{thermal limits, voltage limits, transient stability limits}].$$

Transmission reliability margin (TRM) is the amount of transfer capability necessary to ensure that the interconnected transmission network is secure under a reasonable range of system conditions. This measure reflects the effect of various sources of uncertainty in the system and in system operating conditions. It also captures the need for operational flexibility in ensuring reliable and secure system operations.

Capacity benefit margin (CBM) is the amount of transfer capability reserved by load-serving entities to ensure access to generation from interconnected neighboring systems to meet generation reliability requirements.

ATC then is defined to be a measure of the transfer capability remaining in the physical transmission network for further commercial activity over and above already committed uses. It is given by the relationship

$$ATC = TTC - TRM - CBM - \text{existing transmission commitments.}$$

FERC requirements call for the calculation and posting (on the OASIS) of continuous ATC information for the next hour, month, and for the following 12 months.

Several regions have already started calculating and posting ATCs. However, there are certain inconsistencies in their calculation that need to be addressed:

- Use of real-time conditions for ATC calculation.
- Use of accurate power flow methods to calculate ATC rather than DC power flows or linear interpolation techniques.
- Power system flows are not independent.
- Individual users do not coordinate their transmission system use.
- Limiting effects on a transmission system are nonlinear functions of the power flow.

2.12.2 ATC Calculation

The ATC problem is the determination of the largest additional amount of power above some base case value that can be transferred in a prescribed manner between two sets of buses: the source, in which power injections are decreased by an offsetting amount. Increasing the transfer power increases the loading in the network, and at some point causes an operational or physical limit to be reached that prevents further increase. The effects of contingencies are taken into account in the determination. The largest value of transfer power that causes no limit violations, with or without a contingency, is used to compute the TTC and ATC.

Limits checked for the base case are normal branch flow limits, normal corridor transfer limits, bus voltage limits, and voltage collapse condition. For contingency cases, the limitations checked are the emergency branch flow limits, emergency corridor transfer limits, bus voltage limits, bus change limits, and voltage collapse condition.

The following is a summary of the steps for determining the ATC for a list of source/sink transfer cases for one time period, usually 1 h.

1. Establish and solve the base case power flow for the time period.
2. Select a transfer case.
3. Use continuation power flow (CPF) to make a step increase in transfer power.
4. Establish a power flow problem consisting of the base case modified by the cumulative increases in transfer power from step 3.
5. Solve the power flow problem of step 4.
6. Check the solution of step 5 for violations of operational physical limits.
7. If there are violations, decrease the transfer power to the minimum amount necessary to eliminate them.
8. Solve the power flow problem of steps 4–6 for each listed contingency.
9. If there are violations, decrease the transfer power by the minimum amount necessary to eliminate them. This is the maximum transfer power for the case.
10. Compute the ATC from the interface flows in the adjusted solution of step 7.
11. Return to step 2 for the next transfer case.

2.13 Illustrative Examples

Example 2.13.1

The system in Figure 2.20 is known to be balanced and is a negative sequence network. Assuming that

$$Z_{line} = 10.0\angle-15° \ (\Omega) \quad \text{and} \quad V_{ac} = 208\angle-120° \text{ V}$$

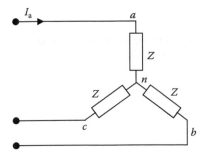

FIGURE 2.20
Y-connected network for Example 2.13.1.

find

1. Voltages V_{ac}, V_{bc}, V_{bn}, V_{cn}
2. Currents I_a, I_b, and I_c
3. Total three-phase power $S_{3\phi}$

SOLUTION

1. Line voltages are

$$\overline{V}_{ca} = -\overline{V}_{ac} = 208\angle-120° + 180 = 208 = \angle60° \text{ V}$$
$$\overline{V}_{ab} = \overline{V}_{ac}\angle120° = 208\angle60° + 120° = 208\angle180° \text{ V}$$
$$\overline{V}_{bc} = \overline{V}_{ab}\angle-240° = 208\angle180° - 240° = 208\angle-60° \text{ V}.$$

Phase voltages are

$$\overline{V}_{an} = \frac{\overline{V}_{ab}}{\sqrt{3}}\angle30° = \frac{208}{\sqrt{3}}\angle180° + 30° = 120\angle210° \text{ V}$$
$$\overline{V}_{bn} = \overline{V}_{an}\angle-240° = 120\angle210° - 240° = 120\angle-30° \text{ V}$$
$$\overline{V}_{cn} = \overline{V}_{bn}\angle120° = 120\angle-30° + 120° = 120\angle90° \text{ V}.$$

2. Line current

For the Y-connected network, the line current = phase current,

$$\overline{I}_a = \frac{\overline{V}_{an}}{Z_{line}} = \frac{120\angle210°}{10\angle-15°} = 12\angle195° \text{ V}$$
$$\overline{I}_b = \overline{I}_a\angle-240° = 12\angle195° - 240° = 12\angle-45° \text{ V}$$
$$\overline{I}_c = \overline{I}_b\angle120° = 12\angle-45° + 120° = 12\angle75° \text{ V}.$$

3. Three-phase power

$$S_{3-\phi} = \sqrt{3}*208*12*\cos(15°) = 4175.9 \text{ W}$$

Example 2.13.2

In Figure 2.21, find the total complex power delivered to the load. Assume that

$$Z_c = -j0.2 \ \Omega$$
$$Z_L = +j0.1 \ \Omega$$
$$R = 10 \ \Omega.$$

SOLUTION

$$Y_{\text{equivalent}} = \frac{1}{R} + \frac{1}{Z_c} + \frac{1}{Z_L} = \frac{1}{10} + \frac{1}{-j0.2} + \frac{1}{j0.1}$$
$$= 0.1 + j5 - j10 = (0.1 - j5) \ \Omega.$$

The total complex power is defined as $S = VI^* = V(VY)^*$. Or we can say $S^* = V^*I = V^*(VY) = V^2Y = (1.0)^2(0.1 - j5.0)$. The total complex power is $S = 0.1 + j5$ VA.

Example 2.13.3

You are given two system areas connected by a tie line with the following characteristics.

Area 1	Area 2
$R = 0.01$ pu	$R = 0.02$ pu
$D = 0.8$ pu	$D = 1.0$ pu
Base MVA $= 2000$	Base MVA $= 1000$

A load change of 100 MW occurs in area 2. Assume both areas were at nominal frequency (60 Hz) to begin

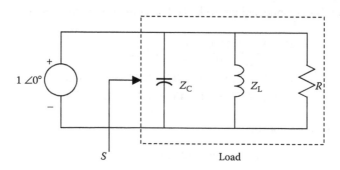

FIGURE 2.21
Composite R–L–C Load for Example 2.13.1

1. What is the new steady-state frequency in Hertz?
2. What is the change in tie-line flow in megawatt?

SOLUTION

Based on the information in Section 2.8, if several generators (each having its own governor and prime mover) were connected to the system, the frequency droop characteristics β_1 and β_2 are given by

$$\beta_1 = \left(\frac{1}{R_1} + D_1\right); \quad \beta_2 = \left(\frac{1}{R_2} + D_2\right)$$

$$\Rightarrow \beta_1 = \frac{1}{0.01} + 0.8 = 100.8 \text{ pu} \quad \text{and} \quad \beta_2 = \frac{1}{0.02} + 1.0 = 51.0 \text{ pu.}$$

Using a common base of 1000 MVA, the frequency droop characteristics becomes

$$\beta_1 = \frac{100.8 \times 2000}{1000} = 201.6 \text{ pu} \quad \text{and} \quad \beta_2 = \frac{51.0 \times 1000}{1000} = 51.0 \text{ pu.}$$

Calculating the change in frequency based on the load change in area 2

$$\Delta\omega = \frac{-\Delta P_{L_2}}{\beta_1 + \beta_2} = \frac{-\Delta P_{L_2}}{(1/R_1) + (1/R_2) + D_1 + D_2}.$$

The change in load is given as $\Delta P_{L_2} = 100 \text{ MW} = 100/1000 = 0.1 \text{ pu.}$

Hence, $\Delta\omega = \dfrac{-0.1}{201.6 + 51.0} = -0.00039588 \text{ pu.}$

(a) We have $\dfrac{\Delta\omega}{\omega_0} = \dfrac{\Delta f}{f_0} = -0.00039588 \text{ pu}$

$$\Rightarrow \Delta f = 60(-0.00039588) = -0.023753 \text{ Hz}$$

and the new frequency becomes

$$f = 60 - 0.023753 = 59.976 \text{ Hz.}$$

(b) Change in tie flow is given by

$$\Delta P_{\text{tie}} = -\beta_1 \Delta\omega = \frac{-\Delta P_{L_2}((1/R_1) + D_1)}{(1/R_1) + (1/R_2) + D_1 + D_2} = -(201.6)(-0.00039588)$$

$$= 0.07981 \text{ pu}$$

$$\Rightarrow \Delta P_{\text{tie}} = 0.07981 \text{ pu} \times 1000 \text{ MW} = 79.81 \text{ MW.}$$

This analysis shows that 79.81 MW of power is being supplied from area 1 to area 2 to meet the 100 MW load increase in area 2.

2.14 Conclusions

The chapter explained major basics for modeling different components in typical electric power systems. First, the concept of electric power was introduced expressed in terms of active, reactive, and apparent power. Second, a review of three-phase systems was introduced. Finally, the synchronous machine model from an electric network standpoint was presented; the reactive capability curves are also examined. Furthermore, the model was extended to handle the excitation system governor models. Static and dynamic load models were also discussed.

2.15 Problem Set

PROBLEM 2.15.1

In Figure 2.22, assume that

$$|V_1| = |V_2| = 1.00 \text{ pu}$$

$$Z_{\text{line}} = 0.1\angle 85°.$$

1. For what nonzero θ_{12} is S_{12} purely real?
2. What is the maximum power, $-P_{12}$, that can be received by V_2, and at what θ_{12} does this occur?
3. When $\theta_{12} = 85°$, what is the active power loss in the line?
4. For what value θ_{12} is $-P_{12} = 1.00$ pu?

PROBLEM 2.15.2

Draw the power circle diagram in the case $|V_1| = 1.05$, $|V_2| = 0.95$, $Z_{\text{line}} = 0.1$ pu at $85°$. Find

1. $P_{12(\text{max})}$
2. θ_{12} at which we get $P_{12(\text{max})}$
3. $-P_{12(\text{max})}$
4. θ_{12} at which we get $-P_{12(\text{max})}$
5. Active power loss in the line when $\theta_{12} = 10°$

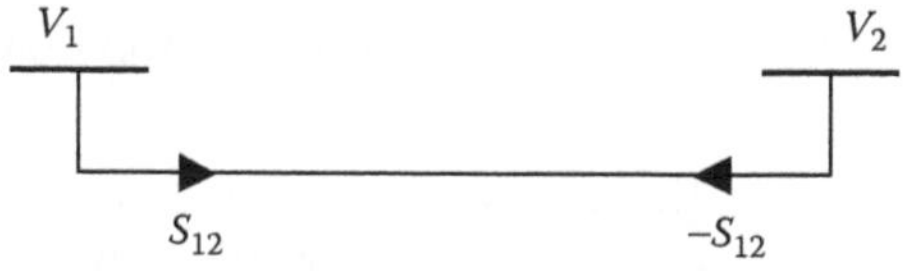

FIGURE 2.22
Simple two-bus power system network for Problem 2.15.1.

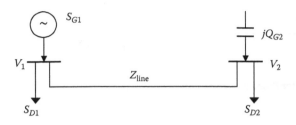

FIGURE 2.23
Power system model for Problem 2.15.3.

PROBLEM 2.15.3

In Figure 2.23, assume that

$$V_1 = 1.00 \angle 0° \text{ pu}$$

Given that

$$Z_{\text{line}} = 0.10 + j0.10$$
$$S_{D1} = 0.50 + j0.50$$
$$S_{D2} = 0.50 + j0.50,$$

select Q_{G2} such that $|V_1| = 1.00$ pu. In this case what are Q_{G2}, S_{G1}, and arg V_2?

PROBLEM 2.15.4

A three-phase, Y-connected synchronous generator is rated for a 600 kVA and 6.9 kV. The reactance of the stator winding is $X_s = j3.97 \ \Omega/\text{phase}$. Express X_s in per units on the machine ratings as its bases.

PROBLEM 2.15.5

A three-phase, 500 kVA, 6.9 Y/1.2 Y–kV transformer has series impedance referred to the low-voltage side of $Z_{T_2} = 0.72 \angle 80° \ \Omega/\text{phase}$. Express Z_T in per unit.

PROBLEM 2.15.6

A three-phase synchronous generator having $R_a = 0$ and $X_S = 1.0$ pu is operating as a generator with $V_t = 1.00$ pu, pf $= 1.0$, and power delivered $= 1.00$ pu.

1. Determine E_{F1} and δ_1, for this operation.
2. Obtain I_a, pf_2, and δ_2 for $E_{F2} = 1.6$ pu with $P = $ constant.

FIGURE 2.24
A two-area power system interface for Problem 2.15.8.

PROBLEM 2.15.7

Consider a load connected across an AC voltage source where the load impedance is given by $R_L + jX_L$ and the series source impedance is given by $R_S + jX_S$. Determine the optimum values for R_L and X_L for maximum power transfer to the load.

PROBLEM 2.15.8

Consider the two area systems shown in Figure 2.24. The system characteristics are given in the figure.

$$P_{G_o}^A = P_L^A = 1500 \text{ MW}$$
$$R_A = 0.015$$
$$P_{G_o}^A = P_L^B = 5000 \text{ MW}$$
$$R_B = 0.008.$$

Assume a load increase of 100 MW occurs in area 1; determine the change in frequency and the ACE in each area.

PROBLEM 2.15.9

Given the block diagram of the two interconnected areas shown in Figure 2.25.

1. Derive the transfer function that relates $\Delta W_1(s)$ and $\Delta W_2(s)$ to load change $\Delta P_L(s)$.

2. For the following data (base power is 1000 MVA)

$$M_1 = 3.8 \text{ pu}, \quad D_1 = 0.9 \text{ pu}, \quad f_1^o = 60 \text{ Hz}$$
$$M_2 = 4.0 \text{ pu}, \quad D_2 = 0.8 \text{ pu}, \quad f_2^o = 60 \text{ Hz}$$
$$T = 7.6 \text{ pu}, \quad \Delta P_L = 200 \text{ MW}, \quad \text{and} \quad R = 0.05 \text{ pu},$$

calculate the final frequency.

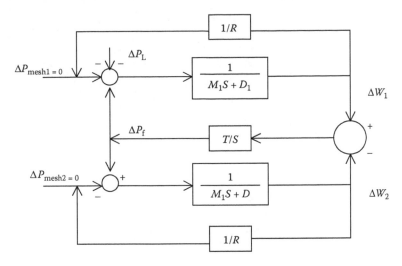

FIGURE 2.25
Block diagram for the control of two interconnected areas for Problem 2.15.9.

References

1. Bergen, A. R. and Vittal, V. *Power Systems Analysis*, 2nd edn., Prentice-Hall, Upper Saddle River, NJ, 1999.
2. Elgerd, O. I. *Electrical Energy Systems Theory—An Introduction*, McGraw-Hill, New York, 1982.
3. El-Hawary, M. E. *Electric Power Systems*, 2nd edn., IEEE Press, New York, 1995.
4. Fizgerald, A., Kingsley, L., and Urans, S. *Electric Machinery*, McGraw-Hill, New York, 1983.
5. Fouad, A. A. and Vittal, V. *Power System Transient Stability; Analysis in Transient Energy Function Method*, Prentice-Hall, Englewood Cliffs, NJ, 1992.
6. Momoh, J. A. and El-Hawary, M. E. *Electrical Systems, Dynamics and Stability with Artificial Intelligence Applications*, Marcel Dekker, New York, 1999.
7. Sauer, P. W. and Pai, M. A. *Power System Dynamics and Stability*, Prentice-Hall, Upper Saddle River, NJ, 1998.

3

Power-Flow Computations

3.1 Introduction

Power flow (PF) is an evaluation tool for operation and planning for determining the steady condition of operation [1,2]. As there is a lot of enhancement of the computation for power system performance, we need to assure the capability of PF to be guaranteed, stressed nature for congestion stresses. The feasibility of PF solution is a necessary condition for pricing theory and stability analysis as well as further robust optimal power flow (OPF) program. Therefore stability for necessity of the performance of PF tools becomes very important.

We must monitor and provide measures to achieve convergence through

- Appropriate selection of initial condition
- Handling of negative impedances
- Location of the swing bus
- Computation of losses
- Managing congestion and power balance under different operating conditions
- Error handling and alert messages
- Final interpretation of results, especially for nonconvergence of the PF methods

It should be noted that PF is a well-seasoned tool used by the industry and researchers. Several packages are available in the market for easy access and study both for industry and education [3].

PF studies are routinely used in planning, control, and operations of existing electric power systems as well as planning for future expansion. Satisfactory operation of power systems depends upon knowing the effects of adding interconnections, connecting new loads, introducing new generating stations, or constructing new transmission lines before they are installed. PF studies also allow us to determine the best size and the most favorable locations for power capacitors both for improving the power factor and also raising the bus voltages of the electrical network. PF studies help us determine the best location as well as the optimal capacity of proposed generating stations, substations, or new lines.

In power system operations, a PF study is used to evaluate the state of the system under a given operational condition and to allow the operator to detect overloads and evaluate transfer limits. Contingency analysis relies on load-flow studies designed to test for the effects of line or generator outages. The information obtained from the load-flow studies includes the magnitude and phase angle of voltages at each bus and the active and reactive PF in each line.

The extensive calculations required for both PF voltage determination requires the use of specialized computing tools [3–6]. This led to the design of a special purpose analog computer called the AC network analyzer in 1929. An AC network analysis was capable of simulating the operation of the power system under existing conditions as well as proposed future expansion. Programming digital computers for PF studies gained importance, and the first computer-based planning studies were completed in 1956. This change from the network analyzer to the digital computer has resulted in greater flexibility, economy, accuracy, and faster operation.

3.2 Types of Buses for PF Studies

The buses in an electric power system network are generally divided into three categories: generation bus, load bus, and slack bus, and two of the following quantities are specified at each bus.

1. Magnitude of the voltage, $|V|$
2. Phase angle of the voltage, ϕ
3. Active or real power, P
4. Reactive power, Q

The quantities specified at each of the bus types are:

1. Generation bus (or voltage-controlled bus): This is also called the P–V bus, where the voltage magnitude $|V|$ and real power P are specified.
2. Load bus: This is also called the P–Q bus, where the real power P and reactive power Q are specified.
3. Slack or swing bus: This is also known as the reference bus where the voltage magnitude $|V|$ and phase angle ϕ are specified. This bus is selected to provide additional real and reactive power to supply transmission losses since these are unknown until the final solution is obtained. If the slack bus is not specified, then a generation bus (usually with maximum real power P) is taken as the slack bus. There can be more than one slack bus in a given system.

TABLE 3.1

Typical Load-Flow Bus Specifications

Bus Classification	Real Power P_i	Reactive Power Q_i	Voltage Magnitude $|V|$	Voltage Argument Φ	Comments
Slack or reference bus			$\checkmark$		Must adjust net power to hold voltage constant
Voltage controlled or P–V bus			$\checkmark$		$Q_{min} \leq Q \leq Q_{max}$
Load or P–Q bus	$\checkmark$				Constant power, constant impedance, and constant current

Table 3.1 shows the load-flow bus specifications. The checked quantities are the boundary conditions for the particular bus.

PF analysis is concerned not only with the actual physical mechanism that controls the PF in the network, but also with how to select a best or optimum flow configuration from among the myriad possibilities [1–6]. The following points summarize the more important features incorporated in PF analysis:

1. Transmission links can carry only certain amounts of power; therefore, links must not be operated too close to their stability or thermal limits.

2. It is necessary to keep the voltage levels of certain buses within close tolerances. This can be achieved by proper scheduling of reactive power levels.

3. If the power system is part of a larger pool, it must fulfill certain contractual commitments via its tie-lines to neighboring systems.

The overall PF problem can be divided into the following:

1. Formulation of a suitable mathematical network model: The model must describe adequately the relationships among voltages and powers in the interconnected system.

2. Specification of the power and voltage constraints that must apply to the various buses of the network.

3. Numerical computation of the PF equations (PFEs) subject to specified constraints. These computations provide, with sufficient accuracy, the values of all bus voltages.

4. When all bus voltages have been determined, the actual PFs in all transmission links are computed.

3.3 General Form of the PFEs

To understand PF in an electric power system, a mathematical model from which the flow picture can be predicted must be established. This model is the PFE. In the general case, consider the general bus i shown in Figure 3.1. Its generation and load are assumed to be equal, S_{Gi} and S_{Di}, respectively. The bus power S_i is thus given by

$$S_i = S_{Gi} - S_{Di} = P_{Gi} - P_{Di} + j(Q_{Gi} - Q_{Di}). \tag{3.1}$$

Transmission lines connect bus i to other buses k in the system. At most there can be $n - 1$ such outgoing lines, each of which can be represented by a π equivalent, with series and parallel admittances y_{sik} and y_{pik}, respectively. If a line does not exist, its admittances simply are set at zero. The current balanced equation is

$$I_i = \frac{S_i^*}{V_i^*} = V_i \sum_{\substack{k=1 \\ k \neq 1}}^{n} y_{pik} + \sum_{\substack{k=1 \\ k \neq 1}}^{n} y_{sik}(V_i - V_k)$$

$$= V_i \sum_{\substack{k=1 \\ k \neq 1}}^{n} y_{sik}(V_i - V_k) + \sum_{\substack{k=1 \\ k \neq 1}}^{n} (-y_{sik})V_k, \quad \text{for } i = 1, 2, 3, \ldots, n. \tag{3.2}$$

These equations can obviously be written in the form

$$I_i = \frac{S_i^*}{V_i^*} = Y_{i1}V_1 + Y_{12}V_2 + \cdots Y_{ii}V_i + \cdots Y_{in}V_n, \quad \text{for } i = 1, 2, \ldots, n, \tag{3.3}$$

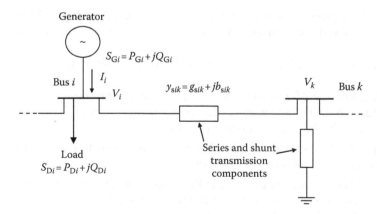

FIGURE 3.1
General bus with generation, load, and outgoing lines.

where

$$Y_{ii} \triangleq \sum_{\substack{k=1 \\ k \neq 1}}^{n} (y_{pik} + y_{sik}) \tag{3.4}$$

and

$$Y_{ik} = Y_{ki} \triangleq -\frac{1}{2} y_{sik}. \tag{3.5}$$

The n-dimensional bus current and bus voltage vectors are then introduced:

$$\mathbf{I}_{bus} = \begin{bmatrix} I_1 \\ \vdots \\ I_n \end{bmatrix} \tag{3.6}$$

$$\mathbf{V}_{bus} = \begin{bmatrix} V_1 \\ \vdots \\ V_n \end{bmatrix}$$

and the $n \times n$-dimensional bus admittance and bus impedance matrices are

$$\mathbf{Y}_{bus} = \left[Y_{ij} \right]_{n \times n}; \quad \mathbf{Z}_{bus} = \mathbf{Y}_{bus}^{-1} = \left[Z_{ij} \right]_{n \times n}. \tag{3.7}$$

3.3.1 PF Control by Transformer Regulation

To employ transformer ratio or angle setting for PF control we proceed to choose transformer ratios and angular settings that appear to best meet the specified requirements. By analyzing the PF results, one finds a basic difference between this type of control and source control.

Adjusting transformer settings will affect the $\mathbf{Y}_{bus}$ matrix, while changes in generation level will not change the $\mathbf{Y}_{bus}$ matrix. Consider a two-bus system, with a regulating transformer (RT) at bus 2 as shown in Figure 3.2a. The RT has a complex transformation ratio

$$a = |a| \angle \alpha. \tag{3.8}$$

RT is represented as an ideal transformer and its effects upon the $\mathbf{Y}_{bus}$ matrix are calculated. From Figure 3.2b, it can be observed that the voltage and current on the primary side of RT are equal to aV_2 and I_2/a^*, respectively. V_2 and I_2 are the secondary variables. The current balance at the two buses requires that

$$I_1 = \frac{S_1^*}{V_1^*} = V_1 Y_p + (V_1 - aV_2)Y_s \tag{3.9}$$

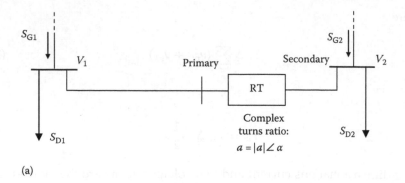

(a)

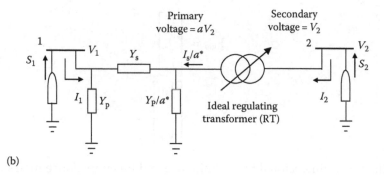

(b)

FIGURE 3.2
Two-bus example system demonstrating the use of the RT.

and

$$\frac{I_2}{a^*} = \frac{S_2^*}{(aV_2)^*} = aV_2 Y_{\mathrm{p}} + (aV_2 - V_1)Y_{\mathrm{s}}.$$

We rewrite Equation 3.9 and obtain the pair

$$\begin{aligned} I_1 &= (Y_{\mathrm{p}} + Y_{\mathrm{s}})V_1 + (-aY_{\mathrm{s}})V_2 \\ I_2 &= (-a^*Y_{\mathrm{s}})V_1 + aa^*(Y_{\mathrm{p}} + Y_{\mathrm{s}})V_2. \end{aligned} \tag{3.10}$$

As a result, we conclude that the addition of RT changes the $\mathbf{Y}_{\mathrm{bus}}$ to a new value

$$\mathbf{Y}_{\mathrm{bus}} = \begin{bmatrix} Y_{\mathrm{p}} + Y_{\mathrm{s}} & -aY_{\mathrm{s}} \\ -a^*Y_{\mathrm{s}} & aa^*(Y_{\mathrm{p}} + Y_{\mathrm{s}}) \end{bmatrix}. \tag{3.11}$$

The addition of the RT has added the two additional variables $|a|$ and α to the previous ones and the linearized power balance equations will contain the added difference variables $\Delta|a|$ and $\Delta\alpha$. In a practical situation, the RT is designed for either voltage magnitude or phase control. In the former case $\alpha = 0$ and $|a|$ can be changed in discrete steps $\Delta|a|$. In the latter case $|a|$ is constant and α can be changed in steps $\Delta\alpha$.

3.4 Practical Modeling Considerations

The power system model presented earlier is a simplified version of the actual state of the operational network. There are some additional complexities that deserve further consideration and more detailed modeling.

3.4.1 Generation Subsystem

The generator bus model presented thus far is a *P–V* model in which the turbine controls hold the generated active power at a fixed level and the machine voltage regulator maintains the bus voltage magnitude fixed. This model may be unrealistic if the voltage regulator is unable to set the field current to produce the desired bus voltage. Figure 3.3 shows a simplified circuit diagram of the generator. To increase $|V_t|$, the field current is increased such that $|E_f|$ increases. Under constant generated power conditions, the torque angle δ decreases slightly, but the phase of the armature current I_a will rotate such that the power factor angle ϕ increases. Thus more reactive power is injected into the bus. The limit on the control of $|V_t|$ occurs primarily due to the limit on the field current (and hence the limit on Q). In most cases, generators inject Q into the system since most loads have a lagging power factor. Thus, the salient limitation is related to the upper field current limit. In cases where the generator absorbs reactive power, the field current must be decreased. The limitations in this region are twofold: limits imposed by the maximum value of $|I_a|$ and practical stability limits associated with low field currents. In the latter case the dominant limitation may depend on the circuit power factor which always depends on the generator control designs. In addition, the generator complex volt–ampere rating $|S|$ warrants further

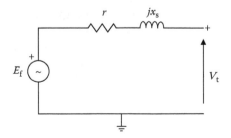

FIGURE 3.3
Circuit diagram for a synchronous generator.

attention. The generator $|S|$ rating implies possible further limitation of Q depending on the active power setting.

A simplified approach to generator field controller limits is usually followed for PF studies: a single upper limit on the injected Q for each generator is imposed. Under the assumption of an ideal P–V bus, the reactive power mismatch is not calculated since there is no specified reactive power at a generator bus. However, the Q generated by the machine must equal the total line flow reactive powers,

$$Q_{\text{generated at } i} = \sum_{j=1}^{N} |Y_{ij}| v_j v_i \sin(-\theta_{ij} + \delta_i - \delta_j).$$

To check whether the generated Q is below the upper limit,

$$Q_{\text{generated at } i} < Q_{\text{limit } i}^{u}$$

must hold. Similarly, the lower limit ($Q_{\text{limit } i}^{l} < 0$) must be checked

$$Q_{\text{limit } i}^{l} < Q_{\text{generated at } i} < Q_{\text{limit } i}^{u}. \tag{3.12}$$

If Equation 3.12 is violated at the upper or lower limit, the P–V bus model must be modified to reflect the physical reality. This usually entails conversion of the P–V bus to a P–Q bus where the specified Q is set at $Q_{\text{limit } i}^{l}$ or $Q_{\text{limit } i}^{u}$ as appropriate. The origin of Q limits of machines comes from the several operating conditions cited earlier, including limits on armature current, limits on field current and voltage, and operational stability requirements.

The concept and treatment of the slack bus requires further discussion. The PF formulation precludes specifying active and reactive power at all buses since the power lost in the transmission network is unknown. A formulation in which P is specified at all buses is overspecified and will yield results inconsistent with transmission losses. This is a mathematical consideration that translates to an operating condition: one machine in the system is usually operated such that the active power generated is set to hold the area error active power at zero. The term area error refers to the generation error in a given system, which causes a frequency error. If the net generation in a given system (an area) is greater than the load plus losses, the excess power injected will result in a net injected energy. This excess energy causes the machines in the system to accelerate. In other words, the integral of the error power (i.e., the error energy) becomes rotating kinetic energy. Similarly, if the net generation is too low, the system kinetic energy (and, hence, frequency) will decrease. The operational turbine setting is obtained at the (slack) swing machine from the area error: a raise power signal is required when the system frequency falls below the standard and a low power signal is required when the system frequency is too high.

The concept of a generalized swing machine was introduced in order to distribute area power error over several machines. Thus, the swing power is made up not at a single machine, but at several generators. Usually, a production computer program still uses a single bus as a swing bus for load-flow studies. One may encounter several swing buses in one load-flow study involving more than one area. (For purposes of this discussion, the term "area" refers to a coordinated power transmission network and generation in which a single swing bus is present.) There are several techniques for considering multiple swing buses, perhaps the simplest being separate load flows in which intertie buses are separated. It is possible to model the swing power in a distributed fashion (i.e., at several generators) by employing several swing machines.

To incorporate the transformer model into the PF formula we describe the various forms of the PF formulation, namely, the rectangular and the polar forms.

The matrix $\mathbf{Y}_{bus}$ represents the model of the passive portion of the n-bus network, in system form. From Equations 3.4 and 3.5 we conclude the following simple rules for finding the elements of the $\mathbf{Y}_{bus}$ matrix:

1. Diagonal element y_{ii} is obtained as the algebraic sum of all admittances incident to node i.

2. Off-diagonal elements $y_{ik} = y_{ki}$ are obtained as the negative of the admittance connecting nodes i and k.

It is usually the case that powers rather than currents are known. Then the PFEs have the nonlinear form

$$S_i^* = P_i - jQ_i = V_i^* \sum_{k=1}^{n} Y_{ik} V_k, \quad \text{for } i = 1, 2, \ldots, n. \tag{3.13}$$

This is the general form of the PFE. It should be noted that the n PFEs (Equation 3.13) are complex. These n equations therefore represent $2n$ real equations:

$$P_i = \sum_{k=1}^{n} |Y_{ik}||V_i||V_k| \cos(\delta_k - \delta_i + \gamma_{ik}) \underline{\triangle} f_{ip}, \quad \text{for } i = 1, 2, \ldots, n \tag{3.14}$$

$$Q_i = -\sum_{k=1}^{n} |Y_{ik}||V_i||V_k| \sin(\delta_k - \delta_i + \gamma_{ik}) \underline{\triangle} f_{iq}, \quad \text{for } i = 1, 2, \ldots, n. \tag{3.15}$$

Equations 3.14 and 3.15 express the balance of real and reactive powers at bus i, respectively. The nodal current can be written in the power form:

$$P_i - jQ_i = \left(V_i^*\right) \sum_{j=1}^{n} (Y_{ij} V_j). \tag{3.16}$$

Equation 3.16 can be expanded into various forms. Below are the commonly encountered rectangular and polar formulations.

3.4.1.1 Rectangular Formulation

The bus voltages on the right-hand side can be substituted by using the rectangular form $V_i = e_i + jf_i$. Choose the rectangular form; then we have by substitution

$$P_i = e_i \left(\sum_{j=1}^{n} (G_{ij}e_j - B_{ij}f_j) \right) + f_i \left(\sum_{j=1}^{n} (G_{ij}f_j - B_{ij}e_j) \right) \tag{3.17}$$

$$Q_i = f_i \left(\sum_{j=1}^{n} (G_{ij}e_j - B_{ij}f_j) \right) - e_i \left(\sum_{j=1}^{n} (G_{ij}f_j + B_{ij}e_j) \right), \tag{3.18}$$

where the admittance is expressed in the rectangular form

$$Y_{ij} = G_{ij} + jB_{ij}. \tag{3.19}$$

3.4.1.2 Polar Formulation

Using the notation

$$V_i = |V_i|e_i^{j0}$$
$$= |V_i|\angle\theta_i, \tag{3.20}$$

we obtain the following representation

$$P_i = |V_i| \sum_{j=1}^{n} |Y_{ij}||V_j| \cos(\theta_i - \theta_j - \psi_{ij}) \tag{3.21}$$

$$Q_i = |V_i| \sum_{j=1}^{n} |Y_{ij}||V_j| \sin(\theta_i - \theta_j - \psi_{ij}), \tag{3.22}$$

where the admittance is expressed in the polar form: $Y_{ij} = |Y_{ij}|\psi_{ij}$.

3.5 Iterative Techniques for PF Solution

PFEs are large-scale nonlinear sets of equations that require using iterative techniques to obtain their solution. The following major methods are discussed: Gauss–Seidel (G–S) iterative technique, Newton–Raphson method (N–R), and fast-decoupled Newton method.

3.5.1 G–S Iterative Technique

This is a simple iterative method that was very popular in the early days of digital computer-based PF analysis. The more powerful N–R method dominates the field today. The G–S method is still used for small power systems where program simplicity is more important than computing costs, and, in many cases, it is used in large-scale systems to obtain an "initial solution" for the N–R program [2].

3.5.1.1 G–S Algorithm

We consider solving an n-dimensional equation of the type

$$F(x) = 0. \tag{3.23}$$

We introduce the method by first solving the following scalar equation iteratively.

$$f(x) = 0. \tag{3.24}$$

We need to reshape the given function $f(x)$ into the alternative form

$$x = F(x). \tag{3.25}$$

For a given function f, it is always possible to find a function F. As an example, consider the second-order equation

$$f(x) = x^2 - 2x + 5 = 0.$$

The function $f(x)$ represents a parabola. We can obviously write $f(x)$ as

$$x = \underbrace{\frac{1}{2}x^2 + \frac{5}{2}}_{F(x)},$$

which is of the form of Equation 3.23.

Note that $F(x)$ is not unique. In the present example we could also have written

$$x = \underbrace{\frac{2x + 5}{x}}_{F(x)}.$$

The algorithm is based on the following empirical reasoning.

The equations $x - F(x) = 0$ and $f(x) = 0$ must have identical roots. The function $x - F(x)$ represents a "boxed in" region between the sloping line x

and the function $F(x)$. As this region "thins out" to zero close to a root, that root can be reached by a zigzag search process.

3.5.1.2 G–S Method Applied to the PFEs

3.5.1.2.1 Load Buses

For this case, real power P and reactive power Q are given. We assume an n-node system including the slack bus S where both V and ϕ are specified and remain fixed throughout. Since $|P|$ and $|Q|$ are given for all buses except the slack bus, we have for any bus K,

$$I_K = \frac{P_K - jQ_K}{V_K^*}, \quad \text{for} \begin{cases} K = 1, 2, \ldots, n \\ K \neq S \end{cases}, \tag{3.26}$$

where S is the slack bus.

Now the performance equation in the bus frame of reference using Y_{bus} where ground is included as a reference node will be

$$I_{\text{BUS}} = [Y_{\text{BUS}}]\overline{V}_{\text{BUS}}. \tag{3.27}$$

For an n-node system, there are $n - 1$ linear independent equations to solve. Expanding Equation 3.27, we get

$$I_1 = Y_{11}V_1 + Y_{12}V_2 + \cdots$$

and hence for Kth bus

$$I_K = Y_{K1}V_1 + Y_{K2}V_2 + \cdots = \sum_{q=1}^{n} Y_{Kq}V_q; \tag{3.28}$$

that is,

$$I_K = \sum_{q=1}^{n} Y_{Kq}V_q, \quad K = 1, \ldots, n; \quad K \neq S$$

$$= Y_{KK}V_K + \sum_{\substack{q=1 \\ q \neq K}}^{n} Y_{Kq}V_q. \tag{3.29}$$

Thus

$$V_K = \frac{1}{Y_{KK}} \left[I_K - \sum_{\substack{q=1 \\ q \neq K}}^{n} Y_{Kq}V_q \right], \quad \text{for } K = 1, \ldots, n; \quad K \neq S.$$

Substituting for I_K from Equation 3.26, we have

$$V_K = \frac{1}{Y_{KK}} \left[\frac{P_K - jQ_K}{V_k^*} - \sum_{\substack{q=1 \\ q \neq K}}^{n} Y_{Kq} V_q \right], \tag{3.30}$$

where
$K = 1, \ldots, n$
$K \neq S$

In the direct Gauss method, we assume guess values for voltage for all buses except the slack bus where the voltage magnitude and phase angle are specified and remain fixed. Normally, we set the voltage magnitude and phase angle of these buses to be equal to that of the slack bus and work in the per unit system (i.e., we may take in pu, the voltage magnitude and phase angle as $1 \angle 0$). The assumed bus voltage and the slack bus along with P and Q are substituted in the right-hand side of Equation 3.30 to obtain a new and improved set of bus voltages. After the entire iteration is completed, the new set of bus voltages is again substituted along with the specified slack bus voltage in the right-hand side of Equation 3.30 to obtain a new set of bus voltages. The process is continued until

$$|V_i^{k+1} - V_i^k| \leq \varepsilon, \tag{3.31}$$

where
k is an iteration count
ε is a very small number, which depends upon the system accuracy and
is normally equal to 10^{-3} in the per unit system

The iterations are continued until the node of the bus voltage obtained at the current iteration minus the value of the bus voltage at the previous iteration is less than a chosen very small number and in this way we obtain the solution for $|V|$ and ϕ.

3.5.1.3 *G–S Iterative Technique*

In the G–S method, the value of a bus voltage calculated for any bus immediately replaces the previous values in the next step. This contrasts the case of the Gauss method where the calculated bus voltages replace the earlier value only at the end of the complete iteration. Normally the G–S method converges much faster than the Gauss method; that is, the number of iterations needed to obtain the solution is much less in the G–S method, as compared to the Gauss method.

3.5.1.3.1 *Generator Bus*

This is the voltage-controlled bus where $|P|$ and $|V|$ are specified. However, usually there is a limit on reactive power; that is, Q_{max} and Q_{min} to hold the generation voltage within limits are also given. For any bus p we have

$$P_p - jQ_p = V_p^* I_p \tag{3.32}$$

and

$$I_p = \sum_{q=1}^{n} Y_{pq} V_q. \tag{3.33}$$

Then substituting Equation 3.33 into Equation 3.32, we get

$$P_p - jQ_p I_p = V_p^* \sum_{q=1}^{n} Y_{pq} V_q. \tag{3.34}$$

As a result

$$Q_p = \text{Imaginary} \left\{ V_p^* \sum_{q=1}^{n} Y_{pq} V_q \right\}. \tag{3.35}$$

The iterative process starts by assuming the values of bus voltages and the phase angles, say $|V|$ and ϕ, from which we determine the real and imaginary components (e.g., e and f) of the bus voltages.

$$e_p = V_p \cos \phi_p$$
$$f_p = V_p \sin \phi_p. \tag{3.36}$$

For any bus p, V_p is specified and, therefore, the selected value of e_p and f_p must satisfy Equation 3.37 approximately.

$$e_p^2 + f_p^2 \simeq [|V_p| \text{ scheduled}]^2, \tag{3.37}$$

where V_p (scheduled) is the specified bus voltage for any bus p.

We now substitute the assumed values of e_p and f_p in Equation 3.35 to calculate the reactive power Q. If the calculated reactive power exceeds $Q_{\max}$ or is below $Q_{\min}$, then we do the following:

$$\text{If } Q > Q_{\max}, \text{ set } Q = Q_{\max} \quad \text{or} \quad \text{if } Q < Q_{\min}, \text{ set } Q = Q_{\min}, \tag{3.38}$$

and treat this bus as a load bus to find the voltage solution. If Equation 3.38 is not true, then use the phase angle of this assumed bus voltage (i.e., ϕ_p) to recalculate e_p and f_p for any bus p.

We know that $\phi_p^k = \tan^{-1}\left(f_p^k / e_p^k \right)$, where k is the iteration count. Assuming this phase angle ϕ_p also to be that of the scheduled bus voltage V_p (for this bus, ϕ_p is unknown but V_p is given), we get

$$e_p^k (\text{new}) = |V_p \text{ scheduled}| \cos \phi_p^k$$
$$f_p^k (\text{new}) = |V_p \text{ scheduled}| \sin \phi_p^k. \tag{3.39}$$

We substitute the e_p^k and f_p^k from Equation 3.39 in Equation 3.30 to recalculate bus voltages and the process is continued until the process converges. However, at every step the reactive power is also calculated to check whether the calculated Q is within the limit.

3.5.1.3.2 Acceleration Factor

We use acceleration factors to increase the rate of convergence. The choice of an acceleration factor depends upon the system and appropriate values normally lie within 1.4–1.6. Then, after calculating e_p^{k+1} and f_p^{k+1} at the $k+1$st iteration and knowing the acceleration factor, say α and β, we calculate the new estimate for the bus voltages given as

$$e_p^{k+1} \text{ (accelerated)} = e_p^k + \alpha \left[e_p^{k+1} - e_p^k \right]$$
$$f_p^{k+1} \text{ (accelerated)} = f_p^k + \beta \left[f_p^{k+1} - f_p^k \right] \tag{3.40}$$

and this new estimate replaces the calculated values e_p^{k+1} and f_p^{k+1}.

3.5.1.4 Line Flow and Losses

After calculating bus voltages and their phase angles for all the buses, the line flow and line losses are calculated. We assume the normal π representation of the transmission line, as shown in Figure 3.4. Having found the solution of bus voltages (both $|V|$ and ϕ for load buses and only ϕ for a generation bus), we calculate the line flows between any buses p and q.

Let i_{pq} be the current flow from bus p toward q; then

$$i_{pq} = [V_p - V_p]Y_{pq} + V_p \frac{Y'_{pq}}{2}, \tag{3.41}$$

where V_p and V_q are the bus voltages at the buses p and q that have been calculated from the load-flow studies.

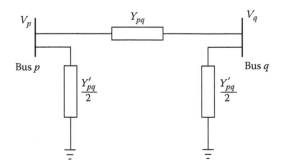

FIGURE 3.4
π-representation of transmission line.

The PF in the line p–q at the bus p is given by

$$P_{pq} - jQ_{pq} = V_p^* i_{pq}$$

$$= V_p^*[V_p - V_q]Y_{pq} + V_p^* V_p \frac{Y_{pq}'}{2}. \qquad (3.42)$$

Similarly the line flow in the line p–q at bus q is given by

$$P_{pq} - jQ_{pq} = V_q^*[V_q - V_p]Y_{pq} + V_q^* V_q \frac{Y_{pq}'}{2}. \qquad (3.43)$$

The algebraic sum of Equations 3.42 and 3.43 is the line losses in the element p–q.

3.5.2 N–R Method

The G–S iterative algorithm (Figure 3.5) is very simple but convergence becomes increasingly slow as the system size grows. The N–R technique (Figure 3.6) converges fast in less than 4–5 iterations regardless of system size. It is popular for large system studies.

3.5.2.1 N–R Algorithm in the Scalar Case

In this method, we start with the scalar Equation 3.21. Assume an initial solution $x^{(0)}$. We try to evaluate the error $\Delta x^{(0)}$ associated with our guess. As $x^{(0)} + \Delta x^{(0)}$ by definition is the correct root then we must require that

$$f(x^{(0)}) + \Delta x^{(0)} = 0. \qquad (3.44)$$

If we expand $f(x)$ in a Taylor series around the initial guess value, we obtain

$$f(x^{(0)}) + \Delta x^{(0)} \left(\frac{df}{dx}\right)^{(0)} + \frac{1}{2}(\Delta x^{(0)})^2 \left(\frac{d^2 f}{d^2 x}\right)^{(0)} + \cdots = 0. \qquad (3.45)$$

All derivatives are computed for $x = x^{(0)}$.

Assuming that the error is relatively small, then the higher-order terms can be neglected and we have

$$f(x^{(0)}) + \Delta x^{(0)} \left(\frac{df}{dx}\right)^{(0)} \approx 0, \qquad (3.46)$$

from which we then compute an approximate value for the error

$$\Delta x^{(0)} = -\frac{f(x^{(0)})}{(df/dx)^{(0)}}. \qquad (3.47)$$

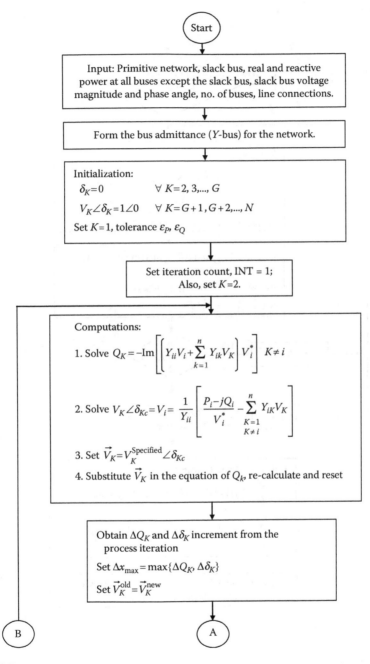

FIGURE 3.5
G–S method.

(*continued*)

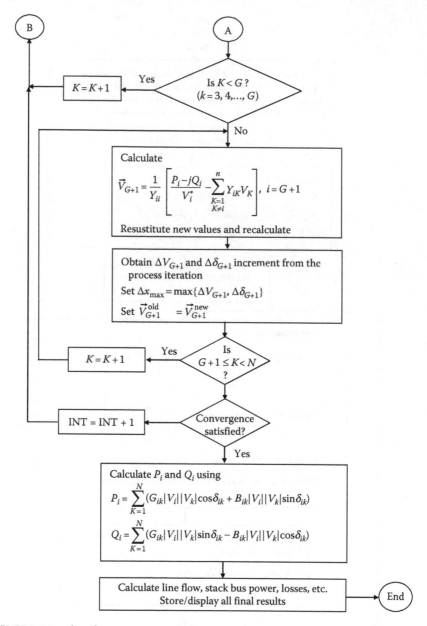

FIGURE 3.5 (continued)

We add this error to the original guess and should thus obtain an improved value $x^{(1)}$ for the root, that is,

$$x^{(1)} = x^{(0)} + \Delta x^{(0)} = x^{(0)} - \frac{f(x^{(0)})}{(df/dx)^{(0)}}.$$

Repeated use of this procedure thus yields the N–R algorithm:

$$x^{(v+1)} = x^{(v)} - \frac{f(x^{(v)})}{(df/dx)^{(v)}}. \tag{3.48}$$

We illustrate the method by finding the larger root of the quadratic equation in x: $f(x) = x^2 - 5x + 4$. Use the initial guess $x^{(0)} = 6$.

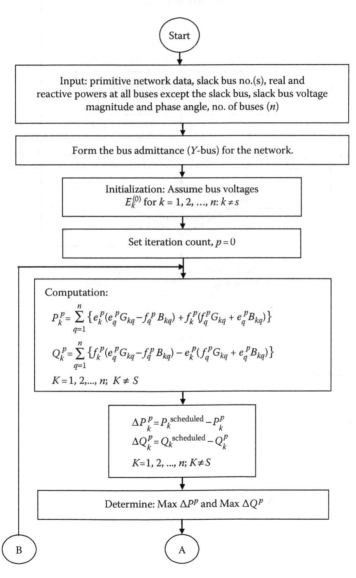

FIGURE 3.6
N–R method.

(continued)

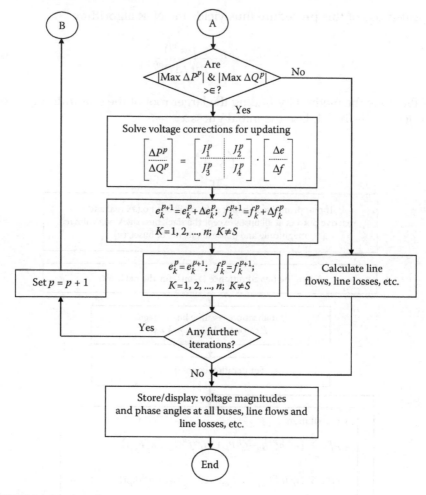

FIGURE 3.6 (continued)

We have

$$\frac{df}{dx} = 2x - 5.$$

The computations proceed according to the following steps:

1. $f(x^{(0)}) = f(6) = 6^2 - 5 \times 6 + 4 = 10.$
2. $\left(\frac{df}{dx}\right)^{(0)} = 2 \times 6 - 5 = 7.$
3. $\Delta x^{(0)} = -\,10/7 = -1.429.$
4. $x^{(1)} = 6 - 1.429 = 4.571.$

Two additional iterations yield

$$x^{(2)} = 4.079$$
$$x^{(3)} = 4.002.$$

The correct value is 4 and we have thus arrived within 0.05% in only three iterations. As in the case of the G–S algorithm, one does not have a full convergence guarantee, unless the initial point is suitably chosen. This, of course, is more important when attempting to solve large-scale, nonlinear or multidimensional system of equations.

3.5.2.2 N–R Algorithm in the n-Dimensional Case

Consider now the solution of the n-dimensional Equation 3.21. By expanding each equation in a Taylor series around the initial guess

$$x^{(0)} = \begin{bmatrix} x_1^{(0)} \\ \vdots \\ x_n^{(0)} \end{bmatrix} \tag{3.49}$$

and upon retaining only the first derivative terms we have

$$f_1(x^{(0)}) + \left(\frac{\partial f_1}{\partial x_1}\right)^{(0)} \Delta x_1 + \cdots + \left(\frac{\partial f_1}{\partial x_n}\right)^{(0)} \Delta x_n \approx 0$$

$$\vdots \qquad\qquad\qquad \vdots \tag{3.50}$$

$$f_n(x^{(0)}) + \left(\frac{\partial f_n}{\partial x_1}\right)^{(0)} \Delta x_1 + \cdots + \left(\frac{\partial f_n}{\partial x_n}\right)^{(0)} \Delta x_n \approx 0.$$

We can write this system of n linear equations as follows:

$$\begin{bmatrix} f_1(x^{(0)}) \\ \vdots \\ f_n(x^{(0)}) \end{bmatrix} + \begin{bmatrix} \left(\frac{\partial f_1}{\partial x_1}\right)^{(0)} & \cdots & \left(\frac{\partial f_1}{\partial x_n}\right)^{(0)} \\ \vdots & & \vdots \\ \left(\frac{\partial f_n}{\partial x_1}\right)^{(0)} & \cdots & \left(\frac{\partial f_n}{\partial x_n}\right)^{(0)} \end{bmatrix} \begin{bmatrix} \Delta x_1 \\ \vdots \\ \Delta x_n \end{bmatrix} \approx \begin{bmatrix} 0 \\ \vdots \\ 0 \end{bmatrix}.$$

Or, in compact matrix–vector notation,

$$\mathbf{f}(\mathbf{x}^{(0)}) + \mathbf{J}^{(0)} \Delta \mathbf{x}^{(0)} \approx 0. \tag{3.51}$$

From this last equation, we then solve for the error vector

$$\Delta \mathbf{x}^{(0)} \approx -[\mathbf{J}^{(0)}]^{-1} \mathbf{f}(\mathbf{x}^{(0)}). \tag{3.52}$$

Upon adding this vector to the original guess and at the same time indicating a repetitive process we obtain the N–R algorithm

$$\mathbf{x}^{(v+1)} = \mathbf{x}^{(v)} - [\mathbf{J}^{(v)}]^{-1}\mathbf{f}(\mathbf{x}^{(v)}). \tag{3.53}$$

The $n \times n$ matrix $\mathbf{J}$, having as its elements the partial derivatives $\partial f_i / \partial x_j$, is referred to as a Jacobian. Use of this algorithm obviously necessitates the need of inverting an $n \times n$ matrix in each iteration.

3.5.2.3 N–R Algorithm Applied to the PFEs

The solution of the power system problem of a nonlinear set of power equations is popularly accomplished by using the N–R technique, which is applicable to the rectangular and polar formulations of the problem. The two cases are presented below.

3.5.2.3.1 Case A: Formulation of the N–R Method in Rectangular Form

Based on the help of the N–R iterations, the general nonlinear algebraic equations of power are transformed into a set of linear algebraic equations relating the changes in power (i.e., error in power) to the change in real and reactive components of bus voltages with the help of the Jacobian matrix. This is actually done by expanding the function by a Taylor's series and neglecting higher-order derivatives and higher power. It is necessary that the initial guess be close to the solution point.

$$
\begin{bmatrix} \Delta P_1 \\ \cdots \\ \Delta P_{n-1} \\ \cdots \\ \Delta Q_1 \\ \cdots \\ \Delta Q_{n-1} \end{bmatrix}
=
\begin{bmatrix}
\dfrac{\partial P_1}{\partial e_1} & \dfrac{\partial P_1}{\partial e_{n-1}} & \dfrac{\partial P_1}{\partial f_1} & \dfrac{\partial P_1}{\partial f_{n-1}} \\
\dfrac{\partial P_{n-1}}{\partial e_1} & \dfrac{\partial P_{n-1}}{\partial e_{n-1}} & \dfrac{\partial P_{n-1}}{\partial f_1} & \dfrac{\partial P_{n-1}}{\partial f_{n-1}} \\
\dfrac{\partial Q_1}{\partial e_1} & \dfrac{\partial Q_1}{\partial e_{n-1}} & \dfrac{\partial Q_1}{\partial f_1} & \dfrac{\partial Q_1}{\partial f_{n-1}} \\
\dfrac{\partial Q_{n-1}}{\partial e_1} & \dfrac{\partial Q_{n-1}}{\partial e_{n-1}} & \dfrac{\partial Q_{n-1}}{\partial f_1} & \dfrac{\partial Q_{n-1}}{\partial f_{n-1}}
\end{bmatrix}
\begin{bmatrix} \Delta e_1 \\ \vdots \\ \Delta e_{n-1} \\ \vdots \\ \Delta f_1 \\ \vdots \\ \Delta f_{n-1} \end{bmatrix}
\tag{3.54}
$$

where the nth bus is the slack bus. Equation 3.54 can be expressed as

$$
\begin{bmatrix} \Delta P \\ \cdots \\ \Delta Q \end{bmatrix}
=
\begin{pmatrix} J_1 & \vdots & J_2 \\ \cdots & \cdots & \cdots \\ J_3 & \vdots & J_4 \end{pmatrix}
\begin{bmatrix} \Delta e \\ \cdots \\ \Delta f \end{bmatrix},
\tag{3.55}
$$

where J_1, J_2, J_3, and J_4 are the elements of the Jacobian matrix that are calculated from the expression of power as follows.

Elements of J_1
Off-diagonal elements

$$\partial P_p / \partial e_q = e_p G_{pq} - f_p B_{pq}, \quad \text{for } q = p. \tag{3.56}$$

Diagonal elements

$$\frac{\partial P_p}{\partial e_p} = 2e_p G_{pp} - f_p B_{pp} - f_p B_{pp}$$

$$= \sum_{\substack{q=1 \\ q \neq p}}^{n} [e_q G_{pq} - f_q B_{pq}]. \tag{3.57}$$

Elements of J_2
Off-diagonal elements

$$\partial P_p / \partial f_q = e_p B_{pq} + f_p G_{pq} \quad q \neq p. \tag{3.58}$$

Diagonal elements

$$\partial P_p / \partial f_p = e_p B_{pp} + 2f_p G_{pp} - e_p B_{pp}$$

$$+ \sum_{\substack{q=1 \\ q \neq p}}^{n} [f_q G_{pq} - e_q B_{pq}]. \tag{3.59}$$

Elements of J_3
Off-diagonal elements

$$\partial Q_p / \partial e_q = f_p G_{pq} + e_p B_{pq}, \quad \text{for } q \neq p. \tag{3.60}$$

Diagonal elements

$$\partial Q_p / \partial e_p = f_p G_{pp} - f_p G_{pp} + 2e_p B_{pp} - \sum_{\substack{q=1 \\ q \neq p}}^{n} [f_p G_{pq} - e_q B_{pq}]. \tag{3.61}$$

Elements of J_4
Off-diagonal elements

$$\partial Q_p / \partial f_q = f_p B_{pq} - e_p B_{pq}, \quad \text{for } q \neq p. \tag{3.62}$$

Diagonal elements

$$\frac{\partial Q_p}{\partial f_p} = e_p G_{pp} + 2f_p B_{pp} - e_p G_{pp} + \sum_{\substack{q=1 \\ q \neq p}}^{n} [e_q G_{pq} - f_q B_{pq}]. \tag{3.63}$$

The following steps are performed in sequence to obtain the PF solution by N–R method:

1. For load buses where P and Q are given, we assume the bus voltages' magnitude and phase angle for all the buses except the slack bus where $|V|$ and ϕ are specified. Normally, we use a flat voltage start; that is, we set the assumed bus voltage magnitude and its phase angle (in other words the real and imaginary component e and f of the bus voltages) equal to the slack bus values.

2. Substituting these assumed bus voltages (i.e., e and f), we calculate the real and reactive components of power (i.e., P_p and Q_p) for all the buses $p = 1, \ldots, n-1$ except the slack bus.

3. Since P_p and Q_p for any bus p are given, the error in the power will be

$$\Delta P_p^k = P_p \text{ (scheduled)} - P_p^k$$
$$\Delta Q_p^k = Q_p \text{ (scheduled)} - Q_p^k, \tag{3.64}$$

   where
   k is an iteration count
   P_p^k and Q_p^k are the active and reactive power values calculated with the latest value of bus voltages at any iteration k

4. Elements of the Jacobian matrix (J_1, J_2, J_3, and J_4) are calculated with the latest bus voltages and calculated power equations.

5. We then solve the linear set of Equations 3.54 by either an iterative technique or by the method of elimination (normally by the Gaussian elimination method) to determine the voltage correction, that is Δe_p and Δf_p at any bus p.

6. This value of voltage correction is used to determine the new estimate of bus voltages as follows:

$$e_p^{k+1} = e_p^k + \Delta e_p^k$$
$$f_p^{k+1} = f_p^k + \Delta f_p^k,$$

   where k is an iteration count.

7. Now this new estimate of the bus voltage e_p^{k+1} and f_p^{k+1} is used to recalculate the error in power and, thus, the entire algorithm starting from step 3 as given earlier is repeated.

During each iteration, the elements of the Jacobian are calculated since they depend upon the latest voltage estimate and calculated power. The process is continued until the error in power becomes very small, that is,

$$|\Delta P| < \varepsilon \quad \text{and} \quad |\Delta Q| < \varepsilon, \tag{3.65}$$

where ε is a very small number. This method converges faster than the G–S method and exhibits quadratic convergence properties. Moreover, while the number of iterations to obtain the nominal solutions increases with the problem size in the case of the G–S method, the number of iterations to obtain a converged solution is nearly constant in the N–R method. Later, we show that with the modified Newton method, nearly 5–6 iterations are needed to obtain a solution although the time taken to complete an iteration is nearly seven times as that of the G–S method.

Example 3.5.1

Solve the following two equations in x_1 and x_2 using the N–R method:

$$F_1(x) = x_1^2 + x_2^2 - 4x_1 = 0$$
$$F_2(x) = x_1^2 + x_2^2 - 8x_1 + 12 = 0.$$

1. Find expressions for the elements of the Jacobian matrix and find the correction increments Δx_1, Δx_2.
2. Calculate the first five iterations to find estimates of the solution using the following initial guesses:

(a) $x_1 = 2, \quad x_2 = 4.$
(b) $x_1 = -5, \quad x_2 = -5.$
(c) $x_1 = -0.1, \quad x_2 = 1.$

SOLUTION

The Jacobian elements are as follows:

$$\frac{\partial F_1}{\partial x_1} = 2x_1 - 4; \quad \frac{\partial F_1}{\partial x_2} = 2x_2; \quad \frac{\partial F_2}{\partial x_1} = 2x_1 - 8; \quad \frac{\partial F_2}{\partial x_2} = 2x_2.$$

Now, using the formulation for the typical N–R problem, we have

$$\begin{bmatrix} \dfrac{\partial F_1}{\partial x_1} & \dfrac{\partial F_1}{\partial x_2} \\ \dfrac{\partial F_2}{\partial x_1} & \dfrac{\partial F_2}{\partial x_2} \end{bmatrix} \begin{bmatrix} \Delta x_1 \\ \Delta x_2 \end{bmatrix} = \begin{bmatrix} -F_1(x) \\ -F_2(x) \end{bmatrix}$$

$$\begin{bmatrix} 2x_1 - 4 & 2x_2 \\ 2x_1 - 8 & 2x_2 \end{bmatrix} \begin{bmatrix} \Delta x_1 \\ \Delta x_2 \end{bmatrix} = \begin{bmatrix} -(x_1^2 + x_2^2 - 4x_1) \\ -(x_1^2 + x_2^2 - 8x_1 + 12) \end{bmatrix}.$$

The solution for Δx_1, Δx_2 is given by

$$\begin{bmatrix} \Delta x_1 \\ \Delta x_2 \end{bmatrix} = \begin{bmatrix} 2x_1 - 4 & 2x_2 \\ 2x_1 - 8 & 2x_2 \end{bmatrix}^{-1} \begin{bmatrix} -(x_1^2 + x_2^2 - 4x_1) \\ -(x_1^2 + x_2^2 - 8x_1 + 12) \end{bmatrix}$$

$$\begin{bmatrix} \Delta x_1 \\ \Delta x_2 \end{bmatrix} = \begin{bmatrix} 3 - x_1 \\ -(x_1^2 + x_2^2 - 8x_1 + 12) \quad 12x_2 \end{bmatrix}.$$

As a result the new estimates of the solution are given by

$$x_i^{n+1} = x_i^n + \Delta x_i^n$$
$$x_i^{n+1} = x_i^n + (3 - x_1^n)$$
$$x_i^{n+1} = 3.$$

Note that x_1 is independent of the starting point. In addition, the iterative formula for x_2 is

$$x_2^{(n+1)} = x_2^{(n)} + \frac{x_1^{2(n)} - x_2^{2(n)} - 6x_1^{(n)} + 12}{2x_2^n}$$
$$= \frac{x_1^{2n} + x_2^{2n} - 6x_1^n + 12}{2x_2^n},$$

with $x_i^n = 3$,

$$x_2^{(n+1)} = \frac{x_2^{2n} + 3}{2x_2^n}.$$

Tables 3.2 through 3.4 give different values of x_1 and x_2 at different iterations.

TABLE 3.2

Iteration Process with Start Point (2, 4)

Iteration	x_1	x_2
0	2	4
1	3	2.5
2	3	1.85
3	3	1.74
4	3	1.7321
5	3	1.73205

TABLE 3.3

Iteration Process with Start Point (−5, −5)

Iteration	x_1	x_2
0	−c5	−5
1	3.0	−9.2
2	3.0	−4.763
3	3.0	−2.6964
4	3.0	−1.9045
5	3.0	−1.7399
6	3.0	−1.7321
7	3.0	−1.73205

TABLE 3.4

Iteration Process with Start Point (−0, 1)

Iteration	x_1	x_2
0	−0.1	1
1	3.0	6.805
2	3.0	3.623
3	3.0	2.225
4	3.0	1.787
5	3.0	1.7329
6	3.0	1.73205

Example 3.5.2

Use the N–R method to find the roots of the equation

$$F(x) = x^3 - 6x^2 + 11x - 6.$$

Assume the following initial guesses: $x = 0$, 0.5, 1.5, 2.5, and 4.

SOLUTION

$$F(x) = x^3 - 6x^2 + 11x - 6$$
$$F'(x) = 3x^2 - 12x + 11$$

$$\Delta x = -\frac{F}{F'} = -\frac{x^3 + 6x^2 - 11x + 6}{3x^2 - 12x + 11}$$

New estimates are given by

$$x^{n+1} = x^n + \Delta x.$$

For the different initial points the solution is developed in Table 3.5. Note that the equation has roots at the points where $x = 1.00$, 2.00, and 3.00. We converged on the first and the third roots depending on the starting estimate. The proximity of the initial guess to the solution affects the N–R solutions of the problem.

TABLE 3.5

Iteration Process of Example 3.5.2 with Different Initial Points

Iteration	Value of the Variable x at the End of Each Iteration for the Five Initial Values Used				
0 (initial)	0.0000	0.5000	1.5000	2.5000	4.0000
1	0.5450	0.8260	3.0000	1.0000	3.4500
2	0.8499	0.9677	—	—	3.1500
3	0.9747	0.9985	—	—	3.0300
4	0.9909	1.0000	—	—	3.0009
5	1.0000	—	—	—	3.0000

Example 3.5.3

For the network shown in Figure 3.7, do the following:

1. Compute the Y-bus admittance matrix.
2. Write down the load-flow equations of the problem using the rectangular formulation.

The nodal admittances are given as

$$Y_{11} = 4 - j5, \quad Y_{22} = 4 - j10, \quad Y_{33} = 8 - j15$$
$$Y_{12} = 0, \quad Y_{13} = -4 + j5, \quad Y_{23} = -4 + j10.$$

We also have

$$|V_1| = 1, \quad \theta_1 = 0$$
$$P_2 = 1.7, \quad |V_2| = 1.1249$$
$$P_3 = -2, \quad Q_3 = -1.$$

SOLUTION

The load-flow equations are obtained utilizing the following formula:

$$P_i - jQ_i = V_i^* \sum Y_{ij} V_j.$$

Recall Equation 3.23. To obtain the rectangular forms, we have for bus 1

$$P_1 - 1Q_1 = (1 + j0[(4 - j5)(1 + j0) + (-4 + j5)(e_3 + jf_3)]).$$

This yields

$$P_1 = 4 - e_3 - 5f_3.$$
$$-Q_1 = -5 + 5e_3 - 4f_3.$$

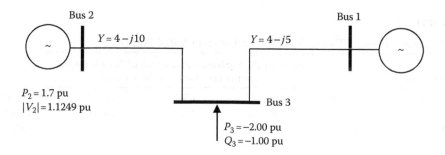

FIGURE 3.7
Single-line diagram for Example 3.5.3.

For bus 2 we have

$$1.7 - jQ_2 = (e_2 - jf_2)[(4 - j10)(e_2 + jf_2) + (-4 + j10)(e_3 + jf_3)].$$

This reduces to

$$1.7 - jQ_2 = (4 - j10)(e_2^2 + jf_2^2) + (e_2 - jf_2)(-4 + j10) * (e_3 + jf_3).$$

Using

$$|V_2|^2 = e_2^2 + f_2^2 = (1.1249)^2,$$

we obtain

$$1.7 - jQ_2 = 5.0616 - j12.654 + (-4 + j10)[(e_2 e_3 + f_1 f_3) + j(f_3 e_2 - f_2 e_3)].$$

Separating real and imaginary parts, we get

$$1.7 = 5.0616 - 4(e_2 e_3 + f_2 f_3) - 10(f_3 e_2 - f_2 e_3)$$
$$-Q_2 = -10.654 + 10(e_2 e_3 + f_2 f_3) - 4(f_3 e_2 - f_2 e_3).$$

We should include

$$(1.1249)^2 = e_2^2 + f_2^2.$$

For bus 3 we have

$$-2 + j1 = (e_3 - jf_3)[(-4 + j5) + (-4 + j10)(e_2 + jf_2) + (8 - j15)(e_3 + jf_3)].$$

Separating real and imaginary parts, we get

$$-2 = -4e_3 + 5f_3 - e_3(4e_3 + 10f_2) + f(10e_2 - 4f_2) + 8(e_3^2 + f_3^2)$$
$$1 = 5e_3 + 4f_3 + e_3(10e_2 - 4f_2) + f_3(4e_2 - 10f_2) - 15(e_3^2 + f_3^2).$$

3.5.2.3.2 Case B: Formulation of the N–R Method in Polar Form

The PF problems using the N–R method can be formulated in polar coordinates. For any bus p, we have

$$V_p = |V_p|e^{j\delta p}, \quad \text{then } V_p^* = |V_p|e^{-j\delta p} \tag{3.66}$$

$$V_q = |V_q|e^{j\delta p}, \quad \text{then } Y_{pq} = |Y_{pq}|e^{-j\theta pq}, \tag{3.67}$$

where
   δ is the phase angle of the bus voltages
   θ_{pq} is the admittance angle

As a result, for any bus p,

$$P_p - jQ_p = V_p^* \sum_{q=1}^{n} Y_{pq} V_q. \tag{3.68}$$

Substituting Equation 3.105 into Equation 3.106, we have

$$P_p - jQ_p = \sum_{q=1}^{n} |V_p V_q Y_{pq}| e^{-j(\theta_{pq} + \delta_p - \delta_q)}. \tag{3.69}$$

Thus

$$\begin{aligned}
P_p &= \text{Real}\left[V_p^* \sum_{q=1}^{n} Y_{pq} V_q \right] \\
&= \sum_{q=1}^{n} |V_p V_q Y_{pq}| \cos(\theta_{pq} + \delta_p - \delta_q) \\
&= |V_p V_p Y_{pp}| \cos(\theta_{pp}) + \sum_{\substack{q=1 \\ q \neq p}}^{n} |V_p V_q V_{pq}| \cos(\theta_{pq} + \delta_p - \delta_q) \tag{3.70}
\end{aligned}$$

and

$$\begin{aligned}
Q_p &= \text{Imaginary}\left[V_p^* \sum_{q=1}^{n} Y_{pq} V_q \right] \\
&= \sum_{q=1}^{n} |V_p V_q Y_{pq}| \sin(Q_{pq} + \delta_p - \delta_q) \\
&= |V_p V_p Y_{pp}| \sin\theta_{pp} + \sum_{\substack{q=1 \\ q \neq p}}^{n} |V_p V_q Y_{pq}| \sin(\theta_{pq} + \delta_p - \delta_q), \tag{3.71}
\end{aligned}$$

for $p = 1, \ldots, n-1$ as the nth bus is a slack bus.

Now the linear N–R Equation in polar form becomes

$$\begin{bmatrix} \Delta P \\ \Delta Q \end{bmatrix} = \begin{bmatrix} J_1 & \vdots & J_2 \\ \cdots & \cdots & \cdots \\ J_3 & \vdots & J_4 \end{bmatrix} \begin{bmatrix} \Delta \delta \\ \Delta |V| \end{bmatrix}, \tag{3.72}$$

where $J_1, J_2, J_3,$ and J_4 are the elements of Jacobian $\mathbf{J}$ which can be calculated from the power Equations 3.70 and 3.71 as follows:

Elements of J₁
Off-diagonal elements

$$\frac{\partial P_p}{\partial \delta_p}\bigg|_{p \neq q} = |V_p V_q Y_{pq}| \sin(\theta_{pq} + \delta_p - \delta_q). \tag{3.73}$$

Diagonal elements

$$\frac{\partial P_p}{\partial \delta_p} = -\sum_{\substack{q=1 \\ q \neq p}}^{n} |V_p V_q Y_{pq}| \sin(\theta_{pq} + \delta_p - \delta_q). \tag{3.74}$$

Elements of J₂
Off-diagonal elements

$$\frac{\partial P_p}{\partial |V_q|}\bigg|_{p \neq q} = |V_p Y_{pq}| \cos(\theta_{pq} + \delta_p - \delta_q). \tag{3.75}$$

Diagonal elements

$$\frac{\partial P_p}{\partial |V_p|}\bigg| = 2|V_p Y_{pp}| \cos \theta_{pp} + \sum_{\substack{q=1 \\ q \neq p}}^{n} |V_q Y_{pq}| \cos(\theta_{pq} + \delta_p - \delta_q). \tag{3.76}$$

Elements of J₃
Off-diagonal elements

$$\frac{\partial Q_p}{\partial \delta_q} = -|V_p V_q Y_{pq}| \cos(\theta_{pq} + \delta_p - \delta_q). \tag{3.77}$$

Diagonal elements

$$\frac{\partial Q_p}{\partial \delta_p} = -\sum_{\substack{q=1 \\ q \neq p}}^{n} |V_p V_q P_{pq}| \cos(\theta_{pq} + \delta_p - \delta_q). \tag{3.78}$$

Elements of J₄
Off-diagonal elements

$$\frac{\partial Q_p}{\partial |V_q|}\bigg|_{q \neq p} = |V_p Y_{pq}| \sin(\theta_{pq} + \delta_p - \delta_q). \tag{3.79}$$

Diagonal elements

$$\frac{\partial Q_p}{\partial |V_p|} = 2|V_p Y_{pp}| \sin \theta_{pp} + \sum_{\substack{q=1 \\ q \neq p}}^{n} |V_q Y_{pq}| \sin(\theta_{pq} + \delta_p - \delta_q). \tag{3.80}$$

The elements of the Jacobian are calculated based on the latest voltage estimate and calculated power. However, the algorithm is the same as that applied with the rectangular coordinates. Notably, the formulation in the polar coordinates takes less computational effort and also requires less memory space.

The real power P is less sensitive to changes in the voltage magnitude $\Delta|V|$ and similarly the reactive power Q is less sensitive to the changes in the phase angle δ. We can thus approximate Equation 3.72 as follows:

$$\begin{bmatrix} \Delta P \\ \Delta Q \end{bmatrix} = \begin{bmatrix} J_1 & \vdots & 0 \\ \cdots & \vdots & \cdots \\ 0 & \vdots & J_4 \end{bmatrix} \begin{bmatrix} \Delta \delta \\ \Delta|V| \end{bmatrix}. \tag{3.81}$$

In the case of a generator bus P and $|V|$ are given. The real power P for any bus p is given by

$$P_p = \text{Real} \left[V_p^* \sum_{q=1}^{n} Y_{pq} V_q \right]. \tag{3.82}$$

and also for bus p, we have

$$|V_p|^2 = e_p^2 + f_p^2, \tag{3.83}$$

where
  V_p is the voltage magnitude
  e_p and f_p are its real and imaginary components, respectively

The matrix equations relating the changes in bus powers and square of the bus voltage magnitudes to the changes in the real and imaginary components of the voltages are

$$\begin{bmatrix} \Delta P \\ \cdots \\ \Delta Q \\ \cdots \\ \Delta|V|^2 \end{bmatrix} \begin{bmatrix} J_1 & \vdots & J_2 \\ \cdots & \cdots & \cdots \\ J_3 & \vdots & J_4 \\ \cdots & \cdots & \cdots \\ J_5 & \vdots & J_6 \end{bmatrix} \begin{bmatrix} \Delta e \\ \cdots \\ \Delta f \end{bmatrix}, \tag{3.84}$$

where

$$\Delta|V_p^k|^2 = \left[|V_p \text{ (scheduled)}|^2 - |V_p^k|^2 \right]$$

(V_p^k is the calculated bus voltage after the kth iteration.)

The elements of Jacobian are calculated as follows:

Elements of J_5
Off-diagonal elements

$$\frac{\partial |V_p|^2}{\partial e_q} = 0, \quad \text{for } q \neq p. \tag{3.85}$$

Diagonal elements

$$\frac{\partial |V_p|^2}{\partial e_p} = 2e_p. \tag{3.86}$$

Elements of J_6
Off-diagonal elements

$$\frac{\partial |V_p|^2}{\partial f_q} = 0, \quad \text{for } q \neq p. \tag{3.87}$$

Diagonal elements

$$\frac{\partial |V_p|^2}{\partial f_p} = 2f_p, \tag{3.88}$$

where
V_p^k is the bus voltage calculated at the kth iteration
V_p (scheduled) is the voltage given (i.e., specified) at any bus p as it is
  the generation bus

Calculations for the elements J_1, J_2, J_3, and J_4 were discussed earlier. After obtaining bus voltages, PF and line losses are calculated. We next give a modified Newton method, which has several benefits over the conventional N–R method.

Example 3.5.4

For the network shown in Figure 3.8, do the following:

1. Compute the Y-bus admittance matrix and state the initial bus voltages in polar form.
2. Write down the load-flow equations of the problem using the polar formulation.

SOLUTION
  1. We have in polar form:
     The diagonal terms of the Y_{bus}

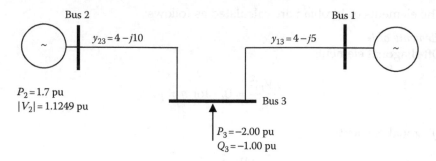

FIGURE 3.8
Single-line diagram.

$$Y_{11} = 6.4031\angle-51.34°$$
$$Y_{22} = 10.77\angle-68.199°$$
$$Y_{33} = 17.00\angle-61.928°.$$

The off-diagonal terms of the Y_{bus}

$$Y_{12} = 0$$
$$Y_{13} = 6.4031\angle128.66°$$
$$Y_{23} = 10.77\angle111.80°.$$

Furthermore, $Y_{ij} = Y_{ji}$ since the admittance matrix is inherently symmetrical. The bus voltages are represented as

$$V_1 = 1\angle0$$
$$V_2 = 1.1249\angle\theta_2$$
$$V_3 = |V_3|\angle\theta_3.$$

1. We now can write the PFE for the three-bus system shown in Figure 3.8. For bus 1, we have

$$P_1 - jQ_1 = V_1^*(Y_{11}V_1 + Y_{12}V_2 + Y_{13}V_3)$$
$$\Rightarrow P_1 - jQ_1 = 6.4031\angle-51.34° + 6.4031|V_3|\angle128.66 + \theta_3.$$

For bus 2, we have

$$P_2 - jQ_2 = V_2^*(Y_{12}V_1 + Y_{22}V_2 + Y_{23}V_3)$$
$$\Rightarrow 1.7 - jQ_2 = 13.628\angle-68.199° + 12.115V_3\angle118 + \theta_3 - \theta_2.$$

For bus 3, we have

$$P_3 - jQ_3 = V_3^*(Y_{13}V_1 + Y_{23}V_2 + Y_{33}V_3)$$
$$\Rightarrow -2 + j1 = V_3\angle-\theta_3[6.4031\angle128.66° + 10.77\angle111.8*1.1249$$
$$\angle\theta_2 + 17\angle-61.928^*V_3\angle\theta_3].$$

This reduces to

$$-2 + j1 = 6.4031V_3\angle128.66° + 12.11V_3\angle111.8 - \theta_2 - \theta_2$$
$$+ 17.00V_3\angle-61.928°.$$

We now separate the real and imaginary parts of the above equations such that:

For bus 1

$$P_1 = 4 + 6.4031V_3\cos(128.66 + \theta_3)$$
$$-Q_1 = -5 + 6.4031V_3\sin(128.66 + \theta_3).$$

For bus 2

$$1.7 = 5.6612 + 12.115V_3\cos(111.8 + \theta_3 - \theta_2)$$
$$-Q_2 = -12.653 + 12.115V_3\sin(111.8 + \theta_3 - \theta_2).$$

For bus 3

$$-2 = 6.4031V_3\cos(128.66 - \theta_3) + 12.115V_3\cos(111.8 + \theta_2 - \theta_3)$$
$$+ 8V_3^2$$
$$1 = 6.4031V_3\sin(128.6 - \theta_3) + 12.115V_3\sin(111.8 + \theta_2 - \theta_3)$$
$$- 15V_3^2.$$

The above six equations define the PF problems in polar form.

3.5.3 Fast-Decoupled PF Method

The fast-decoupled PF method is a very fast technique for computing the PF solution, in which both speed and sparsity are exploited. The technique is an extension of Newton's method formulated in polar coordinates with certain approximations.

The earlier PFE (i.e., Equation 3.72) using the N–R method can be expressed in polar coordinates as

$$\begin{bmatrix} \Delta P \\ \Delta Q \end{bmatrix} = \begin{bmatrix} H & N \\ M & L \end{bmatrix} \begin{bmatrix} \Delta\delta \\ \dfrac{\Delta|V|}{V} \end{bmatrix}, \tag{3.89}$$

where H, N, M, and L are the elements (viz., J_1, J_2, J_3, and J_4) of the Jacobian matrix. Since changes in real power (i.e., ΔP) are less sensitive to the changes in voltage magnitude (ΔV) and changes in reactive power (i.e., ΔQ) are less sensitive to the changes in angle ($\Delta\delta$), Equation 3.89 is written as

$$\begin{bmatrix} \Delta P \\ \Delta Q \end{bmatrix} = \begin{bmatrix} H & O \\ O & L \end{bmatrix} \begin{bmatrix} \Delta\delta \\ \dfrac{\Delta|V|}{V} \end{bmatrix}. \tag{3.90}$$

Equation 3.90 consists of two decoupled equations:

$$[\Delta P] = [H][\Delta\delta] \tag{3.91}$$

and

$$[\Delta Q] = [L]\frac{[\Delta V]}{V}. \tag{3.92}$$

The elements of the Jacobian (Equation 3.90) are defined again (Equation 3.88),

$$H_{pq} = \frac{\delta P_p}{\delta\delta_q} \quad \text{and} \quad L_{pq} = \frac{\delta Q_p}{\delta E_q}|E_q|.$$

Equations 3.91 and 3.92 for calculating the elements of the Jacobian (H and L) are

$$P_p = \sum_{q=1}^{n} |V_p V_q Y_{pq}| \cos[\theta_{pq} + \delta_p - \delta_q]$$

$$= |V_p V_p Y_{pp}| \cos\theta_{pp} + \sum_{q=1}^{n} |V_p V_q Y_{pq}| \cos[\theta_{pq} + \delta_p - \delta_q] \tag{3.93}$$

and

$$Q_p = |V_p V_p Y_{pp}| \sin\theta_{pp} + \sum_{\substack{q=1 \\ q\neq p}}^{n} |V_p V_q Y_{pq}| \sin[\theta_{pq} + \delta_p - \delta_q]. \tag{3.94}$$

The off-diagonal elements of H are

$$H_{pq} = \frac{\delta P_p}{\delta\delta_q} = |V_p V_q Y_{pq}| \sin[\theta_{pq} + \delta_p - \delta_q]$$

$$= |V_p V_q Y_{pq}| [\sin[\theta_{pq}] \cos[\delta_p - \delta_q] + \cos\theta_{pq} \sin[\delta_p - \delta_q]]$$

$$= |V_p V_q| [Y_{pq} \sin\theta_{pq} \cos[\delta_p - \delta_q] + Y_{pq} \cos\theta_{pq} \sin[\delta_p - \delta_q]]$$

$$= |V_p V_q| [-B_{pq} \cos[\delta_p - \delta_q] + G_{pq} \sin[\delta_p - \delta_q]]. \tag{3.95}$$

Similarly, the off-diagonal elements of L are

$$L_{pq} = \frac{\delta Q_p |V_q|}{\delta V_q} = |V_p V_q Y_{qp}| \sin[\theta_{pq} + \delta_p - \delta_q]$$

$$= |V_p V_q| [G_{pq} \sin[\delta_p - \delta_q] - B_{pq} \cos[\delta_p - \delta_q]]. \tag{3.96}$$

From Equations 3.95 and 3.96, it is evident that

$$H_{pq} = L_{pq} = |V_p V_q|[G_{pq} \sin[\delta_p - \delta_q] - B_{pq} \cos[\delta_p - \delta_q]]. \tag{3.97}$$

The diagonal elements of H are given by

$$H_{pp} = \frac{\delta P_p}{\delta \delta_p} = -\sum_{\substack{q=1 \\ q \neq p}}^{n} |V_p V_q Y_{pq}| \sin[\theta_{pq} + \delta_p - \delta_q]$$

$$= -\left[\sum_{q=1}^{n} |V_p V_q Y_{pq}| \sin[\theta_{pq} + \delta_p - \delta_q] - |V_p V_q Y_{pp}| \sin \theta_{pp}\right]$$

$$= [+Q_p + V_p^2 B_{pp}]$$

$$= -V_p^2 B_{pp} - Q_p \tag{3.98}$$

and the elements of L are given by

$$L_{pp} = \frac{\delta Q_p |V_p|}{\delta V_p}$$

$$= |2V_p^2 Y_{pp}| \sin \theta_{pp} + \sum_{\substack{q=1 \\ q \neq p}}^{n} |V_p V_q Y_{pq}| \sin[\theta_{pq} + \delta_p - \delta_q]$$

$$= |2V_p^2 Y_{pp}| \sin \theta_{pp} + Q_p - |V_p^2 Y_{pp}| \sin \theta_{pp}$$

$$= Q_p + = |V_p^2 Y_{pp}| \sin \theta_{pp}$$

$$= -V_p^2 B_{pp} + Q. \tag{3.99}$$

In the case of fast-decoupled PF formulation, the following approximations are made.

$$\cos[\delta_p - \delta_q] \approx 1$$

$$G_{pq} \sin[\delta_p - \delta_q] \ll B_{pq}$$

and

$$Q_p \ll B_{pp} V_p^2. \tag{3.100}$$

These approximations are made for the calculations of elements of Jacobian H and L, to yield

$$H_{pq} = L_{pq} = -|V_p||V_q|B_{pq} \quad \text{for } q \neq p$$

and

$$H_{pp} = L_{pp} = -B_{pq}|V_p^2|. \tag{3.101}$$

Equations 3.91 and 3.92 with the substitutions of elements of Jacobian Equation 3.99 take the form

$$\Delta P = H\Delta\delta,$$

that is,

$$[\Delta P_p] = [V_p][V_q][B'_{pq}][\Delta\delta_q] \qquad (3.102)$$

and similarly

$$[\Delta Q_p] = [V_p][V_q][B''_{pq}][\Delta V], \qquad (3.103)$$

where B'_{pq} and B''_{pq} are the elements of the $[-B]$ matrix.

Further decoupling and the final algorithm for fast-decoupled PF studies are obtained by the following:

1. Omit from B' the representation of those network elements that affect MVAr flows, that is, shunt reactances and off-nominal in-phase transformer taps.
2. Omit from B'' the angle-shifting affects of phase-shifters.
3. Divide Equations 3.102 and 3.103 by V_p set $V_q = 1$ pu, and also neglect the series resistances in calculating the elements of B'.

With these assumptions, Equations 3.102 and 3.103 take the following final form:

$$\left[\frac{\Delta P_p}{V_p}\right] = [B'][\Delta\delta] \qquad (3.104)$$

and

$$\left[\frac{\Delta Q_p}{V_p}\right] = [B'][\Delta V], \qquad (3.105)$$

where both $[B']$ and $[B'']$ are real and sparse and have the same structure as H and L, respectively. Since they contain network admittances, they are constant and need to be triangularized only once at the beginning of the iterations. This algorithm results in a very fast solution of $\Delta\delta$ and ΔV.

3.5.4 Linearized (DC) PF Method

The resistance of transmission lines is rather small compared to the reactances of those lines. Then the resistances are neglected to simplify the solutions. Considering the nominal π model of a medium transmission line, we can represent the line with its inductive reactance only. Then the transmission line parameters ($ABCD$) will be

$$
\begin{aligned}
A &= D = 1\angle 0 \\
B &= jX_{ik}, \; C = 0
\end{aligned}
\tag{3.106}
$$

and the PFs through the line are

$$
\begin{aligned}
P_S &= -\frac{V_S V_R}{X} \cos(90° + \delta) = \frac{V_S V_R}{X} \sin(\delta) \\
P_R &= -\frac{V_S V_R}{X} \cos(90° - \delta) = \frac{V_S V_R}{X} \sin(\delta)
\end{aligned}
\tag{3.107}
$$

for small angles, $\sin(\delta) = \delta$. Hence, the line flows become

$$
\begin{aligned}
P_{ik} &= \frac{V_i V_k}{X_{ik}} \delta_{ik}, \\
\delta_{ik} &= \delta_i - \delta_k.
\end{aligned}
\tag{3.108}
$$

Since all bus voltages of a power system are around 1 pu, let

$$
\begin{aligned}
V_i &= V_k = 1 \, \text{pu}, \\
b_{ik} &= \frac{-1}{X_{ik}}.
\end{aligned}
\tag{3.109}
$$

Then Equation 3.108 becomes

$$
P_{ik} = -b_{ik}\delta_{ik} = \frac{\delta_i - \delta_k}{X_{ik}}.
\tag{3.110}
$$

Now, the bus power at any bus is the sum of the PFs in the lines connected to that bus. Hence

$$
P_i = \sum_{k=1}^{N} P_{ik} = \sum_{k=1}^{N} -b_{ik}\delta_{ik} \quad i = 1, 2, \ldots, N
\tag{3.111}
$$

or in matrix form

$$
\begin{bmatrix} P_1 \\ P_2 \\ \vdots \\ P_N \end{bmatrix} = - \begin{bmatrix} b_{11} & b_{12} & \cdots & b_{1N} \\ b_{21} & b_{22} & \cdots & b_{2N} \\ \vdots & \vdots & \vdots & \vdots \\ b_{N1} & b_{N2} & \cdots & b_{NN} \end{bmatrix} \begin{bmatrix} \delta_1 \\ \delta_2 \\ \vdots \\ \delta_N \end{bmatrix}.
\tag{3.112}
$$

which can be abbreviated as

$$[P] = [b][\delta],$$

$$b_{ik} = \frac{-1}{X_{ik}}, \quad b_{ii} = \sum_{k=1}^{N}(-b_{ik}). \tag{3.113}$$

That is, the matrix

$$[\delta] = [b]^{-1}[[P]]. \tag{3.114}$$

Until now, we have kept the system ground as the reference bus. However, since we have adopted all shunt branches in simplifying things, we have lost our reference. This means that the matrix $[b]$ of Equation 3.112 obtained by Equation 3.113 will be a singular matrix. Hence $[b]^{-1}$ does not exist. To overcome this difficulty, we select one of the buses as reference and assign zero radian to its angle (as in the swing bus for ac PF). Then, all the calculated angles will be referred to this bus, and the row and column corresponding to this bus in $[b]$ will be dropped to produce a nonsingular $[b]$ matrix.

Note that since the system is linearized, the solution is direct and there is no need for an iterative procedure.

3.6 Practical Applications of PF Studies

The way in which PF studies are used encompasses two system areas: system planning and system operation. System planning has the objective of designing a system capable of providing reliable bulk electric supply. The common tasks of the system planner include transmission planning (the design, analysis, and sizing of future transmission circuits), interchange studies, generation adequacy studies, and cost-to-benefit analyses of system additions. Generation planning is also considered as part of system planning. System operation aims at setting, adjusting, and operating the system to produce a reliable and economical electric energy supply. Included in system operation tasks are economic dispatch of the generating stations (i.e., calculating the power levels at each generating unit so that the system is operated most economically), contingency analysis (analysis of outages and other forced operating conditions), and studies that ensure power pool coordination.

In most phases of each of these areas, load-flow studies are used to assess system performance under given operating conditions. An important application of PF studies involves transmission planning. Models of the future system are prepared and, typically, peak load conditions are used to run a study. The intention is to find the proper size components (conductors, transformers, reactors, and shunt capacitors), and, in addition, transmission facility siting may be considered. In some cases, series capacitors, phase-shifters, and tap-changing under loads (TCULs) are studied where control of PF is important. A second general area of system planning in which PF studies are indispensable is interchange studies. Interties with neighbors are planned to meet predicted needs of the system consistent with reliability requirements. Load-flow studies identify line loads and bus voltages out of range, inappropriately large bus phase angles (and the potential for stability problems), component loads (principally transformers), proximity to Q-limits at generation buses, and other parameters that have the potential of creating operating difficulties. While peak load studies are frequently performed, sometimes intermediate loads and off-peak (minimum) load conditions are used. Off-peak loads may result in high-voltage conditions that are not identified during the peak load.

3.6.1 Case Study Discussion

System operation tasks that utilize PF studies retaining all bus voltage and transmission component loads within range involve loss calculations, area coordination, calculation of fixed-tap settings, and various types of contingency checks. The latter checks the effects of line and component outages (sometimes multiple contingencies are considered). Contingency studies are usually done offline (i.e., in advance of operating the system under the conditions considered). A typical line outage contingency study consists of a base-case PF study (with all lines in service) followed by contingency cases in which key lines are outaged. Key lines are identified in a variety of ways: lines that are heavily loaded or have a small load margin (rating minus operating load), lines with high-angular difference in terminal bus voltage phases, extra high-voltage circuits, or lines that are known to the operators to be essential to system integrity. The latter come from system operating experience and are often the most reliable means of identifying key lines. In some cases, approximate PF studies may be sufficient to identify problems in contingency studies—several suitable approximate methods are presented in Chapter 4. Contingency studies are particularly important to maintain reliable service. If a line outage causes other circuits to become heavily loaded, those circuits may trip out by the action of protective relays. These additional outages can result in further outages in an uncontrolled way. The term cascade tripping applies to this undesirable operating condition.

3.7 Illustrative Examples

Example 3.7.1

Find all the solutions to the following two nonlinear simultaneous equations

$$4x_1 + 3x_2^2 = 0$$
$$6x_1 x_2 + 2x_2 = 0.$$

SOLUTION

To find all the solutions, we can use the first equation to define x_1 as a function of x_2, $x_1 = ((-3)/4)x_2^2$. By substituting in the second equation, we get

$$6\left(\frac{-3}{4}x_2^2\right)x_2 + 2x_2 = 0.0 \Rightarrow 2x_2\left(1 - 9x_2^2\right) = 0.0,$$

which can be factored as follows

$$2x_2\left[(1 - 3x_2)(1 + 3x_2)\right] = 0.0,$$

which has three possible solutions, namely

$$x_2 = 0.0, \quad x_2 = \frac{1}{3}, \quad \text{and} \quad x_2 = \frac{-1}{3}.$$

The available solutions are

1. $x_2 = 0.0$, $x_1 = 0.0$

2. $x_2 = \frac{1}{3}$, $x_1 = \frac{-3}{4}\left(\frac{1}{9}\right) = \frac{-1}{12}$

3. $x_2 = \frac{-1}{3}$, $x_1 = \frac{-3}{4}\left(\frac{1}{9}\right) = \frac{-1}{12}$

Then the solution points are

$$(0.0)', \quad \left(\frac{-1}{12}, \frac{1}{3}\right)', \quad \text{and} \quad \left(\frac{-1}{12}, \frac{-1}{3}\right)'.$$

Example 3.7.2

The circuit elements in the 138 kV circuit in Figure 3.9 are in per unit on a 100 MVA base with the nominal 138 kV voltage as base. The $(P+jQ)$ load is scheduled to be 170 MW and 50 MVAr.

1. Write the Y matrix for the two-bus system.
2. Assume that bus 1 is a reference bus and set up the G–S correction equation for bus 2.
3. Use a flat start on bus 2. Carry out two or three iterations and show that you are converging.

SOLUTION

1. The line impedance of this line is $z_{12} = 0.01 + j0.04$ (pu). Then the line admittance can be expressed as

$$y_{12} = 1/z_{12} = \frac{1}{(0.01 + j0.04)} = 5.88 - j23.53 \text{ (pu)}.$$

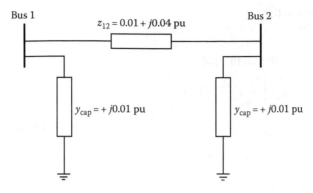

FIGURE 3.9
One-line diagram for Example 3.7.2.

The bus admittance matrix components can be expressed as follows.

$$Y_{11} = y_{12} + y_{10} = (5.88 - j23.53) + (j0.01) = 5.88 - j23.54$$
$$Y_{22} = y_{12} + y_{20} = (5.88 - j23.53) + (j0.01) = 5.88 - j23.54$$
$$Y_{12} = y_{21} + -y_{12} = -(5.88 - j23.53).$$

Then the Y-matrix can be expressed as

$$Y_{bus} = \begin{bmatrix} 5.88 - j23.54 & -5.88 + j23.53 \\ -5.88 + j23.53 & 5.88 - j23.54 \end{bmatrix}.$$

2. Taking bus 1 as reference, $V_1 = 1.0\angle 0.0$ (pu).
   The G–S equation for bus 2 is

$$V_2^{(n+1)} = \frac{1}{Y_{22}} \left[\frac{P_2 - jQ_2}{V_2^{*(n)}} - Y_{21} V_1^{(n)} \right].$$

We have the load $P_2 - jQ_2 = (1.7 - j0.5)$ (pu).

First iteration
   Taking as an initial value $V_2^{(0)} = 1.0\angle 0.0$ (pu),

$$V_2^{(1)} = \frac{1}{Y_{22}} \left[\frac{P_2 - jQ_2}{V_2^{*(0)}} - Y_{21} V_1 \right]$$

$$V_2^{(1)} = \frac{1}{(5.88 - j23.54)} \left[\frac{(1.7 - j0.5)}{1.0\angle 0.0} - (-5.88 + j23.53)(1.0\angle 0.0) \right]$$

$$= 1.039\angle 3.48° \text{ (pu)}.$$

Second iteration
 Now

$$V_2^{(1)} = 1.039\angle 3.48° \text{ (pu)}$$

$$V_2^{(2)} = \frac{1}{Y_{22}}\left[\frac{P_2 - jQ_2}{V_2^{*(1)}} - Y_{21}V_1\right]$$

$$V_2^{(2)} = \frac{1}{(5.88 - j23.54)}\left[\frac{(1.7 - j0.5)}{(1.039)\angle -3.48°} - (-5.88 + j23.53)(1.0\angle 0.0)\right]$$

$$= 1.04\angle 3.21° \text{ (pu)}.$$

Third iteration
 Now

$$V_2^{(2)} = 1.04\angle 3.21° \text{ (pu)}.$$

$$V_2^{(3)} = \frac{1}{Y_{22}}\left[\frac{P_2 - jQ_2}{V_2^{*(2)}} - Y_{21}V_1\right]$$

$$V_2^{(2)} = \frac{1}{(5.88 - j23.54)}\left[\frac{(1.7 - j0.5)}{1.04\angle -3.21°} - (-5.88 + j23.53)(1.0\angle 0.0)\right]$$

$$= 1.0402\angle 3.22° \text{ (pu)}.$$

As a result, we can say $V_2^{(1)} \approx V_2^{(2)}$.
 Then $V_2 = 1.0402\angle 3.22°$ (pu).

Example 3.7.3

Consider the simple electric power system shown in Figure 3.10 and carry out the following calculations. Write down the elements of the bus admittance matrix Y.

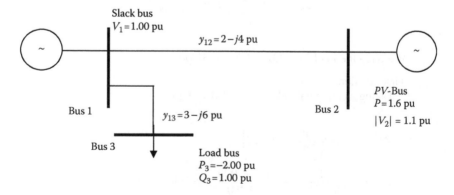

FIGURE 3.10
System for illustrative Example 3.7.3 (Section 3.7).

1. Using the active power equation at bus 2. Calculate the phase angle θ_2.
2. Using the active and reactive power equations at bus 3. Calculate $|V_3|$ and hence θ_3
3. Calculate the active real power at bus 1.
4. Find the total active power loss in the system.

SOLUTION

1. Calculation of the terms of the Y_{BUS} yields

$$Y_{11} = 5 - j10, \qquad Y_{22} = 2 - j4$$
$$= 11.18\angle-63.43, \qquad = 4.47\angle-63.43$$

$$Y_{33} = 3 - j6, \qquad Y_{12} = -2 + j4$$
$$= 6.71\angle-63.43, \qquad = 4.47\angle-116.57$$

$$Y_{13} = -3 + j6$$
$$= 6.71\angle116.57.$$

At bus 2

$$P_2 = |V_2|[Y_{12}V_1 \cos(\theta_2 - \theta_1 - \Psi_{12}) + Y_{22}V_2 \cos(\theta_2 - \theta_2 - \Psi_{22})$$
$$+ Y_{23}V_3 \cos(\theta_2 - \theta_3 - \Psi_{23})]$$

and thus

$$1.6 = 1.1[(4.47)(1.0)\cos(\theta_2 - 116.57) + (4.47)(1.1)\cos(-63.43)].$$

As a result

$$\cos(\theta_2 - 116.57) = -0.1669$$
$$\theta_2 - 116.57 = \pm99.59535$$
$$\theta_2 = (216.16°) \text{ or } (16.97465°).$$

Take $\theta_2 = 16.97465°$.
For bus 3, we have

$$P_2 = |V_3|[|Y_{31}||V_1|\cos(\theta_3 - \theta_1 - \Psi_{31}) + |Y_{33}||V_3|\cos(-\Psi_{33})].$$

By substituting, we get

$$-2 = |V_3|[6.71*1*\cos(\theta_3 - 116.57) + 6.71V_3 \cos(63.43)].$$

Thus

$$\frac{-2}{6.71} = V_3^2 \cos 63.43 + V_3 \cos(\theta_3 - 116.57).$$

Also we have

$$Q_3 = |V_3|[|Y_{31}||V_1|\sin(\theta_3 - \theta_1 - \Psi_{31}) + |Y_{33}||V_3|\sin(-\Psi_{33})]$$
$$1 = |V_3|[6.71\sin(\theta_3 - 116.57) + 6.71V_3\sin(63.43)]$$
$$\frac{1}{6.71} = |V_3|^2\sin 63.43 + |V_3|\sin(\theta_3 - 116.57).$$

Combining Equations (a) and (b),

$$\left[\frac{2}{6.71} + |V_3|^2\cos(63.43)\right]^2 + \left[\frac{1}{6.71} + |V_3|^2\sin(63.43)\right]^2 = |V_3|^2$$

or

$$|V_3|^4 + \frac{4}{6.71}|V_3|^2[\cos(63.43) + 0.5\sin(63.43)] + \frac{5}{(6.71)^2} = |V_3|^2;$$

this gives

$$|V_3|^4 - |V_3|^2 + \frac{1}{a} = 0$$
$$|V_3|^2 = \frac{1 \pm \sqrt{5/3}}{2}.$$

Take the positive sign: $|V_3|^2 = 0.8727$.
The solution for $|V_3|$ is $|V_3| = 0.9342$.
As a result $-2/6.71 = 0.8727\cos(63.43) + 0.9342\cos(\theta_3 - 116.57)$. Thus $\cos(\theta_3 - 116.57) = -0.7369014307$, or $\theta_3 - 116.57 = \pm137.468$. As a result $\theta_3 = -20.898$. We now obtain P_1 as

$$P_1 = |V_1|[|Y_{11}||V_1|\cos(-\Psi_{11}) + |Y_{12}||V_2|\cos(\theta_1 - \theta_2 - \Psi_{12})$$
$$+ |Y_{13}||V_3|\cos(\theta_1 - \theta_3 - \Psi_{13})]$$
$$P_1 = 0.9937.$$

3.8   Conclusion

The chapter discussed the PF problem as an important tool for the operation of the existing power system as well as planning for expected expansion. It started by classifying the different buses in an electric power system network and defined the quantities associated with each of the specific types in Section 3.2. Section 3.3 presented the general forms of the PFEs in an electric power system. Section 3.4 gave insights on potential modeling considerations in the generation subsystem, transmission subsystems, and load modeling. The formulation was extended to handle transformer and phase-shifter inclusion in the PF model; the formulation in both polar and rectangular coordinates was also presented in this section. The different iterative techniques used to solve the PF problem including the G–S iterative technique and the N–R method were presented in Section 3.5. Special forms of load-flow solution methods such as fast-decoupled PF were presented at the end of this section. Some practical applications of the power-flow studies and some illustrative examples to the power-flow problem were presented in Section 3.5. The formulation in both polar and rectangular coordinates was also presented followed by some unsolved exercises.

3.9 Problem Set

PROBLEM 3.9.1

Given the following equations

$$f_1(x) = x_1 - x_2^2 + x_2 = 0$$
$$f_2(x) = 2x_2^2 - x_1 x_2 - 4 = 0.$$

Obtain their solutions using the Jacobian method?

PROBLEM 3.9.2

Consider the following nonlinear simultaneous equations

$$f_1(x) = x_1^2 + x_2 = 0$$
$$f_2(x) = x_1 + x_2 = 0.$$

1. Solve the equations graphically.
2. Solve f_1 for x_2 and f_2 for x_1, solve the equations by the Jacobian method with $x_1 = 0.5$ as an initial guess.
3. Show how Part 2 converges graphically.

PROBLEM 3.9.3

A one-line diagram of a three-bus power system is given in Figure 3.11, where the reactances are in per units on a 100 MVA base. Use G–S iteration to obtain the PF solution within a tolerance of 0.01 pu,

$$V_1 = 1.0\angle 0 \text{ pu}$$
$$V_2 = 1.0\angle 0 \text{ pu}$$
$$P_2 = 0.6 \text{ pu}$$
$$P_3 = -1.0 \text{ pu}$$
$$Q_3 = -0.75 \text{ pu (lagging).}$$

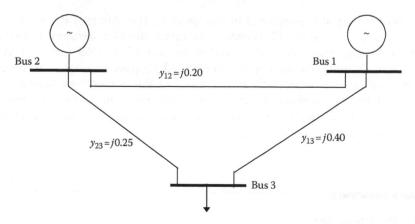

FIGURE 3.11
One-line diagram for Problem 3.9.3.

PROBLEM 3.9.4

Do one N–R iteration of the power system of Problem 3.9.3?

PROBLEM 3.9.5

Using the solution to Problem 3.3, calculate the PF through lines 1–3 at both ends?

PROBLEM 3.9.6

Consider a six-bus power system as shown in Figure 3.12. The line impedances are in per units on a 100 MVA, 138 kV system base and the generator at the slack or reference bus operates at rated voltage. Table 3.6 shows the system load for the six-bus power system.

1. Solve the PF problem for the given system using the G–S approach. Calculate the total system losses in the transmission network.
2. Consider the loss of line L5–6. Resolve the PF problem and the transmission losses.
3. Compare the results of Part (1) and Part (2). Explain the effect of the line outage on the voltage profile of the system.

PROBLEM 3.9.7

Given the network shown in Figure 3.13 (base 100 MVA), do the following:

1. Calculate the phase angles for the set of power injections?
   $P_1 = 100$ MW generation
   $P_2 = 120$ MW load
   $P_3 = 150$ MW generation
   $P_4 = 200$ MW load.

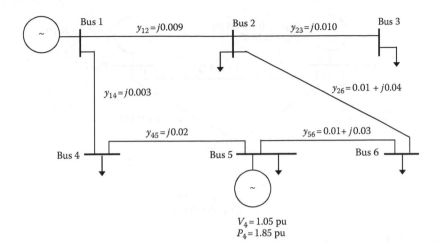

FIGURE 3.12
Simplified six-bus power system.

TABLE 3.6

Load Profile for the Six-Bus Power System

Bus ID	Load	
	Real Power, P_i (MW)	Reactive Power, Q_i (MVAr)
1	0	0
2	85	60
3	15	8.0
4	60	37
5	10	−4.0
6	36	24

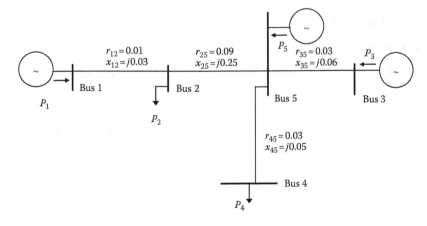

FIGURE 3.13
Four-bus network for Problem 3.9.7.

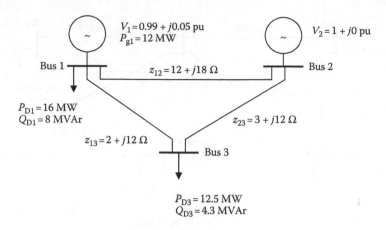

FIGURE 3.14
Three-bus network for Problem 3.9.8.

2. Calculate P_5 according to the decoupled load flow?
3. Calculate all PFs on the system using the phase angles found in Part I?
4. Optional: Calculate the reference bus penalty factors for buses through 4. Assume all bus voltage magnitudes are 1.0 pu?

PROBLEM 3.9.8

For the network shown in Figure 3.14 solve the PF problem using the N–R method.

This part of the network is working on 138 kV.

References

1. Elgerd, O. I., *Electric Energy Systems Theory—An Introduction*, McGraw-Hill, New York, 1982.
2. El-Hawary, M. E., *Electric Systems, Dynamics and Stability with Artificial Intelligence Applications*, Marcel Dekker, New York, 1999.
3. Grainger, J. J. and Stevenson, W. D., Jr., *Power System Analysis*, McGraw-Hill, New York, 1994.
4. Gross, C. A., *Power Systems Analysis*, Wiley, New York, 1979.
5. Pai, M. A., *Computer Techniques in Power System Analysis*, Tata McGraw-Hill, New Delhi, India, 1979.
6. Weedy, B. M., *Electric Power Systems*, 3rd edn., Wiley, London, 1979.

4

Constrained Optimization and Applications

4.1 Introduction

Practical system problems are very often formulated with some constraints imposed on their variables. Optimization of such problems must be carried out within the limits of these constraints. The basic difference between the constrained and the unconstrained problems is the selection of admissible points that are eligible for optimization. The so-called admissible points are those vectors that satisfy the constraints of the problem. An optimum of the constrained problem is judged by the comparison of admissible points. Constrained problems are assumed to have at least one admissible point.

Lagrange multipliers are known as an effective means of dealing with constraints. Two theorems for functions are given. The first theorem in each case gives a necessary condition and the second theorem gives a sufficient condition for minimization of constrained problems. Again, we apply the sufficiency theorems to test for maximum of $f(y)$ by minimizing $f(y)$. Procedures for solving the constrained problems by using the theorems are listed and also utilized in the solutions of the illustrative problems [1,2].

Before we continue further, I wish to introduce some necessary background in analytic computation that will be useful in optimization work. First, it includes matrix notation and some special properties that have significant applications in operation research and power systems in particular. Basic foundations in matrix algebra and functional calculus are required by the reader and the author has introduced standard background operations to test negative and positive semidefinite conditions on Jacobian and Hessian matrices.

Now, consider the n-dimensional before given by $y(t) = [y(t)_1, y(t)_2, \ldots, y(t)_n]^T$ where each element $y_i(t)$ is continuous in time and y is special scalar when $n = 1$. In this chapter, we limit our attention to case when y is independent of t. Then, $n \times m$ matrix $\mathbf{M}$ is now defined as $\mathbf{M} = [\mathbf{y}_j]$ where the vector construct of the matrix is $\mathbf{M} = [\mathbf{y}_j \mid j = 1, m] = [y_{i,j}]_{n \times m}$.

By definition, if A is square and Hermitian such that its $\mathbf{M} = (\mathbf{M}^T)^* = \mathbf{M}^\dagger$, then $\mathbf{M}$ is positive definite if, for all nonzero n-dimensional vectors $\mathbf{x}$, the inner product is strictly positive. That is to say $\langle \mathbf{Mx}, \mathbf{x} \rangle > 0$. And, $\mathbf{M}$ is positive semidefinite if, under the same conditions, $\langle \mathbf{Mx}, \mathbf{x} \rangle \geq 0$. (Reversing the signs concludes a parallel definition for negative definite and negative semidefinite conditions, respectively.)

Alternatively, $\mathbf{M}$ is said to be positive semidefinite if all its eigenvalues are positive and $\langle \mathbf{Mx,x} \rangle \equiv \mathbf{x^T Mx} > 0$. This definition is particularly important in characterizing the convex behaviors of many nonlinear systems, and in particular, those with quadratic models.

Other computational considerations of matrices in vector notation used in the text include the formulations of the following:

1. $n \times n$ Jacobian matrix or $J(\mathbf{f(x)})$

   The Jacobian matrix is the array of all first-order partial derivatives of a vector-valued function such that:

   $$J(x) = \nabla(f(x)) = \left[\frac{\partial f_i}{\partial x_j}\right]_{n \times n}, \quad \text{where } i \in \{1,n\} \text{ and } i \in \{1,n\}$$

2. Development of the Hessian $H(\mathbf{x})$ matrix

   Consider the matrix product given by

   $$\mathbf{y^T} J(\mathbf{f(x)})\mathbf{y} = \sum_{i=1}^{n} \sum_{j=1}^{n} \left(\frac{\partial^2 f}{\partial x_i x_j} y_i y_j \right)$$

   (quadratic form of the matrix product)

   The form is positive definite if $\mathbf{y^T} J(x)\mathbf{y} > 0$ for all real values of z_i and z_j. The test for this condition can be found by evaluating a set of determinants of $J(\mathbf{y(x)})$. The Hessian is now defined at the determinant of the Jacobian matrix J and is the nth principal determinant of J given by

   $$H(\mathbf{f(x)}) = |J(\mathbf{f(x)})| = \left\lVert \left[\frac{\partial^2 f}{\partial x_i x_j}\right] \right\rVert_{n \times n}, \quad \text{where } i \in \{1,n\} \text{ and } i \in \{1,n\}$$

   (The corresponding principal subdeterminants are called sub-Hessians.)

   The symmetric matrix $\mathbf{A}$ is said to be negative definite if its corresponding form is negative definite $\mathbf{A} = \mathbf{y^T} J(x)\mathbf{y} < 0$. In general, the positive (or negative) definite tests on the symmetric Hessian matrix are used in the criteria for local minima (or maxima) to be discussed later in subsequent sections of the chapter.

4.2 Theorems on the Optimization of Constrained Functions [5,6]

The constrained problem is defined to optimize the scalar function $f(y)$ subject to constraints

$$f_i(y) = K_i, \quad \text{for } i = 1, 2, \ldots, m, \tag{4.1}$$

where all $f_i(y)$ are real, scalar, and single-valued functions of y. Although the problem is formulated with equality constraints, problems with inequality constraints can be solved by varying the parameters K_i after the optimization.

The Lagrange function for the problem is defined to be the scalar function

$$L(y) = f(y) + \lambda F(y), \tag{4.2}$$

where

$$\lambda = [\lambda_1, \lambda_2, \ldots, \lambda_m] \tag{4.3}$$

and

$$F(y) = [f_1(y), \ldots, f_m(y)]^T. \tag{4.4}$$

The components λ_i of the row vector λ are constant and are known as the Lagrange multipliers. Note that the Lagrange function is a function of y and λ.

4.2.1 Continuity Assumption

All the functions involved in the following theorems are assumed to be continuous at x which is an optimum under consideration. For simplicity, the arguments of the partial derivatives of the Lagrange and constraint functions evaluated at x are dropped. That is, $L_x = L_x(x)$, $L_{xx} = L_{xx}(x)$, and $F_x = F_x(x)$ are used in the statement without proof of the theorems.

4.2.2 Theorems

THEOREM 4.2.1

If x is an optimum of the constrained problem, then it is necessary that $L_x = 0$ for some λ.

$$L_x = f_x + \lambda F_x = 0. \tag{4.5a}$$

The Lagrange multipliers designate the coefficients of linear combination.

THEOREM 4.2.2

If x is a candidate solution as determined by the necessary conditions of Theorem 4.2.1, then the sufficiency conditions that must be satisfied are

$$T = L_{xx}(x,\lambda) + \beta F_x^{\mathrm{T}}(x)F_x(x) > 0 \qquad (4.5\text{b})$$

for some positive semidefinite values of β (i.e., $\beta \geq 0$).

The significance of the theorem lies in the facts that (i) the constrained problem can be extremized without considering the constraints and (ii) the nonnegative definite matrix $F_x^{\mathrm{T}}F_x$ that is added to L_{xx} in the second theorem enables one to test for sufficiency without considering the constraints.

4.3 Procedure for Optimizing Constrained Problems (Functions)

There are four steps involved in solving constrained problems by applying the theorems:

Step 1: Formulation of the Lagrange function
The inequality constraint $K_{i1} \leq f_i(y) \leq K_{i2}$ is assumed to be $f_i(y) = K_i$ with $K_{i1} \leq K_i \leq K_{i2}$ considered as constant during the optimization. Form $F(y) = [f_1(y), f_2(y), \ldots, f_m(y)]^{\mathrm{T}}$ and $L = f(y) + \lambda F(y)$, where $f(y)$ is to be optimized.

Step 2: Determination of optimum candidate(s)
Set $L_x = 0$ (Theorem 4.2.1), which means replace y by x and then differentiate with respect to x. Here it is understood that L_x is continuous at x. There are altogether $n+m$ equations; n from $L_x = 0$ and m from the constraints that can be obtained to solve for the optimum candidate x and Lagrange multipliers λ_i.

Step 3: Sufficiency test
Assume that all the functions involved in the following are continuous at x. Then the candidate x obtained in step 2 is a minimum (Theorem 4.2.2) if the square matrix

$$L_{xx} + \beta F_x^{\mathrm{T}}F_x$$

is positive definite for $\beta \geq 0$. It is a maximum if the matrix is positive definite after replacing L_{xx} by $-L_{xx}$ (actually L by $-L$).

Step 4: Further optimization
The function $f(x)$ is now a function of K_i. Further optimization is to be sought with respect to K_i in the interval $K_{i1} \leq K_i \leq K_{i2}$ for all i. The optimum of K_i may occur at the boundaries of the constraints.

4.4 Karush–Kuhn–Tucker Condition

Also known as Khun–Tucker or KKT conditions, they are developed to providing unique tests of the necessary and sufficiency conditions of optimality. It is well developed for nonlinear programming that involves the use of the Lagrangian approach. A special treatment of the concepts to be described shortly is also applicable to linear programming problems by taking advantage of duality theorems [6–10].

Recall the generalized NLP

Min **f(x)**

subject to

$$g_i(x) = 0 \quad \text{and} \quad h_j(x) \geq 0,$$

where $f(x)$ is the scalar function of the n-vector x to be minimized. And, for the set of equality constraints, $i \in \{1, p\}$; for the set of inequality constraints, $j \in \{1, m\}$. We now summarize the following conditions of KKT.

KKT conditions: If their should exist a point x^* at a local minimum of $f(x)$, there exist constants λ_i and μ_j where $i \in \{1, p\}$ and $j \in \{1, m\}$, respectively, such that

1. $\nabla f(x^*) - \sum_{i=1}^{p} \lambda_i \nabla g_i(x^*) - \sum_{j=1}^{m} \mu_j \nabla h_j(x^*) = 0$
2. $g_i(x^*) = 0$ for all $i \in \{1, p\}$
3. $h_j(x^*) \geq 0$ for all $j \in \{1, m\}$
4. $\mu_j \geq 0$ for all $j \in \{1, m\}$
5. $\mu_j h_j(x^*) = 0$ for all $j \in \{1, m\}$

This KKT conditions present a new Lagrange multiplier μ_j that corresponds to the dual variables in linear programming with economic implications based on the problem being solved. It should be noted that condition 1 is the necessary condition for x^* to be a candidate, conditions 2 and 3 are the constraints qualifications, and conditions 4 and 5 are eliminated nonoptimal solutions to guarantee feasibility, if they exist.

4.5   Illustrative Problems

PROBLEM 4.5.1

Given a function $f(y) = y_1 - y_2^2$, find an optimum subject to constraint $y_1 y_2 \geq 2$.

Solution

1. $f_1(y) = y_1 y_2 = K$ with $2 \leq K \leq \infty$ and $L = y_1 - y_2^2 + \lambda y_1 y_2$.
2. $L_x = [1 + \lambda x_2, \quad -2x_2 + \lambda x_1] = 0$ gives $1 + \lambda x_2 = 0$ and $-2x_2 + \lambda x_1 = 0$.

   Solution of these yields $x_1 = -2\lambda^{-2}$, $x_2 = -\lambda^{-1}$, and $K_1 = 2\lambda^{-3}$.

3. $L_{xx} + \beta F_x^T F_x = \begin{bmatrix} 0 & \lambda \\ \lambda & -2 \end{bmatrix} + \beta \begin{bmatrix} x_2 \\ x_1 \end{bmatrix} [x_2 \quad x_1]$

   $\qquad = \begin{bmatrix} \beta x_2^2 & \beta x_1 x_2 + \lambda \\ \beta x_1 x_2 + \lambda & \beta x_1^2 - 2 \end{bmatrix}.$

The matrix has a negative determinant for any $\beta \geq 0$ and, hence, it cannot be positive definite. Let us try for maximum by replacing L_{xx} by $-L_{xx}$. That is

$$L_{xx} + \beta F_x^T F_x = \begin{bmatrix} \beta x_2^2 & \beta x_1 x_2 - \lambda \\ \beta x_1 x_2 - \lambda & \beta x_1^2 + 2 \end{bmatrix}$$

$$\qquad = \begin{bmatrix} \beta \lambda^{-2} & 2\beta \lambda^{-3} - \lambda \\ 2\beta \lambda^{-3} - \lambda & 4\beta \lambda^{-4} + 2 \end{bmatrix}.$$

It is positive definite for $\beta > \lambda^4/6$ and, therefore, the solution of step 2 is a maximum.

4. Maximum function is

$$f(x) = -2\lambda^{-2} - \lambda^{-2} = -3\lambda^{-2} = -3\left(\frac{1}{2}K\right)^{2/3},$$

where $2 \leq K \leq \infty$. Further maximization yields $K = 2$ for which $f(x) = -3$, $x_1 = -2$, and $x_2 = -1$.

PROBLEM 4.5.2

Optimize $f(y) = y^T P y + C^T y$ subject to linear constraints $Ay = d$ where P is a symmetric n-square and nonsingular matrix. The matrix A is of $m \times n$ with full rank and C and d are n- and m-vectors.

Solution

1. Constraints are $Ay = d$ and, hence

   $$L = y^T P y + C^T y + \lambda Ay.$$

2. $L_x = 2x^T P + C^T + \lambda A = 0$. The solution of x and λ is

$$x = -\frac{1}{2} P^{-1}(A^T \lambda^T + C),$$

where $\lambda^T = -(AP^{-1}A^T)^{-1}(AP^{-1}C + 2d)$.

3. $L_{xx} + \beta F_x^T F_x = 2P + \beta A^T A$. The solution of step 2 is a minimum if $2P + \beta A^T A > 0$ and is a maximum if $2P + \beta A^T A > 0$ for some $\beta \geq 0$.

4. $f(x) = \frac{1}{4}(\lambda A - C^T) P^{-1}(A^T \lambda^T + C)$.

   (a) Minimum norm of $Ay = d$ with $m \leq m$. The problem is a special case of $P = 1$ and $C = 0$. It follows from the result of step 2 that

   $$\lambda^T = -2(AA^T)^{-1} d \quad \text{and} \quad x = A^T(AA^T)^{-1} d.$$

   The sufficiency for the minimum is assured by choosing $\beta = 0$. The minimized norm is $f(x) = x^T x = d^T(AA^T)^{-1} d$.

   (b) Least square approximation for $Ax = d$ with mn. The problem is to minimize, without constraint,

   $$f(y) = e^T e = (Ax - d)^T(Ax - d)$$
   $$= x^T A^T A x - 2d^T A x + d^T d.$$

   Let $P = A^T A$ and $C = -2A^T d$; then the result of step 2 yields

   $$x = (A^T A)^{-1} A^T d \quad \text{and} \quad e^T e = d^T [I - A(A^T A)^{-1} A^T] d.$$

   The norm is a minimum because A is of rank n and hence $A^T A > 0$. The sufficiency is assured again by choosing $\beta = 0$.

4.5.1 Nonpower Systems Application Examples

PROBLEM 4.5.3

Consider the function

$$f(x_1, x_2, x_3) = x_1 + 2x_3 + x_2 x_3 - x_1^2 - x_2^2 - x_3^2.$$

Solution

Applying the necessary condition $\nabla f(X_0) = 0$, this gives

$$\frac{\partial f}{\partial x_1} = 1 - 2x_1 = 0.$$

$$\frac{\partial f}{\partial x_2} = x_3 - 2x_2 = 0.$$

$$\frac{\partial f}{\partial x_3} = 2 + x_2 - 2x_3 = 0.$$

The solution of these simultaneous equations is given by

$$X_0 = \left(\frac{1}{2}, \frac{2}{3}, \frac{4}{3}\right).$$

Another way to check the sufficiency condition is to check the Hessian matrix $\mathbf{H}$ for positive or negative definiteness. Thus

$$H|x_0 = \begin{bmatrix} \dfrac{\partial^2 f}{\partial x_1^2} & \dfrac{\partial^2 f}{\partial x_1 \partial x_2} & \dfrac{\partial^2 f}{\partial x_1 \partial x_3} \\[2mm] \dfrac{\partial^2 f}{\partial x_2 \partial x_1} & \dfrac{\partial^2 f}{\partial x_2^2} & \dfrac{\partial^2 f}{\partial x_2 \partial x_3} \\[2mm] \dfrac{\partial^2 f}{\partial x_3 \partial x_1} & \dfrac{\partial^2 f}{\partial x_3 \partial x_2} & \dfrac{\partial^2 f}{\partial x_3^2} \end{bmatrix}_{X_0}.$$

$$= \begin{bmatrix} -2 & 0 & 0 \\ 0 & -2 & 1 \\ 0 & 1 & -2 \end{bmatrix}$$

The principal minor determinants $H|x_0$ have the values -2, 4, and -6, respectively. Thus, $H|x_0$ is negative definite and

$$X_0 = \left(\frac{1}{2}, \frac{2}{3}, \frac{4}{3}\right)$$

represents a maximum point.

4.6   Power Systems Application Examples

4.6.1   Optimal Operation of an All-Thermal System: Equal Incremental Cost-Loading

A simple, yet extremely useful problem in optimum economic operation of electric power systems is treated here. Consider the operation of m thermal generating units on the same bus as shown in Figure 4.1 [4]. Assume that the variation of the fuel cost of each generator (F_i) with the active-power output (P_i) is given by a quadratic polynomial. The total fuel cost of the plant is the sum of the individual unit cost converted to $/h$:

$$F = \sum_{i=1}^{m} \alpha_i + \beta_i P_i + \gamma_i P_i^2, \tag{4.6}$$

where α_i, β_i, and γ_i are assumed available.

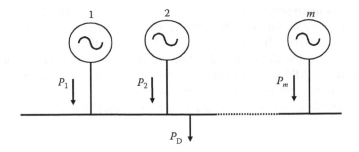

FIGURE 4.1
Units on the same bus.

We wish to determine generation levels such that F is minimized while simultaneously satisfying an active-power balance equation. This utilizes the principle of power-flow continuity. Here, the network is viewed as a medium of active-power transfer from the generating nodes to the load node. Only one equation is needed.

The first active-power balance equation model neglects transmission losses and, hence, we can write

$$P_D = \sum_{i=1}^{m} (P_i) \tag{4.7}$$

with P_D being a given active-power demand for the system.

The demand P_D is the sum of all demands at load nodes in the system. The model is useful in the treatment of parallel generating units at the same plant since in this case the negligible transmission losses assumption is valid.

We write the constraint equation (Equation 4.7)

$$P_D - \sum_{i=1}^{m} (P_i) = 0. \tag{4.8}$$

The technique is based on including Equation 4.8 in the original cost function by use of a Lagrange multiplier, say λ, which is unknown at the outset. Thus

$$L = F_T + \lambda \left[P_D - \sum_{i=1}^{m} (P_i) \right], \tag{4.9}$$

where

$$F_T = \sum_{i=1}^{m} [F_i(P_i)].$$

Note that λ is to be obtained such that Equation 4.8 is satisfied. The idea here is to penalize any violation of the constraint by adding a term corresponding to the resulting error. The Lagrange multiplier is, in effect, a conversion factor that accounts for the dimensional incompatibilities of the cost function (\$/h) and constraints (MW). The resulting problem is unconstrained, and we have increased the number of unknowns by one.

The optimality conditions are obtained by setting the partial derivatives of L with respect to P_i to 0. Thus

$$\frac{\partial F_i}{\partial P_i} - \lambda = 0. \tag{4.10}$$

Note that each unit's cost is independent of the generations of other units.

The expression obtained in Equation 4.10 leads to the conclusion that

$$\lambda = \frac{\partial F_1}{\partial P_1} = \frac{\partial F_2}{\partial P_2} = \cdots . \tag{4.11}$$

The implication of this result is that for optimality, individual units should share the load such that their incremental costs are equal. We can see that the λ is simply the optimal value of incremental costs at the operating point. Equation 4.10 is frequently referred to as the equal incremental cost-loading principle.

Implementing the optimal solution is straightforward for the quadratic cost case where we have

$$F_i(P_i) = \alpha_i + \beta_i P_i + \gamma_i P_i^2.$$

Our optimality conditions from Equation 4.10 reduce to

$$\beta_i + 2\gamma_i P_i - \lambda = 0. \tag{4.12}$$

The value of λ is determined such that Equation 4.8 is satisfied. This turns out to give

$$\lambda = \frac{2P_\mathrm{D} + \sum_{i=1}^{m} (\beta_i/\gamma_i)}{\sum_{i=1}^{m} \gamma_i^{-1}}. \tag{4.13}$$

Finally, using Equation 4.12 the optimal generations are obtained as

$$P_i = \frac{\lambda - \beta_i}{2\gamma_i}. \tag{4.14}$$

4.6.2 Optimal Operation of an All-Thermal System, Including Losses

We are interested in minimizing the total cost given by Equation 4.6 while satisfying the active-power balance equation including losses. Thus

$$P_D = \sum_{i=1}^{m} (P_i) - P_L, \tag{4.15}$$

where P_L is the active-power loss considered as a function of the active-power generation alone as outlined in the previous section. Following our treatment for the loss-free case, we form the augmented cost function:

$$\hat{F} = F_T + \lambda \left[P_D + P_L - \sum_{i=1}^{m} (P_i) \right]. \tag{4.16}$$

The optimality conditions are obtained using the same arguments as before and are

$$\frac{\partial F_i}{\partial P_i} + \lambda \left(\frac{\partial P_L}{\partial P_i} - 1 \right) = 0. \tag{4.17}$$

Note that with negligible transmission losses, the above expression reduces to Equation 4.19.

It is convenient to transform the obtained optimality expression into an equivalent form. This is done by defining the factors L_i:

$$L_i = \left(1 - \frac{\partial P_L}{\partial P_i} \right)^{-1}. \tag{4.18}$$

We can write Equation 4.17 as

$$L_i \frac{\partial F_i}{\partial P_i} = \lambda \ (i = 1, \dots, m). \tag{4.19}$$

This is of the form of Equation 4.11 except for the introduction of the new factors L_i, which account for the modifications necessitated by including the transmission loss. These are traditionally called the penalty factors to indicate that plant costs (F_i) are penalized by the corresponding incremental transmission losses ($\partial P_L / \partial P_i$).

Examination of Equation 4.19 reveals that the optimal generations are obtained when each plant is operated such that the penalized incremental costs are equal.

Flowcharts for the two cases of economic dispatch are shown in Figure 4.2.

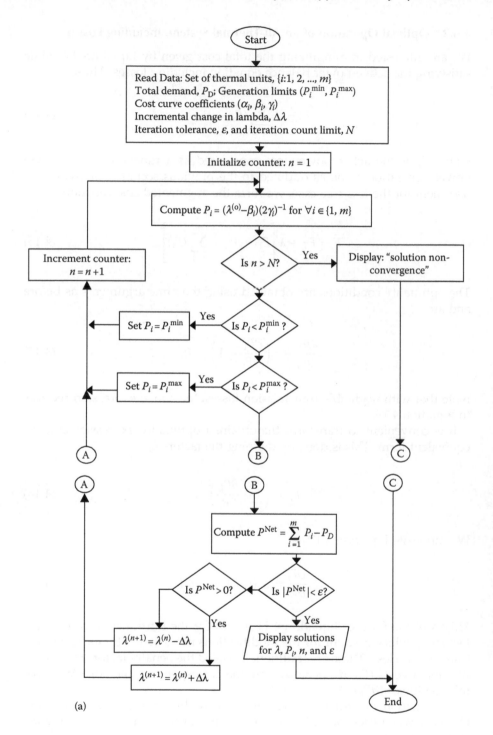

FIGURE 4.2
Economic dispatch flow chart. (a) Neglecting the contributions of transmission losses, and

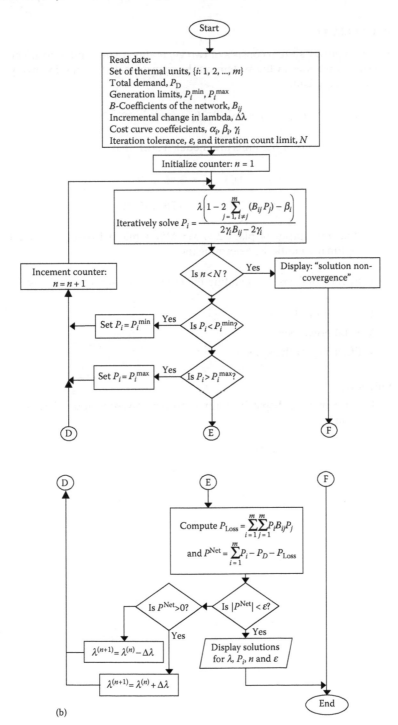

FIGURE 4.2 (continued)
(b) accounting for the effects of transmission losses.

PROBLEM 4.6.1

A certain power system consists of two generating plants and a load. The transmission losses in the system can be approximated by the following quadratic equation.

$$P_{\text{Loss}} = 0.606 \times 10^{-3}P_1^2 + 0.496 \times 10^{-3}P_2^2.$$

The cost functions of the units are modeled such that

$$\beta_1 = 10.63 \quad \gamma_1 = 3.46 \times 10^{-3}$$

$$\beta_2 = 12.07 \quad \gamma_2 = 3.78 \times 10^{-3}.$$

Given that the output of plant 2 is 445 MW while it is being operated under optimal conditions, determine the

1. Incremental cost of operating the units, assuming the equal cost sharing
2. Output of plant 1
3. Total power losses
4. Efficiency of the system

Solution

1. For economic dispatch (optimal power flow, see Figure 4.2) at plant 2,

$$\frac{\partial F_i(P_i)}{\partial P_i} = \lambda \left(1 - \frac{\partial P_{\text{Loss}}}{\partial P_i}\right)\bigg|_{i=2},$$

where

$$F_2(P_2) = \alpha_2 + \beta_2 P_2 + \gamma_2 P_2^2,$$

$$\beta_2 + 2\lambda_2 P_2 = \lambda \left(1 - \frac{\partial P_{\text{Loss}}}{\partial P_2}\right)$$

$$\Rightarrow \lambda = \frac{\beta_2 + 2\lambda_2 P_2}{(1 - (\partial P_{\text{Loss}}/\partial P_2))}$$

$$= \frac{12.07 + 2(3.78 \times 10^{-3})(445)}{1 - 2(0.496 \times 10^{-3})(445)}$$

$$\lambda = 27.63 \ (\$/\text{MW h}).$$

2. Similarly, for the optimal operation of plant 1:

$$\beta_1 + 2\lambda_1 P_1 = \lambda \left(1 - \frac{\partial P_{\text{Loss}}}{\partial P_1}\right)$$

$$\beta_1 + 2\lambda_1 P_1 = \lambda(1 - 2\beta_{11}P_1)$$

$$\Rightarrow P_1 = \frac{\lambda - \beta_1}{2(\gamma_1 + \lambda\beta_{11})}$$

$$= \frac{27.63 - 10.63}{2(0.00346 + (27.63)(0.606 \times 10^{-3}))}$$

$$= 421 \text{ MW}.$$

3. Hence, the total power losses due to transmission are

$$P_{\text{LOSS}} = 0.606 \times 10^{-3} \times (421)^2 + 0.496 \times 10^{-3} \times (445)^2$$

$$= 107 + 98.2$$

$$= 107 + 98.2$$

$$= 206 \text{ MW}.$$

4. Efficiency of the system is given as

$$\eta = \frac{P_{\text{output}}}{P_{\text{input}}}$$

$$= 1 - \left(\frac{P_{\text{LOSS}}}{\sum_i P_i}\right)$$

$$= 1 - \left(\frac{206 \text{ MW}}{(421 + 445) \text{ MW}}\right)$$

$$= 0.763\% \text{ or } 76.3\%.$$

4.7 Illustrative Examples

Example 4.7.1

Find the local and global minima of the function

$$f(x) = f(x_1, x_2) = x_1^2 + x_2^2 - 2 \times 1 - 4x_2 + 5|x_1|, \ |x_2| \le 3.$$

SOLUTION

$$\min F(x) = x_1^2 + x_2^2 - 2x_1 - 4x_2 + 5$$

$$|x_1| \le 3 \rightarrow -3 \le x_1 \le 3$$

s.t.

$$|x_2| \le 3 \rightarrow -3 \le x_2 \le 3.$$

We can change the constraint to be an equality constraint

$$x_1 - k_1, \quad -3 \leq k_1 \leq 3$$
$$x_2 = k_2, \quad -3 \leq k_2 \leq 3.$$

1. From the Lagrange function.

$$L = x_1^2 + x_2^2 - 2x_1 - 4x_2 + 5 + \lambda_1 x_1 + \lambda_2 x_2.$$

2. Determine optimum candidates:

$$\frac{\partial L}{\partial x_i} = 0$$

$$\frac{\partial L}{\partial x_1} = 2x_1 - 2 + \lambda_1 = 0 \rightarrow x_1 = 1 - 0.5\lambda_1$$

$$\frac{\partial L}{\partial x_2} = 2x_2 - 4 + \lambda_2 = 0 \rightarrow x_2 = 2 - 0.5\lambda_2.$$

3. Sufficiency test.

$$L_{xx} + \beta F_x^T F_x > 0 \text{ for some positive } \beta,$$

$$\begin{bmatrix} 2 & 0 \\ 0 & 2 \end{bmatrix} + \beta \begin{bmatrix} 1 \\ 1 \end{bmatrix} [1 \ 1] = \begin{bmatrix} 2+\beta & \beta \\ \beta & 2+\beta \end{bmatrix} = 4 + 4\beta > 0 \text{ for any } \beta.$$

Then it is at a minimum.

4. Further optimization.

$$K_1 = K_2 = K, \quad \text{as} -3 \leq K \leq 3$$

$$F(K) = K^2 - 2K - 4K + 5 = 2K^2 - 2K + 5.$$

For minimization $F_K = 4K - 2 \rightarrow K = 0.5$.

$$F_{KK} = 2 \text{ (i.e., it is minimum)}$$

$$K = 0.5 = X_1 = X_2, \quad f_{min} = 4.5$$
$$f(K) = 4.5.$$

Example 4.7.2

Two thermal units at the same station have the following cost models.

$$F_1 = 793.22 + 7.74P_1 + 0.00107P_1^2$$

$$F_2 = 1194.6 + 7.72P_2 + 0.00072P_2^2$$

$$100 \leq P_2 \leq 800 \text{ MW} \quad \text{and} \quad 100 \leq P_1 \leq 600 \text{ MW}.$$

Find the optimal power generated P_1 and P_2 and the incremental cost of power delivered for demands of 400, 600, and 1000 MW, respectively.

SOLUTION

$$\frac{\partial F_1}{\partial P_1} = 7.74 + 0.00214 P_1$$

$$\frac{\partial F_2}{\partial P_2} = 7.72 + 0.00144 P_2.$$

For optimality

$$\frac{\partial F_1}{\partial P_1} = \frac{\partial F_2}{\partial P_2}$$

$$7.74 + 0.00214 P_1 = \lambda \rightarrow P_1 = 467.29(\lambda) - 3616.8$$

$$7.72 + 0.00144 P_2 = \lambda \rightarrow P_2 = 694.44(\lambda) - 5361.1$$

$$P_D = P_1 + P_2.$$

1. At $P_D = 400$ MW, $P_1 + P_2 = 400$.

$$400 = 467.29\lambda - 3626.8 + (694.44\lambda - 5361.1)$$

$$9377.91 = 1161.73\lambda \rightarrow \lambda = 8.072$$

$$P_1 = 155.3 \text{ MW}, \quad P_2 = 244.7 \text{ MW};$$

both of them are within the desired limit.

2. At $P_D = 600$ MW, $P_1 + P_2 = 600$ MW.

$$600 = 467.29(\lambda) - 3616.8 + (694.44(\lambda) - 5361.1)$$

then

$$\lambda = 8.245, \quad P_1 = 235.78 \text{ MW}, \quad P_2 = 364.22 \text{ MW};$$

both of them are within the desired limit.

3. At $P_D = 1000$ MW, $P_1 + P_2 = 1000$ MW.

$$1000 = 467.29(\lambda) - 3616.8 + (6944(\lambda) - 5361.1)$$

then

$$\lambda = 8.589, \quad P_1 = 396.7 \text{ MW}, \quad P_2 = 603.33 \text{ MW};$$

both of them are in the desired limit.

Example 4.7.3

The following fuel cost equations model a plant consisting of two thermal units.

$$F_1 = 0.00381P_1^2 + 9.2P_1 + 70.5 \ (\$)$$

$$F_2 = 0.00455P_2^2 + 6.0P_2 + 82.7 \ (\$).$$

1. For a load of 600 MW, determine the optimal power generated by each unit and the equal incremental fuel cost λ at which it operates.
2. For the same load as in Part 1, given that the generation is constrained as

$$80.0 \le P_1 \le 250 \text{ MW}$$

$$120 \le P_2 \le 400 \text{ MW}$$

at what values of λ should the units be operated?

SOLUTION

1. From the given data,

$$\lambda_1 = 0.00381, \quad \beta_1 = 9.2, \quad \alpha_1 = 70.5$$
$$\lambda_2 = 0.00455, \quad \beta_2 = 6.0, \quad \alpha_2 = 82.7.$$

By the formula for the incremental fuel cost,

$$\lambda = \frac{2P_D + \sum (\beta_i/\gamma_i)}{\sum (\gamma_i)^{-1}},$$

where

$$\sum \frac{\beta_i}{\lambda_i} = \frac{9.2}{0.00381} + \frac{6.0}{0.00455} = 3.733 \times 10^4$$

$$\sum \frac{1}{\lambda_i} = \frac{1}{0.00381} + \frac{1}{0.00455} = 4.822 \times 10^2$$

$$\therefore \lambda = \frac{2(600) + 3.733 \times 10^4}{4.822 \times 10^2}$$
$$= 10.23 \ \$ \text{ MW h,}$$

at which point, $(\partial F_i/\partial P_i) = \lambda$, for all i, such that

$$2(0.00381)P_1 + 9.2 = 10.23$$
$$\Rightarrow P_1 = 135 \text{ MW.}$$

Similarly, $2(0.00455)P_2 + 6.0 = 10.23 \Rightarrow P_2 = 465$ MW.
Notably, $P_D = P_1 + P_2$ (the power balance equation).

2. For equal incremental fuel costing, there is a violation on the given constraint. Therefore, since P_2 is greater than P_{2max}, let P_2 assume its upper bound 200 MW. Thus

$$P_1 = P_D - P_2 = (600 - 400)$$
$$= 200 \text{ MW,}$$

which is within the desired limits.

The incremental cost for each unit is calculated from $(\partial F_1 / \partial P_i) = \lambda_i$, for all is

$$\lambda_1 = 0.00762P_1 + 9.2 = (0.00762 \times 200) + 9.2 = 10.72 \text{ \$/MW h}$$

and

$$\lambda_2 = 0.00910P_2 + 6.0 = (0.00910 \times 400) + 6.0 = 9.64 \text{ S/MW h.}$$

Example 4.7.4

Consider a power system that is modeled by two generating plants such that their cost function parameters are

$$\beta_1 = 5.93 \quad \gamma_1 = 4.306 \times 10^{-3}$$
$$\beta_2 = 6.02 \quad \gamma_2 = 4.812 \times 10^{-3}.$$

Given also that the network has the following B-coefficients,

$$B_{11} = 3.95 \times 10^{-4}$$
$$B_{22} = 4.63 \times 10^{-4},$$

the system load is 700 MW and the constraints on the generation are

$$100 \leq P_{gl} \leq 500 \text{ MW}$$
$$80 \leq P_{gl} \leq 500 \text{ MW.}$$

1. Calculate the incremental cost and the optimal value for the plants' output.
2. Neglecting the transmission losses, repeat Part 1 while considering the following generation constraints.

SOLUTION

1. Generally, the quadratic approximation of the cost-function modeling the ith plant is

$$F_i(P_i) = \alpha_i + \beta_i P_i + \gamma_i P_i^2,$$

and the transmission losses can be expressed as

$$P_{\text{Loss}} = \sum_{i=1}^{m} \sum_{j=1}^{m} (P_i B_{ij} P_j),$$

where
m is the number of units
B_{ij} is the loss coefficients

For economic dispatch of power from the units with equal cost sharing:

$$\frac{\partial F_i(P_i)}{\partial P_i} = \lambda \left(1 - \frac{\partial P_{\text{Loss}}}{\partial P_i} \right).$$

From the data given, we observe that

$$P_{\text{Loss}} = B_{11} P_1^2 + B_{22} P_2^2; \quad B_{12} = 0.$$

Therefore, for plant 1: $-\beta_1 + 2\gamma_1 P_1 = \gamma(1 - 2B_{11}P_1)$ and for plant 2: $\beta_2 + 2\gamma_2 P_2 = \lambda(1 - 2B_{22}P_2)$,

$$\Rightarrow \frac{\beta_1 + 2\gamma_1 P_1}{\beta_2 + 2\gamma_2 P_2} = \frac{1 - 2B_{11}P_1}{1 - 2B_{22}P_2}.$$

Using $P_2 = P_D - P_1$, further manipulation of the loss equation with substitution of the constants gives

$$P_1^2 + 7.5897 \times 10^4 P_1 - 2.86767 \times 10^7 = 0 \Rightarrow P_1 = 376 \text{ MW}$$

$$\therefore P_2 = P_D - P_1 = 700 - 376 = 324 \text{ MW}.$$

Now, from the equation for plant 1

$$\beta_1 + 2\gamma_1 P_1 = \lambda(1 - 2B_{11}P_1) \Rightarrow \lambda = \frac{\beta_1 + 2\gamma_1 P_1}{1 - 2B_{11}P_1}$$

$$= \frac{5.93 + (2 \times 4.306 \times 10^{-3} \times 3760)}{1 - (2 \times 3.95 \times 10^{-4} \times 376)} = \frac{9.16811}{0.70296} = 13.042 \ \$ \text{ MW h.}$$

The incremental cost for the plants is 13.042 $ MW.

2. Neglecting the transmission losses implies that $(\partial P_{\text{Loss}}/\partial P_i) = 0$. Hence, for optimal power flow (economic dispatch) at each plant

$$\frac{\partial F_T}{\partial P_i} = 0$$

$$\beta_i + 2\gamma_i P_i = \lambda_i, \quad \text{for } \vee\, i$$

$$\therefore \beta_1 + 2\gamma_1 P_1 = \lambda_2 P_2 = \lambda_2, \quad \text{and} \quad P_2 = P_D - P_1.$$

Assuming that $\lambda_1 = \lambda_2 = \lambda$, then

$$\beta_1 + 2\gamma_1 P_1 = \beta_2 + 2\gamma(P_D - P_1)$$

$$\Rightarrow \lambda = \frac{\beta_2 - \beta_1}{2(\gamma_1 + \gamma_2)} = \frac{6.01 - 5.93}{2(4.306 - 4.812) \times 10^{-3}}$$

$$= 655 \text{ MW}.$$

But $P_{1\max} = 500$ MW; therefore we must set P_1 to 500 MW. It follows that the incremental cost for each unit must then be recalculated.

Now, $P_2 = P_D - P_1 = 700 - 500 = 200$ MW, a value within the desired limits of P_2. Therefore, $\lambda_1 = \beta_1 + 2\gamma_1 P_1 = 5.93 + (2)(4.306 \times 10^{-3})$ (500) $= 10.236$ \$ MW h and $\lambda_2 = \beta_2 + 2\gamma_2 P_2 = 6.01 + (2)(4.812 \times 10^{-3})$ (200) $= 7.935$ \$ MW h.

This is an example of a much-simplified iterative process whereby the constraints impose limitations on the desired values of λ_i for each unit.

4.8 Conclusion

This chapter handled practical system problems formulated as constrained optimization problems. The definition of admissible points was presented as those vectors that satisfy the constraints of the problems. It was shown that constrained problems are assumed to have at least one admissible point. The so-called Lagrange multipliers were used as an effective way of dealing with constraints. In Section 4.8.2 theorems on the optimization of constrained functions were presented and necessary and sufficient conditions were defined. In Section 4.8.3 a procedure for optimizing constrained problems was stated in the form of sequential steps. Section 4.8.4 presented some solved problems for illustration and validation purposes. Finally, Section 4.8.5 presented power system application examples such as optimal operation of all thermal system incremental cost-loading.

4.9   Problem Set

PROBLEM 4.9.1

Consider the following problem.

Maximize $x_1^2 - 4x_1x_2 + x_2^2$

Subject to $x_1 + x_2^2 = 1$.

1. Using the Kuhn–Tucker (K–T) conditions, find an optimal solution to the problem.
2. Test for the second-order optimality condition.

Does the problem have a unique optimal solution?

PROBLEM 4.9.2

Consider the following problem:

Maximize $3x_1 - x_2 + x_2^2$

Subject to $x_1 + x_2^2 + x_3 \leq 0$

$-x_1 - 2x_2 + x_3^2 = 0$.

1. Using the Lagrange multiplier technique, write the K–T optimal conditions.
2. Test for the second-order optimality conditions.
3. Argue why the problem is unbounded.

PROBLEM 4.9.3

Find all local minimum and maximum points of the following functions for the four indicated domains. Be sure to list each of the specified points along with the corresponding functional value at the particular point.

Functions

1. $F(x,y) = x^4 + 6x^2 y^2 + y^4 - 2x^2 - 2y^2$.
2. $F(x,y) = x^3 - 3x^2 + 4x + y^2$.
3. $F(x,y) = x^4 + y^4$.

Four indicated domains of the permissible constraints set for each of the objective function.

1. $\{(x,y): x^2 + y^2 \leq 1\}$.
2. $\{(x,y): x^2 + y^2 \geq 2\}$.

3. $\{(x,y): |x| \le 1 \text{ and } |y| \le 1\}$.
4. $\{(x,y): |x| \le 2 \text{ and } |y| \le 2\}$.

PROBLEM 4.9.4

Find the local and global maxima and minima of the function

$$f(x) = f(x_1, x_2, x_3) = x_1^2 + 7x_1 + x_1x_2 + x_2^3 + x_2x_3^2$$

$$|x_1|, |x_2|, |x_3| \le 4.$$

PROBLEM 4.9.5

Two electrical generators are interconnected to provide total power to meet the load. Each generator's cost is a function of the power output (Figure 4.3). All power costs are expressed on a per unit basis. The total power need is at least 60 units. Formulate a minimum cost design problem and solve it graphically. Verify K–T conditions at the solution points and show gradients of cost and constraint functions on the graph.

PROBLEM 4.9.6

Repeat Example 4.7.3 for a transmission loss equation given by

$$P_L = 0.08P_2.$$

All other data are unchanged.

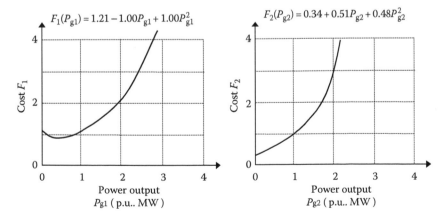

FIGURE 4.3
Power generator cost curves for Problem 4.9.5.

References

1. Bertsekas, D., *Constrained Optimization and Lagrange Multiplier Methods*, Academic Press, New York, 1982; republished by Athena Scientific, 1996.
2. Heterse, M. R., *Optimization Theory, the Finite Dimensional Case*, Wiley, New York, 1975.
3. Hillier, F. S. and Lieberman, G. J., *Introduction to Operations Research*, 4th edn., Holden-Day, San Francisco, CA, 1986.
4. Mulvey, J. M., Vanderbei, R. J., and Zenios, S. A., Robust optimization of large-scale systems, *Operations Research*, 43:264–281, 1995.
5. Luenberger, D. G., *Optimization by Vector Space Methods*, Wiley, New York, 1969.
6. Luenberger, D. E., *Introduction to Linear and Nonlinear Programming*, Addison-Wesley, Reading, MA, 1975.
7. Mokhtar, S., Bazaraa, C., and Shetty, M., *Nonlinear Programming, Theory and Algorithm*, Wiley, New York, 1979.
8. Pierre, D. A., *Optimization Theory with Application*, Wiley, New York, 1969.
9. Potter, W. A., *Modern Foundation of Systems Engineering*, Macmillan, New York, 1966.
10. Zangwill, W. I., Minimizing a function without calculating derivations, *Computer Journal*, 10:293–296, 1967.

5

Linear Programming and Applications

5.1 Introduction

While linear programming (LP) has been a very useful tool developed and used by operation research, power systems, and industry, such as aerospace, for power system, it is fully used in operation and planning of determining resource allocation and value. In recent years, it is a major tool for pricing measured resources [3,8]. It has been of major use in prediction and estimation of data from measurements. The operation research environment and economics have to rely on linear program as a tool for pricing and resources allocation. Due to capability challenge of handling of data new advances to improve computation speed have been introduced namely, interior point method is the application we plan to introduce the concept of LP formulation in matrix form and develop solution requirements based on graphical, simplex (recommended), and interior point method. The application of the methods is numerous in large-scale systems.

Additionally, integer programming (IP) to account for discrete decision variables is introduced in this chapter. The sensitivity method that reduces computational burden by accounting for small perturbation in right-hand side variables will also be found useful to the reader. There are other different techniques, such as calculation for steady-state-based Lagrange technique, which are discussed.

Research papers that apply LP-based optimal power flow (OPF) for minimal losses, cost, etc. use LP to solve optimal operating points of voltage, power as constraints in order to determine the minimized objective values and determine the optimal objectives. The implementation of intermitted resources from distributed generation (DG) is also studied in OPF based on LP [14] provides examples of well-written research work done in the area of LP applications.

The most general description of LP problem [6,7,10–12] is given as the problem of allocating a number m of resources among $1, 2, \ldots, n$ activities in such a way as to maximize the worth from all the activities. The term "linear" refers to the fact that all the mathematical relationships among the decisions (variables) to allocate resources to activities and the various restrictions applicable therein (constraints), as well as the criterion (objective function) are

TABLE 5.1

Notations Commonly Used in LP

Resources	Activity				Total Resources
	1	2	...	n	
1	a_{11}	a_{12}	...	a_{1n}	b_1
2	a_{21}	a_{22}	...	a_{2n}	b_2
.	.	.	...	.	.
.	.	.	...	.	.
.	.	.	...	.	.
M	a_{m1}	a_{m2}	...	a_{mn}	b_m
ΔP/unit of activity	c_1	c_2	...	c_n	—
Level of activity	x_1	x_2	...	x_n	—

devoid of any nonlinearity. The objective function is some measure of the overall performance of the activities (e.g., cost, profit, net worth, system efficiency, etc.). Standard notation for LP is summarized in Table 5.1.

For activity j, c_j ($j = 1, \ldots, n$) is the increase in P (i.e., ΔP), that would result from each unit of increase in x_j (the level of activity j). For resource i, $i = 1, \ldots, m$, b_i is the total amount available for allocation to all the activities. The coefficient a_{ij} denotes the amount of resource i consumed by activity j. The set of inputs (a_{ij}, b_i, c_j) constitutes the parameters of the LP model.

5.2 Mathematical Model and Nomenclature in LP

The conventional LP model reduces to the following equation:

$$\text{Maximize } P = c^T x \tag{5.1}$$

Subject to

$$Ax \leq b \tag{5.2}$$

$$x_j \geq 0, \quad \forall j \in \{1, n\}, \tag{5.3}$$

where the following vectors are defined

Decision matrix: $x = [x_1, x_2, \ldots, x_n]^T$.

Cost coefficient array: $c^T = [c_1, c_2, \ldots, c_n]$.

Constant array: $b = [b_1, b_2, \ldots, b_m]^T$.

System or state matrix:

$$A = \begin{bmatrix} a_{11} & a_{12} & \cdots & a_{1n} \\ a_{21} & a_{22} & \cdots & a_{2n} \\ \vdots & \vdots & \ddots & \vdots \\ a_{m1} & a_{m2} & \cdots & a_{mn} \end{bmatrix}.$$

The following are important terminologies used in LP:

Objective functions and constraints. The function $P(x)$ being maximized is called the objective or goal function subject to the restriction sets of constraints given by Equations 5.2 and 5.3. Equation 5.2 represents a restriction set that is often referred to as the functional constraints and Equation 5.3 is termed as the nonnegativity constraints.

Feasible solution and region. Any specification of the variable x_j is called a solution. A feasible solution is one that satisfies all constraints. The feasible region is the collection of all feasible solutions. If the problem does not have any feasible solution, it is called an infeasible problem.

Optimal solution. An optimal solution corresponds to the minimum or maximum value of the objective function. The problem is either one of minimization or maximization depending on the nature of the objective function under consideration, that is, cost and profit, respectively.

Multiplicity in solution. There can be multiple optimal solutions in cases where a number of combinations of the decision variables give the same maximum (or minimum) value.

Unbounded solutions. There may also be unbounded solutions in that the LP problem objective function could be infinitely low or high depending on minimizing or maximizing cases, respectively.

5.2.1 Implicit Assumptions in LP

There are certain assumptions utilized in LP models that are intended to be only idealized representations of real problems. These approximations simplify the real-life problem to make it tractable. Adding too much detail and precision can make the model too unwieldy for useful analysis. All that is needed is that there be a reasonably high correlation between the prediction of the model and what would actually happen in reality.

The four assumptions are as follows:

1. *Proportionality.* This relates to the assumption of constant-cost coefficients c_j irrespective of the level of x_j. Stated alternatively, there can be correlation between the coefficients c_j and x_j. For example, the incremental cost associated with additional units of activity may be

lower, which is referred to as "increasing marginal return" in the theory of economics. The proportionality assumption is needed essentially to avoid nonlinearity.

2. *Additivity.* Apart from the proportionality assumption, the additivity assumptions are needed to avoid cross-product terms among decision variables. This assumption amounts to state that the total contribution from all activities can be obtained by adding individual contributions from respective activities.

3. *Divisibility.* Decision variables can take any fractional value between specified limits. In other words, the variables cannot be restricted to take some discrete values or integer values.

4. *Certainty.* All the parameters in the model are assumed to be known constants with no uncertainty about the values that these parameters may assume.

It is important in most cases to study and analyze the disparities due to these assumptions by checking with more complex models. The more complex alternative models are a result of relaxing the assumptions regarding nonlinear programming models due to relaxing assumption (1) and/or assumption (2), IP models by relaxing assumption (3), and stochastic programming models by ignoring assumption (4). More complex models are obtained by dropping various combinations of assumptions (1)–(4).

5.3   LP Solution Techniques

Various approaches have been developed over the years to solve LP problems. The commonly encountered techniques that have gained wide attention from engineers, mathematicians, and economists are the graphical approach, the simplex method, the revised simplex method, and the tableau approach. In the text, we omit the treatment of the tableau approach as its flexibility in the development of programs for more complex algorithms outdoes its ability to solve large-scale problems [1–5].

5.3.1   Graphical Method

In the LP problem, the optimal solution (if it exists) lies at one of the corner points of the polytope formed by the boundary conditions of the functional and nonnegativity constraints of the problem. In the graphical method, each corner-point solution, which is also a feasible solution, is checked using the objective function. The solution that yields the greatest improvement to the objective value is the optimal solution to the problem. The graphical technique for solving LP problems is demonstrated by the following example.

Consider the following two-variable LP problem.

Maximize $P = 3x_1 + 5x_2$

Subject to

$x_1 \leq 4$

$2x_2 \leq 12$

$3x_1 + 2x_2 \leq 18$

$x_j \geq 0, \quad \forall j \varepsilon \{1,2\}.$

A two-dimensional figure can be constructed corresponding to the two variables x_1 and x_2 (Figure 5.1). The nonnegativity constraints automatically imply that the search is restricted to the positive side of the x_1 and x_2 axes. Next, it should be observed that the feasible solution cannot lie to the right of the line $x_1 = 4$, because of the restriction $x_1 \leq 4$. In the same way, the restrictions $2x_2 \leq 12$, and $3x_1 + 2x_2 \leq 18$ provide the other two cuts to generate the feasible region in Figure 5.1 as indicated by the shaded area OABCD.

The vertices A, B, C, D, and the origin of the polytope are the corner-point solutions to this problem and at least one of such points represents the optimal solution set for x_1 and x_2. We must now select the point in this region that maximizes the values of $P = 3x_1 + 5x_2$. Different lines corresponding to different values of P are drawn to check the point beyond which there is no point of the feasible region, which lies on the line $P = 3x_1 + 5x_2$. This is

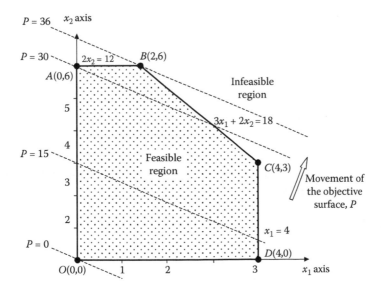

FIGURE 5.1
Graphical representation of the two-dimensional problem.

achieved by drawing a line for $P = 0$ and then progressively increasing it to 15, 30, and so on. It is found that the optimal solution is

$$x_1^* = 2$$
$$x_2^* = 6$$

and the maximum value of P is

$$p^* = 36.$$

This is shown in Figure 5.1 as vertex B of the polytope.

5.3.2 Matrix Approach to LP

The matrix approach is quite effective for the computer-based solution of LP problems. It generally first requires a matrix inversion. The simplex method equipped with an artificially constructed identity matrix for initialization is called the revised simplex method. The revised simplex method and a flowchart for computer programming are discussed with illustrative problems. The IP problem is to be solved as an LP in conduction with Gomory cut, which eliminates the nonintegral optimal solution step by step.

The purpose of LP is to solve the problem formulated in the following standard form:

Maximize $p = c^T x$ (5.4)

Subject to

$Ax = b$ (5.5)

and $x \geq 0$,   (5.6)

where

$b_i > 0$,   for all $i = 1, 2, \ldots, m$.

$x_j \geq 0$,   for all $j = 1, 2, \ldots, n$.

We now define the following arrays in preparation for formulating the LP problem in this chapter:

1. *m*-vectors

$$y = \begin{bmatrix} y_1 \\ y_2 \\ \vdots \\ y_m \end{bmatrix}, \quad b = \begin{bmatrix} b_1 \\ b_2 \\ \vdots \\ b_m \end{bmatrix}, \quad x_B = \begin{bmatrix} x_{B1} \\ x_{B2} \\ \vdots \\ x_{Bm} \end{bmatrix}.$$

2. *n*-vectors

$$u = \begin{bmatrix} u_1 \\ u_2 \\ \vdots \\ u_n \end{bmatrix}, \quad x = \begin{bmatrix} x_1 \\ x_2 \\ \vdots \\ x_n \end{bmatrix}, \quad c = [c_1, c_2, \ldots, c_n].$$

3. $n \times m$ vectors for the case when $n > m$,

$$Z = [z_1, z_2, \ldots, z_{n-m}].$$

4. Matrices

$$A = [a_{ij}]_{m \times n} = \begin{bmatrix} a_{11} & a_{12} & \cdots & a_{1n} \\ a_{21} & a_{22} & \cdots & a_{2n} \\ \cdots & \cdots & \cdots & \cdots \\ a_{m1} & a_{m2} & \cdots & a_{mn} \end{bmatrix}.$$

Now let B represent an $m \times m$ matrix and $\bar{B}$ represent an $m \times (n-m)$ matrix such that $A = [B, \bar{B}]$. Matrix A is now said to be augmented into two components.

The state vector x is a feasible solution (or simply called a solution) if it satisfies the constraint Equations 5.5 and 5.6. It is a nondegenerate basic feasible solution (or simply called a basic solution) if it contains exactly m positive components with all others equal to zero. It is a maximum feasible solution (or simply called a maximum solution) if it is a solution and also maximizes the objective function P.

In practical applications, a given problem may not appear the same as the standard form. For example, the state variable x_1 may be bounded, the constraints may be inequalities, or the objective function may require minimization. In most cases, the following rules can be useful in converting a practical problem to the standard form:

Rule 1: Change of constraints on state variables
The bounded state variable of the following forms can be converted to the inequality form (Equation 5.6):

1. If $x_i \leq d_i$, then add a slack variable x_s to make $x_i + x_s = d_i$ or replace x_i by $d_i - x_s$ where x_s is nonnegative.
2. If $x_i \geq d_i$, then add a slack variable x_s to make $x_i - x_s = d_i$ or replace x_i by $x_s + d_i$.
3. If $-\infty < x_i < \infty$, then replace x_i by $x_s - x_{s+1}$.
4. If $d_i \leq x_i \leq h_i$, then the constraint is equivalent to $0 \leq x_i - d_i \leq h_i - d_i$ and hence x_i is to be replaced by $x_s + d_i$ and $x_s + x_{s+1} = h_i - d_i$ is a new constraint.

Rule 2: Conversion from inequality to equality constraint

We use $(A_x)_i$ to denote the ith row of the vector (Ax) in the following:

1. If $d_i \le (Ax)_i \le h_i$, then two slack variables are needed to make $(Ax)_i - x_s = d_i$ and $(Ax)_i + x_{s+1} = h_i$.
2. If $(Ax)_i = b_i \le 0$, then multiply both sides by -1 when $b_i < 0$, and create a new constraint by adding it to another constraint when $b_i = 0$.

Rule 3: Modification of objective function

1. Minimization of P. Any solution for max(–P) is the same as that for min(P), but min(P) = –max(–P).
2. Maximization of |P|. Find both max(P) and min(P) subject to the same constraint and then select the larger of |max(P)| or |min(P)|.

5.3.3 Simplex Method

The geometric method cannot be extended to large-scale problems for which the number of decision variables can run into several thousands. The simplex method, developed by Dantzig in 1947 [13], and its variants have been widely used to date and a number of commercial solvers have been developed based on it. The first step in setting up the simplex method converts the inequality constraints into equivalent equality constraints by adding slack variables.

Recall the following example:

Maximize $P = 3x_1 + 5x_2$

Subject to

$2x_2 \le 12$

$3x_1 + 2x_2 \le 18$

$x_j \ge 0, \quad \forall j \in \{1,2\}.$

This example illustrative problem can be converted into the following equivalent form by adding three slack variables x_3, x_4, and x_5. Thus, we obtain

Maximize $P = 3x_1 + 5x_2$

Subject to

$x_1 + x_3 = 4$

$2x_1 + x_4 = 12$

$3x_1 + 2x_2 + x_5 = 18$

$x_j \ge 0, \quad \forall j \in \{1,2\}.$

This is called the augmented form of the problem and the following definitions apply to the newly formulated LP problem.

An augmented solution is a solution for the original variables that has been augmented by the corresponding values of the slack variables.

A basic solution is an augmented corner-point solution.

A basic feasible solution is an augmented corner-point feasible solution.

For example, augmenting the solution (3, 2) in the example yields the augmented solution (3, 2, 1, 8, 5); and the corner-point solution (4, 6) is translated as (4, 6, 0, 0, −6).

The main difference between a basic solution and a corner-point solution is whether the values of the slack variables are included. Because the terms basic solution and basic feasible solution are integral parts of LP techniques, their algebraic properties require further clarification. It should be noted that there are five variables and only three equations in the present problem. This implies that there are two degrees of freedom in solving the system since any two variables can be chosen to be set equal to any arbitrary value in order to solve the three equations for the three remaining variables. The variables that are currently set to zero by the simplex method are called nonbasic variables, and the others are called basic variables. The resulting solution is a basic solution. If all the basic variables are nonnegative, the solution is called a basic feasible solution. Two basic feasible solutions are adjacent if all but one of their nonbasic variables are the same.

The final modification needed to use the simplex method is to convert the objective function itself in the form of a constraint to get

$$P - 3x_1 - 5x_2 = 0.$$

Obviously, no slack variables are needed since it is already in the form of equality. Also, the goal function P can be viewed as a permanent additional basic variable.

The feasibility of achieving a solution for the LP problem as described by Equations 5.4 through 5.6 relies on a set of optimality conditions. In solving the problem, the optimality conditions are forced to be satisfied by means of the simplex method, involving a systematic process of selecting a basis to increase the objective function P until it can no longer be increased. We assume that the problem is nondegenerate; that is, all solutions are basic. The following notations are used in the statement and the theorem in this chapter:

B	m-square and nonsingular matrix formed by m columns of A
x_B	m-vector of which the components are the positive components of a basic solution x corresponding to B; that is, $x_B = B^{-1}b$
C_B	m-row vector formed by m components of C that correspond to x_B; that is, $P = Cx = C_B x_B$
$\bar{C}_B$	$(n - m)$-row vector makes $C = (C_B, \bar{C}_B)$
$\bar{c}_{BK}$	kth component of $\bar{C}_B$
$\bar{a}_k$	kth column of the matrix $\bar{B}$
Z	$(n - m)$-row vector defined by $Z = C_B B^{-1} \bar{B}$
z_k	kth component of Z

THEOREM 5.3.1

The objective function P can be increased if $z_k - \bar{c}_{Bk} < 0$ for some k, and the components of x_B are the positive components of a maximum solution if $Z - \bar{C}_B \leq 0$.

The theorem gives a condition of optimality that suggests a systematic approach to a maximum solution of the LP problem.

It is assumed in the simplex method that initially there exists a basis, which yields a basic solution. In practical problems, such a basis is hard if not impossible to find, especially for large m. To overcome this difficulty, an identity matrix is created here for the initialization of the simplex method. The simplex method requires that the basis matrix be inverted at each iteration. We note that there is only a one column change between two consecutive bases. This enables us to apply a lemma of matrix inversion to obtain the inverse matrix for the new basis from that of the old one.

Consider the following example:

$$x_1 + x_2 + x_3 + x_4 = b_1$$

$$x_1 + 2x_2 + x_3 = b_2$$

$$2x_1 - x_2 - x_3 = b_3.$$

Introduce artificial variables x_5 and x_6 such that

$$x_1 + x_2 + x_3 + x_4 = b_1$$

$$x_1 + 2x_2 + x_3 + x_5 = b_2$$

$$2x_1 - x_2 - x_3 + x_6 = b_3.$$

Then, the matrix A of Equation 5.5 has the form

$$A = \begin{bmatrix} 1 & 1 & 1 & 1 & 0 & 0 \\ 1 & 2 & 1 & 0 & 1 & 0 \\ 1 & -1 & -1 & 0 & 0 & 1 \end{bmatrix}$$

and, hence, an identity matrix is created by the last three columns. The basic solution corresponding to the basis of the identity matrix is, therefore, $x_4 = b_1$, $x_5 = b_2$, and $x_6 = b_3$. There are no more than m artificial variables because some state or slack variables can be utilized as artificial variables, such as x_4 in this example.

5.3.4   Lemma of Matrix Inversion

The equation

$$[P^{-1} + H_1 Q H_2]^{-1} = P - P H_1 [H_2 P H_1 + Q^{-1}]^1 H_2 P \tag{5.7}$$

is true if the inverses exist. The lemma can be applied generally to the n-square matrix P, m-square matrix Q, $n \times m$ matrix H_1, and $m \times n$ matrix H_2. The integer m is practically less than n.

We make use of the identity to generate a recursive formula of matrix inversion that is useful to LP. To this end, we assume that $p = B^{-1}$, $Q = 1$, $H_1 = \bar{a}_k - a_r$, and $H_2 = J_r^T$, where J_r is the column matrix with rth component equal to one but all others equal to zero. Thus, the following result can be obtained for nonsingular matrix B:

$$
\begin{aligned}
\text{(a) } 1 + H_2 B^{-1} H_1 &= 1 + J_r^T B^{-1}(\bar{a}_k - a_r) \\
&= 1 + J_r^T y - J_r^T B^{-1} a_r \\
&= 1 + y_r - J_r^T J_r = y_r
\end{aligned}
$$

where
$$B^{-1}\bar{a}_k = y = [y_1, y_2, \ldots, y_m]^T$$
a_r is the rth column of the B matrix

$$
\begin{aligned}
\text{(b) } B^{-1} H_1 H_2 B^{-1} &= B^{-1}(\bar{a}_k - a_r)J_r^T B^{-1} \\
&= (y - J_r)J_r^T B^{-1} \\
&= (y_1, y_2, \ldots, y_{r-1}, \ldots, y_m)^T (b_{r1}, b_{r2}, \ldots, b_{rm}).
\end{aligned}
$$

Substitutions of (a) and (b) into the lemma give

$$
(B + H_1 H_2)^{-1} = B^{-1} - \frac{1}{y_r}(y_1, y_2, \ldots, y_{r-1}, \ldots, y_m)^T
$$

$$
(b_{r1}, b_{r2}, \ldots, b_{rm}).
$$

Let b_{ij} and b'_{ij} be the general elements of the matrices B^{-1} and $(B + H_1 H_2)^{-1}$, respectively. Then, it follows from the above equations that

$$
b'_{ij} = b_{ij} - \frac{y_i}{y_r} b_{rj}; \quad \text{for } i \neq r \tag{5.8}
$$

$$
= \frac{b_{rj}}{y_r}; \quad \text{for } i = r \tag{5.9}
$$

As such, the matrix inversion for the new basis can be derived from that of the old one according to Equations 5.8 and 5.9. Now, the simplex method can be broken down into three stages: namely initialization, iterative procedure, and optimality checking.

1. *Initialization step*: The simplex method can start at any corner-point feasible solution; thus a convenient one such as (0, 0) is chosen for the original variables. Therefore the slack variables become the basic variables and the initial basic feasible solution is obtained at (0, 0, 4, 12, 18).

2. *Iterative procedure*: The iterative procedure involves moving from one corner-point solution to an adjacent basic feasible solution. This movement involves converting one nonbasic variable into a basic variable and simultaneously converting a basic variable into a nonbasic one. The former is referred to as the entering variable and the latter as the leaving variable.

The selection of the nonbasic variable to be converted into a basic variable is based on improvement in the objective function P. This can be calculated from the new representation of the objective function in the form of a constraint. The nonbasic variable, which contributes to the largest increment in the P value, is the entering variable. For the problem at hand, the two nonbasic variables x_1 and x_2 add per-unit contribution to the objective function as 3 and 5, respectively (the coefficients c_j), and, hence, x_2 is selected as the entering variable.

The adjacent basic feasible solution is reached when the first of the basic variables reaches a value of zero. Thus, once the entering variable is selected, the leaving variable is not a matter of choice. It has to be the current basic variable whose nonnegativity constraint imposes the smallest upper bound on how much the entering basic variable can be increased. The three candidate leaving variables are x_3, x_4, and x_5. Calculation of the upper bound is illustrated in Table 5.2.

Since x_1 remains a nonbasic variable, $x_1 = 0$ which implies x_3 remains nonnegative irrespective of the value of x_2, $x_4 = 0$, when $x_2 = 6$, and $x_5 = 0$, when $x_2 = 9$, indicate that the lowest upper bound on x_2 is 6, and is determined when $x_4 = 0$. Thus, x_4 is the current leaving variable.

3. *Optimality test*: To determine whether the current basic feasible solution is optimal, the objective function can be rewritten further in terms of the nonbasic variables:

$$P = 30 + 3x_1 - \frac{5}{2}x_4.$$

Increasing the nonbasic variables from zero would result in moving toward one of the two adjacent basic feasible solutions. Because x_1 has a positive coefficient increasing x_1 would lead to an adjacent basic feasible solution that is better than the current basic feasible solution; thus, the current solution is

TABLE 5.2

Calculation of Upper Bound

Basic Variable	Equation	Upper Bound for x
x_3	$x_3 = 4 - x_1$	No limit
x_4	$x_4 = 12 - 2x_2$	$x_2 \leq \frac{12}{2} = 6 \leftarrow$ min
x_5	$x_5 = 18 - 3x_1 - 2x_2$	$x_2 \leq \frac{18}{2} = 9$

not optimal. Speaking generally, the current feasible solution is optimal if and only if all of the nonbasic variables have nonpositive coefficients in the current form of the objective function. The flowchart of the simplex method is shown in Figure 5.2.

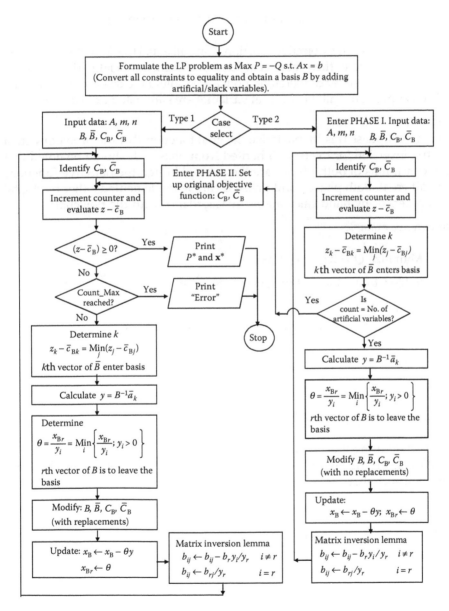

FIGURE 5.2
Flowchart showing the three stages of the simplex method.

5.3.5 Revised Simplex Method

The so-called revised simplex method consists of two phases:

Phase I

This first phase starts with an identity matrix for the basis (an artificial basis) and the maximization of x_0, where

$$x_0 = -(\text{sum of all the artificial variables}).$$

The coefficients of the objective function x_0 are all equal to -1. If the maximized x_0 is less than zero, then there is at least one of the artificial variables that is different from zero and, therefore, no feasible solution exists for the problem. We enter phase II if all the artificial variables are eliminated or $\max(x_0 = 0)$.

Phase II

The aim of phase II is to maximize the goal function P of the problem with a matrix inversion for the basis inherited from phase I. It should be noted that the constraints of both phases are the same although the objective functions are different. Both phase I and phase II can be implemented either by digital computers or by hand calculations in accordance with the flowchart.

Illustrative example

Minimize $Q = 2x_1 + x_2 + x_3$

Subject to the inequality constraints given by

$3x_1 + 5x_2 + 2x_3 \geq 16$

$4x_1 - 2x_2 + x_3 \geq 3$ and $x_i \geq 0$,   for all $i \in \{1,2,3\}$

using the

1. Simplex method
2. Revised simplex method

1. Solution using the simplex method

We convert the minimization problem to a maximization one by changing the sign such that

$$P = -2x_1 - x_2 - x_3.$$

Utilizing two slack variables x_4 and x_5 and two artificial variables x_6 and x_7 we obtain the following equations:

$$3x_1 + 5x_2 + 2x_3 - x_4 + x_6 = 16$$
$$4x_1 - 2x_2 + x_3 - x_5 + x_7 = 3.$$

In matrix notation

$$A = [\bar{B} \quad : \quad B] = \begin{bmatrix} 3 & 5 & 2 & -1 & 0 & \vdots & 1 & 0 \\ 4 & -2 & 1 & 0 & -1 & \vdots & 0 & 1 \end{bmatrix}$$

$$x = [x_1, x_2, x_3, x_4, x_5, x_6, x_7]^T$$

$$C = \begin{bmatrix} \bar{C}_B & : & C_B \end{bmatrix} = [-2 \quad -1 \quad -1 \quad 0 \quad 0 \quad \vdots \quad 0 \quad 0]$$

and

$$b = [16, 3]^T.$$

Notably, $B = B^{-1} = I$. Thus, $x_B = B^{-1}b = Ib = [16, 3]^T$.

First iteration

$$Z - C_B = C_B B^{-1}\bar{B} - \bar{C}_B = -\bar{C}_B$$
$$= [2 \quad 1 \quad 1 \quad 0 \quad 0],$$

which is definitely positive.

$$P = C_B x_B = C_B B^{-1} = C_B Ib$$
$$= [0 \quad 0] I \begin{bmatrix} 16 \\ 3 \end{bmatrix} = 0$$

and

$$x_B = \begin{bmatrix} 16 \\ 3 \end{bmatrix} = \begin{bmatrix} x_6 \\ x_7 \end{bmatrix}$$

$$x_1 = x_2 = x_3 = 0$$

$\max P = 0 \Rightarrow \min Q = 0$.

NOTE: This solution is infeasible as the second constraint $4x_1 - 2x_2 + x_3 \geq 3$ is not valid here as it is zero. The revised simplex method shows this result.

2. Solution using the revised simplex method

Phase I

$$\max(x_0) = -(x_6 + x_7)$$
$$C_B = [-1, -1]$$
$$\bar{C}_B = [0 \quad 0 \quad 0 \quad 0 \quad 0]$$
$$x_B = B^{-1}b = \begin{bmatrix} 16 \\ 3 \end{bmatrix} = \begin{bmatrix} x_6 \\ x_7 \end{bmatrix}.$$

First iteration

Step 1

$$Z - \bar{C}_B = C_B B^{-1} \bar{B} - \bar{C}_B$$

$$= [-1 \quad -1] I \begin{bmatrix} 3 & 5 & 2 & -1 & 0 \\ 4 & -2 & 1 & 0 & -1 \end{bmatrix}$$

$$= [-7 \quad -3 \quad -3 \quad 1 \quad 1].$$

The minimum of the previous row is -7; then the first vector of $\bar{B}$, x_1 enters the basis $k = 1$.

Step 2

Calculate

$$y = B^{-1}\bar{a}_k = I\bar{a}_1 = \begin{bmatrix} 3 \\ 4 \end{bmatrix}$$

$$\theta = x_{Br} = \min_i \left\{ \frac{x_{Bi}}{y_i}; \ y_i > 0 \right\}$$

$$= \min\left\{ \frac{16}{3}, \frac{3}{4} \right\} = \frac{3}{4} \quad \text{and} \quad r = 2.$$

Now, $r = 2$ implies that the second vector of B is to leave the basis.

Step 3

The new basic solution is

$$x_B = \begin{cases} x_{Br} = \theta, & i = r \\ x_{Bi} - \theta y, & i \ne r \end{cases}$$

$$x_B = \begin{cases} x_1 = \frac{3}{4} \\ x_7 = 3 - (4(3/4)) = \begin{Bmatrix} \frac{3}{4} \\ 0 \end{Bmatrix}. \end{cases}$$

Step 4

Now we calculate the inverse of the new basic

$$B^{-1} = \begin{bmatrix} b'_{ij} \end{bmatrix}, \quad \text{for } i = r = 2, \quad b'_{ij} = \frac{b_{rj}}{y_r}$$

$$b'_{21} = \frac{b_{21}}{y_2} = 0 \quad \text{and} \quad b'_{22} = \frac{b_{22}}{y_2} = \frac{1}{4}$$

$$\text{for } i \ne r \Rightarrow i \ne 2 \quad b'_{ij} = b_{ij} - \frac{y_i}{y_r} b_{rj}$$

$$b'_{11} = b_{11} - \left(\frac{y_1}{y_2}\right) b_{21} = 1 - 0 = 1$$

$$b_{21} = b_{21} - \left(\frac{y_1}{y_2}\right) b_{22} = 0 - \left(\frac{3}{4}\right)(1) = \frac{-3}{4}$$

$$[\bar{B}] = \begin{bmatrix} 5 & 2 & -1 & 0 \\ -2 & 1 & 0 & -1 \end{bmatrix}$$

$$[\bar{C}_B] = [-1 \quad -1 \quad 0 \quad 0]$$

$$\bar{x}_b = [x_2 \quad x_3 \quad x_4 \quad x_5]$$

$$[B] = \begin{bmatrix} 1 & 3 \\ 0 & 4 \end{bmatrix} \Rightarrow [B^{-1}] = \frac{1}{4}\begin{bmatrix} 4 & -3 \\ 0 & 1 \end{bmatrix}$$

$$[C_B] = [0 \quad -2] x_b = [x_b \quad x_1] = \left[\frac{3}{4} \quad 0\right].$$

Second iteration

Step 1

$$C_B B^{-1} = \frac{1}{4}[0 \quad -2]\begin{bmatrix} 4 & -3 \\ 0 & 1 \end{bmatrix}$$

$$= \frac{1}{4}[0 \quad -2]$$

$$Z - \bar{C}_B = C_B B^{-1} \bar{B} - \bar{C}_B$$

$$= \frac{1}{2}[0 \quad -1]\begin{bmatrix} 5 & 2 & -1 & 0 \\ -2 & 1 & 0 & -1 \end{bmatrix} - \bar{C}_B$$

$$= \frac{1}{2}[2 \quad -1 \quad 0 \quad 1] - [-1 \quad -1 \quad 0 \quad 0]$$

$$= \left[2 \quad \tfrac{1}{2} \quad 0 \quad \tfrac{1}{2}\right],$$

which is positive definite; then

$$\max P = C_B x_B = [0 \quad -2]\begin{bmatrix} \frac{3}{4} \\ 0 \end{bmatrix} = 0$$

and all the artificial variables are not eliminated from the basis. Therefore we cannot enter phase II of the calculations as the function has an infeasible solution as shown in the simplex method.

5.4 Duality in LP

LP problems exhibit an important property known as duality. The original problem is therefore referred to as the primal and there exists a relationship between the primal of the LP problem and its duality. Now, from Equations 5.1 and 5.2, the general LP problem in its primal form may be expressed as

Maximize $P(x) = c^T x$

Subject to the constraints:

$Ax \leq b$

and

$x_j \geq 0$

$\forall j \in \{1,n\}$,

where
  A is an $m \times n$ matrix
  x is a column n-vector
  c^T is a row n-vector
  b is a column m-vector

The LP model above may also be written as

$$\text{Maximize } P(x) = \sum_{j=1}^{n} c_j^T x_j \tag{5.10}$$

Subject to the constraints:

$$\sum_{i}^{m} \left(\sum_{j=1}^{n} a_{ij} x_j \leq b_i \right) \tag{5.11}$$

and

$x_j \geq 0$

$\forall i \in \{1,m\}$ and $\forall j \in \{1,n\}$.

Notably, the slack variables have not yet been introduced in the inequality constraints.

By the definition of the dual of the primal LP problem expressed as Equations 5.12 through 5.14, we obtain the following model:

$$\text{Minimize } Q(y) = b_D y \tag{5.12}$$

Subject to the constraints:

$$A_D y \geq c_D^T \tag{5.13}$$

and

$$y_j \geq 0 \tag{5.14}$$

$\forall j \in \{1,m\}$,

where
A_D is an $n \times m$ matrix
y is a column m-vector
c_D is a row m-vector
b_D is a column n-vector

Similar to the model for the primal case, the dual LP model above may also be written as

$$\text{Minimize } Q(y) = \sum_{i=1}^{m} b_{D_i} y_i \tag{5.15}$$

Subject to the constraints:

$$\sum_{j}^{n} \left(\sum_{i=1}^{m} a_{ij} y_j \geq c_j \right) \tag{5.16}$$

$$y_i \geq 0$$

$$\forall i \in \{1,m\} \quad \text{and} \quad \forall j \in \{1,n\}.$$

The dual problem can now be solved using any of the previously discussed solution techniques such as the revised simplex method. In addition, the duality model exhibits special properties that can be summarized using the following well-known duality theorems.

THEOREM 5.4.1

The dual of the dual linear programming model is the primal.

THEOREM 5.4.2

The value of the objective function P(x) for any feasible solution of the primal is greater than or equal to the value of the objective function Q(y) for any feasible solution of the dual.

THEOREM 5.4.3

The optimum value of the objective function P(x) of the primal, if it exists, is equal to the optimum value of the objective function Q(y) of the dual.

THEOREM 5.4.4

The primal has an unbounded optimum if and only if the dual has no feasible solution. The converse is also true.

Finally, we make note of the fact that the duality principle, when applied correctly, can reduce the computation needed to solve multidimensional problems. This is attributed to the fact that the large number of constraints are now modeled as the objective function and the old objective with smaller dimension has become the new constraints. Also, the duality concept can be applied to problems that are associated with the inverse of the A matrix in the primal model of the LP problem.

5.5 Khun–Tucker Conditions in LP

In this section, selected formulations for computing the Lagrange multipliers in LP are summarized using the standard notations. Here, we investigate Min/Max LP problems with inequality and equality constraints. In order to determine the Lagrange multipliers associated with each case, the concepts of duality, Khun–Tucker (K–T) conditions, and Extended K–T (EKT) conditions are applied.

In general, the solution of the dual LP problem results in the computation of the desired Lagrange vectors for any equality constraints that are present. Primal LP with both equality constraints $(A_1x = b_1)$ and inequality constraints $(A_2x \leq b_2)$ must be transformed into a purely equality using nonnegative slack variables. This is because the dual of mixed equality and inequality constraints does not exist.

A brief summary of common cases of LP and the corresponding duality conditions now give rise to the appropriate use of complementary slackness and are presented next. Complementary slackness is a term used to describe the orthogonal nature of the dual and primal shadow prices or Lagrange vectors in LP. Also, a simple example is used to demonstrate how Lagrange multipliers for LP problems are determined.

5.5.1 Case 1: LP and KKT Conditions for Problems with Equality Constraints

Recall the formulation
  Min c^Tx s.t.

$$Ax \geq b$$

$$x \geq 0,$$

where

  c^T and b are n- and m-vectors, respectively

  A is an $m \times n$ matrix

The Karush–Khun–Tucker (KKT) conditions requiring optimal solutions $[x^*, \lambda^*, \mu^*]^T$ are

$$Ax \geq b, \quad x \geq 0$$

$$A^T\lambda + \mu = c, \quad \lambda \geq 0, \quad \mu \geq 0$$

$$\lambda^T(Ax - b) = 0, \quad \mu^T x \geq 0,$$

Condition 1: Candidate point must be feasible (primal feasibility of x).

Condition 2: Dual feasibility of λ and μ must be met.

Condition 3: Complementary slackness conditions must be satisfied.

In these conditions for optimality in LP, λ is the vector of Lagrange multipliers associated with $Ax \geq b$ and μ is the vector of Lagrange multipliers associated with $x \geq 0$.

Illustrative example

Compute the shadow prices or Lagrange multipliers for the problem given by

  Min $f(x) = -2x_1 - 4x_2$

  subject to

  $x_1 - 3x_2 \geq -3$

  $-2x_1 - 2x_2 \geq -5$

  $x_1, x_2 \geq 0$.

Step 1: Determine c^T, $\mathbf{A}$, and $\mathbf{b}$ that correspond to the standard LP formulation Min $c^T x$ s.t. $Ax \geq b$ with $x \geq 0$.

  Therefore

$$c^T = [c_1, c_2]^T = [-2, -4]^T$$

$$b = [b_1, b_2]^T = [-3, -5]^T$$

$$x = [x_1, x_2]^T$$

$$A = \begin{bmatrix} +1 & -3 \\ -2 & -2 \end{bmatrix}.$$

Step 2: Solve the LP problem using the simplex or revised simplex method. Solution yields primal solution and objective values of

$$x^* = [x_1^*, x_2^*]^T = [1.12, 1.38]^T \quad \text{and} \quad f^* = -7.75$$

Step 3: Apply KKT

KKT #1: Primal feasibility was satisfied since

$$Ax^* = \begin{bmatrix} +1 & -3 \\ -2 & -2 \end{bmatrix} \begin{bmatrix} 1.12 \\ 1.38 \end{bmatrix} \cong \begin{bmatrix} -3.03 \\ -5.00 \end{bmatrix} = b.$$

KKT #2: Dual feasibility to compute λ and μ.

$$\text{Recall } A^T \lambda + \mu = c, \quad \lambda \geq 0, \quad \mu \geq 0$$

$$\Rightarrow \begin{bmatrix} +1 & -3 \\ -2 & -2 \end{bmatrix} \begin{bmatrix} \lambda_1 \\ \lambda_2 \end{bmatrix} + \begin{bmatrix} \mu_1 \\ \mu_2 \end{bmatrix} = \begin{bmatrix} -2 \\ -4 \end{bmatrix}.$$

KKT #3: Complementary slackness conditions

$$\lambda^T(Ax - b) = 0, \quad \mu^T x \geq 0,$$

But since $x^* \neq 0$, then from $\mu^T x \geq 0$, this implies that $\mu^T = 0$. Therefore

$$\begin{bmatrix} +1 & -3 \\ -2 & -2 \end{bmatrix} \begin{bmatrix} \lambda_1 \\ \lambda_2 \end{bmatrix} + \begin{bmatrix} 0 \\ 0 \end{bmatrix} = \begin{bmatrix} -2 \\ -4 \end{bmatrix}$$

$$\begin{bmatrix} \lambda_1 \\ \lambda_2 \end{bmatrix} = \begin{bmatrix} +1 & -3 \\ -2 & -2 \end{bmatrix}^{-1} \begin{bmatrix} -2 \\ -4 \end{bmatrix} = \begin{bmatrix} 0.50 \\ 0.75 \end{bmatrix}.$$

Therefore, the desired shadow prices or Lagrange multipliers for this problem are $\lambda_1 = 0.50$ and $\lambda_2 = 0.75$.

5.5.2   Case 2: KKT Applied to the Dual LP Problem

This approach is computationally efficient as the Lagrange multipliers are the direct solutions of the Dual LP problem. Furthermore, KKT allows us to compute the optimal value of x without necessarily re-solving the primal LP problem.

Now, by the duality transformation, we have

Primal LP

Min $c^T x$ s.t. $Ax \geq b, x \geq 0$

Dual LP

Max $b^T \lambda$ s.t. $A^T \lambda \leq c, \lambda \geq 0$.

The KKT conditions requiring optimal solutions $[x^*, \lambda^*, \mu^*]^T$ for the Dual problem are given as

Condition 1: Dual feasibility of λ and μ must be met.

$$A^T\lambda \leq c, \quad \lambda \geq 0.$$

Condition 2: Candidate point must be feasible (primal feasibility of x).

$$Ax + \mu = b, \quad \lambda \geq 0, \quad \mu \geq 0.$$

Condition 3: Complementary slackness conditions must be satisfied.

$$x^T(A^T\lambda - c) = 0, \quad \mu^T x = 0.$$

Illustrative example
Recall the problem

Min $f(x) = -2x_1 - 4x_2$

subject to

$x_1 - 3x_2 \geq -3$

$-2x_1 - 2x_2 \geq -5$

$x_1, x_2 \geq 0.$

Solution steps are as follows:

Step 1: Transform the problem to its dual form

$$\text{Max } f_D(x) = b^T\lambda = [-3, \ -5]\begin{bmatrix} \lambda_1 \\ \lambda_2 \end{bmatrix} \text{ subject to}$$

$$A^T\lambda \leq c \text{ and } \lambda \geq 0 \text{ such that}$$

$$\begin{bmatrix} +1 & -2 \\ -3 & -2 \end{bmatrix}\begin{bmatrix} \lambda_1 \\ \lambda_2 \end{bmatrix} \leq \begin{bmatrix} -2 \\ -4 \end{bmatrix} \text{ and } \begin{bmatrix} \lambda_1 \\ \lambda_2 \end{bmatrix} \geq \begin{bmatrix} 0 \\ 0 \end{bmatrix}.$$

Step 2: Solve the LP problem using the simplex or revised simplex method. Solution yields candidate dual solution and objective values of

$$\lambda^* = [\lambda_1^*, \lambda_2^*]^T = [0.50, 0.75]^T \quad \text{and} \quad \text{Max } f_D^* = -7.75.$$

Step 3: Apply KKT

KKT #1: Dual feasibility is established since

$$A^T \lambda^* = \begin{bmatrix} +1 & -2 \\ -3 & -2 \end{bmatrix} \begin{bmatrix} 0.50 \\ 1.25 \end{bmatrix} = \begin{bmatrix} -2.0 \\ -4.0 \end{bmatrix} = \begin{bmatrix} c_1 \\ c_2 \end{bmatrix}$$

Therefore $A^T \lambda^* \leq c$ holds true.

KKT #2: Primal feasibility
Recall $Ax + \mu = b, \ \mu \geq 0$

$$\Rightarrow \begin{bmatrix} +1 & -3 \\ -2 & -2 \end{bmatrix} \begin{bmatrix} x_1 \\ x_2 \end{bmatrix} + \begin{bmatrix} \mu_1 \\ \mu_2 \end{bmatrix} = \begin{bmatrix} -3 \\ -5 \end{bmatrix}.$$

KKT #3: Complementary slackness conditions

$$x^T(A^T\lambda - c) = 0, \quad x^T\mu = 0,$$

But since $A^T\lambda - c = 0$, then x^T and hence $\mu^T = 0$.

$$\text{Therefore } \begin{bmatrix} +1 & -3 \\ -2 & -2 \end{bmatrix} \begin{bmatrix} x_1 \\ x_2 \end{bmatrix} + \begin{bmatrix} 0 \\ 0 \end{bmatrix} = \begin{bmatrix} -3 \\ -5 \end{bmatrix}$$

$$\begin{bmatrix} x_1 \\ x_2 \end{bmatrix} = \begin{bmatrix} +1 & -3 \\ -2 & -2 \end{bmatrix}^{-1} \begin{bmatrix} -3 \\ -5 \end{bmatrix} = \begin{bmatrix} 1.125 \\ 1.375 \end{bmatrix}.$$

This is the optimal solution, x^* and the minimal value function is
$\text{Min } f(x) = f^* = c^T x^* = [-2, \ -4] \begin{bmatrix} 1.125 \\ 1.375 \end{bmatrix} = -7.75$, which is Max f_{Dual}.

5.5.3 Case 3: KKT Applied to LP Problems with Equality Constraints

Primal LP
Min $c^T x$ s.t. $Ax = b$ with $x \geq 0$. The primal KKT conditions requiring optimal solutions x^*, λ^*, and μ^* are

Condition 1: Candidate point must be feasible (primal feasibility of x).

$$Ax = b, \quad x \geq 0.$$

Condition 2: Dual feasibility of λ and μ must be met.

$$A^T\lambda + \mu = c, \quad \lambda \text{ is unrestricted}, \quad \mu \geq 0.$$

Condition 3: Complementary slackness conditions must be satisfied.

$$\lambda^T(Ax - b) = 0, \quad \mu^T x \geq 0.$$

Dual LP

Max $b^T\lambda$ s.t. $A^T\lambda \leq c$ with λ is unrestricted. The dual KKT conditions requiring optimal solutions x^*, λ^*, and μ^* are

Condition 1: Dual feasibility of λ and μ must be met.

$$A^T\lambda + \mu = c, \quad \lambda \text{ is unrestricted}, \quad \text{and } \mu \geq 0.$$

Condition 2: Candidate point must be feasible (primal feasibility of x).

$$Ax = b, \quad x \geq 0.$$

Condition 3: Complementary Slackness conditions must be satisfied.

$$x^T(A^T\lambda - c + \mu) = 0, \quad \mu^T x = 0.$$

Primal LP

Min $c^T x$ s.t. $Ax \leq b$ with x is unrestricted. The primal KKT conditions requiring optimal solutions x^*, λ^*, and μ^* are

Condition 1: Primal feasibility

$$Ax \leq b, \quad x \text{ is unrestricted}.$$

Condition 2: Dual feasibility

$$A^T\lambda + \mu = c, \quad \lambda \geq 0, \quad \mu \geq 0.$$

Condition 3: Complementary Slackness

$$\lambda^T(Ax - b) = 0, \quad \mu^T x \geq 0.$$

Dual LP

Max $b^T\lambda$ s.t. $A^T\lambda = c$ and $\lambda \geq 0$. The dual KKT conditions requiring optimal solutions x^*, λ^*, and μ^* are

Condition 1: Dual feasibility

$$A^T\lambda = c, \quad \lambda \geq 0.$$

Condition 2: Primal feasibility

$$Ax + \mu = b, \quad x \text{ is unrestricted} \quad \text{and } \mu \geq 0.$$

Condition 3: Complementary slackness

$$x^T(A^T\lambda - c) = 0, \quad \mu^T x = 0.$$

5.6 Mixed-Integer Programming

Integer and mixed-IP (MIP) problems are special classes of LP where all or some of the decision variables are restricted to integer values. There are many practical examples where the divisibility assumption in LP needs to be dropped and some of the variables can take up only discrete values. However, even greater importance can be attributed to problems where the discrete values are restricted to zero and one only, that is, "yes" or "no" decisions or binary decision variables. In fact, in many instances, MIP problems can be reformulated to have only binary decision variables that are easier to handle. The occurrence of binary variables may be due to a variety of decision requirements, the most common among which are the following.

1. *ON/OFF decisions*: The most common type of binary decision falls into this category for engineering optimization problems. This decision variable can also have alternative representation of GO/NO GO, BUILD/NOT BUILD, OR SCHEDULE/NOT SCHEDULE, and so on, depending on the specific application under consideration in short, medium, and long-term planning contexts.

2. *Logical either-or/and constraints*: Binary variables can also indirectly handle mutual inclusive or exclusive restrictions. For example, there might be cases where a choice can be made between two constraints, so that only one must hold. Also, there could be cases where process B must be selected if process A has already been selected.

3. *K out of N constraints must hold*: Consider the case where the overall model includes a set of N possible constraints such that only some K of these constraints must hold (assuming $K < N$). Part of the optimization task is to choose which combination of K constraints permits the objective function to reach its best possible value. In fact, this is nothing but a generalization of the either-or constraints, and can handle a variety of problems.

4. *Function with N-possible values*: In many real-life problems, the functions do not have smooth continuous properties, but can take up only a few discrete values. For example, consider the following case

$$f(x_1, \ldots, x_n) = d_1 \quad \text{or} \quad d_2, \ldots, d_N. \tag{5.17}$$

The equivalent IP formulation would be

$$f(x_1, \ldots, x_n) = \sum_{i=1}^{N} d_i y_i \tag{5.18}$$

$$\sum_{i=1}^{N} y_i = 1 \tag{5.19}$$

and $Y_i =$ binary (0 or 1) for $i = 1, \ldots, N$.

5. *Fixed-charge problem*: In most problems, it is common to incur a fixed-cost/set-up charge when undertaking a new activity. In a process-engineering context, it might be related to the set-up cost for the production facility to initiate a run. A typical power system example is the start-up cost of a thermal-generating unit. This fixed charge is often independent of the length or level of the activity and, hence, cannot be approximated by allocating it to the (continuous) level of activity variables.

Mathematically, the total cost comprising fixed and variable charges can be expressed as

$$f_j(x_j) = \begin{cases} K_j + C_j x_j & \text{if } x_j > 0 \\ 0 & \text{if } x_j = 0. \end{cases} \tag{5.20}$$

The MIP transformation would look like

$$\text{Minimize } Z = \sum_{j=1}^{n} (C_j x_j + K_j y_j), \tag{5.21}$$

where

$$y_j = \begin{cases} 1 & \text{if } x_j > 0 \\ 0 & \text{if } x_j = 0. \end{cases}$$

Pure integer or MIP problems pose a great computational challenge. While there exist highly efficient LP techniques to enumerate the basic LP problem at each possible combination of the discrete variables (nodes), the problem lies in the astronomically large number of combinations to be enumerated. If there are N discrete variables, the total number of combinations becomes 2^N! The simplest procedure one can think of for solving an integer or MIP problem is to solve the LP relaxation of the problem (i.e., allowing the discrete variables to take continuous value so that the MIP reduces to non-linear programming) and then rounding the noninteger values to the closest integer solution. There are, however, major pitfalls:

1. Resulting integer solution may not be feasible in the first place.
2. Even if the rounding leads to a feasible solution, it may, in fact, be far from the optimal solution.

Algorithmic development for handling large-scale integer or MIP problems continues to be an area of active research. There have been exciting algorithmic advances during the middle and late 1980s. The most popular method to date has been the branch-and-bound technique and related ideas to implicitly enumerate the feasible integer solutions.

5.6.1 Branch-and-Bound Technique for Binary Integer Programming Problems

The basic philosophy in the branch-and-bound procedure is to divide the overall problem into smaller and smaller subproblems and enumerate them in a logical sequence. The division procedure is called branching and the subsequent enumeration is done by bounding to check how good the best solution in the subset can be, and then discarding the subset if its bound indicates that it cannot possibly contain an optimal solution for the original problem.

The general structure of the MIP problem is

$$\text{Maximize } P(x) = \sum_{j=1}^{n} c_j x_j. \tag{5.22}$$

Subject to the constraints:

$$\sum_{i=1}^{m} \left(\sum_{i=1}^{n} a_{ij} x_j \le b_i \right) \tag{5.23}$$

and

$$x_j \ge 0 \tag{5.24}$$

$$\forall j \in \{1, n\},$$

and x_j is an integer $\forall i \in \{1, I\}$.

Assume for simplicity of notation that the first I variables are the integer decision variables.

Before a formal description of the branch-and-bound procedure is given, the basic procedures of branching, bounding, and fathoming are illustrated using a simple numerical example.

Illustrative example

Consider the following IP problem:

$$\text{Maximize } P = 9x_1 + 5x_2 + 6x_3 + 4x_4$$

Subject to

$$6x_1 + 3x_2 + 5x_3 + 2x_4 \le 0$$
$$x_3 + x_4 \le 1$$
$$-x_1 + x_3 \le 0$$
$$-x_2 + x_4 \le 0$$
$$\forall x_j \in (0, 1).$$

This is a pure integer problem and except for its small size is typical of many of the practical decision-making problems.

Branching

Branching involves developing subproblems by fixing the binary variables at 0 or 1. For example, branching on x_1 for the example problem gives

Subproblem 1. $(x_1 = 0)$

     Maximize $P = 5x_2 + 6x_3 + 4x_4$

     Subject to

     $3x_2 + 5x_3 + 2x_4 \leq 10$

     $x_3 + x_4 \leq 1$

     $x_3 \leq 0$

     $-x_2 + x_4 \leq 0.$

Subproblem 2. $(x_1 = 1)$

     Maximize $P = 9 + 5x_2 + 6x_3 + 4x_4$

     Subject to

     $3x_2 + 5x_3 + 2x_4 \leq 4$

     $x_3 + x_4 \leq 1$

     $x_3 \leq 1$

     $-x_2 + x_4 \leq 0.$

The procedure may be repeated at each of the two subproblem nodes by fixing additional variables such as x_2, x_3, and x_4. Thus, a tree structure can be formulated by adding branches at each iteration, which is referred to as the solution tree. The variable used to do this branching at any iteration by assigning values to the variable is called the branching variable.

Bounding

For each of the subproblems, a bound can be obtained to determine how good its best feasible solution can be. Consider first the relaxed LP formulation for the overall problem, which yields the following solution:

$$(x_1, x_2, x_3, x_4) = \left(\frac{5}{6}, 1, 0, 1\right)$$

with $P = 16\frac{1}{2}$.

Therefore, $P \leq 16.5$ for all feasible solutions for the original problem. This bound can be rounded off to 16, because all coefficients in the objective function are integers; hence, all integer solutions must have an integer value for P. The bound for whole problem is $P \leq 16$.

In the same way, the bounds for the two subproblems are obtained:

Subproblem 1. $(x_1, x_2, x_3, x_4) = (0, 1, 0, 1)$
with $P = 9$.

Subproblem 2. $(x_1, x_2, x_3, x_4) = \left(1, \frac{4}{5}, 0, \frac{4}{5}\right)$
with $P = 16.2$.

Therefore, the resulting bounds are

Subproblem 1. $P \leq 9$,
Subproblem 2. $P \leq 16$.

Fathoming

If a subproblem has a feasible solution, it should be stored as the first incumbent (the best feasible solution found so far) for the whole problem along with its value of P. This value is denoted P^*, which is the current incumbent for P.

A subproblem is said to be fathomed, that is, dismissed from further consideration, if

Test 1. Its bound is less than or equal to P.

Test 2. Its LP relaxation has no feasible solutions.

Test 3. Optimal solution for its LP relaxation is an integer; if this solution is better than the incumbent, it becomes the new incumbent, and the test is reapplied to all unfathomed subproblems with the new larger P^*.

Optimality Test

The iterative procedure is halted when there are no remaining subproblems. At this stage, the current incumbent for P is the optimal solution. Otherwise, we return to perform one more iteration. The solution tree of the current example is provided in Figure 5.3.

The markings $F(1)$, $F(2)$, and $F(3)$ on Figure 5.3 indicate that the node has been fathomed by Tests 1, 2, and 3.

For the general branch-and-bound approach in MIP problems, some deviations are necessary to improve the efficiency of the algorithm. These include

1. *Choice of branching variable*: The variables, which have a noninteger solution in the LP relaxation, are selected for branching.

2. *Values assigned to the branching variable for creating subproblems*: Create just two new subproblems by specifying two ranges of values for the variable.

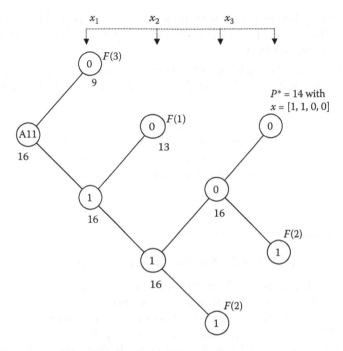

FIGURE 5.3
Solution tree diagram for the IP problem.

3. *Bounding step*: The bound of P is the optimal value of P itself (without rounding) in the LP relaxation.

4. *Fathoming test*: Only the integer decision variables need to be checked for integer solution to decide the fathoming node.

The branch-and-bound procedure is summarized in the following important steps:

Step 1. Initialization: Set $P^* = -\infty$. Apply the bounding step, fathoming step, and optimality test described below to the whole problem. If not fathomed, classify this problem as the one "remaining" subproblem for performing the first full iteration below.

Step 2. Branching: Among the unfathomed subproblems, select the one that was created most recently (breaking ties according to which has the larger bound). Among the integer restricted variables that have a noninteger value in the optimal solution for the LP relaxation of the subproblem, choose the first one in the natural ordering of the variables to be the branching variable. Let x_j be this variable and x_j^* its value in this solution. Branch from the node for the subproblem to create two new subproblems by adding the respective constraints, $x_j \leq [x_j^*]$ and $x_j \geq [x_j^*] + 1$, where $[x_j^*] = $ greatest integer $\leq x_j^*$.

Step 3. Bounding: For each new subproblem, obtain its bound by applying the simplex method to its LP relaxation and using the value of the P for the resulting optimal solution.

Step 4. Fathoming and optimality test: For each new subproblem, apply the three fathoming tests and discard those subproblems that are fathomed by any of the tests.

These are the fundamental steps in the branch-and-bound technique that is applicable to a wide range of MIP problems. For each subproblems that is created, the LP algorithm can be applied in the constrained problem in its pure linear form. We now turn our attention to sensitivity methods in LP.

5.7 Sensitivity Methods for Postoptimization in LP

In many applications, both nonpower and power system types, we often encounter practical problems. In one instance, we seek the optimal solution and in another, we wish to know what happens when one or more variables are changed. In order to save computational effort, it is desirable not to resolve the problem if small perturbations are made to the variables. Sensitivity analysis is the study used to compute such solutions [7,9,10].

Now, recall the LP problem given by the following model as shown in Equations 5.10 and 5.11:

$$\text{Maximize } P(x) = \sum_{j=1}^{n} c_j^{\text{T}} x_j \tag{5.25}$$

Subject to the constraints:

$$\sum_{i}^{m} \left(\sum_{j=1}^{n} a_{ij} x_j \le b_i \right) \tag{5.26}$$

and

$$x_j \ge 0 \tag{5.27}$$

$$\forall i \in \{1, m\} \quad \text{and} \quad \forall j \in \{1, n\}.$$

Here, we observe that changes in the system can be attributed to the modifications such as

1. Perturbation in the parameters, b_i
2. Perturbation in the cost coefficients, c_j
3. Perturbation in the coefficient, a_{ij}

4. Injection of new constraints
5. Injection of new variables

We discuss the effect of each case as it applies to sensitivity analysis and further expound on the first class in more detail in a subsequent section.

5.7.1 Case 1: Perturbation in the Parameters b_1

Let the optimal basis solution for the problem in its primal form be

$$x_B = [x_1, x_2, \ldots, x_m]^T, \tag{5.28}$$

where

$$x_b \geq 0.$$

Since the nonbasic variables are zero, then we can write

$$Bx_B = b \tag{5.29}$$

$$\Rightarrow x_B = B^{-1}b, \tag{5.30}$$

where B is an m-square and nonsingular matrix formed by m columns of A. Let b change to $b + \Delta b$, where $\Delta b = [\Delta b_1, \Delta b_2, \ldots, \Delta b_m]'$, and everything else in the problem remains the same. Then

$$x_B + \Delta x_B = B^{-1}(b + \Delta b), \tag{5.31}$$

given the new values $x_B + \Delta x_B$ of the variables that were the original optimal basic variables.

If $B^{-1}(b + \Delta b) \geq 0$, then the variables continue to be basic feasible. They would also continue to be optimal if the relative cost coefficients given by $C_B B^{-1}\bar{B} - \bar{C}_B$ continued to be nonnegative; that is

$$Z - \bar{C}_B = C_B B^{-1}\bar{B} - \bar{C}_B \geq 0. \tag{5.32}$$

The optimum value of $P(x)$ changes can be calculated with the new values of the variables given by $x_B + \Delta x_B$, or by using the following equation,

$$P(x) = C_B B^{-1}(b + \Delta b). \tag{5.33}$$

5.7.2 Case 2: Perturbation in the Cost Coefficients C_j

If C_j are changed to C'_j, everything else in the problem remaining the same, the relative cost coefficients are given by

$$C'_B B^{-1}\bar{B} - \bar{C}'_B. \tag{5.34}$$

These may not all be nonnegative. For some j, some of them are negative. This would mean that the basic feasible solution that was optimal for C_j is not optimal for C'_j. So from this point onwards further iterations may be done with new values C'_j to obtain a new optimal solution.

If, however, C'_j are such that $Z - \bar{C}'_B = C'_B B^{-1} \bar{B} - \bar{C}'_B \geq 0$, then the original optimal basis still remains optimal, and the value of the optimal basic variables also remains unchanged. The optimum value of $P(x)$ is given by

$$P(x) = C'_B B^{-1} b. \tag{5.35}$$

5.7.3   Case 3: Perturbation in the Coefficient a_{ij}

If the changes are in a_{ik}, where x_k is the nonbasic variable of the optimal solution, then we get:

$$C'_B B^{-1} \bar{B}_{(ik)} - \bar{C}'_B, \tag{5.36}$$

where $\bar{B}_{(ik)}$ means that the value of a_{ik} in $\bar{B}$ is changed.

If $C'_B B^{-1} \bar{B}_{(ik)} - \bar{C}'_B \geq 0$, then the original optimal basis still remains optimal. If not, further iterations with the new values of $C'^{B-1}_B \bar{B}_{(ik)} - \bar{C}'_B$ and a_{ik} may be done.

5.7.4   Case 4: Injection of New Constraints

Generally, if the original optimal solution satisfies the new constraints that are added to the system, then that solution will still be an optimal solution. However, if some of the injected constraints are violated by the original optimal solution, then the problem must be solved taking into account the new constraints to the system. The new initial point may constitute the old basic variables of the original optimal solution along with one additional basic variable that is associated with each added constraint.

5.7.5   Case 5: Injection of New Variables

Since the number of constraints remains the same, the number of basic variables remains the same. Therefore, the original optimal solution along with zero values of the new variables would result in a basic feasible solution for the new problem. That solution would remain optimal if the newly introduced cost coefficients corresponding to them are nonnegative.

5.7.6   Sensitivity Analysis Solution Technique for Changes in Parameters b_i

We consider the well-known LP problem defined by

$$\min (P = c^T x:\ \text{s.t. } Ax = b;\ x \geq 0), \tag{5.37}$$

where
  c and x are n-vectors
  b is an m-vector

It is assumed that for a given b, the LP problem has been solved with an optimal basis that yields an optimal solution which may be nondegenerate or degenerate. Then the vector b is subject to change with an increment Δb. The postoptimization problem is to further optimize P with respect to Δb under the condition that the optimal basis remains unchanged.

Now let B and X_B be the optimal basis and the associated optimal solution of Equation 5.37. Then it is clear from Equation 5.37 that both P and X_B change as b does. We define the rate of change of P with respect to b as the sensitivity denoted by

$$S_b^P = \frac{\partial P}{\partial b} \quad \text{(a row } m\text{-vector)}. \tag{5.38}$$

Thus, the new objective function becomes

$$P + \Delta P = P + \frac{\partial P}{\partial b}\Delta b \tag{5.39}$$

$$\Delta P = S_b^P \Delta b. \tag{5.40}$$

The new optimal solution is

$$x_B' = B^{-1}(b + \Delta b) = x_B + H\Delta b \geq 0, \tag{5.41}$$

where $H = B^{-1} = [h_{ij}]$.

In practical applications, only some components of b are subject to change and the changes are usually bounded. If J is the index set that contains j for which b_j changes with increment Δb_j, then

$$f_j \leq \Delta b_j \leq g_j. \tag{5.42}$$

In order to ensure that $\Delta b_j = 0$ is feasible, we impose the condition

$$f_j \leq 0 \leq g_j, \quad \text{for } j \in J. \tag{5.43}$$

The trivial condition does not affect the practicality of the postoptimization, yet it guarantees a solution. It can be shown that the sensitivity S_b^P is the dual solution of Equation 5.37. For this reason, we use the conventional notation

$$y^T = S_b^P, \tag{5.44}$$

where y is the dual solution and a column vector.

The problem is to minimize ΔP with respect to Δb by keeping B unchanged. That is,

Minimize $\Delta P = y^T \Delta b = \sum_j y_j \Delta b_j$ (5.45)

Subject to

$x_{Bi} + h_{ij}\Delta b_j \geq 0$ (5.46)

and

$f_i \leq \Delta b_j \leq g_j$ (5.47)

where $j \in J$ and $i = 1, 2, \ldots, m$.

5.7.6.1 Solution Methodology

We wish to utilize the solution of Equation 5.37 to find the sensitivity. Let C_B be associated with the optimal basis B and solution X_B. That is,

$$P = C_B^T X_B$$ (5.48)

and

$$B X_B = b.$$ (5.49)

A new vector is defined here to satisfy

$$B^T y = c_B.$$ (5.50)

Such an m-vector y is unique since B is nonsingular. In view of Equations 5.49 and 5.50, we know that both P and X_B change if the vector b changes in order to maintain the optimality. But B may or may not change due to the insensitive nature of B to b. We consider here the case for which B remains unchanged when b changes. Thus, the vector y of Equation 5.41 is constant when b changes. It follows from Equations 5.48 through 5.50 that

$$S_b^P = \frac{\partial(C_B^T X_B)}{\partial b} = C_B^T \frac{\partial X_B}{\partial b} = y^T B \frac{\partial X_B}{\partial b}$$
$$= y^T \frac{\partial(B X_B)}{\partial b} = y^T \frac{\partial b}{\partial b} = y^T I = y^T.$$ (5.51)

This result shows that the sensitivity is indeed the y that satisfies Equation 5.50. Importantly, it is a constant and, hence, ΔP as given by Equation 5.40 is exact without any approximation. Since y is also the dual solution of Equation 5.37 for the nondegenerate case, it can be obtained together with B.

We do not impose the condition of nondegeneracy but only require an optimal basis B which may yield a degenerate ($X_B \geq 0$) or nondegenerate ($X_B > 0$) solution. In any case, y can be found from Equation 5.50 by B^{-1} which is also needed in the algorithm.

It is intended to solve the problem by changing one component of Δb at a time. By looking at the sign of y_j, one may choose a feasible Δb_j in such a way that

$$y_j \Delta b_j = -|y_i \Delta b_j| \leq 0, \tag{5.52}$$

which is equal to zero only when $\Delta b_j = 0$.

A process is employed to change Δb_j step by step with the resetting of necessary quantities. In the process, $j \in J$ advances from the first one to the last and then back to the first, and so on, until all $\Delta b_j = 0$ (steady state). Since ΔP is decreasing from one step to another unless $\Delta b_j = 0$, the steady state is reachable if a solution exists.

The method can be implemented by using a series of logical decisions. For Δb_j with $j \in J$, Equation 5.52 requires that $h_{ij} \Delta b_j \geq -X_{Bi}$ where $i = 1, 2, \ldots, m$ and h_{ij} is the ith row and jth column of the matrix H. The above inequality can be fulfilled by

$$L_j \leq \Delta b_j \leq R_j. \tag{5.53}$$

The bounds are to be determined as follows:

$$L_j = \max[-X_{Bi}/h_{ij}; \ h_{ij} > 0]$$
$$= -\infty, \quad \text{if all } h_{ij} \leq 0 \tag{5.54}$$

$$R_j = \min[-X_{Bi}/h_{ij}; \ h_{ij} < 0]$$
$$= \infty, \quad \text{if all } h_{ij} \geq 0, \tag{5.55}$$

where the maximization and minimization are taken over $i = 1, 2, \ldots, m$. (Note that $L_j \leq 0$ and $R_j \geq 0$ are always true as evidenced by $X_B \geq 0$.) Chosen according to Equation 5.53, Δb_j is feasible for the constraint Equation 5.41. To satisfy Equation 5.42 and make the smallest $y_i b_j < 0$, one must select Δb_j as follows:

$$\Delta b_j = 0, \quad \text{if } y_j = 0 \tag{5.56}$$

$$= \max[f_i, L_j], \quad \text{if } y_j > 0 \tag{5.57}$$

$$= \min[g_j, R_j], \quad \text{if } y_j < 0. \tag{5.58}$$

Note that the maximal selection is always negative but the minimal one is always positive.

The steady state may be replaced by a simpler expression $\Delta P = 0$. They are equivalent because

$$\Delta P = \sum_j y_j \Delta b_j = -\sum |y_j \Delta b_j| = 0, \tag{5.59}$$

which implies that

$$y_j \Delta b_j = 0, \quad \text{for all } j \in J. \tag{5.60}$$

This is true only when $\Delta b_j = 0$ for all $j \in J$ due to the selection rule that $\Delta b_j = 0$ when $y_j = 0$.

5.7.6.2 Implementation Algorithm

The algorithm for sensitivity analysis with changes in the parameter b_i can be implemented by using the following steps:

Step 1. Obtain B, X_B, C, and b from the LP problem, Equation 5.10.

Step 2. Calculate $P = C_B^T C_B$, $H = B^{-1} = [h_{ij}]$, and then $y = H^T C_B$.

Step 3. Identify J and $[f_j, g_j]$ from Equation 5.13 of the problem.

Step 4. Set $\Delta P = 0$.

Step 5. Do the following Steps 5.1 through 5.3. For $j \in J$ and $i = 1, 2, \ldots, m$

    5.1 If $y_j = 0$, then set $\Delta b_j = 0$.

    5.2 If $y_j > 0$, then set

$$L_j = \max\,[-X_{Bi}/h_{ij};\ h_{ij} > 0]$$
$$= -\infty \quad \text{if all } h_{ij} \leq 0$$

    Then set $\Delta b_j = \max[f_j, L_j]$.

    5.3 If $y_j < 0$, then set

$$R_j = \min\,[-X_{Bi}/h_{ij};\ h_{ij} < 0]$$
$$= \infty \quad \text{if all } h_{ij} \geq 0,$$

    Then set $\Delta b_j = \min[g_j, R_j]$.

Step 6. Update

$$\Delta P = \Delta P + y_j \Delta b_j$$
$$b_j = b_j + \Delta b_j$$
$$f_j = f_j - \Delta b_j$$
$$g_j = g_j - \Delta b_j.$$

Step 7. Update

$$X_{Bi} = X_{Bi} + h_{ij} \Delta b_j, \quad \text{for all } i = 1, 2, \ldots, m.$$

Step 8. Update

$$P = P + \Delta P.$$

Step 9. Go to step 10 if $\delta P = 0$; go back to step 4 otherwise.

Step 10. Stop with X_B, b, and P as a solution for the postoptimization as formulated.

Finally, it is easy to understand the update performed in steps 6 and 7 except for the intervals. The new intervals must be shifted by the amount of Δb_j to the left if $\Delta b_j > 0$ and to the right if $\Delta b_j < 0$. The update in steps 6 and 7 may be exempted if $\Delta b_j = 0$.

Illustrative example

Minimize $[P = X_1 + X_2 + X_3]$

Subject to

$X_1 + X_3 + X_4 - X_5 = 3 = b_1$

$X_2 - X_3 + X_4 + X_6 = 1 = b_2$

and

$\forall i, \quad X_i \geq 0.$

It is required to further minimize P for $-1 \leq \Delta b_1 \leq 1$ and $0 \leq \Delta b_2 \leq 2$ after a minimized solution has been achieved for the problem. This is a problem of postoptimization and can be solved as follows:

Solution

Step 1. From the primal problem

$$B = \begin{pmatrix} 1 & 1 \\ -1 & 1 \end{pmatrix}, \quad X_B = \begin{bmatrix} X_3 \\ X_4 \end{bmatrix} = \begin{bmatrix} 1 \\ 2 \end{bmatrix}, \quad C_B = \begin{bmatrix} 1 \\ 0 \end{bmatrix} \quad \text{and} \quad b = \begin{bmatrix} 3 \\ 1 \end{bmatrix}.$$

Step 2. $P = X_3 = 1.$

$$H = \begin{pmatrix} 1 & 1 \\ -1 & 1 \end{pmatrix}^{-1} = \frac{1}{2} \begin{pmatrix} 1 & -1 \\ 1 & 1 \end{pmatrix}$$

$$y = \frac{1}{2} \begin{pmatrix} 1 & 1 \\ -1 & 1 \end{pmatrix} \begin{bmatrix} 1 \\ 0 \end{bmatrix} = \frac{1}{2} \begin{bmatrix} 1 \\ -1 \end{bmatrix}.$$

Step 3. $J = \{1,2\}.$

$$f_1 = -1, \quad f_2 = 0$$
$$g_1 = 1, \quad g_2 = 2.$$

Step 4. $\Delta P = 0.$

Step 5. With $j=1$, then with

Step 5.2 with $y_1 = 0.5$, we get

$$L_1 = \max\left[-\frac{1}{0.5}, -\frac{2}{0.5}\right] = -2$$

$$\Delta b_1 = \max[-1, -2] = -1$$

$$\Delta P = 0 + \left(\frac{1}{2}\right)(-1) = -\frac{1}{2}$$

$$b_1 = 3 + (-1) = 2$$

$$f_1 = -1 - (-1) = 0$$

$$g_1 = 1 - (-1) = 2$$

with Step 5.3,

$$X_{B1} = 1 + 0.5(-1) = 0.5$$

$$X_{B2} = 2 + 0.5(-1) = 1.5.$$

Step 6. With $j=2$, then

$$y_2 = -0.5$$

$$R_2 = \min\begin{bmatrix}-0.5\\-0.5\end{bmatrix} \quad \text{and} \quad \Delta b_2 = \min[2, 1] = 1.$$

Step 7. $\Delta P = -0.5 + (-0.5)\,(1) = -1$

$$b_2 = 1 + 1 = 2, \quad g_2 = 2 - 1 = 1$$

$$f_2 = 0 - 1 = -1.$$

Step 8. $X_{B1} = 0.5 + (-0.5)\,(1) = 0$

$$X_{B2} = 1.5 + (0.5)(1) = 2.$$

Step 9. $P = 1 + (-1) = 0$.

Step 10. $\Delta P = -1 \neq 0$.

      10.1 $\Delta P = 0$.

      10.2 $j = 1$, $y_1 = \frac{1}{2}$: $\Delta b_1 = \max[0, 0] = 0$, no update.

      10.3 $j = 2$, $y_2 = -\frac{1}{2}$: $\Delta b_2 = \min[1, 0] = 0$, no update.

      Therefore, $\Delta P = 0$.

Step 11. The optimal solution to the problem is

$$X_B = \begin{bmatrix} 0 \\ 2 \end{bmatrix}, \quad b = \begin{bmatrix} 2 \\ 2 \end{bmatrix}, \quad \text{and} \quad P = 0.$$

Notably, if we change b_2 first and then b_1 or $J = \{2, 1\}$, the solution becomes

$$X_B \begin{bmatrix} 0 \\ 3 \end{bmatrix}, \quad b = \begin{bmatrix} 3 \\ 3 \end{bmatrix}, \quad \text{and} \quad P = 0.$$

In fact, the problem has an infinite number of solutions:

$$X_B = \begin{bmatrix} 0 \\ \beta \end{bmatrix}, \quad b = \begin{bmatrix} \beta \\ \beta \end{bmatrix}, \quad \text{and} \quad P = 0,$$

where $2 \leq \beta \leq 3$.

5.7.6.3 Duality in Postoptimal Analysis

The dual of the primal LP problem given by Equations 5.12 through 15.14 can be restated in the form:

$$\max \{D = b^T y \text{ s.t. } A^T y \leq C; \, y \geq 0\}. \tag{5.61}$$

Let the vectors in A and components in C be arranged as

$$A = [B, N] \tag{5.62}$$

and

$$C = \begin{bmatrix} C_B \\ C_N \end{bmatrix}. \tag{5.63}$$

Then, it follows that

$$AX = [B, N] \begin{bmatrix} X_B \\ 0 \end{bmatrix} = BX_B = b \tag{5.64}$$

and

$$P = C^T X = [C_B^T, C_N^T] \begin{bmatrix} X_B \\ 0 \end{bmatrix} = C_B^T X_B. \tag{5.65}$$

Introduce a new m-vector u such that

$$Bu = a_k(A_k), \tag{5.66}$$

where a_k is a vector in N.

The vector y in Equation 5.62 is feasible for the dual problem if

$$N^T y \leq C_N. \tag{5.67}$$

This is true because

$$A^T y = \begin{bmatrix} B^T y \\ N^T y \end{bmatrix} \leq \begin{bmatrix} C_B \\ C_N \end{bmatrix} = C. \tag{5.68}$$

To show Equation 5.69, we assume the contrary to have

$$a_k^T y - C_{Nk} > 0. \tag{5.69}$$

Multiply Equation 5.67 by a positive number θ and then subtract it from Equation 5.65:

$$B(X_B - \theta u) + \theta a_k = b. \tag{5.70}$$

We obtain from Equations 5.66 and 5.67 that

$$y^T B u = y^T a_k = C_B^T u = a_k^T y$$

and hence

$$a_k^T y - C_{NK} = C_B^T u - C_{NK}. \tag{5.71}$$

Multiply Equation 5.71 by θ and then subtract it from Equation 5.66:

$$P - \theta(a_k^T y - C_{NK}) = C_B^T(X_B - \theta u) + \theta C_{Nk}. \tag{5.72}$$

Choose

$$\theta = \frac{X_{Br}}{u_r} = \min\left\{ \frac{X_{Bi}}{u_i} : u_i > 0 \right\} \tag{5.73}$$

which always exists since $X_B > 0$ (nondegenerate). With θ so chosen, Equation 5.71 demonstrates that a new feasible basis is formed by replacing a_r in B with a_k in N. On the other hand, Equation 5.72 indicates that the new P is decreased from the old value by

$$\theta(a_k^T y - C_{NK}) > 0. \tag{5.74}$$

This contradicts the fact that B is a minimal basis.

For any feasible X of Equation 5.10 and its dual y, we have

$$P = C^T X \geq (A^T y)^T X = y^T A X = y^T b = D. \tag{5.75}$$

But, for $X = X_B$, it becomes

$$P = C_B^T X_B = (B^T y)^T X_B = y^T B X_B \tag{5.76}$$

$$= y^T b = D. \tag{5.77}$$

Therefore y is also a maximum of the dual problem.

5.8 Power Systems Applications

Consider a subtransmission system in which certain bus voltages of interest form the vector $|V|$. Assume that the increase in bus voltage magnitude is linearly proportional to injected reactive power at several buses, denoted by the vector ΔQ. The dimensions of vectors $|V|$ and ΔQ may not be the same since the capacitor placement may occur at a different number of buses compared to the buses at which $|V|$ is supported. Then

$$|V| = B\Delta Q, \tag{5.78}$$

where B involves elements from the inverse of the $\partial Q / \partial |V|$ portion of the Jacobian (under the assumptions of superposition and decoupled power flow). To ensure high enough bus voltage,

$$|V| + \Delta |V| \geq |V_{min}|, \tag{5.79}$$

where $|V_{min}|$ is a vector of minimum bus voltage magnitudes and $|V|$ is the "base case" (i.e., no capacitive compensation) bus voltage profile. The $\Delta |V|$ term stems from capacitive compensation. As a result

$$|\Delta V| = B\Delta Q \geq |V_{min}| - |V|. \tag{5.80}$$

Again the concept of a vector inequality in Equation 5.68 is said to hold when each scalar row holds. Furthermore, it is desired to minimize c_q,

$$c_q = c^t \Delta Q. \tag{5.81}$$

c^t is a row vector of 1s commensurate dimension with ΔQ. The cost function, c_q, is a scalar. The minimization of c_q subject to Equation 5.68 is accomplished by LP. In most LP formulations, the inequality constraints are written with the solution vector appearing on the "smaller than" side (i.e., opposite to inequality Equation 5.70), and the index that is extremized is maximized rather than minimized. Both problems are avoided by working with $\Delta Q'$, rather than ΔQ, where

$$\Delta Q' = K - \Delta Q. \tag{5.82}$$

In this discussion, ΔQ entries are assumed to be positive for shunt capacitive compensation.

It is possible also to introduce capacitor costs, which depend on the size of the unit. This is done by allowing other than unity weighting in c^t. Upper limits may also be introduced, but these are usually not needed.

5.9   Illustrative Examples

Example 5.9.1

A subsystem has two generators. There are four key lines with limits given by

$$P_{L1}^{max} = 12 \text{ MW}, \quad P_{L2}^{max} = 12 \text{ MW}$$
$$P_{L3}^{max} = 18 \text{ MW}, \quad P_{L4}^{max} = 18 \text{ MW}.$$

The sensitivity relation between the key lines and the generators is given below:

$$P_{L1} = 2P_{G1} + 3P_{G2}, \quad P_{L2} = 3P_{G1} + 2P_{G2}$$
$$P_{L3} = 6P_{G1} + P_{G2}, \quad P_{L4} = P_{G1} + 6P_{G2}.$$

The system benefit function is $F = P_{G1} + 2P_{G2}$. Find the maximum benefit value of the system using LP.

SOLUTION

$$P_{L1}^{max} = 12, \quad P_{L2}^{max} = 12$$
$$P_{L3}^{max} = 18, \quad P_{L4}^{max} = 18$$
$$P_{L1} = 2P_{G1} + 3P_{G2} \rightarrow 2P_{G1} + 3P_{G2} \leq 12$$
$$P_{L2} = 3P_{G1} + 2P_{G2} \rightarrow 3P_{G1} + 2P_{G2} \leq 12$$
$$P_{L3} = 6P_{G1} + P_{G2} \rightarrow 6P_{G1} + P_{G2} \leq 18$$
$$P_{L4} = P_{G1} + 6P_{G2} \rightarrow P_{G1} + 6P_{G2} \leq 18$$
$$F = P_{G1} + 2P_{G2} \rightarrow \text{maximize.}$$

If we change the variables to be $P_{G1} \rightarrow x_1$, $P_{G2} \rightarrow x_2$ then the problem will be

Maximize $F = x_1 + 2x_2$

Subject to

$2x_1 + 3x_2 \le 12$

$3x_1 + 2x_2 \le 12$

$6x_1 + x_2 \le 18$

$x_1 + 6x_2 \le 18.$

SOLUTION

Change the inequality constraints to be equality by adding a slack variable to each inequality constraint. Now the problem will be

Maximize $F = x_1 + 2x_2$

Subject to

$2x_1 + 3x_2 + x_3 = 12$

$3x_1 + 2x_2 + x_4 = 12$

$6x_1 + x_2 + x_5 = 18$

$x_1 + 6x_2 + x_6 = 18$

$$C = [1 \quad 2 \quad | \quad 0 \quad 0 \quad 0 \quad 0] = \lfloor \bar{C}_B \mid C_B \rfloor$$

$$A = \begin{bmatrix} 2 & 3 & | & 1 & 0 & 0 & 0 \\ 3 & 2 & | & 0 & 1 & 0 & 0 \\ 6 & 1 & | & 0 & 0 & 1 & 0 \\ 1 & 6 & | & 0 & 0 & 0 & 1 \end{bmatrix} = [\bar{B} \mid B]$$

$$x = [x_1 \quad x_2 \quad | \quad x_3 \quad x_4 \quad x_5 \quad x_6]^T$$

$$x_B = [x_3 \quad x_4 \quad x_5 \quad x_6]^T = [12 \quad 12 \quad 18 \quad 18]^T$$

$$Z - \bar{C}_B = C_B B^{-1} \bar{B} - \bar{C}_B = [0 \quad 0] - [1 \quad 2] = [-1 \quad -2].$$

First iteration: $k = 2$ implies that the second vector of $\bar{B}$, x_2 is to enter the basis

$$y = B^{-1}\bar{a}_2 = j\begin{bmatrix} 3 \\ 2 \\ 1 \\ 6 \end{bmatrix}$$

$$\theta = \min\left\{\frac{x_{Bi}}{y_i}, \ y_i > 0\right\} = \min\left\{\frac{12}{3}, \frac{12}{2}, \frac{18}{1}, \frac{18}{6}\right\} = 3,$$

$r = 4$, $\theta = 3$; therefore, the fourth vector X_6 is to leave the basis. The new basis after replacement X_2 and X_6 is

$$x_{B\ new} = [x_B - \theta y, x_{Br} = \theta] = \{12 - 3 \times 3, 12 - 2 \times 3, 18 - 3 \times 1, 3\}^T$$
$$= [6, 6, 15, 3]^T.$$

New values

$$B = \begin{array}{c} \begin{array}{cccc} x_3 & x_4 & x_5 & x_2 \end{array} \\ \begin{bmatrix} 1 & 0 & 0 & 3 \\ 0 & 1 & 0 & 2 \\ 0 & 0 & 1 & 1 \\ 0 & 0 & 0 & 6 \end{bmatrix} \end{array} \qquad \bar{B} = \begin{array}{c} \begin{array}{cc} x_1 & x_6 \end{array} \\ \begin{bmatrix} 2 & 0 \\ 3 & 0 \\ 6 & 0 \\ 1 & 1 \end{bmatrix} \end{array}$$

$$C_B = [0 \quad 0 \quad 0 \quad 2]$$
$$x_B = [6, 6, 15, 3]^T$$
$$\bar{C}_B = [1 \quad 0].$$

Components of B^{-1}, $r = 4$, $y_r = 6$,

$$b_{ij} = b_{ij} - \frac{y_i}{y_r} b_{4j}, \quad i \neq 4$$

$$b_{ij} = \frac{y_i}{y_4} b_{4j}, \quad i = 4$$

$$b_{11} = b_{11} - \frac{y_1}{y_4} b_{41} = 1$$

$$b_{12} = b_{12} - \frac{y_1}{y_4} b_{42} = 0$$

$$b_{13} = b_{13} - \frac{y_1}{y_4} b_{43} = 0$$

$$b_{14} = b_{14} - \frac{y_1}{y_4} b_{44} = -\frac{3}{6}(1) = -0.5$$

$$b_{21} = b_{21} - \frac{y_2}{y_4} b_{41} = 0$$

$$b_{22} = b_{22} - \frac{y_2}{y_4} b_{42} = 1$$

$$b_{23} = b_{23} - \frac{y_2}{y_4} b_{43} = 0$$

$$b_{24} = b_{24} - \frac{y_2}{y_4} b_{44} = 0 - \frac{2}{6}(1) = -\frac{1}{3}$$

$$b_{31} = b_{31} - \frac{y_3}{y_4} b_{41} = 0$$

$$b_{32} = b_{32} - \frac{y_3}{y_4} b_{42} = 0$$

$$b_{33} = b_{33} - \frac{y_3}{y_4} b_{43} = 1 - \frac{1}{6}(0) = 1.0$$

$$b_{34} = b_{34} - \frac{y_3}{y_4} b_{44} = 0 - \frac{1}{6}(1) = -\frac{1}{6}$$

$$b_{41} = \frac{b_{41}}{y_4} = 0, \quad b_{42} = 0, \quad b_{43} = 0, \quad b_{44} = \frac{1}{6}.$$

$$B^{-1} = \begin{bmatrix} 1 & 0.0 & 0.0 & -0.5 \\ 0 & 1 & 0 & -\frac{1}{3} \\ 0 & 0 & 1 & -\frac{1}{6} \\ 0 & 0 & 0 & \frac{1}{6} \end{bmatrix}.$$

Second iteration

$$Z = C_B B^{-1} \bar{B} = [0 \ 0 \ 0 \ 2] \begin{bmatrix} 1 & 0 & 0.0 & -0.5 \\ 0 & 1 & 0 & -\frac{1}{3} \\ 0 & 0 & 1 & -\frac{1}{6} \\ 0 & 0 & 0 & \frac{1}{6} \end{bmatrix} \begin{bmatrix} 2 & 0 \\ 3 & 0 \\ 6 & 0 \\ 1 & 1 \end{bmatrix}$$

$$= \begin{bmatrix} 0 & 0 & 0 & \frac{1}{3} \end{bmatrix} \begin{bmatrix} 2 & 0 \\ 3 & 0 \\ 6 & 0 \\ 11 & 1 \end{bmatrix} = \begin{bmatrix} \frac{1}{3} & \frac{1}{3} \end{bmatrix}$$

$$Z - \bar{C}_B = \begin{bmatrix} \frac{1}{3} & \frac{1}{3} \end{bmatrix} - [1 \ 0] = \begin{bmatrix} -\frac{2}{3} & \frac{1}{3} \end{bmatrix}.$$

Then x_1 is to enter the basis

$$y = B^{-1} \bar{a} = \begin{bmatrix} 1 & 0 & 0.0 & -0.5 \\ 0 & 1 & 0 & -\frac{1}{3} \\ 0 & 0 & 1 & -\frac{1}{6} \\ 0 & 0 & 0 & \frac{1}{6} \end{bmatrix} \begin{bmatrix} 2 \\ 3 \\ 6 \\ 1 \end{bmatrix} = \begin{bmatrix} 2 \\ 6 \\ 6 \\ \frac{1}{6} \end{bmatrix}$$

$$\theta = \min\left\{ \frac{x_{Bi}}{y_i}, y_i > 0 \right\} = \min\left\{ \frac{6}{2} \ \frac{6}{6} \ \frac{15}{6} \ 18 \right\} = 1,$$

$r = 2$, $\theta = 1$. Therefore, the second column (x_4) is to leave the basis. x_1 replaces x_4, $y_r = y_2 = 6$.
The new basis is

$$X_{B \text{ new}} = \{X_B - \theta y, \ X_{B_2} = \theta\} = \left\{ 6 - 2, \ 1, \ 15 - 6, \ 3 - \frac{1}{6} \right\} = \left\{ 4, 1, 9, \frac{17}{6} \right\}.$$

B^{-1} components are

$$b_{ij} = b_{ij} - \frac{y_i}{y_r} b_{rj}, \quad i \neq r, \quad b_{ij} = b_{ij} - \frac{y_i}{y_2} b_{2j},$$

$$b_{ij} = \frac{y_{rj}}{y_r}, \quad b_{ij} = \frac{b_{2j}}{y_2},$$

$$b_{11} = b_{11} - \frac{y_1}{y_2} b_{21} = 1 - \frac{2}{6}(0) = 1$$

$$b_{12} = b_{12} - \frac{y_1}{y_2} b_{22} = 0 - \frac{2}{6}(1) = -\frac{1}{3}$$

$$b_{13} = b_{13} - \frac{y_1}{y_2} b_{23} = 0$$

$$b_{14} = b_{14} - \frac{y_1}{y_2} b_{24} = 3.0 - \frac{2}{6}(2) = \frac{7}{3}$$

$$b_{21} = \frac{b_{21}}{y_2} = 0, \quad b_{22} = \frac{b_{22}}{y_2} = \frac{1}{6}, \quad b_{23} = 0$$

$$b_{24} = \frac{b_{24}}{y_2} = \frac{2}{6} = \frac{1}{3}$$

$$b_{31} = b_{31} - \frac{y_3}{y_2} b_{21} = 0$$

$$b_{32} = b_{32} - \frac{y_3}{y_2} b_{22} = 0 - \frac{6}{6}(1) = -1$$

$$b_{33} = b_{33} - \frac{y_3}{y_2} b_{23} = 1 - \frac{6}{6}(0) = 1.0$$

$$b_{34} = b_{34} - \frac{y_3}{y_2} b_{24} = 1 - (1)2 = -1$$

$$b_{41} = b_{41} - \frac{y_4}{y_2} b_{21} = 0 - \frac{1/6}{6}(0) = 0$$

$$b_{42} = b_{42} - \frac{y_4}{y_2} b_{22} - \frac{1/6}{6}(1) = -\frac{1}{36}$$

$$b_{43} = b_{43} - \frac{y_4}{y_2} b_{23} = 0, \quad b_{44} = b_{44} - \frac{y_4}{y_2} b_{24} = 6 - \frac{1}{36}(2) = \frac{107}{18}$$

$$B^{-1} = \begin{matrix} & x_3 & x_1 & x_5 & x_2 & \\ & \begin{bmatrix} 1 & -\frac{1}{3} & 0 & \frac{7}{3} \\ 0 & \frac{1}{6} & 0 & \frac{1}{3} \\ 0 & -1 & 1 & -1 \\ 0 & -\frac{1}{36} & 0 & \frac{107}{18} \end{bmatrix} & x_4\ x_6 \end{matrix}$$

$$B = \begin{bmatrix} 1 & 2 & 0 & 3 \\ 0 & 3 & 0 & 2 \\ 0 & 6 & 1 & 6 \\ 0 & 1 & 0 & 1 \end{bmatrix} \quad \bar{B} = \begin{bmatrix} 0 & 0 \\ 1 & 0 \\ 0 & 0 \\ 0 & 1 \end{bmatrix}$$

$$x_b = \left[4, 1, 9 \ \frac{17}{6} \right]^T$$

$$C_B = \begin{bmatrix} 0 & 1 & 0 & 2 \end{bmatrix}$$

$$\bar{C}_B \begin{bmatrix} 0 & 0 \end{bmatrix}.$$

Third iteration

$$Z = C_B \bar{B}^1 \bar{B} = \begin{bmatrix} 0 & 1 & 0 & 2 \end{bmatrix} \begin{bmatrix} 1 & -\frac{1}{3} & 0 & \frac{7}{3} \\ 0 & \frac{1}{6} & 0 & \frac{1}{3} \\ 0 & -1 & 1 & -1 \\ 0 & -\frac{1}{36} & 0 & \frac{107}{18} \end{bmatrix} \begin{bmatrix} 0 & 0 \\ 1 & 0 \\ 0 & 0 \\ 0 & 1 \end{bmatrix}$$

$$= \begin{bmatrix} 0 & \frac{2}{28} & 0 & \frac{110}{9} \end{bmatrix} \begin{bmatrix} 0 & 0 \\ 1 & 0 \\ 0 & 0 \\ 0 & 1 \end{bmatrix} = \begin{bmatrix} \frac{1}{9} & \frac{110}{9} \end{bmatrix}$$

$$Z - \bar{C}_B = \begin{bmatrix} \frac{1}{9} & \frac{110}{9} \end{bmatrix} - \begin{bmatrix} 0 & 0 \end{bmatrix} = \begin{bmatrix} \frac{1}{9} & \frac{110}{9} \end{bmatrix} > 0,$$

which means that we reached the optimum point,

$$\min P = C_B x_B = \begin{bmatrix} 0 & 1 & 0 & 2 \end{bmatrix} \begin{bmatrix} 4 \\ 1 \\ 9 \\ 17/6 \end{bmatrix} = \frac{20}{3},$$

$$x_1 = 1, \quad x_2 = (17/6).$$

Then as a result, $P_{G1} = 1$ and $P_{G2} = (17/6)$.

Example 5.9.2

(a) Use LP to minimize $F = 4x_1 + 5x_2$ subject to

$$x_1 + x_2 \leq 6$$
$$x_1 + 3x_2 \leq 15$$
$$2x_1 + x_2 \leq 12.$$

(b) Determine the solution of if the RHS vector is changed to $\begin{bmatrix} 10 & 21 & 16 \end{bmatrix}^T$.

SOLUTION TO PART (a)

Changing inequality constraints to be equality by adding slack:

Min $F = 4x_1 + 5x_2$

s.t. $x_1 + x_2 + x_3 = 6$

$x_1 + 3x_2 + x_4 = 15$

$2x_1 + x_2 + x_5 = 12$

$$C = \begin{bmatrix} 4 & 5 & | & 0 & 0 & 0 \end{bmatrix} = \begin{bmatrix} \bar{C}_B & | & C_B \end{bmatrix}$$

$$A = \begin{bmatrix} 1 & 1 & | & 1 & 0 & 0 \\ 1 & 3 & | & 0 & 1 & 0 \\ 2 & 1 & | & 0 & 0 & 1 \end{bmatrix} = [\bar{B} \mid B]$$

$$x = [x_1 \quad x_2 \quad | \quad x_3 \quad x_4 \quad x_5] \, x_B = [x_3 \quad x_4 \quad x_5]^T = [6 \quad 15 \quad 12]^T$$

$$Z - \bar{C}_B = C_B B^{-1} \bar{B} - \bar{C}_B = [-4 \quad -5].$$

First iteration

Now $k = 2$ implies the second vector of $\bar{B}$ and x_2 enters bases

$$y = B^{-1} \bar{a}_2 = \begin{bmatrix} 1 \\ 3 \\ 1 \end{bmatrix}$$

$$\theta = \min \left\{ \frac{x_{Bi}}{y_i}, \, y_i > 0 \right\} = \min \left\{ \frac{6}{1}, \frac{15}{3}, \frac{12}{1} \right\} = [1 \quad 5 \quad 7]^T.$$

New values

$$B = \begin{matrix} & \begin{matrix} x_2 & x_1 & x_4 \end{matrix} \\ & \begin{bmatrix} 1 & 1 & 0 \\ 0 & 3 & 0 \\ 0 & 1 & 1 \end{bmatrix} \end{matrix} \qquad \bar{B} = \begin{matrix} & \begin{matrix} x_1 & x_4 \end{matrix} \\ & \begin{bmatrix} 1 & 0 \\ 1 & 1 \\ 2 & 0 \end{bmatrix} \end{matrix}$$

$$C_B = [0 \quad 5 \quad 0] \quad \bar{C}_B = [4 \quad 0]$$

$$x_B = [1 \quad 5 \quad 7]^T.$$

Components of B^{-1} were calculated from the matrix inversion lemma to get:

$$B^{-1} = \begin{bmatrix} 1 & 0 & 0 \\ 0 & 1 & 0 \\ 0 & 0 & 1 \end{bmatrix}.$$

Second iteration

$$Z = C_B B^{-1} \bar{B} = [0 \quad 5 \quad 0] \begin{bmatrix} 1 & 0 & 0 \\ 0 & 1 & 0 \\ 0 & 0 & 1 \end{bmatrix} \begin{bmatrix} 1 & 0 \\ 1 & 1 \\ 2 & 0 \end{bmatrix} = [5 \quad 5].$$

Since $Z - \bar{C}_B = [1 \quad 5] > 0$, we have reached optimum point at

$$\begin{bmatrix} x_1 \\ x_2 \end{bmatrix} = \begin{bmatrix} 1 \\ 5 \end{bmatrix}$$

and the corresponding objective value is

$$\text{Min } F = C_B x_B = [0 \quad 5 \quad 0] \begin{bmatrix} 1 \\ 5 \\ 7 \end{bmatrix} = 25.$$

SOLUTION TO PART (b)

Using perturbation for $b = \begin{bmatrix} 10 \\ 21 \\ 16 \end{bmatrix}$

given $B = \begin{bmatrix} 1 & 1 & 0 \\ 0 & 3 & 0 \\ 0 & 1 & 1 \end{bmatrix}$, $x_B = \begin{bmatrix} 1 \\ 5 \\ 7 \end{bmatrix}$, $C_B = \begin{bmatrix} 0 \\ 5 \\ 0 \end{bmatrix}$

$$P = C_B^T x_B = \begin{bmatrix} 0 & 5 & 0 \end{bmatrix} \begin{bmatrix} 1 \\ 5 \\ 7 \end{bmatrix} = 25$$

$$H = B^{-1} = \begin{bmatrix} 1 & 0 & 0 \\ 0 & 1 & 0 \\ 0 & 0 & 1 \end{bmatrix} \text{ (from Part (a))}$$

$$y = H^T C_B = \begin{bmatrix} 1 & 0 & 0 \\ 0 & 1 & 0 \\ 0 & 0 & 1 \end{bmatrix} \begin{bmatrix} 0 \\ 5 \\ 0 \end{bmatrix} = \begin{bmatrix} 0 \\ 5 \\ 0 \end{bmatrix}$$

$$J = \{1, 2, 3\}$$

$$\Delta b = b_2 - b_1 = \begin{bmatrix} 10 \\ 21 \\ 16 \end{bmatrix} - \begin{bmatrix} 1 \\ 5 \\ 7 \end{bmatrix} = \begin{bmatrix} 9 \\ 16 \\ 9 \end{bmatrix}$$

$f_1 \leq 9 \leq g_1;\quad f_1 = 5;\quad g_1 = 10$

$f_2 \leq 16 \leq g_2;\quad f_2 = 0;\quad g_2 = 10;$

$f_3 \leq 9 \leq g_3;\quad f_3 = 1;\quad g_3 = 10$

$$\Delta P = 0$$

with $j = 1$, then $y_1 = 0$, $\Delta b_1 = 0$
with $j = 2$, then $y_2 = 5$, we get

$$L_2 = \left[\frac{-x_{Bi}}{h_{i2}}, h_{i2} > 0 \right] = \left[\frac{-5}{1} \right] = -5$$

$$\Delta b_2 = -5$$

$$\Delta P = \Delta P + y_j \Delta b_j = 0 + y_2 \Delta b_2 = -25$$

$$b_2 = b_2 + \Delta b_2 = 21 + (-5) = 16$$

With $f_j = f_j - \Delta b_j, \quad f_2 = 0 - (-5) = 5$

With $g_j = g_j - \Delta b_j, \quad f_2 = 0 - (-5) = 5$

$$x_{B1} = x_{B1} + h_{12} \Delta b_2 = 1 + 0(-5) = 1$$

$$x_{B2} = x_{B2} + h_{22} \Delta b_2 = 5 + 1(-5) = 0$$

$$x_{B3} = x_{B3} + h_{32} \Delta b_2 = 7 + 0(-5) = 7$$

With $j = 3$, $y_3 = 0$ therefore $\Delta b_3 = 0$, then

updating $P = P + \Delta P = 25 + (-25) = 0$.

Second iteration

$$x_B = \begin{bmatrix} 1 \\ 0 \\ 7 \end{bmatrix}, \quad b = \begin{bmatrix} 10 \\ 16 \\ 16 \end{bmatrix},$$

$$f_1 = 5, \quad g_1 = 10$$
$$f_2 = 5, \quad g_2 = 21$$
$$f_3 = 1, \quad g_3 = 10$$

$$P = \bar{C}_B x_B = \begin{bmatrix} 0 & 5 & 0 \end{bmatrix} \begin{bmatrix} 1 \\ 0 \\ 7 \end{bmatrix} = 0$$

$$C_B = \begin{bmatrix} 0 \\ 5 \\ 0 \end{bmatrix}$$

with $j = 1$, then $y_1 = 0$ therefore $\Delta b_1 = 0$
with $j = 2$, then $y_2 = 5$, we get

$$L_2 = \text{Max}\left[\frac{-x_{Bi}}{h_{i2}}, h_{i2} > 0\right] = \text{Max}\left[\frac{-0}{1}\right] = 0$$

Therefore $\Delta b_2 = 0$.

With $j = 3$, $y_3 = 0$ therefore $\Delta b_3 = 0$. And since $\Delta b_1 = \Delta b_2 = \Delta b_3 = 0$, no further update is required.

Therefore $\Delta P = 0$ and the optimal solution to the problem is $x_B = \begin{bmatrix} 1 \\ 0 \\ 7 \end{bmatrix}$ and $P = 0$.

5.10 Conclusion

This chapter covered LP, one of the most famous optimization techniques for linear objectives and linear constraints. In Section 5.1, the formulation of the natural model associated with the basic assumptions was presented and a graphical solution of the LP problem demonstrated. In Section 5.2 the simplex algorithm was presented and supported with illustrative problems together with a summary of the computational steps involved in the algorithms. The matrix approach solution to the LP problem was presented in Section 5.3 where the formulation of the problem and the revised simplex algorithm were also presented. Duality in LP was presented in Section 5.4.

For those cases where the variables take either integer or continuous values, MIP was presented in Section 5.5 as a way of solving such problems. The branch-and-bound technique for solving this problem was explained. In Section 5.6, the sensitivity method for analysis of postoptimization of LP was presented supported with a method of solution and detailed algorithms. In Section 5.7, a power system application was presented where improvement of a voltage profile using reactive power resources installed in the system was shown. The construction of the method and the solution technique were explained.

5.11   Problem Set

PROBLEM 5.11.1

Given the objective function

$$f(x) = x_1 + 2x_2, \quad x_1, x_2 \geq 0$$

subject to the constraints:

$$\phi_1(x) = 2x_1 + 3x_2 - 12 \leq 0$$
$$\phi_2(x) = 3x_1 + 2x_2 - 12 \leq 0$$
$$\phi_3(x) = 6x_1 + x_2 - 18 \leq 0$$
$$\phi_4(x) = x_1 + 6x_2 - 18 \leq 0,$$

find the point (x) by obtaining the candidate points by finding all possible solutions to be the boundary equations implied by the constraints and testing to satisfy the domain of feasibility.

PROBLEM 5.11.2

Given the objective function

$$f(x) = 4.25x_1 + 4.00x_2, \quad x_1, x_2 \geq 0$$

subject to the constraints:

$$5.00x_1 + 3.00x_2 \leq 15.00$$
$$3.50x_1 + 5.00x_2 \leq 17.50$$
$$4.00x_1 + 3.50x_2 \leq 14.00$$

find the point (x) that maximizes f. Determine (x) to two decimal places.

PROBLEM 5.11.3

Solve the following LP.

Maximize $z = 4x_1 + 6x_2 + 2x_3$

Subject to

$4x_1 - 4x_2 \leq 5$

$-x_1 + 6x_2 \leq 5$

$-x_1 - 3x_2 + 2x_3 \leq 3.$

$x_1, x_2,$ and x_3 are nonnegative integers.
    Compare the rounded optimal solution and the integer optimal solution.

PROBLEM 5.11.4

Convert the following problem to standard form and solve.

Maximize $x_1 + 4x_2 + x_3$

Subject to

$2x_1 - 2x_2 + x_3 = 4$

$x_1 - x_3 = 1$

$x_2 \geq 0, \quad x_3 \geq 3.$

PROBLEM 5.11.5

Consider the problem:

Maximize $z = x_1 + x_2$

Subject to

$2x_1 + 5x_2 \leq 16$

$6x_1 + 5x_2 \leq 30,$

where x_1 and x_2 are nonnegative integers.
    Find the optimal noninteger solution graphically. By using IP, show graphically the successive parallel changes in the value that will lead to the optimal integer solution.

PROBLEM 5.11.6

Consider a power system with two generators with cost functions given by

Generator 1: $F_1(P_1) = 80 + 7.2P_1 + 0.00107P_1^2$ ($)

Generator 2: $F_2(P_2) = 119 + 7.2P_2 + 0.00072P_2^2$ ($).

Where generators 1 and 2 are limited to producing 400 and 600 MW of power, respectively. Given that the system load is 500 MW, then

1. Formulate the problem into LP form.
2. Calculate the optimal generation.
3. Determine the optimal generation cost.

PROBLEM 5.11.7

Consider the problem

Maximize $z = x_1 + x_2$

Subject to

$2x_1 + 5x_2 \leq 16$

$6x_1 + 5x_2 \leq 30,$

where x_1 and x_2 are nonnegative integers.

PROBLEM 5.11.8

Consider the following problem:

Maximize $Z = -5x_1 + 5x_2 + 13x_3$

Subject to

$-x_1 + x_2 + 3x_3 \leq 20$

$12x_1 + 4x_2 + 10x_3 \leq 90$

$x_i \geq 0, \quad i = 1, 2, 3.$

Conduct sensitivity analysis by investigating each of the following changes in the original model. Test the solution for feasibility and optimality.

1. Change in the right-hand constraint 1 to

$$b_1 = 30.$$

2. Change in the right-hand constraint to

$$b_2 = 70.$$

3. Change in the right-hand sides to

$$\begin{bmatrix} b_1 \\ b_2 \end{bmatrix} = \begin{bmatrix} 10 \\ 100 \end{bmatrix}.$$

4. Change in the coefficient of x_3 in the objective function to

$$c_3 = 8.$$

5. Change in the coefficient of x_1 to

$$\begin{bmatrix} c_1 \\ a_{11} \\ a_{12} \end{bmatrix} = \begin{bmatrix} -2 \\ 0 \\ 5 \end{bmatrix}.$$

6. Change in the coefficient of x_2 to

$$\begin{bmatrix} c_2 \\ a_{12} \\ a_{22} \end{bmatrix} = \begin{bmatrix} 6 \\ 2 \\ 5 \end{bmatrix}.$$

7. Introduce a new variable x_6 with coefficients

$$\begin{bmatrix} c_6 \\ a_{16} \\ a_{26} \end{bmatrix} = \begin{bmatrix} 10 \\ 3 \\ 5 \end{bmatrix}.$$

8. Introduce a new constraint $2x_1 + 3x_2 + 5x_3 \leq 50$ (denote the slack variables by x_6).
9. Change constraint 2 to $10x_1 + 5x_2 + 10x_3 \leq 100$.

PROBLEM 5.11.9

Consider the following problem:

Maximize $Z = 2x_1 + 7x_2 - 3x_3$

Subject to

$x_1 + 3x_2 + 4x_3 \leq 30$

$x_1 + 4x_2 - x_3 \leq 10$

$x_1 \geq 0, \quad i = 1, 2, 3.$

Reformulate the problem using x_4 and x_5 as slack variables. Conduct sensitivity analysis by investigating each of the following changes in the original model. Test the solution for feasibility and optimality.

1. Change in the right-hand sides to

$$\begin{bmatrix} b_1 \\ b_2 \end{bmatrix} = \begin{bmatrix} 20 \\ 30 \end{bmatrix}.$$

2. Change in the coefficient of x_3 to

$$\begin{bmatrix} c_3 \\ a_{13} \\ a_{23} \end{bmatrix} = \begin{bmatrix} -2 \\ 3 \\ -2 \end{bmatrix}.$$

3. Change in the coefficient of x_1 to

$$\begin{bmatrix} c_1 \\ a_{11} \\ a_{21} \end{bmatrix} = \begin{bmatrix} 4 \\ 1 \\ 2 \end{bmatrix}.$$

4. Introduce a new variable x_6 with coefficients

$$\begin{bmatrix} c_6 \\ a_{16} \\ a_{26} \end{bmatrix} = \begin{bmatrix} 3 \\ 1 \\ 2 \end{bmatrix}.$$

5. Change in the objective function

$$Z = x_1 + 5x_2 - 2x_3.$$

6. Introduce a new constraint

$$3x_1 + 2x_2 + 3x_3 \le 25.$$

7. Change constraint 2 to

$$x_1 + 2x_2 + 2x_3 \le 135.$$

PROBLEM 5.11.10

A manufacturer develops two kinds of products by using three machines types. Each machine has a limited amount of time as shown below:

Machines List	Production Time (h/Unit)		Total Time (h)
	x_1	x_2	
Type I	1	1	6
Type II	1	3	15
Type III	2	1	

1. Solve the problem by using LP method with the objective to minimize total production time.
2. Resolve if the total time is perturbed such that $b = \begin{bmatrix} 7 \\ 16 \\ 12 \end{bmatrix}$.

PROBLEM 5.11.11

Maximize $f(x) = -2x_1 - x_2 - x_3$

Subject to $3x_1 + 5x_2 + 2x_3 \geq 6$

$4x_1 - 2x_2 + x_3 \geq 3.$

Do the LP solution by changing the right-hand side to $\begin{bmatrix} 7 \\ 4 \end{bmatrix}$.

PROBLEM 5.11.12

A problem is to maximize $Z = x_1 + x_2$ subject to the constraints given by $2x_1 + 5x_2 \geq 16$ and $6x_1 + 5x_2 \geq 30$ where $x_i \geq 0$ for $i = 1, 2.$

1. Solve for the decision variables by using the simplex method.
2. If the right-hand side is changed by 10%, what will be the new optimal value of the objective function, Z^* using postoptimal technique?

PROBLEM 5.11.13

Given the primal LP problem to maximize $z = c^T x$ subject to $Ax = b$ with $x \geq 0$, write the complete dual problem formulation. What is the relationship between the values of the objective function in the primal and dual problems?

PROBLEM 5.11.14

Solve the following problem by the revised simplex method:

Maximize $f(x) = 3x_1 - x_2 + 1.5x_3$

Subject to

$2x_1 - x_2 + 2x_3 \leq 2.2$

$x_1 + 4x_3 \leq 4.5$

$x_1, x_2, x_3 \geq 0.$

PROBLEM 5.11.15

Apply the revised simplex method to

Maximize $f(x) = 4x_1 + 2x_2 + 4x_3$

Subject to

$4x_1 + 3x_2 + 8x_3 \leq 10$

$4x_1 + x_2 + 12x_3 \leq 7$

$4x_1 - x_2 + 3x_3 \leq 6$

$x_1, x_2, x_3 \geq 0.$

PROBLEM 5.11.16

Maximize $2x_1 + x_2 + 3x_3 + 4x_4$

s.t. $x_1 + 3x_2 - x_3 + 2x_4 = 6$

$-x_1 + 2x_2 + x_3 + x_4 = 4$

$x_1, x_2, x_3, x_4 \geq 0$

x_1 and x_2 are integers.

PROBLEM 5.11.17

Maximize $z = 2x_1 + 3x_2$

Subject to

$5x_1 + 7x_2 \leq 35$

$4x_1 + 9x_2 \leq 36$

$x_1, x_2 \geq 0$ and integers.

PROBLEM 5.11.18

Maximize $z = 4x_1 + 6x_2 + 2x_3$

Subject to

$4x_1 - 4x_2 \leq 5$

$-x_1 + 6x_2 \leq 5$

$-x_1 + x_2 + x_3 \leq 5$

$x_1, x_2, x_3 \geq 0$ and integers.

Compare the rounded optimal solution and the integer optimal solution.

PROBLEM 5.11.19

Develop and verify the graphical solution to the problem below and determine the existence of an integer solution when the decision variables are discrete:

Maximize $z = 2x_1 + x_2$

Subject to

$10x_1 + 10x_2 \leq 9$

$10x_1 + 5x_2 \geq 1$

$x_1, x_2 \geq 0.$

PROBLEM 5.11.20

Consider the problem

Maximize $z = x_1 + x_2$

Subject to

$2x_1 + 5x_2 \leq 16$

$6x_1 + 5x_2 \leq 30$

$x_1, x_2 \geq 0$ and integers.

Find the optimal noninteger solution graphically. By using the branch-and-bound algorithm, show graphically the successive parallel changes in the objective value that will lead to the optimal integer solution.

References

1. Bialy, H., An elementary method for treating the case of degeneracy in linear programming, *Uternehmensforschung* (Germany), 10(2), 116, 118–123.
2. Bitran, G. R. and Novaes, A. G., Linear programming with a fractional objective function, *Operations Research*, 21(1), 22–29, Jan.–Feb. 1973.
3. Boulding, K. E. and Spivey, W. A., *Linear Programming and the Theory of the Firm*, Macmillan, New York, 1960.
4. Glover, F., A new foundation for a simplified primal integral programming algorithm, *Operations Research*, 16(4), 727–740, July–August 1968.
5. Harris, M. Y., A mutual primal–dual linear programming algorithm, *Naval Research Logistics Quarterly*, 17(2), 199–206, June 1970.
6. Heady, E. O. and Candler, W., *Linear Programming Methods*, Iowa State College Press, Ames, IA, 1958.
7. Hillier, F. S. and Lieberman, G. J., *Introduction to Operations Research*, 4th edn., Holden-Day, San Francisco, CA, 1986.
8. Lavallee, R. S., The application of linear programming to the problem of scheduling traffic signals, *Operations Research*, 3(4), 86–100, 1968.
9. Ravi, N. and Wendell, R. E., The tolerance approach to sensitivity analysis of matrix coefficients in linear programming-I, Working Paper 562, Graduate School of Business, University of Pittsburgh, October 1984.
10. Chieh, H. T., *Applied Optimization Theory and Optimal Control*, Feng Chia University, China, 1990.
11. Luenberger, D. G., *Introduction to Linear and Nonlinear Programming*, Addison-Wesley, Reading, MA, 1984.
12. Hamdy, A., Taha, *Operations Research: An Introduction*, Prentice Hall, Englewood Cliffs, NJ, 8th edn., April 2006.
13. Dantzig, G. B., *Linear Programming and Extension*, Princeton University Press, 1963.
14. Robert Vanderbei, *Linear Programming: Foundations and Extensions*, Springer, 2nd edn., May 2001.

6

Interior Point Methods

6.1 Introduction

Many engineering problems, including the operation of power systems, are concerned with the efficient use of limited resources to meet a specified objective. If these problems can be modeled, they can be converted to an optimization problem of a known objective function subject to given constraints. Most practical systems are nonlinear in nature; however, some approximations are usually tolerable to certain classes of problems. Two methods commonly used are linear and quadratic programming. The former solves those problems where both the objective and constraints are linear in the decision variables [3,4,25]. The quadratic optimization method assumes a quadratic objective and linear constraints.

The well-known simplex method has been used to solve linear programming (LP) problems. In general, it requires burdensome calculations, which hamper the speed of convergence. In an attempt to improve the convergence properties, recent work by Karmarkar [1,2,8,9] on variations of the interior point (IP) method was proposed. The variants include protective, affine-scaling, and path-following methods. Each of these methods solves the LP problem by determining the optimal solution from within the feasible interior (FI) region of the solution space [1,2].

Projective methods are known to require $O(nL)$ iterations. They rely on the projective algorithm, which requires a good scheme. Several schemes have been proposed in the literature. These methods are different from the simplex method, which seeks the optimum solution from a corner point of the solution space.

Affine-scaling methods have no known polynomial time complexity, and can require an exponential number of iterations if they are started close to the boundary of the feasible region. Also, it has been shown that these methods can make it difficult to recover dual solutions and prove optimality when there is degeneracy. However, these methods work well in practice. Very recently, a polynomial time bound for a primal–dual affine method has been obtained.

Path-following methods generally require $O(n0.5L)$ iterations in the worst case, and work by using Newton's method to follow the central path of optimal solutions obtained by a family of problems defined by a logarithmic barrier function. The most popularly used scheme so far is the barrier method developed by Meggido [13], Kojima [23], and Montiero and Adler [24]. They have shown that the algorithm requires $O(n^3L)$ overall time and no IP algorithm has been shown to have a better worst-case complexity bound. McShane et al. [12] have given a detailed implementation of the algorithm. Their results have been recently adapted to power system problems.

Howard University research contract EPRI—RP2436 extends the results of earlier works [14,16] by improving on the starting and terminating conditions of the IP method for solving LP problems and the extension of the algorithm for solving quadratic-type problems.

The optimal power flow (OPF) problem [5,6] has been recently reviewed [7] as a process of determining the state of power systems that guarantee affordability, reliability, security, and dependability. These abilities optimize given objectives that satisfy a set of physical and operating constraints. In general, these objectives are designated as transmission losses, fuel cost, reactive sources allocation, and voltage feasibility. In general, OPF is a large-scale non-LP problem with thousands of input variables and nonlinear constraints. The problem can be formulated as

Minimize $f(z)$

Subject to

$h(z) = 0$,   with $1 \leq z \leq u$.

f and h are continuously differentiate functions in R^n with values in R and R^m, and l and u are vectors in R^n corresponding to lower and upper bounds in the variables, respectively.

The current interest in IP algorithm [12-22] was spanned by searching algorithm for LP which is based on two key ideas:

1. Steepest descent direction is reached more effectively in improving iterate. This is at the center of the polytope forward by the linear constraints than if it were at the boundary.

2. Transformation of the decision space can be found such that it places the iterate at the center of the polytope without altering the problem. Following the work of Karmarkar, several research works have achieved the IP to many variants of IP involving the primal–dual (Mehrotra method). The most popular variant of IP is equivalent to the well-known logarithmic barrier methods which have been discussed in nonlinear programming (NLP). The main idea in all these barrier methods is discussed in this chapter. IP point methods have been proven successful for solving power system OPF problem. It is easily applicable to OPF based on LP technique described in Chapter 5 and is used for quadratic programming and nonlinear-based method derived from the banner method.

6.2 Karmarkar's Algorithm

The new projective-scaling algorithm for LP developed by Karmarkar has caused quite a stir in the optimization community partly because the speed advantage gained by this new method (for large problems) is reported to be as much as 50:1 when compared to the simplex method [8]. This method has a polynomial bound on worst-case running time that is better than ellipsoid algorithms.

Karmarkar's algorithm is significantly different from George Dantzig's simplex method [25] that solves an LP problem starting with one extreme point along the boundary of the feasible region and skips to a better neighboring extreme point along the boundary, finally stopping at an optimal extreme point. Karmarkar's IP rarely visits many extreme points before an optimal point is found. The IP method stays in the interior of the polytope and tries to position a current solution as the "center of the universe" in finding a better direction for the next move. By properly choosing the step lengths, an optimal solution is achieved after a number of iterations. Although this IP approach requires more computational time in finding a moving direction than the traditional simplex method, a better moving direction is achieved resulting in fewer iterations. Therefore, the IP approach has become a major rival of the simplex method and is attracting attention in the optimization community.

Figure 6.1 illustrates how the two methods approach an optimal solution. In this small problem, the projective-scaling algorithm requires approximately the same amount of iterations as the simplex method. However, for a large problem, this method only requires a fraction of the number of iterations that the simplex method would require.

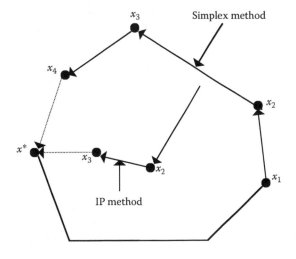

FIGURE 6.1
Illustration of IP and simplex methods.

A major theoretical attraction of the projective-scaling method is its superior worst-case running time (or worst-case complexity). Assume that the size of a problem is defined as the number of bits N required to represent the problem in a computer. If an algorithm's running time on a computer is never greater than some fixed power of N, no matter what problem is solved, the algorithm is said to have polynomial worst-case running time. The new projective-scaling method is such an algorithm.

Due to the results of [10,15], several variants of IPs have been proposed such as the affine-scaling method which is discussed in this chapter. Affine-scaling methods have no known polynomial time complexity, and can require an exponential number of iterations if they are started close to the boundary of the feasible region. It has also been shown that these methods can make it difficult to recover dual solutions and prove optimality when there is degeneracy. However, these methods do work well in practice. Very recently, a polynomial time bound for a primal–dual affine method has been obtained [11].

Path-following methods generally require $O(n0.5L)$ iterations in the worst case, and work by using Newton's method to follow the "central path" of optimal solutions obtained by a family of problems defined by a logarithmic barrier function. Two parameters, the barrier parameter and an underestimated optimal value of the objective function, are the linking parameters between all methods. Barrier methods have been used to construct primal-path-following algorithms, and the method of centers used as a basis for dual algorithms. After scaling has been used to construct both primal and dual algorithms, other variants of barrier methods have been used to construct primal–dual path-following algorithms and an affine variant of the primal–dual algorithms. In general, the above methods, whether projective affine method of centers, or path following, are all simple variants of the algorithm barrier methods applied to the primal, dual, or primal and dual problems together [15].

As mentioned above, since Karmarkar's discovery of the IP method [26,27] and its reported speed advantage obtained over other traditionally used methods, many variants of the IP method have evolved in an attempt to solve the above posed problems. Of these the projective scaling, the dual and primal affine methods, and the barrier function method are the most popular.

These variants of the IP method are presented and evaluated based on their algorithms and the problems they solve.

6.3   Projective-Scaling Method

The projective-scaling algorithm has attracted a great deal of interest due to Karmarkar's ingenious proof that its running time is a polynomial function of the problem size even in the worst case. Karmarkar showed that if n is the number of variables in LP Problems and L is the number of bits used to represent numbers in the computer, the theoretical worst-case running time is $O(n^{3.5}L^2)$. That is, as the problem size increases, the running time tends to be a constant multiple of $n^{3.5}L^2$, which is substantially better than the ellipsoid algorithm's worst-case running time of $O(n^6L^2)$.

The problems solved by the projective algorithm are in the following form:

Minimize $c^T y$

Subject to

$Ay = 0$

$e_y^T = 1, \quad y \geq 0,$

where
A is an m by n matrix
e is a vector of n ones

The main algorithm for the projective method is presented below.

Algorithm for the Projective-Scaling Method

Step 1. *Initialization.*
$k = 0$, $x^0 = c/n$, and let L be a large positive integer.

Step 2. *Optimality check.*
IF $c^T x^k$ is $\leq 2^{-L} c^T e/n$, **THEN** stop with an optimal solution $x^* = x^k$. **Otherwise** go to step 3.

Step 3. *Iterate for a better solution.*
Let $X^T = \text{Diag}(x^k)$
$B^k = \left\{ \frac{AX^k}{e^k} \right\}$, e is a matrix.
$d^k = -[I - B_k^T(B_k B_k^T)^{-1} B_k] X_k c,$
D is the direction of the real line.
$y^{k+1} = \frac{c}{n} + \frac{\alpha}{n} \left(\frac{d^k}{\|d^k\|} \right), 0 \leq \alpha \leq 1$
$x^{k+1} = \frac{(x_k)y^{k+1}}{(e^T x_k)y^{k+1}}.$
Set $k = k+1$. Go to step 2

where
x^k is an interior feasible solution
x^n is an n-dimensional diagonal matrix
B^k is the constrained matrix in Karmarkar's standard form
d^k is a feasible direction of the projective negative direction as defined
above
y^{k+1} is a new interior feasible solution
L is chosen to be the problem size, where $2^{-L} > e, e > 0$

This algorithm terminates in $O(nL)$ iterations. A large value of α tends to speed up the iteration.

6.4 Dual Affine Algorithm

Both the primal and dual interior methods of the variety initiated by Karmarkar can be viewed as special cases of the logarithmic barrier method applied to either the primal or the dual problem. The problems solved by the dual and primal affine methods and their algorithms are presented below. Consider the LP problem given by

Maximize $Z = c^T x$,

Subject to

$Ax \leq b$ and x is unrestricted.

By introducing slack variables into the constraints, the inequality constraints are converted such that we obtain a new formulation given by

Maximize $Z = c^T x$,

Subject to

$Ax + s = b$

$s \geq 0$ and x is unrestricted.

Algorithm

Step 1. *Initialization.*
Set counter value $k=0$ and the tolerance $\in$ (a small positive number).
Obtain a starting solution (x^0, s^0) such that $Ax^0 + s^0 = b$ and $s^0 > 0$.
Set the acceleration constant α, where $0 < \alpha < 1$.

Step 2. *Obtaining the translation direction.*
Computer $d_x^k = (A^T W_k^{-2} A)^{-1} c$, where $W_k = \text{diag}(s_k)$.
Compute the direction vector $d_s^k = -A d_x^k$.

Step 3. *Check for unboundedness.*
IF $d_s^k = 0$, THEN (x^k, s^k) is the dual optimal. Go to step 9.
IF $d_s^k > 0$, THEN the problem is unbounded. Go to step 10.
Otherwise, $d_s^k < 0$. Go to step 4.

Step 4. *Compute the primal estimate y^k.*
$y^k = -W_k^{-2} d_s^k$.

Step 5. *Optimality test.*
IF $y^k \geq 0$ **AND** $(b^T y^k - c^T x^k) \leq \epsilon$, **THEN** (x^k, s^k) is the dual optimal solution and is the primal optimal solution. Go to step 9. **Otherwise**, go to step 6.

Step 6. *Compute the step length* β_k.
$$\beta_k = \min_i \left\{ \frac{\alpha s_i^k}{-(d_s^k)_i} \bigg|_{(d_s^k)_i < 0} \right\}, \quad \text{where } 0 < \alpha < 1.$$

Step 7. *Update the dual variables* (x^k, s^k).
$$x^{k+1} = x^k + \beta_k d_x^k$$
$$s^{k+1} = s^k + \beta_k d_s^k.$$

Step 8. *Increment the counter and check for termination.*
IF $k \geq k_{\max}$, **THEN** flag the user (maximum number of iteration is reached). Go to step 10.
Otherwise, increment the counter
$k = k + 1$ and go to step 2.

Step 9. *Calculate the final objective value and display the results.*
Optimal solution is $x^* = x^k$ and the objective value is $Z^* = c^T x^*$. Display the final solution.

Step 10. Stop.

6.5 Primal Affine Algorithm

Consider the LP problem given by

Maximize $Z = c^T x$
Subject to
$Ax = b$ and $x \geq 0$ (x is unrestricted).

Algorithm

Step 1. *Initialization.*
Set the iteration limit, $k_{\max}$, and initialize the counter value at $k = 0$.
Select tolerance $\in$ (a small positive number).
Set vector $e = [1, 1, \ldots, 1]$.
Obtain a starting solution $x^0 \geq 0$ such that $Ax^2 = b$.
Set the acceleration constant, α where $0 < \alpha < 1$.

Step 2. *Compute the estimate vectors* w^k.
$w^k = (A D_k^2 A^T)^{-1} A D_k^2 c$ where $D_k = \text{diag}(x^k)$.

Step 3. *Compute the reduced cost coefficient vector.*
$r^k = c - A^T w^k.$

Step 4.   *Optimality check.*
IF $r^k = 0$ **AND** $e^T D_k r^k \leq \in s$, **THEN**
x^k is the primal optimal. Go to step 10. **Otherwise**, continue.

Step 5.   *Compute translation direction d_y^k.*
$$d_y^k = -D_k r^k$$

Step 6.   *Check for feasibility and constant objective.*
IF $d_y^k > 0$, **THEN** "the problem is unbounded." Go to step 11.
IF $d_y^k = 0$, **THEN** "the primal optimal solution is x^k."
Go to step 10. **Otherwise**, continue.

Step 7.   *Compute the step length β_k.*
$$\beta_k = \min_i \left\{ \left. \frac{\alpha}{-(d_y^k)_i} \right|_{(d_y^k)_i < 0} \right\}, \quad \text{where } 0 < \alpha < 1.$$

Step 8.   *Update the Primal Variables x^k.*
$$x^{k+1} = x^k + \beta_k D_k d_y^k.$$

Step 9.   *Increment the counter and check for termination.*
IF $k \geq k_{\max}$, **THEN** flag the user (maximum number of iterations is reached). Go to step 11.
Otherwise, increment the counter
$k = k + 1$ and go to step 2.

Step 10.   *Calculate the final objective value and display the results.*
Optimal solution is $x^* = x^k$ and the objective value is $Z^* = c^T x^*$.
Print the solution.

Step 11.   *Stop.*

6.6   Barrier Algorithm

Barrier-function methods treat inequality constraints by creating a barrier function, which is a combination of the original objective function and a weighted sum of functions with a positive singularity at the boundary. As the weight assigned to the singularities approaches zero, the minimum of the barrier-function approaches the minimum of the original constrained problem. Barrier-function methods require a strictly feasible starting point for each minimization, and generate a sequence of strictly feasible iterates. The barrier transformation for linear programs with upper and lower bounds on the variables is given in the following form.

Minimize $c^T x$

Subject to

$Ax = b, \quad 1 \leq x \leq u,$

where A is an m by n matrix with $m \leq n$.

When applying the barrier-function method to the problem described above, the subproblem to be solved takes the following form.

$$\underset{x \in R^n}{\text{Minimize }} F(x) \equiv c^T x - \mu \sum_{j=1}^{n} \ln x_j$$

Subject to

$$Ax = b,$$

where the scalar μ ($\mu > 0$) is known as the barrier parameter and is specified for each subproblem. The equality constraints cannot be treated by a barrier transformation and, thus, are handled directly. The general algorithm is given below.

Algorithm for the Barrier Method (In Brief)

At the start of each iteration, the quantities μ, x, π, η are known, where $\mu > 0$, $x > 0$, $Ax = b$, and $\eta = c - A^T p$. A correction of π is calculated at each stage since a good estimate is available from the previous iteration. The main steps of the algorithm are as follows.

Step 1. Define $D = \text{Diag}(x_j)$ and compute $r = D\eta - \mu e$.
Note that r is a residual from the optimality condition for the barrier subproblem, and hence $||r|| = 0$ if $x = x^*(m)$.

Step 2. Terminate if μ and $||r||$ are sufficiently small.

Step 3. If appropriate, reduce μ and reset r.

Step 4. Solve the least squares problem
$\underset{\delta \pi}{\text{Minimize}} ||r - DA^T \delta \pi||$.

Step 5. Compute the updated vectors
$\pi \leftarrow \pi + \delta \pi$ and $\eta \leftarrow \eta - A^T \delta \pi$.
Set $r = D\eta - \mu e$ (the updated scaled residual) and $\pi = -(1/\mu)Dr$.

Step 6. Find α_M, the maximum value of α such that $x + \alpha p \geq 0$.

Step 7. Determine the step length $\alpha \in (0, \alpha_m)$ at which the barrier function F $(x + \alpha p)$ is suitably less.

Step 8. Update $x \leftarrow x + \alpha p$.

All iterates satisfy $Ax = b$ and $x > 0$. The vectors p and η approximate the dual variables π^* and reduced cost η^* of the original linear program.

6.7 Extended IP Method for LP Problems

The LP problem is formulated in the standard formula of the LP IP method form as follows:

Maximize $P = C^T x$ (6.1)

Subject to

$Ax = b$

$b_i > 0, \quad i = 1, \ldots, m, \quad x_i > 0, \quad i = 1, \ldots, m.$ (6.2)

The IP method involves a sequence that consists of a feasible IP $(Ax = b, x > 0)$ that makes the objective function increase until it reaches its limit. The limit is an optimal solution of the problem. The IP method utilizes all the vectors in A together with the points in the sequence to generate a maximum increase of the objective function. In addition to programming simplicity, it is superior to the simplex method in computation time and convergence for large systems (large $m > n$).

The condition that $b > 0$ as imposed in the simplex method is waived here and the matrix A is only required to be of full rank $m < n$. Also, by using an appropriate conversion of inequality, two-sided constraints can be handled by the proposed IP method.

Finally, to guarantee existence of the feasible IPs, a trivial condition has been imposed in that the problem has no less than two feasible points $(Ax = b, x > 0)$, one of which is a bounded solution for the problem.

6.8   FI Sequence

For convenience, the LP problem can be written as follows:

Maximize $a^T x$ (6.3)

Subject to

$Ax = b,$

Such that $x \geq 0.$ (6.4)

Note C^T is replaced by a^T where a is a column vector to facilitate description of the IP method. The considered FI sequence contains only feasible and IPs of the problem; that is,

$$S = \{x^1, x^2, \ldots, x^k, x^{k+1}\}$$ (6.5)

with

$$Ax^k = b$$ (6.6)

and

$$x^k > 0, \quad \text{for all } k > 0. \tag{6.7}$$

For a known point x^k, a diagonal matrix is formed by

$$D = \text{diag}[x_1^k, x_2^k, \dots, x_n^k], \tag{6.8}$$

where

$$x^k = [x_1^k, x_2^k, \dots, x_n^k]. \tag{6.9}$$

The FI sequence is then generated recursively according to

$$x^{k+1} = x^k + \beta D d_p, \tag{6.10}$$

where
  d_p is an n-vector
  β is a positive number

They are to be chosen in such a way that x^{k+1} is a feasible IP wherever x^k is. As such, S contains all FI points of x.

The objective functions and constraints between two consecutive points are related by

$$a^T x^{k+1} = a^T x^k + \beta d^T d_p \tag{6.11}$$

and

$$A x^{k+1} = A x^k + \beta B d_p, \tag{6.12}$$

where

$$d = Da \tag{6.13}$$

and

$$B = AD. \tag{6.14}$$

D is determined by x^k and it changes from point to point. Let d be orthogonally decomposed into $d = d_p + d_q$, where d_p is the projection on the null space of B and d_q is in the B^T-subspace (spanned by the vectors of B^T). Then, it follows that

$$B d_p = 0, \quad d_q = B^T v, \quad \text{and} \quad d_p^T d_q = 0, \tag{6.15}$$

where v is an m-vector, the coordinates of d_q. Solving v and then d_q, we have

$$d_q = B^T(BB^T)^{-1}Bd \qquad (6.16)$$

and

$$d_p = d - d_q = [I - B^T(BB^T)^{-1}B]d = Du, \qquad (6.17)$$

where

$$U = a - A^Tw \qquad (6.18)$$

with

$$w = (BB^T)^{-1}Bd. \qquad (6.19)$$

To generate the FI sequence S, the LP problem is divided into two cases: trivial and ordinary. The former has the vector a confined in the A^T space while the latter does not. In the trivial case, there exists an m-vector v such that $a = A^Tv$ and hence $a^Tx = v^TAx = v^Tb =$ constant. Thus, all the feasible solutions yield the same objective functions and, hence, there is no optimization involved in the problem. Consequently, it remains only to generate S for the ordinary case.

Starting with a known FI point x^1, S is generated recursively according to Equation 6.10 until reaching a point x at which $d_p = 0$. S contains all the points except x, which is referred to as the limit point of S. The problem is assumed to have no less than two feasible points, and between them there exists a bounded solution of the problem. Since feasible points form a convex set, there are an infinite number of FI points existing within the problem.

For the ordinary case with a bounded solution, one may draw the following conclusions: (1) S contains an infinite number of points and (2) S always has a limit point. To show this, let us assume $d_p = 0$ at a finite k; then $d = d_p + d_q = d_q$ reveals that

$$d = Da = B^T(BB^T)^{-1}Bd = DA^T(BB^T)^{-1}Bd = DA^Tv$$

from which we have $a = A^Tv$ since D is nonsingular. This is a trivial case and, hence, d_p cannot be zero for finite k.

It follows from Equation 6.11 and $d_q^Td_p = 0$ that

$$a^Tx^{k+1} = a^Tx^k + \beta\|d_p\|^2. \qquad (6.20)$$

Summing both sides of Equation 6.20 from $k = 1$ to $k = N$ gives

$$a^Tx^{N+1} = a^Tx^1 + \sum_{k=1}^{N}\beta\|d_p\|^2. \qquad (6.21)$$

To make Equation 6.21 bounded, it is necessary that $d_p = 0$ as $N \to \infty$. Otherwise, the right side becomes positively unbounded and, hence, x^{N+1} (feasible) yields an objective function, which contradicts the assumption of a bounded solution.

Now, it remains to specify β to make x^{k+1} an FI point if x^k is one. We choose for this purpose

$$0 < \beta < \frac{1}{-\gamma'} \tag{6.22}$$

where γ is the smallest component of d_p. It is asserted that $\gamma < 0$ for all points in S. Indeed, if $\gamma \geq 0$ then $d_p \geq 0$ and $x^{k+1} > 0$ for any $\beta > 0$ according to Equation 6.10. The objective function indicated by Equation 6.20 becomes positively unbounded together with β since $d_p \neq 0$ in S. This contradicts the assumption of a bounded solution and, hence, $\gamma < 0$ must be true.

If x^k is an FI point, then the ith component of Equations 6.10 and 6.12 becomes $x_i^{k+1} = x_i^k + \beta x_i^k d_{p_i} = x_i^k(1 + \beta d_{p_i}) \geq x_i^k(1 + \beta\gamma) > x_i^k(1 - 1) = 0$, for all $i = 1, 2, \ldots, n$ and $Ax^{k+1} = Ax^k + 0 = b$. Since x^{k+1} is also an FI point, all the points of S are FI points if x^1 is by induction. Figure 6.2 shows the IP method algorithm.

6.8.1 Optimality Condition

The limit point of the FI sequence is an optimal solution of the LP problem. This can be shown as follows.

First, we demonstrate that $d_p = 0$ at the limit point implies $u \leq 0$. It follows from Equation 6.17 and $d_p = 0$ that $d_{p_i} = x_i u_i = 0$ for all $i = 1, 2, \ldots, m$. We need only consider the case $u_i > 0$ and $x_i = 0$. In such a case, there exists an $\varepsilon > 0$ and integer N such that $u_i^k > \varepsilon$ for $k \geq N$ where u_i^k is the ith component of Equation 6.18 evaluated at x^k. The ith components of Equations 6.10 and 6.17 are

$$x_i^{k+1} = x_i^k + \beta_i x_i^k d_{p_i} \tag{6.23}$$

and

$$d_{p_i} = x_i^k u_i^k. \tag{6.24}$$

Then, we obtain

$$\frac{x_i^{k+1}}{x_i^k} = 1 + \beta_i x_i^k u_i^k. \tag{6.25}$$

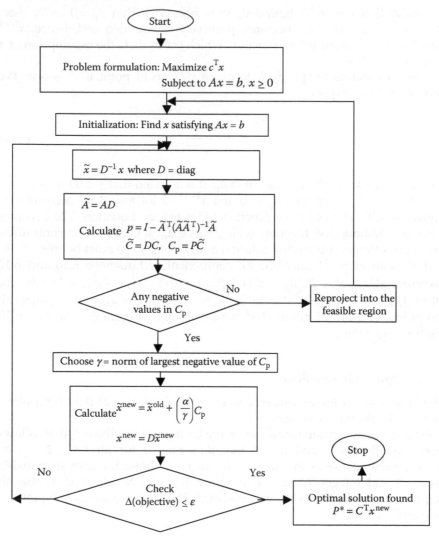

FIGURE 6.2
Algorithm for the IP method.

Multiplying both sides of Equation 6.25 from $k = N$ to $k = K$ gives

$$\frac{x_i^{K+1}}{x_i^N} = \sum_{k=N}^{k=K} \left[1 + \beta_i x_i^k u_i^k\right] > \sum_{k=N}^{k=K} \left[1 + \beta_i x_i^k \varepsilon\right] > 1. \qquad (6.26)$$

Since $x_i^{K+1} > x_i^N > 0$ for any K,

$$x_i = \lim_{x \to \infty} x_i^{K+1} > 0 \qquad (6.27)$$

as K approaches infinity which violates the fact that $x_i = 0$; therefore, $u_i \leq 0$ or $U \leq 0$ must be true.

Using $u \leq 0$ and $d_p = 0$ at x, we obtain $u^T y \leq 0$ and $d_p^T e = 0$, where y is any feasible solution and e is the n-vector with all components equal to one. Substituting Equations 6.17 and 6.18 yields

$$u^T y - d_p^T e = u^T y - u^T De = u^T (y - x)$$
$$= (a^T - w^T A)(y - x)$$
$$= a^T y - a^T x - w^T b + w^T b$$
$$= a^T y - a^T x \leq 0, \tag{6.28}$$

which indicates that x is a maximum solution.

6.9 Extended Quadratic Programming Using IP Method

Extended quadratic programming using the IP method (EQIP) considered here is an extension of the LP version of the IP method developed during the project. The objective function, a quadratic form, is defined by

$$P = \frac{1}{2} x^T Q x + a^T x \tag{6.29}$$

subject to

$$Ax = b \quad \text{and} \quad x \geq 0, \tag{6.30}$$

where Q is any square and symmetric matrix. LP is a special case of the quadratic programming problem when $Q = 0$. The concept for solving the quadratic programming problem is similar to that of LP problems. It is again assumed that the problem has a bounded solution. With A being of full rank m with $m < n$, there are at least two feasible solutions to the problem. The same FI sequence generated below guarantees optimality within the feasible region for quadratic optimization problems.

In order to maintain the solution of the problem of each iteration within the interior feasible region, the algorithm requires the calculation of the initial starting interior feasible point $\tilde{X}^0$; that is, $\tilde{A}\tilde{X}^0 = \tilde{b}$ with $\tilde{x}_j^0 \geq 0$. The initial feasible point can be obtained by introducing the artificial variable x_S. the EQIP obtains the initial feasible point by using an auxiliary problem.

Maximize $[-x_S]$ $\qquad\qquad$ (6.31)

Subject to

$$\tilde{A}\tilde{X} + (\tilde{b} - \tilde{A}e)x_S = \tilde{b},$$
$$\tilde{x}_j \geq 0, \quad j = n+1, \ldots, n+2m, \quad x_S \geq 0. \qquad (6.32)$$

Clearly, any feasible solution of the original problem is a maximum solution $x_S = 0$ for the auxiliary problem and vice versa. Since the latter always has a feasible point at

$$\tilde{X} = e \quad \text{and} \quad x_S = 1,$$

one may use this point as the initial starting point to solve the auxiliary problem by the EQIP with $\tilde{Q} = 0$ to reach a maximum solution and thus obtain a feasible initial point for the original problem. The key point is that the direction vector d_x at each iteration k can be approximately calculated but maintain feasibility of

$$\tilde{X}^{k=1},$$

for example

$$\tilde{A}\tilde{X}^{k+1} = \tilde{b}$$

with

$$\tilde{x}_j^{k+1} \geq 0, \quad j+n+1, \ldots, n+2m.$$

A feasible direction, along with the objective function increases, is found, and then an approximate step length is determined to guarantee the new feasible solution, which is strictly better than the previous one. The stopping criteria are the relative changes in the objective function at iterations; that is

$$|P_{k+1} - P_k| \max\{1, |P_k|\} < \varepsilon_1, \qquad (6.33)$$

or the relative changes in interior feasible solutions in iterations; that is

$$|\tilde{X}^{k+1} - \tilde{X}^k| < \varepsilon, \qquad (6.34)$$

where P_{k+1} and P_k are defined as follows:

$$P_{k+t} = \frac{1}{2}(\tilde{X}^{K+1})^T \tilde{Q}\tilde{X}^{K+1} + \tilde{a}^T \tilde{X}^{K+1}, \qquad (6.35)$$

$$P_k = \frac{1}{2}(\tilde{X}^K)^T \tilde{Q}\tilde{X}^K + \tilde{a}^T \tilde{X}^K. \qquad (6.36)$$

The optimality condition is computed until the maximum is satisfied.

$$P_{k+1} = P_k + \frac{\left\| d_P^k \right\|^4}{-2T}. \tag{6.37}$$

Detailed EQIP Algorithm

A detailed step-by-step description of how the EQIP algorithm solves a quadratic objective function subject to linear constraints is presented below:

Step 1. Identify the problem defined by maximizing:

$$P = \frac{1}{2}x^T Q x + a^T x, \quad \text{subject to } Ax = b$$

$$\text{With } x_i \geq 0, \quad \text{for } i = 1, 2, 3, \ldots, n$$

Step 2. For all i and x_{n+1} is an FI point of the auxiliary problem of maximizing: x_{n+1}

Subject to $Ax + (b - Ae)x_{n+1} = b$

where $e = [1, 1, \ldots, 1]^T$, column vector, $\begin{bmatrix} x \\ x_{n+1} \end{bmatrix} = \begin{bmatrix} e \\ 1 \end{bmatrix}$ is a feasible IP

because $Ax + (b - Ae)x_{n+1} = Ae + b - Ae = b$.

Evidently, $x = e$ is a feasible IP for the diagonal problem

$c = b - Ae = 0$ since $Ae = b$.

Step 3. Construct an updated $B = AD$ for corrected x, which is modified in such a way as to maximize:

$(-x_{n+1})$ in the auxiliary problem.

At the maximum, $-x_{n+1} = 0$ and hence

$Ax + (b - Ae)x_{n+1} = b \Rightarrow Ax + 0 = b$

Thus, x is a feasible IP of the original problem.

Step 4. MN solution for $By = v$ is designed to obtain

$y = B^T (BB^T)^T v$,

where BB^T is nonsingular.

The reason to assume $v = (x_{n+1})C$ is to assure feasibility to be explained in the next step.

Step 5. $\gamma = \text{Min}[\gamma_i]$, $i^{n+1} = i^k(1 + \beta\zeta)$ for all i and $x_{n+1}^{k+1} = x_{n+1}^k(1 - \beta)$

for the auxiliary problem, we have from: $v = (b - Ae)x_{n+1}^k$

$$Ax^{k+1} + (b - Ae)x_{n+1}^{k+1} = A(x^k + \beta Dy) + \beta_1(b - Ae)x_{n+1}^{k+1} + 0 = b$$

$$= Ax^k + \gamma + \beta[By - v]$$

$$= Ax^k + (b - Ae)x_{n+1}^k + 0 = b$$

since $By = v$, thus x^{k+1} is feasible if the objective function is increased by

$(-x_{n+1}^{k+1}) - (-x_{n+1}^k) = \beta x_{n+1}^k > 0$, because $\beta > 0$ is to be chosen.

5(a): $\gamma + 1 > 0$: $\gamma = \text{Min} [y_i] > -1 \rightarrow 1 + y_i > 0$ and

$$x_i^{k+1} = x_i^k(1 + y_i) > 0,$$

since $x_i^k > 0$

$$x_{n+1}^{k+1} = x_{n+1}^k(1 + \beta) = x_{n+1}^k(1 - 1) = 0$$

Because we choose $b = 1$ for such a case. Therefore, x^{k+1} is a feasible IP of the original problem because

$$Ax^{k+1} + (b - Ae)x_{n+1} = Ax^{k+1} + 0$$

5(b): $\gamma + 1 \leq 0$: $\beta = \frac{\alpha}{\gamma} > 1$

$$1 + \beta\xi = 1\left(\frac{\alpha}{-\gamma}\right)y_i > 1 - 1 = 0$$

$$1 - \beta = 1\left(\frac{\alpha}{-\gamma}\right) > 1 - \alpha > 0$$

since $0 < \alpha < 1$; $\gamma \leq -1$, therefore x^{k+1} is an FI point of the auxiliary problem if x^k together with x_{n+1}^k is one.

Step 6.   We exit from the auxiliary problem and now enter the original problem started with the FI point just obtained. For this purpose we construct r, $d = Dg$ and $B = AD$ as required.

Step 7.   Solve for DG from $Bdg = v$ by using the minimum norm program with $v = Bd$.

Step 8.   Replace zero and infinity, respectively, by
ε and M for computer practice.
Assume $M = 10^6$ and $0.001 \leq \varepsilon \leq 0.01$
8(a): $|dp_{ii}| \leq \varepsilon \rightarrow dp_i = 0$ for all i and hence the optimally conditions are reached.
8(b): If a is not met, we will advance the FI point toward the limiting point according to: $x^k + \beta Ddp$, where $D = \text{diag}[x_1^k, x_2^k, \ldots, x_n^k]$
The scalar β must be chosen to ensure FI and also increase P. It will be chosen between β_1 and β_2 in step 2, but β_1 is fixed here as follows.
8(b) (i): $\gamma < 0$: $\beta_1 = \left(\frac{\alpha}{-\gamma}\right)$ with $0 < \alpha < 1$ and
$\gamma = \text{min} [d_{p_i}]$ such as β_1 makes $1 + \beta_1 dp_1 \geq (1 - \frac{\alpha}{\beta})\alpha = 1 - \alpha > 0$
Hence
$x_i^{k+1} = x_i^k(1 + \beta_1 dp_1) > 0$. Thus, x_i^{k+1} is a feasible IP is one
8(b) (ii): $\gamma \geq 0$: x_i^{k+1} is always interior for any $\beta_1 > 0$

We will choose $\beta_1 = M$ and leave β to be constrained by another β_2 to consider in step 9.

Step 9.   It has been shown that the objective function between two consecutive points are related by

$$P_{k-1} = P_k + \beta \|dp\|^2 + \frac{1}{2}\beta^2 T$$

where $T = (Q\,dp)^T Q(D\,dp)$, a scalar. As a function of β,

$$\beta = \beta_2 = \frac{\|dp\|^2}{-T}$$

If $T < 0$ and P_{k+1} increases monotonically with β for $T \geq 0$. Thus, one should choose $\beta = M$ for the latter case and β_2 for the former case. The scalar β_2 is fixed in this step for latter case.

Step 10. Problem does not have a finite maximum if $\beta_1 = M$ and $\beta_2 = M$ at the same time.

Step 11. β is chosen to be smaller one between β_1 and β_2 in order to guarantee that x^{k+1} maximize P_{k+1} and yet preserve FI point status with such β, we form the next point by

$$x_i^{k+1} = x_i^k + \beta x_i^k \, dP_i \quad \text{or} \quad x^{k+1} = x^k[1 + \beta D\,dP]$$

in matrix form as used in step 8.

Step 12. If $\beta = \beta_2$ was chosen, then no adjustment is needed, but we adjust the last point by choosing $\alpha = 1$ if $\beta = \beta_1$ was chosen. Let us consider x^{N+1} in S, that yields: $|dP(1)| \leq \varepsilon$ for a. Then instead of using fractional α, we use $\alpha = 1$ to generate x^N.

$$x^{N+1} = x^N + \frac{\alpha}{-\gamma} D\,dP = x + \frac{1}{\alpha}\Delta x$$

Eliminating x^N we have: $x = x^{N+1} + \delta \Delta x$
where $\delta = \frac{1}{\alpha} - 1 = 0.030927835$ for $x\ 0.97$ and $\Delta x = \frac{\alpha}{-\gamma} D\,dP$
which was computed in step 11.

Step 13. Current x is a solution because the optimal condition is met.

Step 14. No bounded solution exist because if $\beta_1 = M$ and $\beta_2 = M$ may be made infinity by some FI point (infinity).

6.10   Illustrative Examples

Example 6.10.1

Solve the constrained problem using the following:

Maximize $z = x_1 + 2x_2$

Subject to

$x_1 + x_2 + x_3 \leq 8$

$x_j \geq 0$.

1. IP method.
2. Graphical representation.

Based on the algorithm shown in Section 6.6 and following the flowchart in Figure 6.2, we can say $Z = x_1 + 2x_2 = C^t x \Rightarrow C^t = [1 \quad 2 \quad 0]$ as $x = [x_1, x_2, x_3]^t$
Subject to

$$Ax = b \Rightarrow A = [1 \quad 1 \quad 1].$$

We are going to take $\alpha = 0.7$, $\varepsilon = 0.1$.

First iteration

As an initial point, we start by $x = [1, 1, 2]^t$. Substitute in the objective function,

$$Z = C^t x = [1 \quad 2 \quad 0] \begin{bmatrix} 1 \\ 1 \\ 2 \end{bmatrix} = 3.0$$

$$D = \text{diag}(x) = \begin{bmatrix} 1 & 0 & 0 \\ 0 & 1 & 0 \\ 0 & 0 & 2 \end{bmatrix}$$

$$\tilde{x} = D^{-1}x = \begin{bmatrix} 1 & 0 & 0 \\ 0 & 1 & 0 \\ 0 & 0 & 2 \end{bmatrix}^{-1} \begin{bmatrix} 1 \\ 1 \\ 2 \end{bmatrix} = \begin{bmatrix} 1 & 0 & 0 \\ 0 & 1 & 0 \\ 0 & 0 & 0.5 \end{bmatrix} \begin{bmatrix} 1 \\ 1 \\ 2 \end{bmatrix} = \begin{bmatrix} 1 \\ 1 \\ 1 \end{bmatrix}$$

$$\tilde{A} = AD = [1 \quad 1 \quad 1] \begin{bmatrix} 1 & 0 & 0 \\ 0 & 1 & 0 \\ 0 & 0 & 2 \end{bmatrix} = [1 \quad 1 \quad 2].$$

The projection area p,

$$P = I - \tilde{A}^t(\tilde{A}\tilde{A}^t)^{-1}\tilde{A} = \begin{bmatrix} 0.833 & -0.1667 & -0.333 \\ -0.1667 & 0.833 & -0.333 \\ -0.333 & -0.333 & 0.333 \end{bmatrix}$$

$$\tilde{C} = DC = \begin{bmatrix} 1 & 0 & 0 \\ 0 & 1 & 0 \\ 0 & 0 & 2 \end{bmatrix} \begin{bmatrix} 1 \\ 2 \\ 0 \end{bmatrix} = \begin{bmatrix} 1 \\ 2 \\ 0 \end{bmatrix}$$

$$C_p = P\tilde{C} = \begin{bmatrix} 0.5 \\ 1.5 \\ -1 \end{bmatrix}.$$

Then we can get the value of $\gamma = 1$.

$$\tilde{x}^{new} = \tilde{x}^{old} + \left(\frac{\alpha}{\gamma}\right) C_p = \begin{bmatrix} 1.35 \\ 2.05 \\ 0.3 \end{bmatrix}$$

$$x^{new} = D\tilde{x}^{new} = \begin{bmatrix} 1.35 \\ 2.05 \\ 0.6 \end{bmatrix}$$

$$Z^{new} = C'x^{new} = \begin{bmatrix} 1 & 2 & 0 \end{bmatrix} \begin{bmatrix} 2.35 \\ 2.05 \\ 0.6 \end{bmatrix} = 5.45$$

Δ objective $= Z^{new} - Z^{old} = 5.45 - 3.0 = 2.45 > \varepsilon.$

Then we go to the second iteration.

Second iteration

$$x = [1.35, 2.05, 0.6]^t$$

$$D = \text{diag}(x) = \begin{bmatrix} 1.35 & 0 & 0 \\ 0 & 2.05 & 0 \\ 0 & 0 & 0.6 \end{bmatrix}$$

$$\tilde{x} = D^{-1}x = \begin{bmatrix} 1.35 & 0 & 0 \\ 0 & 2.05 & 0 \\ 0 & 0 & 0.6 \end{bmatrix}^{-1} \begin{bmatrix} 1.35 \\ 2.05 \\ 0.6 \end{bmatrix}$$

$$= \begin{bmatrix} (1/1.35) & 0 & 0 \\ 0 & (1/2.05) & 0 \\ 0 & 0 & (1/0.6) \end{bmatrix} \begin{bmatrix} 1.35 \\ 2.05 \\ 0.6 \end{bmatrix} = \begin{bmatrix} 1 \\ 1 \\ 1 \end{bmatrix}$$

$$\tilde{A} = AD = \begin{bmatrix} 1 & 1 & 1 \end{bmatrix} \begin{bmatrix} 1.35 & 0 & 0 \\ 0 & 2.05 & 0 \\ 0 & 0 & 0.6 \end{bmatrix} = [1.35 \quad 2.05 \quad 0.6].$$

The projection area p,

$$P = I - \tilde{A}^t(\tilde{A}\tilde{A}^t)^{-1}\tilde{A} = \begin{bmatrix} 0.7146 & -0.4334 & -0.1269 \\ -0.4334 & 0.3418 & -0.1926 \\ -0.1269 & -0.1926 & 0.9436 \end{bmatrix}$$

$$\bar{C} = DC = \begin{bmatrix} 1.35 & 0 & 0 \\ 0 & 2.05 & 0 \\ 0 & 0 & 0.6 \end{bmatrix} \begin{bmatrix} 1 \\ 2 \\ 0 \end{bmatrix} = \begin{bmatrix} 1.35 \\ 4.1 \\ 0 \end{bmatrix}$$

$$C_p = P\tilde{C} = \begin{bmatrix} -0.8124 \\ 0.8163 \\ -0.9611 \end{bmatrix}.$$

Then we get the value of $\gamma = 0.9611$.

$$\tilde{x}^{new} = \tilde{x}^{old} + \left(\frac{\alpha}{\gamma}\right) C_p = \begin{bmatrix} 0.4083 \\ 0.5945 \\ 0.3 \end{bmatrix}$$

$$x^{new} = D\tilde{x}^{new} = \begin{bmatrix} 0.5512 \\ 3.2688 \\ 0.18 \end{bmatrix}$$

$$Z^{new} = C^t x^{new} \begin{bmatrix} 1 & 2 & 0 \end{bmatrix} \begin{bmatrix} 0.5512 \\ 3.2688 \\ 0.18 \end{bmatrix} = 7.0888$$

Δ objective $= Z^{new} - Z^{old} = 7.0888 - 5.45 = 1.6388 > \varepsilon.$

Then we go to the third iteration.

Third iteration

$$x = \begin{bmatrix} 0.5512 \\ 3.2688 \\ 0.18 \end{bmatrix} \qquad \gamma = 0.5328$$

$$x^{new} = \begin{bmatrix} 0.1654 \\ 3.7383 \\ 0.0963 \end{bmatrix} \qquad Z = 7.6421$$

Δ objective $= Z^{new} - Z^{old} = 7.6421 - 7.0888 = 0.5533 > \varepsilon.$

Then we go to the fourth iteration.

Fourth iteration

$$x = \begin{bmatrix} 0.1654 \\ 3.7383 \\ 0.0963 \end{bmatrix} \qquad \gamma = 0.1923$$

$$x^{new} = \begin{bmatrix} 0.0661 \\ 3.905 \\ 0.0289 \end{bmatrix} \qquad Z = 7.9577$$

$$\Delta \text{ objective} = Z^{\text{new}} - Z^{\text{old}} = 7.9577 - 7.6421 = 0.0816 < \varepsilon.$$

Then we can stop here with a result

$$x_1 = 0.0661, \quad x_2 = 3.9689, \quad x_3 = 0.0112, \quad \text{and} \quad Z = 7.9577.$$

Example 6.10.2

Consider the following problem

Maximize $z = 3x_1 + x_2$

Subject to

$x_1 + x_2 \leq 4$

$x_j \leq 0.$

Starting from the initial point $(1,2)$ solve the problem using the IP algorithm.
  Based on the algorithm shown in Section 6.6 and following the flowchart in Figure 6.2, we can say $Z = 3x_1 + x_2 = C^t x \Rightarrow C^t = [3 \quad 1]$ as $x = [x_1, x_2]^t$
Subject to $Ax = b \Rightarrow A = [1 \quad 1]$.
We are going to take $\alpha = 0.7$, $\varepsilon = 0.01$.

First iteration
  As an initial point, we start with $x = [1, 2]^t$. Substitute in the objective function

$$Z = C^t x = [3 \quad 1]\begin{bmatrix} 1 \\ 2 \end{bmatrix} = 5.0$$

$$D = \text{diag}(x) = \begin{bmatrix} 1 & 0 \\ 0 & 2 \end{bmatrix}$$

$$\bar{x} = D^{-1}x = \begin{bmatrix} 1 & 0 \\ 0 & 2 \end{bmatrix}^{-1}\begin{bmatrix} 1 \\ 2 \end{bmatrix} = \begin{bmatrix} 1 & 0 \\ 0 & 0.5 \end{bmatrix}\begin{bmatrix} 1 \\ 2 \end{bmatrix} = \begin{bmatrix} 1 \\ 1 \end{bmatrix}$$

$$\tilde{A} = AD = [1 \quad 1]\begin{bmatrix} 1 & 0 \\ 0 & 2 \end{bmatrix} = [1 \quad 2].$$

The projection area p

$$P = I - \tilde{A}^t(\tilde{A}\tilde{A}^t)^{-1}\tilde{A} = \begin{bmatrix} 0.8 & -0.4 \\ -0.4 & 0.2 \end{bmatrix}$$

$$\tilde{C} = DC = \begin{bmatrix} 1 & 0 \\ 0 & 2 \end{bmatrix}\begin{bmatrix} 3 \\ 1 \end{bmatrix} = \begin{bmatrix} 3 \\ 2 \end{bmatrix}$$

$$C_p = P\tilde{C} = \begin{bmatrix} 1.6 \\ -0.8 \end{bmatrix}.$$

Then we get the value of $\gamma = 0.8$.

$$\tilde{x}^{new} = \tilde{x}^{old} + \left(\frac{\alpha}{\gamma}\right)C_p = \begin{bmatrix} 2.4 \\ 0.3 \end{bmatrix}$$

$$x^{new} = D\tilde{x}^{new} = \begin{bmatrix} 2.4 \\ 0.6 \end{bmatrix}$$

$$Z^{new} = C^t x^{new} = \begin{bmatrix} 3 & 1 \end{bmatrix} \begin{bmatrix} 2.4 \\ 0.6 \end{bmatrix} = 7.8$$

$$\Delta \text{ objective} = Z^{new} - Z^{old} = 7.8 - 5.0 = 2.8 > \varepsilon.$$

Then we go to the second iteration.

Second iteration

$$x = [2.4, 0.6]^t$$

$$D = \text{diag}(x) = \begin{bmatrix} 2.4 & 0 \\ 0 & 0.6 \end{bmatrix}$$

$$\tilde{x} = D^{-1}x = \begin{bmatrix} 2.4 & 0 \\ 0 & 0.6 \end{bmatrix}^{-1} \begin{bmatrix} 2.4 \\ 0.6 \end{bmatrix} = \begin{bmatrix} 1 \\ 1 \end{bmatrix}$$

$$\tilde{A} = AD = \begin{bmatrix} 1 & 1 \end{bmatrix} \begin{bmatrix} 2.4 & 0 \\ 0 & 0.6 \end{bmatrix} = \begin{bmatrix} 2.4 & 0.6 \end{bmatrix}.$$

The projection area P

$$P = I - \tilde{A}^t(\tilde{A}\tilde{A}^t)^{-1}\tilde{A} = \begin{bmatrix} 0.0588 & -0.2353 \\ -0.2353 & 0.9412 \end{bmatrix}$$

$$\bar{C} = DC = \begin{bmatrix} 7.2 \\ 0.6 \end{bmatrix}$$

$$C_p = P\tilde{C} = \begin{bmatrix} 0.2824 \\ -1.1294 \end{bmatrix}.$$

Then we get the value of $\gamma = 1.1294$.

$$\tilde{x}^{new} = \tilde{x}^{old} + \left(\frac{\alpha}{\gamma}\right)C_p = \begin{bmatrix} 1.2471 \\ 0.0118 \end{bmatrix}$$

$$x^{new} = D\tilde{x}^{new} = \begin{bmatrix} 2.9929 \\ 0.0071 \end{bmatrix}$$

$$Z^{new} = C^t x^{new} = \begin{bmatrix} 3 & 1 \end{bmatrix} \begin{bmatrix} 2.9929 \\ 0.0071 \end{bmatrix} = 8.9859$$

$$\Delta \text{ objective} = Z^{new} - Z^{old} = 8.9859 - 7.8 = 1.1859 > \varepsilon.$$

Then we go to the third iteration.

Third iteration

$$x = [2.9929, 0.0071]^t$$

$$D = \text{diag}(x) = \begin{bmatrix} 2.9929 & 0 \\ 0 & 0.0071 \end{bmatrix}$$

$$\tilde{x} = D^{-1}x = \begin{bmatrix} 1 \\ 1 \end{bmatrix}$$

$$\tilde{A} = AD = [2.9929 \quad 0.0071].$$

The projection area P,

$$P = I - \tilde{A}^t(\tilde{A}\tilde{A}^t)^{-1}\tilde{A} = \begin{bmatrix} 0.0 & -0.0024 \\ -0.0024 & 1.0 \end{bmatrix}$$

$$\tilde{C} = DC = \begin{bmatrix} 8.9788 \\ 0.0071 \end{bmatrix}$$

$$C_p = P\tilde{C} = \begin{bmatrix} 0.00 \\ -0.0142 \end{bmatrix}.$$

Then we get the value of $\gamma = 0.0142$.

$$\tilde{x}^{new} = \tilde{x}^{old} + \left(\frac{\alpha}{\gamma}\right)C_p = \begin{bmatrix} 1.0017 \\ 0.2980 \end{bmatrix}$$

$$x^{new} = D\tilde{x}^{new} = \begin{bmatrix} 2.9979 \\ 0.0021 \end{bmatrix}$$

$$Z^{new} = C^t x^{new} = [3 \quad 1]\begin{bmatrix} 2.9979 \\ 0.0021 \end{bmatrix} = 8.9958$$

$$\Delta \text{ objective} = Z^{new} - Z^{old} = 8.9958 - 8.9859 = 0.0099 < \varepsilon.$$

Then, we stop up to the third iteration

$$x_1 = 2.9979, \quad x_2 = 0.0021, \quad \text{and} \quad Z = 8.9958.$$

Example 6.10.3

The following optimization problem demonstrates the primal affine-scaling algorithm.

Minimize $Z = 2x_1 + x_2 + 4x_3$

Subject to

$x_1 + x_2 + 2x_3 = 3$

$2x_1 + x_2 + 3x_3 = 5$

$x_i \geq 0 \quad (i = 1, 2, 3).$

First iteration
  Let

$$e = \begin{bmatrix} 1 \\ 1 \\ 1 \end{bmatrix}, \quad x^0 = \begin{bmatrix} 1.5 \\ 0.5 \\ 0.5 \end{bmatrix},$$

$$D_0 = \text{diag}(x^0) = \begin{bmatrix} 1.5 & 0.0 & 0.0 \\ 0.0 & 0.5 & 0.0 \\ 0.0 & 0.0 & 0.5 \end{bmatrix}.$$

Therefore, the dual estimate vector is

$$w^0 = (AD_0^2 A^T)^{-1} AD_0^2 c = \begin{bmatrix} 0.8947 \\ 0.5789 \end{bmatrix},$$

and the reduced cost coefficient is

$$r^0 = \begin{bmatrix} -0.0526 \\ -0.4737 \\ 0.4737 \end{bmatrix}.$$

For the optimality check, we calculate

$$e^T = D_0 r^0 = -0.0789 \quad \text{and} \quad d_y^0 = \begin{bmatrix} 0.0789 \\ 0.2368 \\ -0.2368 \end{bmatrix}.$$

The optimality condition is not satisfied but the problem is not unbounded; therefore

$$\beta_0 = 4.1807.$$

The update on the primal variable is

$$x^1 = x^0 + \beta_0 D_0 d_y^0 = \begin{bmatrix} 1.9951 \\ 0.9951 \\ 0.0049 \end{bmatrix}.$$

Second iteration

$$D_1 = \begin{bmatrix} 1.9951 & 0.0000 & 0.0000 \\ 0.0000 & 0.9951 & 0.0000 \\ 0.0000 & 0.0000 & 0.0049 \end{bmatrix}$$

$$w^1 = \begin{bmatrix} 0 \\ 1.0 \end{bmatrix} \quad \text{and} \quad r^1 = \begin{bmatrix} 0 \\ 0 \\ 1 \end{bmatrix}.$$

For the optimality checking, we compute

$$e^T D_1 r^1 = 0.0049$$

$$d_y^1 = \begin{bmatrix} 0.0 \\ 0.0 \\ -0.0049 \end{bmatrix},$$

which implies that the problem is bounded.

Therefore, we compute $\beta_1 = 202.0408$ and update the primal variable to get

$$x^2 = \begin{bmatrix} 2.0 \\ 1.0 \\ 0.0 \end{bmatrix} = x^*,$$

which yields an optimal value of objective value $Z^* = 5$.

Example 6.10.4

The following optimization problem demonstrates the dual affine-scaling algorithm.

Maximize $Z = 15x_1 + 15x_2$

Subject to

$1.5x_1 + 5x_1 = -3.0$

$-1.5x_1 + 1.5x_2 + 5x_2 = 1.5$

$1.5x_3 + s_3 = 0$

$1.5x_4 + s_4 = 0$

$s_i \geq 0 \quad (i = 1, 2, 3, 4).$

First iteration

$$x^0 = \begin{bmatrix} -3 \\ -3 \end{bmatrix}, \quad s^0 = \begin{bmatrix} 1.5 \\ 1.5 \\ 4.5 \\ 4.5 \end{bmatrix},$$

$$W^0 = \text{diag}(s^0) = \begin{bmatrix} 1.5 & 0 & 0 & 0 \\ 0 & 1.5 & 0 & 0 \\ 0 & 0 & 4.5 & 0 \\ 0 & 0 & 0 & 4.5 \end{bmatrix}.$$

Therefore, the direction of translation is

$$d_x^0 = (A^t W_0^{-2} A)^{-1} c = \begin{bmatrix} 23.5321 \\ 34.6789 \end{bmatrix},$$

$$d_s^0 = -A^T d_x^0 = \begin{bmatrix} -35.2982 \\ -16.7202 \\ -35.2982 \\ -52.0183 \end{bmatrix},$$

$$y^0 = \begin{bmatrix} 15.6881 \\ 7.4312 \\ 1.7431 \\ 2.5688 \end{bmatrix}.$$

For the optimality check, we calculate

$$b^T y^0 - c^T X^0 = 46.2385.$$

The optimality condition is not satisfied but the problem is not unbounded; therefore

$$\beta_0 = 0.0421.$$

The update on the primal variable is

$$x^1 = x^0 + \beta_0 d_x^0 = \begin{bmatrix} -2.01 \\ -1.5411 \end{bmatrix}$$

$$s^1 = \begin{bmatrix} 0.015 \\ 0.796 \\ 3.015 \\ 2.312 \end{bmatrix}.$$

Second iteration

$$d_x^1 (A^T W_1^{-2} A)^{-1} c = \begin{bmatrix} 0.0028 \\ 3.7838 \end{bmatrix}$$

$$d_s^1 = -A^T d_x^1 = \begin{bmatrix} -0.0043 \\ -5.6714 \\ -0.0043 \\ -5.6757 \end{bmatrix}$$

$$y^1 = \begin{bmatrix} 18.9374 \\ 8.9378 \\ 0.0005 \\ 1.0622 \end{bmatrix}.$$

For the optimality check, we calculate

$$b^T y^1 - c^T x^1 = 0.3918.$$

The optimality condition is not satisfied but the problem is not unbounded; therefore

$$\beta_0 = 0.1391.$$

The update on the primal variable is

$$x^2 = x^1 + \beta_1 d_x^1 = \begin{bmatrix} -2.0096 \\ -1.0149 \end{bmatrix}$$

$$s^2 = \begin{bmatrix} 0.0144 \\ 0.0079 \\ 3.0144 \\ 1.5224 \end{bmatrix}.$$

The reader may carry out more iterations and verify that the optimal value is assumed at $X^* = (-2, -1)^T$ and $s^* = (0, 0, 3, 1.5)^T$.

Example 6.10.5

Using IP method, solve the following LP problem:

Maximize $\quad 3x_1 + 3x_2 - x_3$
subject to $\quad 2x_1 - 3x_2 + x_3 = 0$
$$x_1 + x_2 + x_3 = 1$$
$$\forall j \in \{1,2,3\}$$

SOLUTION
The problem is formulated in standard form as

Max $z = 3x_1 + 3x_2 - x_3 = C^T x$

$C^T = \begin{bmatrix} 3 & 3 & -1 \end{bmatrix}$ as $x = [x_1, x_2, x_3]^T$

subject to $Ax = b \Rightarrow A = \begin{bmatrix} 2 & -3 & 1 \\ 1 & 1 & 1 \end{bmatrix}$

And using $\alpha = 0.7$ and $\varepsilon = 0.1$

As an initial point, we start by $z = C^T x = \begin{bmatrix} 3 & 3 & -1 \end{bmatrix} \begin{bmatrix} 1 \\ 1 \\ 2 \end{bmatrix} = 4.0, x = [1, 1, 2]^T$

$$z = C^T x = \begin{bmatrix} 3 & 3 & -1 \end{bmatrix} \begin{bmatrix} 1 \\ 1 \\ 2 \end{bmatrix} = 4.0, \quad D = \text{Diag } x = \begin{bmatrix} 1 & 0 & 0 \\ 0 & 1 & 0 \\ 0 & 0 & 2 \end{bmatrix}$$

$$\tilde{x} = D^{-1}x = \begin{bmatrix} 1 & 0 & 0 \\ 0 & 1 & 0 \\ 0 & 0 & 0.5 \end{bmatrix} \begin{bmatrix} 1 \\ 1 \\ 2 \end{bmatrix} = \begin{bmatrix} 1 \\ 1 \\ 1 \end{bmatrix}$$

$$\tilde{A} = AD = \begin{bmatrix} 2 & -3 & 1 \\ 1 & 1 & 1 \end{bmatrix} \begin{bmatrix} 1 & 0 & 0 \\ 0 & 1 & 0 \\ 0 & 0 & 2 \end{bmatrix} = \begin{bmatrix} 2 & -3 & 2 \\ 1 & 1 & 2 \end{bmatrix}$$

The projection matrix is calculated as

$$P = I - \tilde{A}^T(\tilde{A}\tilde{A}^T)^{-1}\tilde{A}$$

$$\Rightarrow P = \begin{bmatrix} 0.0025 & -0.0045 & 0 \\ 0 & -0.023 & 0 \\ 0 & 0 & 1 \end{bmatrix}$$

$$\tilde{c} = DC = \begin{bmatrix} 1 & 0 & 0 \\ 0 & 1 & 0 \\ 0 & 0 & 2 \end{bmatrix} \begin{bmatrix} 3 \\ 3 \\ -1 \end{bmatrix} = \begin{bmatrix} -0.006 \\ -0.069 \\ -2 \end{bmatrix}$$

$$\therefore \gamma = |-2| = 2, \quad \alpha = 0.7$$

$$\therefore \tilde{x}_{new} = \tilde{x}_{old} + \left(\frac{\alpha}{\gamma}\right)C_p = \begin{bmatrix} 0 \\ 1 \\ 0 \end{bmatrix} + \frac{0.7}{2} \begin{bmatrix} -0.006 \\ -0.069 \\ -2 \end{bmatrix} = \begin{bmatrix} -0.0021 \\ 0.97585 \\ -0.7 \end{bmatrix}$$

$$x_{new} = D\tilde{x}_{new} = \begin{bmatrix} 1 & 0 & 0 \\ 0 & 1 & 0 \\ 0 & 0 & 2 \end{bmatrix} \begin{bmatrix} -0.0021 \\ 0.97585 \\ -0.7 \end{bmatrix} = \begin{bmatrix} -0.0021 \\ 0.97585 \\ -1.4 \end{bmatrix}$$

$$\text{Profit}^{new} = C^T x_{new} = [3 \quad 3 \quad -1] \begin{bmatrix} -0.0021 \\ 0.97585 \\ 1.4 \end{bmatrix} = 4.32125$$

After the second iteration, the value of x was found to be $x^* = \begin{bmatrix} -0.021 \\ 0.9758 \\ -0.7 \end{bmatrix}$ and the

profit $P^* = 3.621$.

Example 6.10.6

Recall the linear model

Maximize $f(x) = c^T x$

Subject to $Ax = b$

From the given problem, the following vectors and matrices are identified:

$$x = [x_1 \quad x_2 \quad x_3]^T, \quad c^T = [2 \quad 5 \quad 7], \quad A = [1 \quad 2 \quad 3], \quad \text{and} \quad b = 6.$$

First iteration

Initialize: $x^{(0)} = [1 \quad 1 \quad 1]^T$ such that $Ax = b$

Choosing: $\alpha = 0.95$, where $0 < \alpha < 1$, $\varepsilon = 10^{-3}$

Now, the scaled gradient vector is $\tilde{c} = DC = [2 \quad 5 \quad 7]^T$

And from the problem, $A = [1 \quad 2 \quad 3]$ such that $\tilde{A} = AD = AI = A$

The projection matrix is calculated as

$$P = I - \tilde{A}^T(\tilde{A}\tilde{A}^T)^{-1}\tilde{A} = \begin{bmatrix} 0.9286 & -0.1429 & -0.2143 \\ -0.1429 & 0.7143 & -0.4286 \\ -0.2143 & -0.4286 & 0.3571 \end{bmatrix}$$

The projection gradient is

$$C_p = P\tilde{c} = \begin{bmatrix} 0.9286 & -0.1429 & -0.2143 \\ -0.1429 & 0.7143 & -0.4286 \\ -0.2143 & -0.4286 & 0.3571 \end{bmatrix} \begin{bmatrix} 2 \\ 5 \\ 7 \end{bmatrix} = \begin{bmatrix} -0.3571 \\ +0.2857 \\ -0.0714 \end{bmatrix}$$

Scalar value $\gamma = |\text{Min } C_{p,i}| = |-0.3571| = 0.3571$ for $C_{p,i} < 0$

Updating the scaled solution vector yields:

$$\tilde{x}^{new} = \tilde{x}^{old} + \frac{\alpha C_p}{\gamma}$$

$$\tilde{x}^{new} = \begin{bmatrix} 1 \\ 1 \\ 1 \end{bmatrix} + \frac{0.95}{0.3571} \begin{bmatrix} -0.3571 \\ 0.28570 \\ -0.07140 \end{bmatrix} = \begin{bmatrix} 0.0500 \\ 1.7600 \\ 0.8100 \end{bmatrix}$$

Hence, the scaled solution vector is

$$x^{new} = D\tilde{x}^{old} = \begin{bmatrix} 1 & 0 & 0 \\ 0 & 1 & 0 \\ 0 & 0 & 1 \end{bmatrix} \begin{bmatrix} 0.0500 \\ 1.7600 \\ 0.8100 \end{bmatrix} = \begin{bmatrix} 0.0500 \\ 1.7600 \\ 0.8100 \end{bmatrix}$$

The value of the objective function at this stage is $f(x^{new}) = 14.5700$

After another iteration, the optimal solution was found to be $x^* = \begin{bmatrix} 0.0461 \\ 2.9162 \\ 0.0405 \end{bmatrix}$

and $f^* \approx 14.9567$.

6.11 Conclusions

Variants of IP algorithms were presented. These variants included work by Karmarkar, projection, offline-scaling, and the primal-affine algorithm. These methods were shown in Sections 6.2 through 6.5.

In Section 6.6, the barrier algorithm was presented, where a barrier-function tests hit inequality constraint methods by creating a barrier function which is a combination of the original objective function and a weighted sum of functions with a positive singularity at the boundary. The formulation and algorithm were presented in this section.

In Section 6.7, an extended IP for the LP problem was presented and a discussion of the possible interior sequence was presented in Section 6.8 where the optimality conditions and start and termination of the recursive process were explained. In Section 6.9, an extended quadratic programming algorithm for solving quadratic optimization problems was presented.

6.12   Problem Set

PROBLEM 6.12.1

Solve the unconstrained problem:

$$\text{Minimize } z = \frac{1}{3}x_1^2 + \frac{1}{2}x_2^2 - x_1x_2 - 2x_1,$$

1. Using the IP method
2. Any other method

PROBLEM 6.12.2

Solve the following problem using the quadratic IP method:

$\text{Minimize } z = 2x_1^2 + 3x_2^2 + 5x_3^2 + x_1 + 2x_2 - 3x_3$

Subject to

$x_1 + x_2 = 5$

$x_1 + x_3 = 10$

$x_j \geq 0.$

PROBLEM 6.12.3

Consider the following problem.

$\text{Maximize } z = 2x_1 + 5x_2 + 7x_3$

Subject to

$x_1 + 2x_2 + 3x_3 = 6$

$x_j \geq 0.$

1. Graph the feasible region.
2. Find the gradient of the objective function and then find the projected gradient onto the feasible region.

3. Starting from initial trial solution $(1, 1, 1)$ perform two iterations of the IP algorithm.

4. Perform eight additional iterations.

PROBLEM 6.12.4

Consider the following problem.

Maximize $z = -x_1 - x_2$

Subject to

$x_1 + x_2 \leq 8$

$x_2 \geq 3$

$-x_1 + x_2 \leq 2$

$x_j \geq 0.$

1. Solve this problem graphically.
2. Use the dual simplex method to solve this problem.
3. Trace graphically the path taken by the dual simplex method.
4. Solve this problem using the IP algorithm.

PROBLEM 6.12.5

Maximize $3x_1 + 3x_2 - x_3$

s.t. $2x_1 - 3x_2 + x_3 = 0$

$x_1 + x_2 + x_3 = 1$

PROBLEM 6.12.6

Maximize $4x_1 + 3x_2$

s.t. $3x_1 + 10x_2 \leq 8$

$4x_1 + 2x_2 \leq 15$

PROBLEM 6.12.7

Maximize $10x_1 + 4x_2 + 3x_3$

s.t. $x_1 - x_3 \leq 0$

$x_1 + x_2 \leq 2$

$-x_1 + x_2 + 3x_3 \leq 4$

$x_1, x_2, x_3 \geq 0$

PROBLEM 6.12.8

Maximize $2x_1 + 3x_2 + 4x_3$

Subject to

$x_1 + x_2 + x_3 \geq 5$

$x_1 + 2x_2 = 7$

$5x_1 - 2x_2 + 3x_3 \leq 9$

$x_1, x_2, x_3 \geq 0$

PROBLEM 6.12.9

Maximize $z = 3x_1 + 2x_2$

Subject to

$x_1 + 2x_2 + x_3 = 6$

$2x_1 + x_2 + x_4 = 8$

$-x_1 + x_2 + x_5 = 1$

$x_2 + x_6 = 2$

$x_1, x_2, \ldots, x_6 \geq 0$

References

1. Alder, I., Karmarkar, N., Resende, M. G. C., and Veiga, G., An implementation of Karmarkar's algorithm for linear programming, Working Paper, Operations Research Center, University of California, Berkeley, 1986 (also in *Mathematical Programming*, 44).
2. Alder, I., Resende, M. G. C., Veiga, G., and Karmarkar, N., An implementation of Karmarkar's algorithm, *ORSA Journal on Computing*, 1(2), 1989, 84–106.
3. Anstreicher, K. M. A., Monotonic projective algorithm for fractional linear programming, *Algorithmica*, 1, 1986, 483–498.
4. Barnes, E. R., A variation of Karmarkar's algorithm for solving linear programming problems, *Mathematical Programming 1986 for Computing Projections*, 13th International Mathematical Programming Symposium, Tokyo, August 1988.
5. Carpenter, J., Contribution a l'Etude du Dispatch Economique, *Bulletin de la Société Francaise des Electriciens*, 3, 1962, 431–447.
6. Dommel, H. W. and Tinney, W. F., Optimal power flow solutions, *IEEE Transactions on Power Apparatus and Systems*, 87, 1968, 1866–1878.
7. Galliana, F. D. and Hunneault, M., A survey of the optimal power flow literature, *IEEE Transactions on Power Systems*, 6, 1991, 762–770.
8. Karmarkar, N., New polynomial-time algorithm for linear programming, *Combinatorica*, 4, 1984, 373–397.

9. Karmarkar, N., *A New Polynomial-time Algorithm for Linear Programming*, Combinatorica, 1984, 373–395.
10. Kojima, M., Determining basic variables of optimal solutions in Karmarkar's new LP algorithm, *Algorithmica*, 1, 1986, 499–517.
11. Kozlov, A., The Karmarkar algorithm: Is it for real? *SIAM News*, 18(6), 1987, 1–4.
12. McShane, K. A., Monma, C. L., and Shanno, D., An implementation of a primal–dual interior point method for linear programming, *ORSA J. Computing*, Spring 1989, 70–83.
13. Meggido, N., On finding primal- and dual-optimal bases, Research Report, RJ 6328 (61997), IBM, Yorktown Heights, New York, 1988.
14. Momoh, J. A., Application of quadratic interior point algorithm to optimal power flow, EPRI Final Report RP 2473-36 II, Howard University, March, 1992.
15. Momoh, J. A., Austin, R., and Adapa, R., Feasibility of interior point method for VAR planning, accepted for publication in *IEEE SMC*, 1993.
16. Momoh, J. A., Guo, S. X., Ogbuobiri, E. C., and Adapa, R., The quadratic interior point method solving power system security-constrained optimization problems, Paper No. 93, SM 4T7-BC, Canada, July 18–22, 1993.
17. Ponnambalam, K., New starting and stopping procedures for the dual affine method, Working Paper, Department of Civil Engineering, University of Waterloo, Waterloo, Ontario, Canada, 1988.
18. Reid, G. F. and Hasdorf, L., Economic dispatch using quadratic programming, *IEEE Transactions on Power Apparatus and Systems*, PAS-92, 1973, 2017–2023.
19. Sun, D. I., Ashely, B. B., Hughes, A., and Tinney, W. F., Optimal power flow by Newton method, *IEEE Transactions on Apparatus and Systems*, PAS-103, 1984, 2864–2880.
20. Vanderbei, R. J., Meketon, M. S., and Freedman, B. A., A modification of Karmarkar's linear programming algorithm, *Algorithmica*, 1, 1986, 395–409.
21. Vannelli, A., An adaptation of the interior point method for solving the global routine problem, *IEEE Transactions on Computer-Aided Design*, 10(2) 1991, 193–203.
22. Ye, Y. and Kojima, M., Recovering optimal dual solutions in Karmarkar's polynomial-time algorithm for linear programming, *Mathematical Programming*, 39, 1987, 307–318.
23. Kojima, M., Determining basic variables of optimal solutions in Karmarkar's new LP algorithm, *Algorithmica*, 1(4), 1986, 449–517.
24. Montiero, R. C. and Adler, I., Interior path following primal–dual algorithms. Part I: Linear programming, *Mathematical Programming*, 44, 1989, 27–42.
25. Dantzig, G. F., *Linear Programming and Extensions*, Princeton University Press, Princeton, NJ, 1963.
26. Lustig, I. J., Marsten, R. E., and Shanno, D. F., Interior point methods for linear programming: computational state of the art, *ORSA Journal on Computing*, 6(1), 1994, 1–14.
27. Liu, M. Tso, S. K., and Cheng, Y., An extended nonlinear primal–dual interior-point algorithm for reactive-power optimization of large-scale power systems with discrete control variables, *IEEE Transactions on Power Systems*, 17(4), 2002, 982–991.

7

Nonlinear Programming

7.1 Introduction

While linear programming has found numerous practical applications, the assumptions of proportionality, additivity, and other forms of nonlinearity are common in many engineering applications. Most often the sources of non-linearity are the physical process and the associated engineering principles that are not amenable to linearization. Even though there are linearization schemes, they are subject to large errors in representing the phenomenon.

Nonlinear programming (NLP) aims to solve optimization problems involving a nonlinear objective and constraint functions [1–4,10–13]. The constraints may consist of equality and inequality forms. The inequalities may be specified by two bounds: bounded below and bounded above. There is no generalized approach to solve the NLP problem and a particular algorithm is usually employed to solve the specified type of problem. In other words, it is different from the simplex method which can be applied to any LP problem. However, two methods, namely, sensitivity and barrier, are considered to be quite generalized to be able to successfully solve the NLP. These methods are discussed in detail in this chapter.

Theorems on necessary and sufficient conditions are given for extremizing unconstrained functions and optimizing constrained functions. In conjunction with the necessary condition, the well-known Kuhn–Tucker (K–T) conditions are treated. Finally, based on the K–T conditions, sensitivity and barrier methods are developed for a general approach to solve the NLP problems. The methods are designed for solving NLP involving large numbers of variables such as the power system.

7.2 Classification of NLP Problems

7.2.1 NLP Problems with Nonlinear Objective Function and Linear Constraints

This is a relatively simple problem with nonlinearity limited only to the objective function. The search space is similar to that of the linear programming problem and the solution methods are developed as extensions to the simplex method.

7.2.2 Quadratic Programming

This is a special case of the former where the objective function is quadratic (i.e., involving the square or cross-product of one or more variables). Many algorithms have been developed with the additional assumption that the objective function is convex, which is a direct extension of the simplex method. Apart from being a very common form for many important problems, quadratic programming (QP) is also very important because many of the problems in Section 7.3.1 are often solved as a series of QP or sequential quadratic programming (SQP) problems.

Objective form:

$$f(x) = \sum c_j x_j + \frac{1}{2} \sum_{i=1}^{n} \sum_{j=1}^{n} g_{ij} x_i x_j.$$

Subject to

$$\sum a_{ij} x_j \le b_i$$

$$x_j \ge 0,$$

where

c_j, a_{ij} and b_i and a_{ij} are assumed to be known

$a_{ij} = a_{ji}$ are symmetrical and hence $= a_{ij} = \frac{1}{2}[a_{ij} + a_{ji}]$

7.2.3 Convex Programming

Convex programming arises out of the assumptions of convexity of the objective and constraint functions. Under these assumptions, it can encompass both foregoing problems. The major point of emphasis is that the local optimal point is necessarily for the global optimum under these assumptions.

7.2.4 Separable Programming

Separable programming is a special class of convex programming with the additional assumption that all objective and constraint functions are separable functions; that is, the function can be expressed as a sum of the functions of the individual variables. For example, if $f(x)$ is a separable function it can be expressed as

$$f(x) = \sum_{j=1}^{n} f_j(x_j),$$

where each, $f_j(x_j)$ includes a term involving x_j only.

Subject to

$$\sum g_{ij}(x_j) \leq 0.$$

These problems are called separable programming because the decisive variable approximate separate one in each function g_{ij} in the constraints and one in the each objective function.

7.3   Sensitivity Method for Solving NLP Variables

For simplicity, we hereafter use $i \leq p$ to mean $i = 1, 2, \ldots, p$, $i > p$ to mean $i = P+1, p+2, \ldots, m$, and all i to mean $i = 1, 2, \ldots, m$. Let y be an n-vector and $f(y)$ together with $f_i(y)$ be scalar functions for all i. Then, the NLP problem is defined to

Minimize $f(y)$

Subject to

$$C_i \leq f_i(y) \leq D_i \quad \text{for all } i, \tag{7.1}$$

where

$$C_i = D_i \quad \text{for } i \leq p \quad \text{but } C_i < D_i \quad \text{for } i > p.$$

There are p equality constraints and $m- p$ inequality constraints which are bounded by C_i and D_i for $i > p$. The number of p is less than or equal to n but m may be greater than n. By $C_i = -\infty$ and $D_i = \infty$, we mean that the inequality constraint is bounded above and below, respectively. Any constraint with one bound is thus a special case of the inequality constraint.

The constraints are denoted by the equality form

$$f_i(y) = k_i \tag{7.2}$$

for all i where $K_i = D_i$, when $i \leq p$ and $C_i \leq K_i \leq D_i$, when $i > p$. In matrix form, they are denoted collectively by

$$F(y) = K, \tag{7.3}$$

where

$F(y) = [f_1(y), f_2(y), \ldots, f_m(y)]^T$
$K = [K_1, K_2, \ldots, K_m]^T$

The Lagrange function for this problem is defined as before by the scalar function

$$L(y,\lambda) = f(y) + \lambda F(y), \tag{7.4}$$

where $\lambda = [\lambda_1, \lambda_2, \ldots, \lambda_m]$.

The Lagrange function is assumed to be continuous up to the first partial derivatives at $y = x$ which is a minimum as considered in the theorem.

Extended K–T conditions are considered here to cover constraints bounded below and above. The conditions can be stated in theorem form as follows.

THEOREM 7.3.1

If x is a solution for the NLP problem, then it is necessary that $L_x(x,\lambda) = 0$ and one of the following conditions be satisfied for all $i > p$.

(a) $\lambda_i = 0$ when $C_i < f_i(x) < D_i$.

(b) $\lambda_i \geq 0$ when $f_i(x) = D_i$.

(c) $\lambda_i \leq 0$ when $f_i(x) = C_i$.

7.3.1 Procedure for Solving the NLP Problem

1. Use $L_x(x,\lambda) = 0$ and the equality constraints to find x for cases a, b, and c.

2. Find the smallest $f(x)$ among the three possible x obtained in step 1.

3. Use $T = L_{xx}(x,\lambda) + \beta F_x^T(x)F_x(x) > 0$ for some $\beta \geq 0$ to test sufficiency for x determined in step 2.

The conditions as imposed in the theorem are called here the Extended Kuhn–Tucker (EKT) conditions. They suggest that one can predict the changes of $f(x)$ due to variations of K if λ is known. This fact is utilized in the method to approach the EKT conditions. Note that there may exist multiple sets of EKT conditions.

Transpose Equation 7.4 and then use the column vector z to denote λ^T and U to denote L_x^T:

$$U(x,z) = f_x^T(x) + F_x^T(x)z = 0, \qquad (7.5)$$

where $z_i = \lambda_i$ for all i. Since Equation 7.3 must be satisfied for $y = x$, we have

$$F(x) = K, \qquad (7.6)$$

where $K_i = C_i = D_i$ for $i \leq p$ but is uncertain for $i > p$. The method uses a process that adjusts K_i in $C_i \leq K_i \leq D_i$ for $i > p$ to decrease $f(x)$ and keep Equations 7.4 and 7.6 satisfied.

Consider some x, z, and K that satisfy Equations 7.5 and 7.6 at one step and $x + \Delta x$, $z + \Delta z$, and $K + \Delta K$ at the next. Then, it follows that

$$U(x + \Delta x, z + \Delta z) = f_x^T(x + \Delta x) + F_x^T(x + \Delta x)(z + \Delta z) = 0$$

and $F(x + \Delta x) = K + \Delta K$.

The first-order approximation of the above equations is

$$U(x,z) + U_x(x,z)\Delta x + U_z(x,z)\Delta z = 0 \tag{7.7}$$

and

$$F^{\text{new}}(x) = F^{\text{old}}(x) + F_x(x)\Delta x = K + \Delta K. \tag{7.8}$$

The partial derivatives of $U(x, z)$ can be found from Equation 7.5 as

$$S = U_x(x,z) = f_{xx}(x) + \sum_{i=1}^{m} \lambda_i f_{ixx}(x) \tag{7.9}$$

and

$$U_z(x,z) = F_x^T(x),$$

where f_{xx} and f_{ixx} are, respectively, the second derivatives of $f(x)$ and $f_i(x)$ with respect to x; all of them are assumed to exist.

For simplicity, the augments x and z are omitted here and Equations 7.7 and 7.8 are combined in a matrix form

$$Ay = b, \tag{7.10}$$

where

$$A = \begin{bmatrix} S & F_x^T \\ F_x & 0 \end{bmatrix}, \quad y = \begin{bmatrix} \Delta x \\ \Delta z \end{bmatrix}, \quad b = \begin{bmatrix} -U \\ \Delta K \end{bmatrix}.$$

Note that the condition $U = 0$ may not be true due to first-order approximation but $F = K$ is true since K is calculated from Equation 7.6. Inclusion of U in the vector b would force U to be zero at the next step if it were not zero at the present one. Equation 7.10 shows that any two of the increments may be determined if the third one is specified. However, the change of K not only relates to the constraints but correlates with the objective function as evidenced by Equation 7.6. It is for this reason that ΔK is chosen to be the independent variable. As mentioned earlier, the basic rule of adjusting ΔK is to decrease $f(x)$ without violating the constraints.

The matrix A of Equation 7.10 may not be inverted at each step of the process; this is always the case for $m > n$. We are seeking here the least square solution with minimum norm (LSMN) for Δx and Δz. The solution always exists and is unique as long as A is not a null matrix. Moreover, it reduces automatically to the exact solution if A is nonsingular.

Expression for EKT Conditions (Sensitivity Method)

As it is, the EKT conditions are not suitable for application to a computer-based solution. An alternative expression is sought here for practical applications.

Consider an m-vector J with component J_i ($i = 1, 2, \ldots, m$) defined by

$$T = \min[\lambda_i, (D_i - K_i)] \tag{7.11}$$

and

$$J_i = \max[T, (C_i - K_i)], \tag{7.12}$$

for all $i = 1, 2, \ldots, m$. It can be concluded that a set of EKT conditions is satisfied if and only if $U = 0$ and $J = 0$. This can be shown as follows:

1. *Feasibility $K_i > D_i$:* $J_i = 0$ from Equation 7.12 requires that $T = 0$ which cannot take place in Equation 7.11 since $D_i - K_i < 0$.

   $K_i < C_i$: $J_i = 0$ from Equation 7.12 cannot occur since $C_i - K_i > 0$. Therefore, $C_i \le K_i \le D_i$ must hold when $J_i = 0$.

2. *Optimality $U = 0$* is imposed in the sensitivity method. This fulfills the first part of the EKT conditions: $L_x = 0$.

$C_i < K_i < D_i$: $J_i = 0$ from Equation 7.12 requires that $T = 0$ and hence $\lambda_i = 0$ must hold in Equation 7.11. Conversely, if $\lambda_i = 0$, then $T = 0$ in Equation 7.11 and hence $J_i = 0$ results from Equation 7.12.

$K_i = D_i$: $J_i = 0$ from Equation 7.12 requires that $T = 0$ and hence $\lambda_i \ge 0$ must hold in Equation 7.11. Conversely, if $\lambda_i \ge 0$, then $T = 0$ in Equation 7.11 and hence $J_i = 0$ results from Equation 7.12.

$K_i = C_i$: $J_i = 0$ from Equation 7.12 requires that $T \le 0$ and hence $\lambda_i \le 0$ must hold in Equation 7.11. Conversely, if $\lambda_i \le 0$, then $T \le 0$ in Equation 7.11 and hence $J_i = 0$ results from Equation 7.12.

It follows therefore that $J = 0$ and $U = 0$ are both necessary and sufficient to reach a set of EKT conditions. These conditions are used as the criteria for termination of the method. The process involved in the method may start with any guessed values of x and λ. However, different initial values may lead to different sets of EKT conditions and even divergence.

Then to compute ΔK for the algorithm:

1. Adjustment of ΔK

Consider a change Δx about a known x. The second-order approximation of the objective function can be written as

$$f(x + \Delta x) = f(x) + f_x \Delta x + \frac{1}{2} \Delta x^T f_{xx} \Delta x, \tag{7.13}$$

where the partial derivatives are evaluated at x. It is intended to reduce Equation 7.13 by a proper adjustment of ΔK. To this end, we assume that $\Delta K = qJ$, where J is determined from Equations 7.11 and 7.12. The increments Δx and Δz caused by ΔK satisfy Equation 7.10:

$$A \begin{bmatrix} \Delta x \\ \Delta z \end{bmatrix} = \begin{bmatrix} -U \\ qJ \end{bmatrix}, \tag{7.14}$$

which consists of

$$S\Delta x + F_x^T \Delta x = -U$$

and $F_x \Delta x = \Delta K = qJ$.

We define vectors u and v in such a way that $\Delta x = qu$ and $F_x^T(\Delta z - qv) = (q - 1)U$ where the last equation is satisfied in the sense of LSMN [13]. Then, by eliminating Δx and Δz, we obtain from Equation 7.14 that

$$A \begin{bmatrix} u \\ v \end{bmatrix} = \begin{bmatrix} -u \\ J \end{bmatrix}. \tag{7.15}$$

Substitution of $\Delta x = qu$ into Equation 7.13 gives

$$f(x + \Delta x) = f(x) + qf_x u + \frac{1}{2}q^2 N,$$

where

$$N = u^T f_{xx} u. \tag{7.16}$$

The minimum of $f(x + \Delta x)$ for positive q occurs at

$$q = -\frac{f_x u}{N} \tag{7.17}$$

if $f_x u < 0$ and $N > 0$. We choose q as given by Equation 7.17 only when $f_x u < 0$ and $N > 0$, and choose $\Delta K = J$ or $q = 1$ otherwise. For q not equal to one, ΔK_i is revised for all i, by

$$T = \min[qJ_i, (D_i - K_i)] \tag{7.18}$$

$$\Delta K_i = \max[T, (C_i - K_i)]. \tag{7.19}$$

Using the ΔK, we solve Δx and Δz from Equation 7.10 and then update x and z.

7.4   Algorithm for Quadratic Optimization

Since it is of practical importance, particular attention is given to the NLP that has the objective function and constraints described by quadratic forms [1,7]. This type of problem is referred to as quadratic optimization. The special case where the constraints are of linear forms is known as QP. Derivation of the sensitivity method is aimed at solving the NLP on the computer. An algorithm is generated for this purpose according to the result obtained in the previous section.

Quadratic optimization is involved in power systems [15] for maintaining a desirable voltage profile, maximizing power flow, and minimizing generation cost. These quantities are controlled by complex power generation which is usually bounded by two limits. The first two problems can be formulated with an objective function in a quadratic form of voltage while the last one is in a quadratic form of real power. Formulation of the first problem is given in the last of the illustrative problems.

As usual, we consider only minimization since maximization can be achieved by changing the sign of the objective function. Let the objective and constraint functions be expressed, respectively, by

$$f(x) = \frac{1}{2}x^T Rx + a^T x$$

and

$$f_i(x) = \frac{1}{2}x^T H_i x + b_i^T x \quad \text{for all } i. \tag{7.20}$$

R together with H are n-square and symmetrical matrices, and x, a together with b, are n-vectors. The quadratic functions are now characterized by the matrices and vectors. As defined before, the constraints are bounded by $C_i \le f_i(x) \le D_i$, for all $i = 1, 2, \ldots, m$. Among these i, the first p are equalities ($C_i = D_i$, for $i \le p$).

The matrix A and n-vector U in Equation 7.10 can be found by using

$$F_x^T = [H_1 x + b_1, H_2 x + b_2, \ldots, H_m x + b_m],$$

$$S = R + \sum_{i=1}^{m} \lambda_i H_i, \quad w = a + \sum_{i=1}^{m} \lambda_i b_i,$$

and $U = Sx + w$.

Given below is an algorithm to be implemented in a computer program.

1. *Input Data*
   (a) n, m, p, and ε (to replace zero, usually lies between 10^{-3} and 10^{-5}).
   (b) R, a, H_i, b_i, C_i, and D_i for all $i = 1, 2, \ldots, m$.

2. *Initialization* Set $x_i = 0$ and $z_i = 0$ ($\lambda_i = 0$) for all i or use any other preference.

3. *Testing EKT Conditions*

   (a) Calculate K_i, U_i (the ith component of U) and then J_i from Equations 7.11 and 7.12 for all i.

   (b) A set of EKT conditions is reached if $|U_i| < \varepsilon$ and $|J_i| < \varepsilon$ for all i. Otherwise go to step 4.

4. *Solving for u and v*

   (a) Solve u and v from Equation 7.15 by using LSMN [13].

   (b) Calculate N by Equation 7.16 and then go to step 5 if $N > 0$ and $f_x u < 0$. Go to Part (c) otherwise.

   (c) Update x by $x + u$, and $z + v$, and then go to step 3.

5. *Determining ΔK*

   (a) Calculate q by Equation 7.17 and then find ΔK_i from Equations 7.18 and 7.19 for all i.

   (b) Solve Δx and Δz from Equation 7.10 by using LSMN.

   (c) Update x by $x + \Delta x$ and z by $z + \Delta z$, and then go to step 3.

In using the algorithm, one should discover several set of EKT conditions. This can be done by varying the initial values. Sometimes, intuitive judgment is helpful in deciding if the smallest one is the solution of the problem.

7.5 Illustrative Example (Barrier Method for Solving NLP)

As given by Equation 7.1 in the sensitivity method, the NLP is rewritten here as

$$\left.\begin{array}{l} \text{Minimize } f(x) \\ \text{Subject to } g(x) = 0 \\ \text{and } C \leq h(x) \leq D \end{array}\right\}, \tag{7.21}$$

where
   x is an n-vector
   $f(x)$ is a scalar function

The constraints $g(x)$ and $h(x)$ are, respectively, p- and m-vector functions. The bound vectors C and D are constant. All the functions are assumed to be twice differentiable. It is important to mention that m may be greater than n but p cannot. Any bound imposed on x may be considered as part of $h(x)$.

The problem is solved here by using K–T necessary conditions in conjunction with barrier penalty functions. Involved in the method is a recursive process that solves a set of linear equations at each iteration. The equations are reduced to the least in number. The barrier parameter is generalized to a vector form in order to accommodate discriminatory penalty.

7.5.1 Algorithm for Recursive Process

Newton's numerical method is used in the sequel to approach a solution (if one exists) of the problem. To acquire the K–T conditions, we introduce first nonnegative slack variables to convert the inequalities constraints. That is

$$\left. \begin{array}{l} h(x) + s = D \\ h(x) - r = C \end{array} \right\}, \tag{7.22}$$

where s and r are nonnegative m-vector functions.

The logarithmic barrier function has been used extensively to avoid dealing with the harsh constraint of nonnegativeness on the slack variables; that is, to append $f(x)$ as

$$f_b(x) = f(x) - \sum_{j=1}^{m} U_j \ln s_j - \sum_{j=1}^{m} V_j \ln r_j. \tag{7.23}$$

All the us and vs are specified nonnegative. They may change from one iteration to another in the process. It is known that the optimization of $f_b(x)$ and $f(x)$ subject to the same constraints is the same as the us and vs approach zero. As such, one may optimize $f_b(x)$ by ignoring the nonnegative constraint on the slack variables.

For simplicity, the argument x is dropped from $f(x)$, $g(x)$, and $h(x)$ to form the Lagrange function

$$L = f_b + y^T g + w^T(h + s) - z^T(h - r), \tag{7.24}$$

where y, w, and z are the Lagrange vectors associated with the constraints. Note that they are required to be nonnegative by the K–T conditions for the problem.

Differentiation of L with respect to x, s, and r, and then setting them equal to zero yields the optimally conditions for the appended problem.

To facilitate the derivation, an operator $\nabla = \partial/\partial x$ (read as gradient) is used to mean

$$\nabla L = \left(\frac{\partial L}{\partial x} \right)^T = \left[\frac{\partial L}{\partial x_j} \right]^T \text{ (column } n\text{-vector)},$$

$$\nabla g = \left(\frac{\partial g}{\partial x}\right)^{\mathrm{T}} = \left[\frac{\partial g_i}{\partial x_j}\right]^{\mathrm{T}} \ (n \times p \text{ matrix}),$$

$$\nabla h = \left(\frac{\partial h}{\partial x}\right)^{\mathrm{T}} = \left[\frac{\partial h_i}{\partial x_j}\right]^{\mathrm{T}} \ (n \times m \text{ matrix}),$$

$$\nabla^2 f = \left[\frac{\partial^2 f}{\partial x_i \partial x_j}\right], \quad \nabla^2 g_k = \left[\frac{\partial^2 g_k}{\partial x_i \partial x_j}\right] \quad \text{and}$$

$$\nabla^2 h_k = \left[\frac{\partial^2 h_k}{\partial x_i \partial x_j}\right].$$

Enclosed by the brackets are the entry at the ith row and jth column. T denotes the transpose and the last three n-square matrices are Hessian matrices.

The K–T conditions can be obtained with respect to state and slack variables. For the state vector x, we have

$$\nabla L = \nabla f + (\nabla g)y + (\nabla h)(w - z) = 0. \tag{7.25}$$

Let S, R, W, and Z be the diagonal matrices that contain the elements of the vectors s, r, w, and z, respectively. Then the optimality conditions with respect to s and r are

$$-S^{-1}u + w = 0$$

and

$$-R^{-1}v + z = 0.$$

That is,

$$\left.\begin{array}{l} Sw = u \\ Rz = v \end{array}\right\}. \tag{7.26}$$

The increment equations of Equation 7.26 are

$$\left.\begin{array}{l} (S + \Delta S)(w + \Delta w) = u \\ (R + \Delta R)(z + \Delta z) = v \end{array}\right\}. \tag{7.27}$$

The penalty vectors u and v may alter each iteration but remain constant between iterations. By neglecting the terms $\Delta S \Delta w$ and $\Delta R \Delta z$, we obtain from Equation 7.27,

$$\left.\begin{array}{l} \Delta w = S^{-1}u - w - S^{-1}\Delta Sw \\ \Delta z = R^{-1}v - z - R^{-1}\Delta Rz \end{array}\right\}. \tag{7.28}$$

Using the fact that $\Delta Sw = \overline{W}\Delta s$ and $\Delta Rz = Z\Delta r$, one may express Equation 7.28 as

$$\left.\begin{aligned}\Delta w &= S^{-1}u - w - S^{-1}\overline{W}\Delta s \\ \Delta z &= R^{-1}v - z - R^{-1}Z\Delta r\end{aligned}\right\}. \tag{7.29}$$

Increments of slack variables and state variables are linearly related if high orders of Δx are neglected. It follows from Equation 7.22 that

$$\left.\begin{aligned}\nabla h^{\mathrm{T}}\Delta x + \Delta s &= D - h - s = d_1 \\ \nabla h^{\mathrm{T}}\Delta x - \Delta r &= C - h + r = d_2\end{aligned}\right\}. \tag{7.30}$$

Thus, we can write the relation between the increments as

$$\left.\begin{aligned}\Delta s &= d_1 - \nabla h^{\mathrm{T}}\Delta x \\ \Delta r &= \nabla h^{\mathrm{T}}\Delta x - d_2\end{aligned}\right\}. \tag{7.31}$$

Substitutions of Equation 7.31 into 7.29 give

$$\left.\begin{aligned}\Delta w &= S^{-1}(u - \overline{W}d_1) - w + S^{-1}\overline{W}\nabla h^{\mathrm{T}}\Delta x \\ \Delta z &= R^{-1}(v + Zd_2) - z - R^{-1}Z\nabla h^{\mathrm{T}}\Delta x\end{aligned}\right\}. \tag{7.32}$$

The increment Equation 7.25, after all the variables are augmented, can be similarly determined. By neglecting the high orders of Δx and Δy, we have

$$(\nabla L)_{\mathrm{aug}} = \nabla L + H\Delta x + \nabla g\Delta y + \nabla h(\Delta w - \Delta z), \tag{7.33}$$

where

$$H = \nabla^2 f + \sum_{k=1}^{p} y_k \nabla^2 g_k + \sum_{k=1}^{m} (w_k - z_k)\nabla^2 h_k. \tag{7.34}$$

Substituting equals on Equation 7.32 into 7.33 and then setting Equation 7.33 equal to zero gives

$$(\nabla L)_{\mathrm{aug}} = A\Delta x + \nabla g\Delta y + b = 0, \tag{7.35}$$

where

$$A = H + \nabla h(S^{-1}\overline{W} + R^{-1}Z)\nabla h^{\mathrm{T}} \tag{7.36}$$

and

$$b = \nabla L + \nabla h(S^{-1}(u - \overline{W}d_1) - R^{-1}(v + Zd_2) - (w - z)). \tag{7.37}$$

The linearized equation of the equality constraint is

$$g + \nabla g^{\mathrm{T}} \Delta x = 0. \tag{7.38}$$

Combination of Equations 7.35 and 7.38 makes

$$\begin{bmatrix} A & \nabla g \\ \nabla g^{\mathrm{T}} & 0 \end{bmatrix} \begin{bmatrix} \Delta x \\ \Delta y \end{bmatrix} = - \begin{bmatrix} g \\ g \end{bmatrix}. \tag{7.39}$$

Being symmetrical, the $(n+p)$-square matrix on the left can be inverted by fast means even for large $n+p$. Computation time for Δx and Δy should not cause any problem in the process. Other increments can be readily found from Equations 7.31 and 7.29.

7.5.1.1 Analytical Forms

It is necessary to derive first the n-vector ∇f, $n \times p$ matrix ∇g, $n \times m$ matrix ∇h, and n-square symmetrical matrices $\nabla^2 f$, $\nabla^2 g_k$, and $\nabla^2 h_k$. Then form the n-vector ∇L and n-square matrix H in terms of the Lagrange multipliers according to Equations 7.25 and 7.34:

$$\nabla L = \nabla f + (\nabla g)y + \nabla h(w - z)$$

$$H = \nabla^2 f + \sum_{k=1}^{p} y_k \nabla^2 g_k + \sum_{k=1}^{m} (w_k - z_k) \nabla^2 h_k.$$

A wide class of NLP problems is expressible in the form of quadratic optimization. That is

$$f(x) = \frac{1}{2} x^{\mathrm{T}} Q x + a^{\mathrm{T}} x,$$

$$g_k(x) = \frac{1}{2} x^{\mathrm{T}} G_k x + B_k^{\mathrm{T}} x, \quad k = 1, 2, \ldots, p$$

$$h_k(x) = \frac{1}{2} x^{\mathrm{T}} h_k x + J_k^{\mathrm{T}} x, \quad k = 1, 2, \ldots, m,$$

where
Q, G_k, and H_k are symmetrical
a, B_k, and J_k are n-vectors

The vectors and matrices required by Equation 7.39 are

$$\nabla f = Qx + a$$
$$\nabla g = [G_1 x + B_1, \; G_2 x + B_2, \ldots, \; G_p x + B_p]$$
$$\nabla h = [H_1 x + J_1, \; H_2 x + J_2, \ldots, \; H_m x + J_m]$$
$$\nabla^2 f = Q, \quad \nabla^2 g_k = G_k, \quad \text{and} \quad \nabla^2 h_k = H_k \text{ and}$$
$$H = Q + \sum_{k=1}^{p} y_k G_k + \sum_{k=1}^{m} (w_k - z_k) H_k.$$

7.5.1.2 Penalty Vectors

Each component of u or v may be chosen differently to achieve the discriminative penalty. However, one may choose the equipenalty scheme if there is no preference. That is,

$$u = v = \rho \mu e_m,$$

where $0 < \rho < 1$ and e_m is the m-vector with all elements equal to one. The penalty parameter μ is required to approach zero as the process approaches an optimum.

To meet such a requirement, we choose

$$\mu = \frac{1}{2m}(w^T s + z^T r) = \frac{1}{2m}(w_i s_i + z_i r_i). \tag{7.40}$$

Note that $\mu = 0$ at an optimum according to the K–T condition.

Start of Process. Initial values play an important part in the recursive process. Improper assumptions may cause divergence or convergence to a different solution (if one exists). It is known that the recursive process always converges to a solution if the initial values are close enough to it. If there is no preference, one may consider the following scheme.

State Vector.

1. Assume x to be an estimated solution.
2. Make x satisfy $\nabla f = 0$.
3. Set $x = 0$ if step 1 or 2 fails.

Slack Vectors.
   Use $s = r = 1/2(D - C) > 0$.

Lagrange Vectors. Use $w = z = [1 + \|\nabla f\|] e_m$ and $y = 0$, where $\|\nabla f\|$ is the l_1-norm.

Penalty Parameter. $0.1 \leq \rho \leq 0.5$ may be used. A large number can retard the process and a small number can cause divergence.

Replacement of Zero. Two small numbers ϵ_1 and ϵ_2 are to be used to replace zero for computer implementation. They may be different and exist between 10^{-6} and 10^{-3}.

7.5.2 Computer Implementation

Using relevant equations in the earlier derivations:

1. *Initialization* Assume

$$x, s, r, w, \text{ and } z.$$

$$\rho, \epsilon_1, \text{ and } \epsilon_2.$$

Using current initial conditions,

2. *Computation*

$$\mu = (1/2m)(w^Ts + z^Tr) \quad \text{and} \quad u = v = \rho\mu e_m. \tag{7.40}$$

$$d_1 = D - h - s \quad \text{and} \quad d_2 = C - h + r. \tag{7.30}$$

$$\nabla L = \nabla f + (\nabla g)y + (\nabla h)(w - z) = 0. \tag{7.25}$$

3. *Computation*

$$A = H + \nabla h(S^{-1}\overline{W} + R^{-1}Z)\nabla h^T. \tag{7.36}$$

$$b = \nabla L + \nabla h\big(S^{-1}(u - \overline{W}d_1) - R^{-1}(v + Zd_2) - (w - z)\big). \tag{7.37}$$

4. *Increments*

$$\begin{bmatrix} \Delta x \\ \Delta y \end{bmatrix} = -A^{-1}\begin{bmatrix} b \\ g \end{bmatrix} \tag{7.39}$$

$$\left.\begin{aligned} \Delta s &= d_1 - \nabla h^T \Delta x \\ \Delta r &= \nabla h^T \Delta x - d_2 \end{aligned}\right\} \tag{7.31}$$

$$\left.\begin{aligned} \Delta w &= S^{-1}u - w - S^{-1}\Delta Sw \\ \Delta z &= R^{-1}v - z - R^{-1}\Delta Rz \end{aligned}\right\}. \tag{7.28}$$

5. *Size of increment* Determine two numbers according to

$$N_1 = \min\left[\frac{\Delta s_i}{s_i}, \frac{\Delta r_i}{r_i}, \quad i = 1, 2, \ldots, m\right]$$

$$N_2 = \min\left[\frac{\Delta w_i}{w_i}, \frac{\Delta z_i}{z_i}, \quad i = 1, 2, \ldots, m\right]$$

Then, set

$$\beta_1 = 1 \quad \text{if } N_1 \geq -1$$

$$= (-1/N_1) \quad \text{if } N_1 < -1$$

$$\beta_2 = 1 \quad \text{if } N_2 \geq -1$$

$$= (-1/N_2) \quad \text{if } N_2 < -1$$

$$\Delta x = \beta_1 \Delta x, \quad \Delta s = \beta_1 \Delta s, \quad \Delta r = \beta_1 \Delta r,$$

$$\Delta w = \beta_2 \Delta w, \quad \Delta z = \beta_2 \Delta z, \quad \text{and} \quad \Delta y = \beta_2 \Delta y.$$

6. **Update**

$$\Delta x = \beta_1 \Delta x, \quad \Delta s = \beta_1 \Delta s, \quad \Delta r = \beta_1 \Delta r, \quad \text{and} \quad \Delta y = \beta_2 \Delta y.$$

Note that all the slack variables and Lagrange multipliers are non-negative due to the choice of β_1 and β_2.

7. **Test for Termination**

Compute

$$\mu = \frac{1}{2m}(w^T s + z^T r) \tag{7.40}$$

$$\nabla L = \nabla f + (\nabla g)y + (\nabla h)(w - z) = 0. \tag{7.25}$$

Go to step 9 if both $\mu = \leq \epsilon_1$ and $\|\nabla L\| \leq \epsilon_2$ (l_1-norm) are satisfied. Otherwise, go to step 8.

8. **Adjustment** Go back to step 2 if both $N_1 \geq -0.995$ and $N_2 \geq -0.995$. Otherwise, go back to step 2 after having adjusted all the variables as follows.

$$x = x - 0.005\Delta x, \quad s = s - 0.005\Delta s, \quad r = r - 0.005\Delta r,$$

$$w = w - 0.005\Delta w, \quad z = z - 0.005\Delta z, \quad \text{and} \quad y = y - 0.005\Delta y.$$

9. Stop with solution: $(x, y, s, r, w, \text{and } z)$.

7.6 Illustrative Examples

Example 7.6.1

Minimize $f = (x_1 - 2)^2 + 4$

Subject to

$$x_1^2 + x_2^2 - 1 \leq 0, \quad x_1, x_2 \geq 0.$$

Obtain the solution using the K–T condition.

SOLUTION

We can change the constraint to be two-sided as all the values of x_1, x_2 should be greater than zero. The constraint will be in the form $0 \le x_1^2 + x_2^2 \le 1$.

1. Form the Lagrange function

$$L(x,\lambda) = (x_1 - 2)^2 + 4 + \lambda(x_1^2 + x_2^2).$$

2. Search for the optimum candidates $L_x = 0$:

$$L_{x1} = 2(x_1 - 2) + 2\lambda x_1 = 0$$
$$L_{x2} = 2\lambda x_2 = 0.$$

Solving for the second equation for $L_{x2} = 0$, we have two possible solutions $(x_1 = 0, \lambda = 0)$. But also using the K–T condition, which can be stated as follows,

$$\lambda = 0 \Rightarrow 0 < \left(x_1^2 + x_2^2 < 1\right)$$
$$\lambda \ge 0 \Rightarrow \left(x_1^2 + x_2^2\right) = 1$$
$$\lambda \le 0 \Rightarrow \left(x_1^2 + x_2^2\right) = 0$$

a. At $\lambda = 0$, solving for $L_{x1} = 0$ results in $x_1 = 2$, which means that the main constraint, $0 \le x_1^2 + x_2^2 \le 1$ violates. Then there is no solution at $\lambda = 0$.

b. $\lambda \ge \mathbf{0}$, $x_2 = 0$, based on the K–T condition $\left(x_2^2 + x_2^2\right) = 1$, means that $x_1 = \pm 1$,

$$f(-1,0) = (-1 - 2)^2 + 4 = 13, \quad \lambda = -3 \text{ (out of range)}$$
$$f(1,0) = (1 - 2)^2 + 4 = 5, \quad \lambda = 1 \text{ (in range)}$$

c. $\lambda \le 0$, $x_2 = 0$, based on the K–T condition $\left(x_1^2 + x_2^2\right) = 0$, means that $x_1 = 0$,

$$f(0,0) = (0 - 2)^2 + 4 = 8.$$
$$\text{Then } f_{min} = f(1,0) = 5.$$

Example 7.6.2

Perform 1 iteration in solving the following problem:

Minimize $F = 0.25x_1^2 + x_2^2$

s.t. $1 \le x_1 - x_2 \le 7$

SOLUTION

$$L(x,\lambda) = f(x) + \lambda F(x) = 0.25x_1^2 + x_2^2 + \lambda(x_1 - x_2)$$

$$U(x,z) = f_x^T(x) + zF_x^T(x) = 0$$

$$f_x = x_1 - x_2 \Rightarrow f_{xx} = [1, -1]$$

$$F_x = [0.5x_1 \quad 2x_2]$$

$$S = U_x(x,z) = f_{xx} + \sum_{i=1}^{3} z_i f_{ixx}(x) = \begin{bmatrix} \lambda^T(1 - x_2) + 1 & \lambda^T(x_1 - 1) + 1 \\ \lambda^T(1 - x_2) - 1 & \lambda^T(x_1 - 1) - 1 \end{bmatrix}$$

$$U_z(x,z) = F_x^T(x) = \begin{bmatrix} 0.5x_1 \\ 2x_2 \end{bmatrix}$$

Note $A_y = b$ where $A = \begin{bmatrix} S & F_x^T \\ F_x & 0 \end{bmatrix}$, $y = \begin{bmatrix} \Delta x \\ \Delta z \end{bmatrix}$, and $b = \begin{bmatrix} -u \\ \Delta k \end{bmatrix}$

Using initial solution as $x = [1, 0]$, $\lambda = -2$, and $k = 3$, we get

$$S = \begin{bmatrix} 2x_2 - 1 & 3 - 2x_1 \\ 2x_2 - 3 & 1 - 2x_1 \end{bmatrix} = \begin{bmatrix} -1 & 1 \\ -3 & -1 \end{bmatrix}, \quad S^{-1} = \begin{bmatrix} -\dfrac{1}{4} & -\dfrac{1}{4} \\ \dfrac{3}{4} & -\dfrac{1}{4} \end{bmatrix}$$

$$F_x = [0.5 \quad 0], \quad U = \begin{bmatrix} 0 \\ -1 \end{bmatrix}.$$

If $A_y = b$, then $y = A^{-1}b$

$$\therefore y = \begin{bmatrix} S^{-1} & (F_x^T)^{-1} \\ F_x^{-1} & 0 \end{bmatrix}\begin{bmatrix} -U \\ \Delta K \end{bmatrix} = \begin{bmatrix} S^{-1}(-u) + (F_x^T)^{-1}(\Delta k) \\ F_x^{-1}(-u) + 0 \end{bmatrix}$$

$$S^{-1}(-u) = \begin{bmatrix} -0.25 & -0.25 \\ 0.75 & -0.25 \end{bmatrix}\begin{bmatrix} 0 \\ 1 \end{bmatrix} = \begin{bmatrix} -0.25 \\ -0.25 \end{bmatrix}$$

$$(F_x^T)^{-1}(\Delta k) = \begin{bmatrix} 6 \\ 0 \end{bmatrix}$$

$$(F_x)^{-1}(-u) = [2 \quad 0]\begin{bmatrix} 0 \\ 1 \end{bmatrix} = 0$$

$$y = \begin{bmatrix} \Delta x \\ \Delta z \end{bmatrix} = \begin{bmatrix} [5.25 - 0.25]^T \\ 0 \end{bmatrix}, \quad \Delta x = \begin{bmatrix} 5.25 \\ -0.25 \end{bmatrix}, \quad \Delta z = \Delta\lambda^T = 0$$

using Newton Method: $\begin{bmatrix} x_1 \\ x_2 \end{bmatrix}^{new} = \begin{bmatrix} x_1 \\ x_2 \end{bmatrix}^{old} + \begin{bmatrix} \Delta x_1 \\ \Delta x_2 \end{bmatrix}$

$$= \begin{bmatrix} 1 \\ 0 \end{bmatrix} + \begin{bmatrix} 5.25 \\ -0.25 \end{bmatrix} = \begin{bmatrix} 6.25 \\ -0.25 \end{bmatrix}$$

$$\lambda^{new} = \lambda^{old} + \Delta\lambda = -2 + 0 = -2$$

using $x_1 - x_2 \Rightarrow 6.25 - (-0.25) = 6.25$

$$f_{min}(6.25, -0.25) = 0.25(6.25)^2 + (-0.25)^2 = 9.828 \text{ after 1 iteration.}$$

Example 7.6.3

Minimize $f(x_1,x_2) = (x_1 - 1)^2 + x_2 - 2$

s.t. $h(x_1,x_2) = x_2 - x_1 - 1 = 0$

$g(x_1,x_2) = x_1 - x_2 - 2 \le 0$

SOLUTION
Formulate the Lagrangian as

$$L = f(x_1,x_2) + \lambda_1 h(x_1,x_2) + \lambda_2 g(x_1,x_2)$$
$$= (x_1 - 1)^2 + x_2 - 2 + \lambda_1(x_2 - x_1 - 1) + \lambda_2(x_1 - x_2 - 2)$$

Necessary condition

$$\frac{\partial L}{\partial x_1} = 2(x_1 - 1) - \lambda_1 + \lambda_2 = 0 \Rightarrow x_1 = \frac{\lambda_1 - \lambda_2}{2} + 1 \tag{1}$$

$$\frac{\partial L}{\partial x_2} = 1 + \lambda_1 - \lambda_2 = 0 \Rightarrow \lambda_1 = \lambda_2 - 1 \tag{2}$$

From Equations 1 and 2, we get

$$\therefore x_1 = \frac{\lambda_2 - 1 - \lambda_2}{2} + 1 = 0.5 \tag{3}$$

Substitute Equation 3 into $h(x_1,x_2)$, we obtain $x_2 = 1.5$. Hence $(x_1,x_2) = (0.5, 1.5)$
Sufficiency condition

$$\frac{\partial^2 L}{\partial x_1^2} = 2 \quad \frac{\partial^2 L}{\partial x_2 \partial x_1} = 0$$

$$\frac{\partial L}{\partial x_1 \partial x_2} = 0 \quad \frac{\partial^2 L}{\partial x_2^2} = 0$$

$H = \begin{bmatrix} 1 & 0 \\ 0 & 0 \end{bmatrix}$ is positive definite.

$(x_1,x_2) = (0.5,1.5)$ is the minimum point. $f(0.5,1.5) = (0.5 - 1)^2 + 1.5 - 2 = -0.25$ is minimum value of $f(x_1,x_2)$.

Example 7.6.4

Use the barrier algorithm to minimize the function: $f(x) = x_1 x_2$
     Subject to $1 \leq x_1 - x_2 \leq 2$.

SOLUTION

With the number of variables $n = 2$ and number of constraints $m = 1$,

$$h(x) = x_1 - x_2, \quad D = 2, \quad C = 1.$$

Use the form:

$$f(x) = \frac{1}{2} x^T Q x + a^T x = x_1 x_2.$$

Then

$$Q = \begin{bmatrix} 0 & 1 \\ 1 & 0 \end{bmatrix} \quad \text{and} \quad a = 0$$

Subject to $1 \leq x_1 - x_2 \leq 2$,   then $h(x) = x_1 - x_2$

$$h_k(x) = \frac{1}{2} x^T H_k x + J_k^T x, \quad \text{then } k = 1, H_1 = 0$$

$$J_1 = \begin{bmatrix} 1 \\ -1 \end{bmatrix}$$

$$H = Q = \begin{bmatrix} 0 & 1 \\ 1 & 0 \end{bmatrix}$$

$$\nabla f = Qx + a = \begin{bmatrix} 0 & 1 \\ 1 & 0 \end{bmatrix} \begin{bmatrix} x_1 \\ x_2 \end{bmatrix} + \begin{bmatrix} 0 \\ 0 \end{bmatrix} = \begin{bmatrix} x_2 \\ x_1 \end{bmatrix}$$

$$\nabla h = H_1 x + J_1 = \begin{bmatrix} 1 \\ -1 \end{bmatrix}, \quad \nabla^2 f = Q = \begin{bmatrix} 0 & 1 \\ 1 & 0 \end{bmatrix}$$

$$\nabla L = \nabla f + (w - z)\nabla h = \begin{bmatrix} x_2 \\ x_1 \end{bmatrix} + (w - z) \begin{bmatrix} 1 \\ -1 \end{bmatrix} = \begin{bmatrix} x_2 + w - z \\ x_1 - w + z \end{bmatrix}.$$

First iteration

     1. Initialization

$$s = r = \frac{1}{2}(D - C) = \frac{1}{2}(2 - 1) = 0.5.$$

Set $x_1 = x_2 = 0$ to satisfy $\nabla f = 0$,

$$w = z = [1 + \|\nabla f\|]e_m,$$

where e_m is an m-vector with element 1

$$w = z = (1 + 0)1 = 1$$

Take $\rho = 0.5$, $\varepsilon_1 = 10^{-2}$, and $\varepsilon_2 = 10^{-5}$.

2. Computation

(a) $\mu = \dfrac{1}{2m}(w^T s + z^T r) + \dfrac{1}{2}(0.5 + 0.5) = 0.5$

$u = v = \rho \mu e_m = 0.25.$

(b) $d_1 = D - h - s$, $h = x_1 - x_2 = 0 - 0 = 0$ then

$d_1 = 2 - 0 - 0.5 = 1.5$

$d_2 = C - h + r = 1 - 0 + 0.5 = 1.5.$

(c) $\nabla L = \begin{bmatrix} x_2 + w - z \\ x_1 - w + z \end{bmatrix} = \begin{bmatrix} 0 - 1 + 1 \\ 0 + 1 - 1 \end{bmatrix} = \begin{bmatrix} 0 \\ 0 \end{bmatrix}.$

3. Computation

(a) $A = H + \nabla h(s^{-1}w + r^{-1}z)\nabla h^T$

$= \begin{bmatrix} 0 & 1 \\ 1 & 0 \end{bmatrix} + (2 + 2)\begin{bmatrix} 1 \\ -1 \end{bmatrix}[1 \quad -1] = \begin{bmatrix} 4 & -3 \\ -3 & 4 \end{bmatrix}.$

(b) $b = \nabla L + \nabla h\{s^{-1}(u - wd_1) - r^{-1}(v + zd_2) - (w - z)\}$

$b = \begin{bmatrix} 0 \\ 0 \end{bmatrix} + \begin{bmatrix} 1 \\ -1 \end{bmatrix}\{2(0.25 - 1 \times 1.5) - 2(0.25 + 1 \times 1.5) - 0\}$

$= \begin{bmatrix} -6 \\ 6 \end{bmatrix}.$

4. Increments

$$\begin{bmatrix} \nabla x \\ \nabla y \end{bmatrix} = -A^{-1}\begin{bmatrix} b \\ g \end{bmatrix},$$

$$\Delta x = -A^{-1}b = -\begin{bmatrix} 4 & -3 \\ -3 & 4 \end{bmatrix}^{-1}\begin{bmatrix} -6 \\ 6 \end{bmatrix} = 0.8571\begin{bmatrix} 1 \\ -1 \end{bmatrix}$$

$$\Delta s = d_1 - \nabla h^T \Delta x \nabla h = 1.5 - 0.8571[1 \quad -1]\begin{bmatrix} 1 \\ -1 \end{bmatrix} = -0.2143$$

$$\Delta r = \nabla h^T \Delta x - d_2 = 0.2143$$

$$\Delta w = s^{-1}(u - w\Delta s) - w = 0.1286$$

$$\Delta z = r^{-1}(v - z\Delta r) - z = -0.7286.$$

5. Size of increment

$$N_1 = \min\left\{\frac{\Delta s}{s}, \frac{\Delta r}{r}\right\} = \min\left\{\frac{-0.2143}{0.5}, \frac{0.2143}{0.5}\right\} = -0.42856$$

$$N_2 = \min\left\{\frac{\Delta w}{w}, \frac{\Delta z}{z}\right\} = \min\left\{\frac{0.1286}{1}, \frac{-0.7286}{1}\right\} = -0.7286$$

$$N_1 > -1 \Rightarrow \beta_1 = 1$$

$$N_2 > -1 \Rightarrow \beta_2 = 1.$$

Then there is no change in the previous calculated increments.
6. Update

$$s = s + \Delta s = 0.5 - 0.2143 = 0.2857$$

$$x = x + \Delta x = \begin{bmatrix} 0.85714 \\ -0.85714 \end{bmatrix}$$

$$r = r + \Delta r = 0.5 + 0.2143 = 0.7143$$

$$w = w + \Delta w = 1 + 0.1286 = 1.1286$$

$$z = z + \Delta z = 1 - 0.7286 = 0.2714.$$

7. Test of termination

$$\mu = \frac{1}{2m}(w^T s + z^T r) = \frac{1}{2 \times 1}(1.1286 \times 0.2857 + 0.2714 \times 0.7143)$$

$$= 0.2582 > \varepsilon_1$$

$$\nabla L = \begin{bmatrix} x_2 + w - z \\ x_1 - w + z \end{bmatrix} = \begin{bmatrix} -0.85714 + 1.1286 - 0.2714 \\ 0.85714 - 1.1286 + 0.2714 \end{bmatrix}$$

$$= -0.331 \times 10^{-15} \begin{bmatrix} 1 \\ -1 \end{bmatrix}$$

$$\|\nabla L\| = 4.7103 \times 10^{-16} > \varepsilon_2.$$

8. Adjustment

$$N_1 = -0.4286 > -0.995 \quad \text{and} \quad N_2 = -0.7286 > -0.995.$$

Then go to step 2.

Second iteration
Using current initial conditions,

2. Computation

(a) $$\mu = \frac{1}{2m}(w^T s + z^T r) = 0.2582$$

$$u = v = \rho\mu e_m = 0.1807.$$

(b) $$d_1 = D - h - s, \quad h = x_1 - x_2 = 0.85714 + 0.85714 = 1.71428.$$

Then

$$d_1 = 1.0e - 04$$

$$d_2 = C - h + r = 1 - 1.7142 + 0.7143 = 1.0e - 04.$$

(c) $\qquad \nabla L = \begin{bmatrix} x_2 + w - z \\ x_1 - w + z \end{bmatrix} = \begin{bmatrix} -0.85714 + 1.1286 - 0.2714 \\ 0.85714 - 1.1286 + 0.2714 \end{bmatrix}$

$$= -0.331 \times 10^{-15} \begin{bmatrix} 1 \\ -1 \end{bmatrix}$$

3. Computation

(a) $\qquad A = H + \nabla h (s^{-1} w + r^{-1} z) \nabla h^{\mathsf{T}}$

$$= \begin{bmatrix} 0 & 1 \\ 1 & 0 \end{bmatrix} + \left(\frac{1.1286}{0.2857} + \frac{0.2714}{0.7143} \right) \begin{bmatrix} 1 \\ -1 \end{bmatrix} [1 \quad -1]$$

$$= \begin{bmatrix} 4.3302 & -3.3302 \\ -3.3302 & 4.3302 \end{bmatrix}.$$

(b) $\qquad b = \nabla L + \nabla h \{ s^{-1}(u - w d_1) - r^{-1}(v + z d_2) - (w - z) \}$

$$= \begin{bmatrix} -0.4780 \\ 0.4780 \end{bmatrix}.$$

4. Increments

$$\begin{bmatrix} \nabla x \\ \nabla y \end{bmatrix} = -A^{-1} \begin{bmatrix} b \\ g \end{bmatrix}$$

$$\Delta x = -A^{-1} b = -\begin{bmatrix} 4.3302 & -3.3302 \\ -3.3302 & 4.3302 \end{bmatrix}^{-1} \begin{bmatrix} -0.478 \\ 0.478 \end{bmatrix} = 0.0624 \begin{bmatrix} 1 \\ -1 \end{bmatrix}$$

$$\Delta s = d_1 - \nabla h^{\mathsf{T}} \Delta x \nabla h = -0.1247$$

$$\Delta r = \nabla h^{\mathsf{T}} \Delta x - d_2 = 0.1247$$

$$\Delta w = s^{-1}(u - w \Delta s) - w = -0.0035$$

$$\Delta z = r^{-1}(v - z \Delta r) - z = -0.0658.$$

5. Size of increment

$$N_1 = \min \left\{ \frac{\Delta s}{s}, \frac{\Delta r}{r} \right\} = \min \left\{ \frac{-0.1247}{0.2857}, \frac{0.1247}{0.7143} \right\} = -0.4365$$

$$N_2 = \min \left\{ \frac{\Delta w}{w}, \frac{\Delta z}{z} \right\} = \min \left\{ \frac{-0.0035}{1.1286}, \frac{-0.0658}{0.2714} \right\} = -0.2424$$

$$N_1 > -1 \Rightarrow \beta_1 = 1$$

$$N_2 > -1 \Rightarrow \beta_2 = 1.$$

Then there is no change in the previous calculated increments.

6. Update

$$s = s + \Delta s = 0.2857 - 0.1247 + 0.161$$

$$x = x + \Delta x = \begin{bmatrix} 0.8571 \\ -0.8571 \end{bmatrix} + \begin{bmatrix} 0.0624 \\ -0.0624 \end{bmatrix} = \begin{bmatrix} 0.9195 \\ -0.9195 \end{bmatrix}$$

$$r = r + \Delta r = 0.7143 + 0.1247 = 0.8390$$

$$w = w + \Delta w = 1.1286 - 0.0035 = 1.1251$$

$$z = z + \Delta z = 0.2714 - 0.0658 = 0.2056.$$

7. Test of termination

$$\mu = \frac{1}{2m}(w^T s + z^T r) = \frac{1}{2 \times 1}(1.1251 \times 0.161 + 0.2056 \times 0.8390)$$
$$= 0.1768 > \varepsilon_1$$

$$\nabla L = \begin{bmatrix} x_2 + w - z \\ x_1 - w + z \end{bmatrix} = \begin{bmatrix} -0.9195 + 1.1251 - 0.2056 \\ 0.9195 - 1.1251 + 0.2056 \end{bmatrix} = \begin{bmatrix} 0 \\ 0 \end{bmatrix} < \varepsilon_2.$$

8. Adjustment

$$N_1 = -0.4365 > -0.995 \quad \text{and} \quad N_2 = -0.2424 > -0.995.$$

Then go to step 2.

Repeating the previous step, we get the solution that satisfies the conditions $\varepsilon_1 = 10^{-2}$ and $\varepsilon_2 = 10^{-5}$ after 10 iterations. Table 7.1 tabulates the results of these iterations.

The solution is $x_1 = 0.995$ and $x_2 = -0.995$.

TABLE 7.1

Results of All Iterations for Example 7.6.2

Iteration Number	x_1	x_2	r	s	w	z	μ	$\|\Delta L\|$
1	0.8571	−0.8571	0.7143	0.2857	1.1286	0.2714	0.2585	4.71 E-16
2	0.9195	−0.91915	0.8390	0.1610	1.1251	0.2056	0.1768	0.0
3	0.9416	−0.9416	0.8833	0.1167	1.0782	0.1367	0.1233	1.57 E-16
4	0.9585	−0.9585	0.9170	0.0830	1.0510	0.0925	0.0860	4.79 E-15
5	0.9707	−0.9707	0.9414	0.0586	1.0339	0.0632	0.0601	0.0
6	0.9794	−0.9794	0.9587	0.0413	1.0228	0.0435	0.0420	4.71 E-16
7	0.9855	−0.9855	0.9710	0.0290	1.0156	0.0301	0.0293	1.57 E-16
8	0.9898	−0.9898	0.9796	0.0204	1.0107	0.0209	0.0205	1.57 E-16
9	0.9929	−0.9929	0.9857	0.0143	1.0074	0.0146	0.0144	3.14 E-16
10	0.995	−0.995	0.9900	0.0100	1.0051	0.0102	0.0100	1.884 E-15

7.7 Conclusion

This chapter dealt with optimization techniques that fit most nonlinear engineering applications. It presented several algorithms in NLP that are programmable by scientists and engineers using well developed numerical methods such as in [14]. NLP aims at solving optimization problems involving nonlinear objective and constrained functions [3–9,16]. In Sections 7.1 and 7.2, classification of NLP problems was presented. The classification included quadratic, convex, and separable programming. The sensitivity method for solving NLP variables was presented in Section 7.3. Also, a practical procedure for solving the problem was demonstrated with an alternative expression for the extended K–T condition to provide feasibility and optimality. In Section 7.4 an algorithm for solving the quadratic optimization problem was presented in the form of sequential steps. A technique based on the barrier method for solving NLP problems was presented in Section 7.5, where the recursive process was developed.

7.8 Problem Set

PROBLEM 7.8.1

Solve the following as a separate convex programming problem.

Minimize $Z = (x_1 - 2)^2 + 4(x_2 - 6)^2$

Subject to

$6x_1 + 3(x_2 + 1)^2 \leq 12$

$x_1, x_2 \geq 0$

PROBLEM 7.8.2

Consider the problem

Maximize $Z = 6x_1 + 3x_2 - 4x_1x_2 - 2x_1^2 - 3x_2^2$

Subject to

$x_1 + x_2 \leq 1$

$2x_1 + 3x_2 \leq 4$

$x_1, x_2 \geq 0.$

Show that Z is strictly concave and then solve the problem using the QP algorithm.

PROBLEM 7.8.3

Consider the problem

Minimize $Z = x_1^2 + x_2^2$

Subject to

$2x_1 + x_2 \leq 2$

$-x_1 + 1 \leq 0.$

1. Find the optimal solution to this problem.
2. Formulate a suitable function with initial penalty parameter $\mu = 1$.
3. Starting from the point $(2, 6)$, solve the resulting problem by a suitable unconstrained optimization technique.
4. Replace the penalty parameter μ by 10. Starting from the point obtained in 3, solve the resulting problem.

PROBLEM 7.8.4

Minimize $f = (x_1 + 1)(x_2 - 2)$ over the region $0 \leq x_1 \leq 2, 0 \leq x_2 \leq 1$ by writing the K–T conditions and obtaining the saddle point.

PROBLEM 7.8.5

Minimize $Z = 2x_1 - x_1^2 + x_2$

Subject to

$2x_1 + 3x_2 \leq 6$

$x_1 + x_2 \leq 4$

$x_1, x_2 \geq 0.$

PROBLEM 7.8.6

Minimize $Z = (x_1 - 6)^2 + (x_2 - 8)^2$

Subject to

$x_1^2 - x_2^2 \leq 0.$

Using the auxiliary function $(x_1 - 6)^2 + (x_2 - 8)^2 + \mu \max (x_1^2 - x_2, 0)$, and adopting the cyclic coordinate method, solve the above problem starting from $x_1 = (0, -4)^t$ under the following strategies for modifying μ.

1. Starting from x_1, solve the penalty problem for $\mu_1 = 0.1$ resulting in x_2. Then start from x_2, and solve the problem with $\mu_2 = 100$.
2. Starting from the unconstrained optimal point $(6, 8)$, solve the penalty problem for $\mu_2 = 100$. (This is the limiting case of Part 1 for $\mu_1 = 0$.)
3. Starting from x_1, solve the penalty problem $\mu_1 = 100.0$.

4. Which of the above strategies would you recommend, and why? Also, in each of the above cases, derive an estimate for the Lagrangian multiplier associated with the single constraint.

PROBLEM 7.8.7

Maximize $f(x) = x_1^2 - 2x_2^2$

Subject to ·

$x_1 - x_2 = 2$

$x_2 \leq x_1^2 \leq x_2 + 8.$

PROBLEM 7.8.8

Maximize $f(x) = 3x_1 x_2 - x_1^2 - x_2^2$

Subject to

$x_1 \leq 0$ and $0 \leq x_2 \leq 1.$

PROBLEM 7.8.9

Minimize $x_1^2 + x_2^2$

s.t. $x_1 - x_2^2 - 4 \geq 0$

$x_1 - 10 \leq 0$

PROBLEM 7.8.10

Use KT conditions to solve

Minimize $f(x_1, x_2) = (x_1 - 1)^2 + x_2 - 2$

s.t. $h(x_1, x_2) = x_2 - x_1 - 1 = 0$

$f(x_1, x_2) = x_1 - x_2 - 2 \leq 0$

PROBLEM 7.8.11

Apply the NLP sensitivity method to solve

Minimize $f(x_1, x_2) = 0.5x_1^2 + 2x_2^2$

s.t. $2 \leq 2x_1 - 2x_2 \leq 4$

PROBLEM 7.8.12

Minimize $f(x_1, x_2) = x_1^2 + x_1 x_2 + 2x_2^2 - 6x_1 - 14x_2$

s.t. $3 \leq x_2 + x_1 \leq 6$

$2 \leq -x_1 + 2x_2 \leq 3$

PROBLEM 7.8.13

Minimize $f(x_1,x_2) = 0.25x_1^2 + x_2^2$

s.t. $1 \le x_1 - x_2 \le 2$

PROBLEM 7.8.14

Consider the NLP problem:

Minimize $x_1^2 + x_2^2 + 2x_1x_2 + 2x_1$

s.t. $1 \le x_1 + x_2 \le 2$

Solve by

    a. Sensitivity/Newtonian method
    b. Quadratic interior point method

(Perform at least one iteration for each method)

PROBLEM 7.8.15

Apply K–T conditions to solve the NLP problem given in Problem 7.8.14.

PROBLEM 7.8.16

Maximize $f(x) = -(2x_1 - 5)^2 - (2x_2 - 1)^2$

Subject to

$x_1 + 2x_2 \le 2$

$x_1, x_2 \ge 0$

PROBLEM 7.8.17

Consider the problem defined as Maximize $f(x)$ Subject to $h(x) \le 0$. If the Lagrangian is formulated as $L(x, \lambda, s) = f(x) + \lambda[h(x) + s^2]$, how would this change affect the K–T conditions?

PROBLEM 7.8.18

Consider the following generalized nonlinear problem:

$Max(x)$

Subject to

$g_i(x) \le 0, \quad i = 1, 2, \ldots, r$

$g_i(x) \ge 0, \quad i = r+1, \ldots, p$

$g_i(x) \le 0, \quad i = p+1, 2, \ldots, m$

a. Convert the inequality constraints to equality and construct and appropriate Lagrangian

$$L(x, \lambda, s) = f(x) - \sum_{i=1}^{r} \lambda_i \left[g_i(x) + s^2 \right] - \sum_{i=r+1}^{p} \lambda_i \left[g_i(x) - s^2 \right]$$

$$- \sum_{i=p+1}^{m} \lambda_i g_i(x)$$

b. Develop the K–T conditions to the following Lagrange function

PROBLEM 7.8.19

Determine extreme points for the functions give by

a. $f(x) = x_1^3 + 2x_2^3 - 4x_1 x_2$
b. $f(x) = 2x_1^2 + 3x_2^2 - x_3^2 + 4(x_1 + x_3) + 3x_2 x_3$

PROBLEM 7.8.20

a. Minimize $(x_1 - 2)^4 + (x_1 - 2x_2)^2$
s.t. $x_1^2 + x_2 = 0$
using the Barrier algorithm

b. Using the quadratic interior point (QUIP) method
Maximize $P = -2x_1 - 6x_2 + x_1^2 + x_2^2$
s.t. $x_1 + 2x_2 \leq 5$
$x_1 + x_2 \leq 3$ for all positive x.

PROBLEM 7.8.21

Consider the quadratic problem stated as
Minimize $0.5 \times {}^T Hx + c^T x$: $Ax = b$
Where $A_{m \times n}$ is of rank m and H is positive definite:

a. Show that the matrix $\begin{bmatrix} H & A^T \\ A & 0 \end{bmatrix}$ is nonsingular.
b. Show that the linear equation of the KKT system of the QP yields a unique solution
c. Assume that H is positive definite and hence nonsingular, derive an explicit or closed form expression for the optimal solution to the QP problem.

PROBLEM 7.8.22

a. Determine whether or not the quadratic form

$$Q(x_1,x_2) = 7x_1 + 3x_2 - 4x_1x_2 - 2x_1^2 - 3x_2^2 \text{ is negative-definite.}$$

b. Determine if the quadratic form

$$Q(x_1,x_2,x_3) = 2x_1^2 + 5x_2^2 + 3x_3^2 + 4x_1x_2 + 6x_2x_3 \text{ is positive-definite.}$$

References

1. Hamdy, A. T. *Operations Research: An Introduction*, 8th edn., Prentice-Hall, New Jersey, 2006.
2. Himmelblau, D. M. *Applied Nonlinear Programming*, McGraw-Hill, New York, 1972.
3. Luenberger, D. G. *Introduction to Linear and Nonlinear Programming*, Addison-Wesley, Reading, MA, 1973.
4. Mangasarian, O. L. Nonlinear programming problems with stochastic objective functions, *Management Science*, 10, 1964, 353–359.
5. McCormick, G. P. Penalty function versus non penalty function methods for constrained nonlinear programming problems, *Mathematical Programming*, 1, 1971, 217–238.
6. McMillan, C., Jr. *Mathematical Programming*, Wiley, New York, 1970.
7. Murtagh, B. A. and Sargent, R. W. H. Computational experience with quadratically convergent minimization methods, *Computer Journal*, 13, 1970, 185–194.
8. Pierre, D. A. *Optimization Theory with Applications*, Wiley, New York, 1969.
9. Powell, M. J. D. A method for nonlinear constraints in minimization problems, in: *Optimization*, R. Fletcher (Ed.), Academic Press, London, New York, 1969.
10. Wilde, D. J. *Optimum Seeking Methods*, Prentice-Hall, Englewood Cliffs, NJ, 1964.
11. Wilde, D. J. and Beightler, C. S. *Foundations of Optimization*, Prentice-Hall, Englewood Cliffs, NJ, 1967.
12. Zangwill, W. I. *Nonlinear Programming: A Unified Approach*, Prentice-Hall, Englewood Cliffs, NJ, 1969.
13. Zoutendijk, G. Nonlinear programming, computational methods, in: *Integer and Nonlinear Programming*, J. Abadie (Ed.), North-Holland, Amsterdam, 1970, pp. 37–86.
14. Forsythe, G. E., Malcolm, M. A., and Moler, C.B. *Computer Methods for Mathematical Computations*, Prentice-Hall, Englewood Cliffs, NJ, 1977.
15. Momoh, J. A., Dias, L. G., Guo, S. X., and Adapa, R. A. Economic operation and planning of multi-area interconnected power system, *IEEE Transactions on Power Systems*, 10, 1995, 1044–1051.
16. Vanderbei, R. J. and Shanno, D. F. An interior point algorithm for nonconvex nonlinear programming, Research Report SOR-97-21, Statistics and Operations Research, Princeton University, Princeton, NJ.

8

Dynamic Programming

8.1 Introduction

Dynamic programming (DP) is an optimization approach that transforms a complex problem into a sequence of simpler problems [3–6,9,12–14]; its essential characteristic is the multistage nature of the optimization procedure. More so than the optimization techniques described previously, DP provides a general framework for analyzing many problem types. Within this framework a variety of optimization techniques can be employed to solve particular aspects of a more general formulation. Usually creativity is required before we can recognize that a particular problem can be cast effectively as a dynamic program, and often subtle insights are necessary to restructure the formulation so that it can be solved effectively.

The DP method was developed in the 1950s through the work of Richard Bellman [1,2] who is still the doyen of research workers in this field. The essential feature of the method is that a multivariable optimization problem is decomposed into a series of stages, optimization being done at each stage with respect to one variable only. Bellman [1] gave it the rather undescriptive name of DP. A more significant name would be recursive optimization.

Both discrete and continuous problems can be amenable to this method and deterministic as well as stochastic models can be handled by it. The complexities increase tremendously with the number of constraints. A single-constraint problem is relatively simple, but even more than two constraints can be formidable.

The DP technique, when applicable, represents or decomposes a multi-stage decision problem as a sequence of single-stage decision problems. Thus an N-variable problem is represented as a sequence of N single-variable problems that are solved successively. In most of the cases, these N subproblems are easier to solve than the original problem. The decomposition to N subproblems is done in such a manner that the optimal solution of the original N-variable problem can be obtained from the optimal solutions of the N one-dimensional problems. It is important to note that the particular optimization technique used for the optimization of the N-single-variable problems is irrelevant. It may range from a simple enumeration process to a calculus or a nonlinear programming technique.

Multistage decision problems can also be solved by the direct application of classical optimization techniques. However, this requires the number of variables to be small, the functions involved to be continuous and continuously differentiable, and the optimum points not to lie at the boundary points.

Furthermore, the problem has to be relatively simple so that the set of resultant equations can be solved either analytically or numerically. The nonlinear programming techniques can be used to solve slightly more complicated multistage decision problems. But their application requires the variables to be continuous and for there to be prior knowledge about the region of the global minimum or maximum. In all these cases, the introduction of stochastic variability makes the problem extremely complex and the problem unsolvable except by using some sort of an approximation-like chance constrained optimization.

DP, on the other hand, can deal with discrete variables, and nonconvex, noncontinuous, and nondifferentiable functions [4,7–11]. In general, it can also take into account the stochastic variability by a simple modification of the deterministic procedure. The DP technique suffers from a major drawback, known as the curse of dimensionality. However, in spite of this disadvantage, it is very suitable for the solution of a wide range of complex problems in several areas of decision making.

Optimization overtime in a single or multiple decision process is generally formulated as DP. It involves a large number of variables over different stages. DP is a procedure designed to improve the computation efficiency for solving a set of large-scale problems in time that can be decomposed into smaller and have computationally simple problems. DP solves the optimization in stages with each stage involving exactly one optimization variable. The computations in each stage are linked by the focus called recursive computation in a manner that is feasible in optimization solution. This will be obtained for the entire problem. When the last stage reached this there is an advantage for each stage being optimized. On the other hand, the advantage loss is in the complexity of its solution for a large system, the so-called curse of dimensionality. With this in mind DP, has limited applications and there are many recent works to advance this technology for possible applications in power systems management. These include candidates such as dynamic programming (ADP), artificial neural networks (ANN), and evolutionary programming in the area of genetic algorithms (GA), which are observed in later chapters.

8.2 Formulation of a Multistage Decision Process

8.2.1 Representation of a Multistage Decision Process

Any decision process is characterized by certain input parameters, X (or data), certain decision variables (U), and certain output parameters (T)

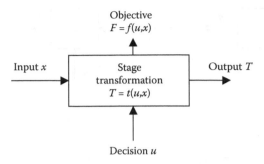

Input x

Objective
$F = f(u,x)$

Stage
transformation
$T = t(u,x)$

Output T

Decision u

FIGURE 8.1
Single-stage decision problem.

representing the outcome obtained as a result of making the decision. The input parameters are called input stage variables, and the output parameters are called output stage variables. Finally there is a return or objective function F, which measures the effectiveness of the decisions for any physical system, represented as a single-stage decision process (shown in Figure 8.1). The output of this single stage is shown in Equations 8.1 and 8.2.

$$x_i = t_i(x_{i+1},u_i) \quad \forall \, i \in \{1,n\} \tag{8.1}$$

$$F_i = f_i(x_{i+1},u_i) \quad \forall \, i \in \{1,n\}, \tag{8.2}$$

where u_i denotes the vector of decision variables at stage i. The objective of a multistage decision process is to find $u_1, u_2, \ldots, u_n$ so as to optimize some function of the individual stage returns, say, $F(f_1, f_2, \ldots, f_n)$ and satisfy Equations 8.1 and 8.2. The nature of the n-stage return function f determines whether a given multistage problem can be solved as a decomposition technique; it requires the separability and monotonicity of the objective function.

In order to have separability of the objective function, we must be able to represent the objective function as the composition of the individual stage returns. This requirement is satisfied for additive objective functions:

$$F = \sum_{i=1}^{n} f_i(u_i,x_{i+1}), \tag{8.3}$$

where u_i are real, and for multiplicative objective functions:

$$F = \prod_{i=1}^{n} f_i(x_{i+1},u_i). \tag{8.4}$$

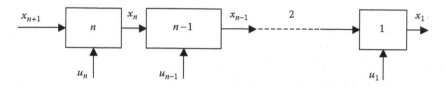

FIGURE 8.2
Multistage initial value problem.

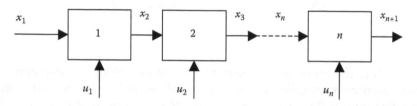

FIGURE 8.3
Multistage final value problem.

8.2.2 Types of Multistage Decision Problems

The serial multistage decision problems can be classified into the following categories.

1. Initial value problem: If the value of the initial state variable x_{n+1} is prescribed, the problem is called an initial value problem (Figure 8.2).
2. Final value problem: If the value of the final state variable x_1 is prescribed, the problem can be transformed into an initial value problem by reversing the directions of u_i, $i = 1$, 2, $n+1$, which is shown in Figure 8.3.
3. Boundary value problem: If the values of both the input and output variables are specified, the problem is called a boundary value problem.

8.3 Characteristics of DP

DP has the following characteristics:

1. Divisible into stages, policy decision is requested at each stage.
2. Each stage has a number of associated states.
3. Efficient policy decision transformation of the current state is needed to associate the next stage.

4. Solution procedure is classified to find the optimum policy for the overall problem.

5. Prescription of optimum policy for the remaining stages, for each possible state.

6. Given the current stage, an optimal policy for the remaining stage is independent of the policy adopted in the previous stage (principle of optimality in DP).

7. By using the recursive relationship, the solution procedure moves backward stage by stage, each time finding the optimum policy for that stage until it finds the optimum policy starting at the initial stage.

8.4 Concept of Suboptimization and the Principle of Optimality

A DP problem can be stated as follows. Find the values of $u_1, u_2, \ldots, u_N$ which optimize $F = \sum_{i=1}^{n} f_i(u_i, x_{i+1})$ and satisfy the design equations $u_i = t_i(x_i, u_{i+1})$. DP makes use of the concept of suboptimization and Bellman's principle of optimality in solving the problem.

Bellman stated the following as part of his principle of optimality.

> An optimal policy (or a set of decisions) has the property that whatever the initial state and initial decisions are, the remaining decisions must constitute an optimal policy with regard to the state resulting from the first decision.

In developing a recurrence relationship, suppose that the desired objective is to minimize the N-stage objective function f, which is given by the sum of the individual stage-returns:

$$\min \; F = f_n(x_{n+1}, u_n) + f_{n-1}(x_n, u_{n-1}) + \cdots + f_1(x_2, u_1) \tag{8.5}$$

and satisfying the design equations $x_i = t_i(x_{i+1}, u_i)$. Figure 8.4 shows how to apply this to DP.

Now consider the first subproblem by starting at the final stage, $i = 1$. If the input to this stage x_2 is specified, then according to the principle of optimality, u_1 must be selected to optimize f_1. Irrespective of what happens to other stages, u_1 must be selected such that $f(u_1, x_2)$ is an optimum for the input x_2. If the optimum is denoted f_1^* we have

$$f_1^*(x_2) = \operatorname*{opt}_{u_1} [f_1(x_2, u_1)]. \tag{8.6}$$

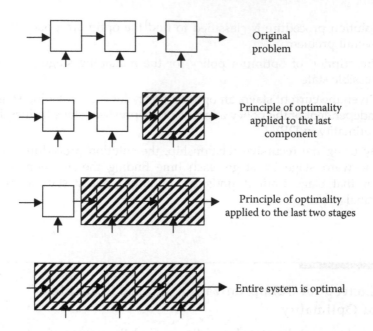

FIGURE 8.4
Illustration of the principle of optimality.

This is called a one-stage policy since once the input stage s_2 is specified, the optimal values of f_1, u_1, and x_2 are completely defined as shown in Equation 8.6.

Next, consider the second subproblem by grouping the last two stages together. If f_2^* denotes the optimum objective value of the second subproblem for a specified value of the input x_3, we have

$$f_2^*(x_3) = \underset{u_2 \cdot u_1}{\text{opt}} \, [f_1(x_2, u_1) + f_2(x_3, u_2)]. \tag{8.7}$$

The principle of optimality requires that u_1 be selected so as to optimize f_1 for a given x_2. Since x_2 can be obtained once u_2 and x_3 are specified, Equation 8.7 can be written as

$$f_2^*(x_3) = \underset{u_2}{\text{opt}} \lfloor f_1^* + f_2(x_3, u_2) \rfloor. \tag{8.8}$$

Thus f_2^* represents the optimal policy for the two-stage problem. Therefore, rewriting Equation 8.8 yields

$$f_2^*(x_3) = \underset{u_2}{\text{opt}} \lfloor f_1^*(t_2(u_2, x_3) + f_2(x_3, u_2)) \rfloor. \tag{8.9}$$

In this form, it can be seen that for a specified input x_3, the optimum is determined solely by a suitable choice of the decision variable u_2. Thus the optimization problem stated in Equation 8.7, in which both u_2 and u_1 are to be simultaneously varied to produce the optimum f_2^*, is reduced to each of these subproblems involving only a single-decision variable, and optimization is, in general, much simpler.

This idea can be generalized and the ith subproblem defined by

$$f_i^*(x_{i+1}) = \operatorname*{opt}_{u_i, u_{i-1}, \dots, u_1} \left[f(u_i, x_{i+1}) + \cdots + f_1(x_2, u_1) \right], \tag{8.10}$$

which can be written as

$$f_i^*(x_{i+1}) = \operatorname*{opt}_{u_i} \left[f_{i-1}(x_i) + f_i(x_{i+1}, u_i) \right], \tag{8.11}$$

where
   f_{i+1}^* denotes the optimal value of the objective function corresponding to the last $i-1$ stages
   s_i is the input to the stage $i-1$

By using the principle of optimality, this problem has been decomposed into i separate problems, each involving only one decision variable.

8.5   Formulation of DP

DP is a transformation from a multiple vector decision process to a series of single-vector decision processes. It converts a problem that contains many vectors for decision processes but contains only one vector to decide at a time. The single vector is usually reduced to contain the minimal number of components for ease of optimization. DP not only reduces the complexity of problems, but also provides a form that can be solved effectively and iteratively on digital computers.

Consider an optimization problem that consists of n decision (control) vectors v_k for $k = 1, 2, \dots, n$ and a scalar objective function $P_n(v_n, v_n, v_{n-1}, \dots, v_1)$ defined as state n. The vectors v_k have r_k components, which may change with the running index k. The decision vectors v_k are constrained in a set S, which is usually described by equations and decision vectors in S for a given state vector y_n.

Now, let $P_k(y_k, v_k, v_{k-1}, \dots, v_1)$ be the objective function at stage k where $k = 1, 2, \dots, n$, and let

$$p_k(y_k) = \text{optimize}[P_k(y_k, v_k, v_{k-1}, \dots, v_1)] = P_k(y_k, u_k, u_{k-1}, \dots, v_1). \tag{8.12}$$

Equation 8.12 is the optimized objective function at stage k. The optimization is carried out with respect to v_1, v_2, v_k in S for a given y_k. The optimal decision or control vectors u, where $i = 1, 2, \ldots, k$, are known vectors and hence the optimized objective function is a function of y_k only as indicated by the left side of Equation 8.12. As such, the problem is equivalent to determining Equation 8.12 for $k = n$. The state vectors y_k have n_k components, which may change with k. By properly choosing the integer r_k, we can always formulate the objective function as follows:

$$P_k(y_k, v_k, v_{k-1}, \ldots, v_1) = f_k(y_k, v_k) \text{ o } P_{k-1}(y_{k-1}, v_k, v_{k-1}, \ldots, v_1), \quad (8.13)$$

for all $k = 2, 3, 4, \ldots, n$ with $P_k(y_1, v_1) = f_1(y_1, v_1)$ for $k = 1$.

The symbol "o" in Equation 8.13 is called an operator, which may be addition, multiplication, or comparison, and it may alter from stage to stage. The scalar function $f_k(y_k, v_k)$ is to be chosen in a manner such that it constitutes a part of the objective function. The relation specifies the state vector y_k,

$$y_{k-1} = T_k(y_k, v_k). \quad (8.14)$$

Both f_k and T_k denote a transformation or mapping that may be described in different forms. There are some known types of constraints that can be replaced by Equation 8.14. For example:

$$g_1(v_1) + g_2(v_2) + \cdots + g_n(v_n) \geq C \quad (\text{or} \leq C) \quad (8.15)$$

$$h_1(v_1)h_2(v_2), \ldots, h_n(v_n) \geq C \quad (\text{or} \leq C), \quad (8.16)$$

where g_s and h_s are scalar functions of the decision vectors. From Equation 8.15, we choose the state Equation 8.14 to be

$$y_{k-1} = y_k - g_k(v_k) \quad \text{for all } k = 1, 2, \ldots, n.$$

Summation of this equation gives

$$g_1(v_1) + g_2(v_2) + \cdots + g_n(v_n) = y_k - g_k(v_k).$$

Hence, $y_n \geq C$ can be concluded if $y_0 = 0$ is assumed.

Similarly, from Equation 8.16, we select the state equation to be

$$y_{k-1} = \frac{y_k}{h_k(v_k)},$$

where

$$h_k(v_k) \neq 0$$
$$k = 1, 2, \ldots, n$$

Multiplication of these equations gives $h_1 v_1, h_2 v_2, \ldots, h_n v_n = y_n / y_0 \geq C$ and hence $y_n \geq C$ follows if $y_0 = 0$ is chosen. Thus, we may choose DP to get $P_n(y_n)$ and then make further optimization for $y_n \geq C$ ($>C$).

We call an optimization problem decomposable if it can be solved by recursive optimization through N stages, at each stage optimization being done over one decision variable. We first define monotonicity of a function, which is used subsequently.

DEFINITION 8.5.1

The function $f(x,y)$ is said to be a *monotonic nondecreasing function* of y for all feasible values of x if $y_1 < y_2 \Rightarrow f(x,y_1) \geq f(x,y_2)$ for every feasible value of x.

Conversely, the function is said to be *monotonic nonincreasing* if $y_1 < y_2 \Rightarrow f(x,y_1) \leq f(x,y_2)$ for every feasible x.

THEOREM 8.5.1

In a serial two-stage minimization or maximization problem if

(a) *Objective function ϕ_2 is a separable function of stage returns $f_1(X_1,U_1)$ and $f_2(X_2,U_{21})$, and*

(b) *ϕ_2 is a monotonic nondecreasing function of f_1 f_0 for every feasible value of f_2, then the problem is said to be decomposable.*

Proof Putting $N = 2$ in the problem, the objective function $\phi_2(f_2,f_1)$ is separable if $\phi_2 = f_2 \circ \phi_1$, $\phi_1 = f_1$. Assuming that this condition holds true, then suppose that ϕ_2 is a monotonic nondecreasing function of f_1 for feasible values of f_2. We prove the theorem for the minimum case. For the maximum case the proof is on identical lines.

Following the notation introduced before, we note that the following expressions are equivalent.

$$F_2(X_2) = \min_{U_1, U_2} \phi_2(X_2, U_2, U_1) \tag{8.17a}$$

$$= \min_{U_1, U_2} \phi_2[f_2(X_2, U_2) \circ f_1(X_1, U_1)] \tag{8.17b}$$

$$= \min_{U_1, U_2} \phi_2[f_2(X_2, U_2) \circ f_1(X_1, U_2, U_1)]. \tag{8.17c}$$

The last form is possible because of the transformation relation

$$X_1 = t_2(X_2, U_2). \tag{8.18}$$

Also

$$F_2(X_2) = \min_{U_1} f_1(X_1,U_1) = \min_{U_1} f_1(X_1,U_2,U_1). \tag{8.19}$$

Let

$$F_2'(X_2) = \min_{U_2} [f_2(X_2,U_2) \circ f_1(X_1)] \tag{8.20}$$

$$= \min_{U_2} \left[f_2(X_2,U_2) \circ \min_{U_2} f_1(X_2,U_2,U_1) \right]. \tag{8.21}$$

Comparing Equations 8.17 and 8.21, we get $F_2'(X_2) \geq F_2(X_2)$.
  Let

$$\min_{U_2} f_1(X_2,U_2,U_1) = f_1(X_2,U_2,U_1^0). \tag{8.22}$$

Then

$$f_1(X_2,U_2,U_1) \geq f_1(X_2,U_2,U_1^0). \tag{8.23}$$

Since ϕ_2 is a monotonic nondecreasing function of f_1, the above inequality implies

$$\begin{aligned} \phi_2(X_2,U_2,U_1) &\geq \phi_2(X_2,U_2,U_1^0) \\ \phi_2(X_2,U_2,U_1) &\leq \min_{U_1} \phi_2(X_2,U_2,U_1^0). \end{aligned} \tag{8.24}$$

Now from Equations 8.21 and 8.23, we observe that

$$\begin{aligned} F_2'(X_2) &= \min_{U_2} \lfloor f_2(X_2,U_2) \circ f_1(X_2,U_2,U_1^0) \rfloor \\ &= \min_{U_2} \phi_2(X_2,U_2,U_1^0) \\ &\leq \min_{U_2} \min_{U_1} \phi_2(X_2,U_2,U_1^0) = sF_2(X_2). \end{aligned}$$

And from Equations 8.15 and 8.17, we get $F_2(X_2) = F_2'(X_2)$ or

$$F_2(X_2) = \min_{U_2} [f_2 \circ F_1(X_1)].$$

The following theorem is an extension of the above to an N-stage optimization problem, treating stages $N-1$ to 1 as a single stage. Theorem 8.5.2 is a direct consequence of Theorem 8.5.1 and needs no further proof.

THEOREM 8.5.2

If the real-valued return function $\phi_N(f_N, f_{N-1}, \ldots, f_1)$ *satisfies*

(a) *Condition of separability, that is*

$$\phi_N(f_N, f_{N-1}, \ldots, f_1) = f_N \circ f_{N-1},$$

where $\phi_N(f_N, f_{N-1}, \ldots, f_1)$ *is real-valued, and*

(b) ϕ_2 *is a monotonic nondecreasing function of* ϕ_{N-1}, *for every* f_N: *then* ϕ_N *is decomposable: that is*

$$\min_{u_N, \ldots, u_1} \phi_N(f_N, \ldots, f_1) = \min_{u_x}\left(f_N \circ \min_{u_{N1}, \ldots, u_1} \phi_{N-1} \right).$$

Theorems 8.5.1 and 8.5.2 indicate that monotonicity is a sufficient condition for decomposability. The converse has not been proved. In fact the monotonicity condition is not necessary.

Often, it is tedious in applications to judge an objective function. Operators that belong to one of the following cases are monotonic and hence the decomposition theorem is applicable.

Case (a): All the operators are additions
  Inequality (Equation 8.23) because

$$P_{k-1}(y_{k-1}) \geq P_{k-1}(y_{k-1}, v_{k-1}, v_{k-2}, \ldots, v_1)$$

for maximization, and the inequality reverses for minimization by definition. Note that abstractions can be converted to additions by absorbing the negative signs.

Case (b): All the operators are multiplications and $f_k(y_k, v_k) \geq 0$ for all $k = 1, 2, \ldots, n$.
  The reason is the same as in case (a).

Case (c): Combination of (a) and (b).
  The reason is the same as case (a).

Case (d): All the operators are comparison.
  In such a case, the maximization is defined as follows.

$$P(y_k) = \max[f_k(y_k v_k), f_{k-1}(y_{k-1}, v_{k-1}), \ldots, f_1(y_1, v_1)],$$

and minimization is defined by replacing maximum by minimum. For maximization, we have max $[f_k(y_k, v_k), P_{k-1}(y_{k-1})] = f_k$. When the first is greater than or equal to the second $= P_{k-1}(y_{k-1})$. Otherwise

$$\max[f_k(y_k, v_k), P_{k-1}(y_{k-1}, v_{k-1}, \ldots, v_1)] = f_k,$$

when the first is greater than or equal to the second, $p_{k-1}(y_{k-1}, v_{k-1}, \ldots, v_1)$ otherwise.

It follows from $P_{k-1}(y_{k-1}) \geq p_{k-1}(y_{k-1}, v_{k-1}, \ldots, v_1)$ that Equation 8.23 holds true. Similarly, Equation 8.23 can be shown valid for minimization.

The task of DP is to carry out the iterative process generated by the decomposition theorem for a monotonic objective function. The advantage of using DP is that only one decision vector is involved in each iteration. As a rule of thumb, it is desirable to reduce the integer r_k as small as possible without violating. This requires more stages but fewer variables in each iteration of the optimization.

8.6 Backward and Forward Recursion

Throughout we have used a recursion procedure in which x_j is regarded as the input and x_{j-1}, as the output for the nth stage, the stage returns are expressed as functions of the stage inputs, and recursive analysis proceeds from stage 1 to stage n. This procedure is called backward recursion because the stage transformation function is of the type $x_{j-1} = t_j(x_j, u_j)$. Backward recursion is convenient when the problem involves optimization with respect to a given input x_n, because then the output x_0 is very naturally left out of consideration.

If, however, the problem is to optimize the system with respect to a given output x_0, it would be convenient to reverse the direction, regard x_j as a function of x_{j-1} and u_j, and put $x_{j-1} = t_j(x_j, u_j)$, $1 \leq j \leq n$ and also express stage returns as functions of stage outputs and then proceed from stage n to stage 1. This procedure is called forward recursion.

In problems where both the input x_N and the output x_0 are given parameters, it is immaterial whether one proceeds in one direction or the other. Both parameters are retained during analysis, and the optimal solution is a function of both. In fact, for most multistage problems there is no essential difference between these two procedures. In mathematical problems inputs and outputs are fictitious concepts and are interchangeable. One can visualize and solve the problem in any direction. It only involves a slight modification in notation.

However, DP is also applicable to nonserial multistage systems, which are important in automatic control systems and in certain process technologies. In such systems stages are not all connected in series but with branches and loops. It is in the application of such systems that the difference between forward and backward recursion procedures becomes not only significant but also crucial. For this reason we proceed to write the recursion formulae for the forward procedure explicitly.

Assume that the return function $\phi_1(x_n, x_0, u_n, \ldots, u_j)$ is a function of the stage returns $f_j(x_j, x_{j-i}u_j)$ in the form

$$\phi_1 = f_n \circ f_{n-1} \circ \cdots \circ f_2 \circ f_j,$$

and further assume that the stage transformation function is given by

$$x_j = i_j(x_{j-1}x_j).$$

Then, by defining

$$F_j(x_{j-1}) = \min_{u_N,\dots,u_f} (f_n \text{ o } f_{n-1}, \dots, f_j)$$

we postulate the forward recursion formulae as

$$F_j(x_{j-1}) = \min_{u_f} \lfloor f_j(x_{j-1},u_j) \text{ o } f_{j+1}(x_j) \rfloor, \quad 1 \le j \le n-1$$

$$F_N(x_{n-1}) = f_n(x_{n-1},u_n).$$

With this notation the optimum value of ϕ_1 which we seek is denoted by $F_1(x_0)$ which is obtained recursively through stages $j = n-1, \dots, 2, 1$.

8.6.1 Minimum Path Problem

Consider the following DP problem where we must find the shortest path from vertex A to vertex B along arcs joining the various vertices lying between A and B (Figure 8.5). The lengths of the paths are as shown.

The vertices are divided into five stages that we denote by subscript j. For $j=0$, there is only one vertex A; also for $j=4$ the only vertex is B. For $j=1,2,3$, there are three vertices in each stage. Each move consists of moving from stage j to stage $j+1$, that is, from any one vertex in stage j to any one vertex in stage $j+1$.

We say that each move changes the state of the system that we denote x_j, and x_0 is the state in which node A lies. Notably, x_0 has only one value,

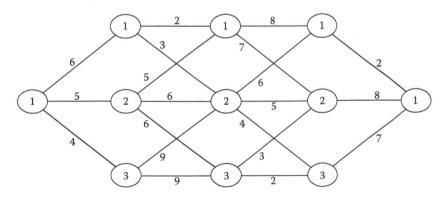

FIGURE 8.5
Minimum path problem.

say $x_0 = 1$. The state x_2 has three possible values, say 1, 2, 3, corresponding to three vertices in stage 2, and so on. We call the possible alternative paths from one stage to the next *decision* variables, and denote by u_j the decision variable that takes us from the state x_{j-1} to state x_j. The return or the gain from the decision u_j we denote by $f_j(u_j)$, return obviously being the function of the decision. In this problem we can identify u_j with the length of the corresponding arc, and so simplify matters by taking $f_j(u_j) = u_j$.

We denote by $F_j(x_j)$ the minimum path from the state, from x_0 any vertex in state x_j. Thus, $F_2(1)$ denotes the minimum path from A to vertex 1 of stage 2. The problem is to determine the minimum path $F_4(x_4)$, and the values of the decision variables u_1, u_2, u_3, and u_4 which yield that path. Let us look at the problem in the following way. The value of u_4 can either be 2, 8, or 7. If $u_4 = 2$ $\Rightarrow x_3 = 1$; similarly $u_4 = 8 \Rightarrow x_3 = 2$, and if $u_4 = 7 \Rightarrow x_3 = 3$, the minimum path from A to B is either through $x_3 = 1$, 2, or 3. With x_3 as 1, 2, or 3, the respective values of u_4 are 2, 8, or 7. Thus

$$F_4(x_4) = \min \begin{cases} 2 + F_3(1) \\ 8 + F_3(2) = \min_{u_4} (u_4 + F_3(x_3)). \\ 7 + F_3(3) \end{cases}$$

Similarly we can argue that

$$F_3(1) = \min \begin{cases} 8 + F_2(1) \\ 6 + F_2(2) \end{cases},$$

or, in general

$$F_3(2) = \min \begin{cases} 7 + F_2(1) \\ 5 + F_2(2) . \\ 3 + F_3(3) \end{cases}$$

Hence

$$F_3(3) = \min \begin{cases} 4 + F_2(2) \\ 2 + F_2(3) \end{cases}.$$

Finally

$$F_3(x_3) = \min_{u_3} (u_3 + F_2(x_2)).$$

We have thus a general recursion formula

$$F_2(x_2) = \min_{u_2} (u_2 + F_1(x_1))$$

that enables us to determine $F_4(x_4)$ recursively.

As we later show, the number of enumerations gone through in this way substantially reduces the total number of enumerations that will have to be gone through if all possible paths are examined.

We now tabulate the information given in the problem and the enumerative steps necessary to find $F_4(x_4)$ with the help of the above recursion formula. We also simultaneously introduce the standard terminology of DP.

Tables 8.1 through 8.4 imply that a function of the type $x_{j-1} = t_j(x_j, u_j)$ exists. It follows from the data of the problem and we call it the stage

TABLE 8.1

Stage 1 for Minimum Path Problem

u_1	6	5	4
x_1	1	2	3

TABLE 8.2

Stage 2 for Minimum Path Problem

u_2 / x_2	x_1				
	2	3	5	6	9
1	1	—	2	—	—
2	—	1	—	2	3
3	—	—	—	2	3

TABLE 8.3

Stage 3 for Minimum Path Problem

u_3 / x_3	x_2						
	2	3	4	5	6	7	8
1	—	—	—	—	2	—	1
2	—	3	—	2	—	1	—
3	3	—	2	—	—	—	—

TABLE 8.4

Stage 4 for Minimum Path Problem

u_4 / x_4	x_3		
	2	7	8
1	1	3	2

transformation function. x_{j-1} may not be defined for all combinations of x_j, u_j. A dash in the tables indicates that the transformation for that pair of values x_j, u_j is not defined and therefore that transformation is not feasible. In Tables 8.5 through 8.8 the recursive operations using the recursive formulae are indicated.

The minimum path from A to B is thus found to be 17. Tracing the minimum path and decisions backwards, the successive decisions are 6, 3, 6, and 2 and the states are $x_0 = 1$, $x_1 = 1$, $x_2 = 2$, $x_3 = 3$, and $x_4 = 1$.

TABLE 8.5

Step 1 for the Recursive Operations

J	x_1	u_1	$F_1(x_1)$
	1	6	6
1	2	5	5
	3	4	4

TABLE 8.6

Step 2 for the Recursive Operations

u_2		$F_1(x_1)$					$F_1(x_1) + u_2$				
x_2	2	3	5	6	9	2	3	5	6	9	$F_2(x_2)$
1	6	—	5	—	—	8	—	10	—	—	8
2	—	6	—	5	4	—	9	—	11	13	9
3	—	—	—	5	4	—	—	—	11	13	11

TABLE 8.7

Step 3 for the Recursive Operations

u_3		$F_2(x_2)$							$F_2(x_2) + u_3$							
x_3	2	3	4	5	6	7	8	2	3	4	5	6	7	8	$F_3(x_3)$
1	—	—	—	—	9	—	8	—	—	—	—	15	—	16	15
2	—	11	—	9	—	8	—	—	—	14	—	14	—	15	—	14
3	11	—	9	—	—	—	—	13	—	13	—	—	—	—	13

TABLE 8.8

Step 4 for the Recursive Operations

u_4		$F_3(x_3)$			$F_3(x_3) + u_4$			
x_4	2	8	7	2	8	7	$F_4(x_4)$
1	15	14	13	17	22	20	17

8.6.2 Single Additive Constraint and Additively Separable Return Problem

Consider the following problem as another illustration of the method of DP.
Find the value of u_j that minimizes

$$z = \sum_{j=1}^{n} f_j(u_j),$$

where $1 \leq j \leq n$, subject to the constraints

$$\sum_{j=1}^{n} a_j u_j \geq b, \ a_j, \ b \in R, \ a_j \geq 0, \ b$$

$$u_j \geq 0.$$

The objective or return function z is a separable additive function of the n variable u_j. We look upon the problem as an n-stage problem, the suffix j indicating the stage. We have to decide about the values of u_j, and so the u_j is called decision variables. With each decision u_j is associated a return function $f_j(u_j)$ which is the return at the jth stage.

Now we introduce the variables $x_0, x_1, x_2, \ldots, x_n$ defined as follows:

$$\left.\begin{array}{l} x_n = a_1 u_1 + a_2 u_2 + \cdots + a_n u_n \geq b \\ x_{n-1} = a_1 u_1 + a_2 u_2 + \cdots + a_{n-1} u_{n-1} = x_n - a_n u_n \\ x_{n-2} = a_1 u_1 + a_2 u_2 + \cdots + a_{N-2} u_{n-2} = x_{n-1} - a_{n-1} u_{n-1} \\ \vdots \qquad\qquad \vdots \\ x_1 = a_1 u_1 = x_2 - a_2 u_2 \end{array}\right\}. \qquad (8.25)$$

These variables are called the state variables. We further notice that $x_{j-1} = t_j(x_j, u_j)$ for $1 \leq j \leq n$. That is, each state variable is a function of the next state and decision variable. This is the stage transformation function. Since x_n is a function of all the decision variables we may denote by $F_n(x_n)$ the minimum value of u for any feasible function of x_n,

$$F_n(x_n) = \min_{u_1, u_2, \ldots, u_n} [f_1(u_1) + f_2(u_2) + \cdots + f_n(u_n)], \qquad (8.26)$$

the minimization being over nonnegative values of u_j subject to $x_n \geq b$.

We select a particular value of u_n and holding u_n fixed, we minimize u over the remaining variables. This minimum is given by

$$f_n(u_n) + \min_{u_1, u_2, \ldots, u_{n-1}} [f_1(u_1) + f_2(u_2) + \cdots + f_{n-1}(u_{n-1})] = f_n(u_n) + F_{n-1}(x_{n-1}).$$

$$(8.27)$$

And the values of $u_1, u_2, \ldots, u_{n-1}$ that would make $\sum_{j=1}^{-1} f_j(u_j)$ minimum for a fixed u_n thus depending upon x_{n-1} which in turn is a function of x_n and u_n. Also, the minimum u over all u_n for feasible x_n would be

$$F_n(x_n) = \min_{u_n} \left[f_n(u_n) + F_{n-1}(x_{n-1}) \right]. \tag{8.28}$$

If somehow $F_{n-1}(x_{n-1})$ were known for all u_n, the above minimization would involve a single variable u_n. Repeating the argument, we get the recursion formula

$$F_j(x_j) = \min_{u_j} \left\lfloor f_j(u_j) + f_{j-1}(x_{j-1}) \right\rfloor, \quad 2 \le j \le n$$

$$F_1(x_1) = f_1(u_1), \tag{8.29}$$

which along with the relation $x_{j-1} = t_j(x_j, u_j)$ defines a typical DP problem.

If we could make a start with $F_1(x_1)$ and recursively go on to optimize to get $F_2(x_2), F_3(x_3), \ldots, F_n(x_n)$, each optimization being done over a single variable, we would get $F_n(x_n)$ for each possible x_n. Minimizing it over x_n we get the solution. The following numerical examples illustrate how this can be done.

8.6.3 Single Multiplicative Constraint, Additively Separable Return Problem

Consider the problem

$$\min \sum_{j=1}^{n} f_j(u_j)$$

subject to the constraints

$$\prod_{j=1}^{n} u_j \ge k > 0, \quad u_j \ge 0.$$

We introduce the state variables x_j defined as follows.

$$\left. \begin{aligned}
x_n &= u_n u_{n-1} = \cdots = u_2 u_1 \ge k, \\
x_{n-1} &= x_n / u_n = u_{n-1} \ldots u_2 u_1, \\
&\cdots \\
x_2 &= x_3 / u_2 = u_2 u_1, \\
x_1 &= x_2 / u_2 = u_1
\end{aligned} \right\}. \tag{8.30}$$

These are the stage transformations of the type

$$x_{j-1} = t_j(x_j, u_j).$$

Denoting by $F_n(x_n)$ the minimum value of the objective function for any feasible x_n, we can get the recursion formula

$$F_j(x_j) = \min_{u_j} \left\lfloor f_j(u_j) + f_{j-1}(x_{j-1}) \right\rfloor, \quad 2 \leq j \leq n$$

$$F_1(x_1) = f_1(u_1) \tag{8.31}$$

which will lead to the solution.

Example 8.6.1 (see Tables 8.9 and 8.10)

Maximize $u_1^2 + u_2^2 + u_3^2$

Subject to

$u_1 u_2 u_3 \leq 6$

and

u_1, u_3, u_3 are positive integers.

SOLUTION

The state variables are

$$x_3 = u_1 u_2 u_3 \leq 6, \quad x_2 = x_3/u_3 = u_1 u_2, \quad x_1 = x_2/u_2 = u_1.$$

TABLE 8.9

Stage 1 for Example 8.6.1

u_j	1	2	3	4	5	6
$f_j(u_j)$	1	4	9	16	25	36

TABLE 8.10

Stage 2 for Example 8.6.1

			u_j			
x_j	1	2	3	4	5	6
1	1	—	—	—	—	—
2	2	1	—	—	—	—
3	3	—	1	—	—	—
4	4	2	—	1	—	—
5	5	—	—	—	1	—
6	6	3	2	—	—	1

The solution is worked out as follows:
Stage return

$$f_j(u_j) = u_j^2; \quad j = 1, 2, 3.$$

Stage transformation

$$x_{j-1}(u_j), \quad j = 2, 3.$$

Recursion operation (see Tables 8.11 through 8.13)
The answer is 38 with $u_3 = 1$, $u_2 = 1$, $u_1 = 6$.

TABLE 8.11

Step 1 for the Recursive Operation

x_1	1	2	3	4	5	6
$F_1(x_1)$	1	4	9	16	25	36

TABLE 8.12

Step 2 for the Recursive Operation

u_2 / x_2	$f_2(u_2)$						$F_1(x_1)$						$F_2(x_2)$
	1	2	3	4	5	6	1	2	3	4	5	6	
1	1	—	—	—	—	—	1	—	—	—	—	—	2
2	1	4	—	—	—	—	4	1	—	—	—	—	5
3	1	—	9	—	—	—	9	—	1	—	—	—	10
4	1	4	—	16	—	—	16	4	—	1	—	—	17
5	1	—	—	—	25	—	25	—	—	—	1	—	26
6	1	4	9	—	—	36	36	9	4	—	—	1	37

TABLE 8.13

Step 3 for the Recursive Operations

u_3 / x_3	$f_3(u_3)$						$F_2(x_2)$						$F_3(x_3)$
	1	2	3	4	5	6	1	2	3	4	5	6	
1	1	—	—	—	—	—	2	—	—	—	—	—	3
2	1	4	—	—	—	—	5	2	—	—	—	—	6
3	1	—	9	—	—	—	10	—	2	—	—	—	11
4	1	4	—	16	—	—	17	5	—	2	—	—	18
5	1	—	—	—	25	—	26	—	—	—	2	—	27
6	1	4	9	—	—	36	37	10	5	—	—	2	38

8.6.4 Single Additive Constraint, Multiplicatively Separable Turn Problem

Consider the problem:

$$\text{Maximize } \prod_{j=1}^{n} f_j(u_j) \tag{8.32}$$

Subject to

$$\sum_{j=1}^{n} a_j u_j = k \tag{8.33}$$

and

$$u_j \geq 0, \quad a_j \geq 0.$$

Here, our state variables are

$$x_n = \sum_{j=1}^{n} a_j u_j = k,$$

$$x_{j-1} = x_j - a_j u_j, \quad 2 \leq j \leq n.$$

Putting

$$F_j(x_j) = \max_{u_1, u_2, \ldots, u_j} \prod_{j=1}^{n} f_j(u_j), \quad 2 \leq j \leq n,$$

the general recursion formula is

$$F_j(x_j) = \max_{u_j} \prod_{j=1}^{n} \left[f_j(u_j) F_{j-1}(x_{j-1}) \right], \quad j - n, n - 1, \ldots, 2 \tag{8.34}$$

$$F_1(x_1) = f_1(u_1).$$

8.7 Computational Procedure in DP

We now discuss the use of the recurrence relationships. DP begins by sub-optimizing the fast component which means the determination of

$$f_1^*(x_2) = \operatorname*{opt}_{u_1} [F_1(u_1, x_2)].$$

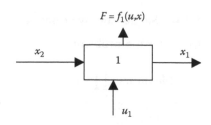

FIGURE 8.6
Suboptimization problem of component 1 for various settings of input state variable x_2.

The best value of the decision variable u_1, denoted u_1^*, is that which makes the objective function f_1, denoted f_1^*, hit its maximum value. Both depend on the condition of the input or feed that component 1 receives from upstream. Figure 8.6 shows a summary of the suboptimization problem of stage 1.

Next, we move up the serial system to include the fast two components. In this two-stage suboptimization, we have to determine

$$f_2^*(x_3) = \underset{u_2,u_1}{\text{opt}} \; [F_1(u_1,x_2) + F_2(u_2,x_3)]$$

since all the information about the first stage has already been calculated. Then the result can be substituted to get the following equation for the second stage

$$f_2^*(x_3) = \underset{u_2}{\text{opt}} \lfloor f_1^*(x_2) + F_2(u_2,x_3) \rfloor,$$

assuming the sequence has been carried on to include $(i-1)$ at the end of the components. Then the solution will be the solution of

$$f_i^*(x_{i+1}) = \underset{u_i,u_{i-1}.....u_1}{\text{opt}} \; [F_i + F_{i-1} + \cdots + F_1].$$

However, all the information regarding the suboptimization of $(i-1)$ end components is known and has been entered in the table corresponding to f_{i-1}^*. Then we can obtain

$$f_i^*(x_{i+1}) = \underset{u_i}{\text{opt}} \; [F_i(u_i,x_{i+1}) + f_{i-1}^*(x_i)];$$

thus the dimension of the ith stage has been reduced to one.

8.8 Computational Economy in DP

DP for discrete problems is an enumerative procedure. A number of alternatives are examined at each stage and some selected for further examination. For large-scale problems, the number of alternatives is very large. Should we examine all the possible alternatives? The question now is if there is any saving in DP. And if so what is the order of the saving.

Consider a problem with n stages and let each decision variable u_j have p possible values and also each state variable u_j have p feasible values and an additive return function. In the direct exhaustive search, a feasible solution is specified by an input value x_n and the values of the decision variables u_j $(j = 1, 2, \ldots, n)$ and each component has p values. Then the total number of feasible solutions is p^{n+1}. To get the objective from each feasible solution, we have to add two at a time involving $(n-1)$ additions; it means $(n-1)p^{n+1}$ additions. Finally to choose the optimum of these we have to make (p^{n+1}) comparisons. The total number of steps is

$$N(\text{DS}) = (n-1)p^{n+1} + p^{n+1} - 1 = np^{n+1} - 1. \qquad (8.35)$$

But for DP, for stages starting at the second one up to n, for every combination, p^2 combinations per stage, we need only one addition. The total number of additions is $(n-1)p^2$ additions. Also at each stage starting from the first one, for each value of x_j, p numbers have to be compared in $(p-1)$ comparisons giving a total of $[np(p-1)]$. To get the optimum value, p numbers have to be compared. Then the total number of additions and comparisons in DP are

$$N(\text{DP}) = (n-1)p^2 + np(p-1) + p - 1 = (2n-1)p^2 - (n-1)p - 1, \qquad (8.36)$$

which is much less than in the exhaustive search case.

8.9 Systems with More than One Constraint

DP can be applied to problems involving many constraints. In multiconstraint problems, there has to be one state variable per constraint per stage. Sometimes it is possible to take advantage of the problem structure to return the number of state variables.

Example 8.9.1

Maximize $u_1^2 + u_2^2 + u_3^2$

Subject to

$u_1 u_2 u_3 \leq 6, \quad u_1 + u_2 + u_3 \leq 6$

and u_1, u_2, u_3 are positive integers.

SOLUTION

The two sets of state variables are

$$x_3 = u_1 u_2 u_3 \qquad y_3 = u_1 + u_2 + u_3$$
$$x_2 = x_3/u_3 = u_1 u_2 \quad y_2 = y_3 - u_3 = u_1 + u_2$$
$$x_1 = x_2/u_2 = u_1 \qquad y_1 = y_2 - u_2 = u_1.$$

The feasible values of u_j are 1, 2, 3, 4. For stage $j=1$, the stage transformation gives the following possible values of x_1, y_1.

$$
\begin{array}{ccccc}
u_1 & 1 & 2 & 3 & 4, \\
x_1 & 1 & 2 & 3 & 4, \\
y_1 & 1 & 2 & 3 & 4.
\end{array}
$$

For $j=2$, 3, Table 8.14 gives the transformations

$$x_{j-1} = t_{j-1}(x_j,u_j), \quad y_{j-1} = t'_{j-1}(y_j,u_j).$$

Because of the constraints we do not need to consider x_j, $y_j > 6$ (see Tables 8.15 through 8.17). Hence max $F_3(x_3,y_3) = 18$ for $(x_3,y_3) = (4,6)$. Tracing back, the optimal decision variables are (1, 1, 4) or (1, 4, 1) or (4, 1, 1).

TABLE 8.14

Transformation for Example 8.9.1

u_j		x_{j-1}, y_{j-1}		
$x_i y_j$	1	2	3	4
1,1	1,0	—,—	—,—	—,—
2,2	2,1	1,0	—,—	—,—
3,3	3,2	—,1	1,0	—,—
4,4	4,3	2,2	—,1	1,0
5,5	5,4	—,3	—,2	—,1
6,6	6,5	3,4	2,3	—,2

TABLE 8.15

Stage 1 of Optimization of Example 8.9.1

u_1	X_1	Y_1	$F(x_1,y_1)$
1	1	1	1
2	2	2	4
3	3	3	9
4	4	4	16

TABLE 8.16

Stage 2, $F_2(X_2,Y_2) = \min\limits_{U_2} \lfloor U_2^2 + F_1(X_1,Y_1) \rfloor$

u_2	x_2	y_2	$F_1(x_1,y_1)$	$u_2^2 + F_1(x_1,y_1)$	x_2	y_2	$F_2(x_2,y_2)$
	1	1	1	2	1	2	2
	2	2	4	5	2	3	5
1	3	3	9	10	3	4	10
	4	4	16	17	4	5	17
	1	1	1	5	2	3	—
2	2	2	4	8	4	4	8
	3	3	9	13	6	5	13
3	1	1	1	10	3	4	—
	2	2	4	13	6	5	—
4	1	1	1	17	4	5	—

TABLE 8.17

Stage 3, $F_3(x_3,y_3) = \max\limits_{U_3} \lfloor u_3^2 + F_2(x_2,y_2) \rfloor$

u_3	x_2	y_2	$F_2(x_2,y_2)$	$[u_3^2 + F_2(x_2,y_2)]$	x_3	y_3	$F_3(x_3,y_3)$
	1	2	2	3	1	3	3
	2	3	5	6	2	4	6
	3	4	10	11	3	5	11
1	4	4	8	9	4	5	9
	4	5	17	18	4	6	18
	6	5	13	14	6	6	14
	1	2	2	6	2	4	—
2	2	3	5	9	4	5	—
	3	4	10	14	6	6	—
3	1	2	2	11	3	5	—
	2	3	5	14	6	6	—
4	1	2	2	18	4	6	—

8.10 Conversion of a Final Value Problem into an Initial Value Problem

Before we discussed DP with reference to an initial value problem. But if the problem is a final value problem as shown in Figure 8.7, it can be converted into an equivalent initial value problem. If the state transformation was given by Equation 8.1

$$x_i = t_i(x_{i+1}, u_i) \quad i = 1, 2, \ldots, n.$$

Assuming the inverse relations exist, we can write

$$x_{i+1} = \bar{t}_i(x_i, u_i) \quad i = 1, 2, \ldots, n,$$

where the input state to stage i is expressed as a function of its output state and its decision variable. Also if the objective function was originally expressed as

$$F_i = f_i(x_{i+1}, u_i) \quad i = 1, 2, \ldots, n,$$

it can be used to express it in terms of the output stage and the decision variable as

$$F_i = f_i[\bar{t}_i(x_i, u_i), x_i] = \bar{f}_i(x_i u_i), \quad i = 1, 2, \ldots, n.$$

Then we can use the same original approach as before in this new problem.

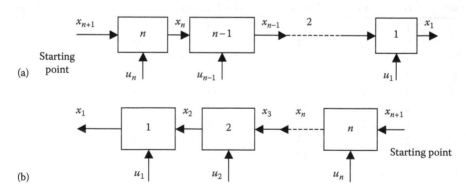

FIGURE 8.7
Conversion of (a) final value problem to (b) initial value problem.

8.11   Illustrative Examples

Example 8.11.1

Maximize $x_1^2 + x_2^2 - x_3^2$

Subject to

$x_1 + x_2 + x_3 \geq 14$

Put

$$u_3 = x_1 + x_2 + x_3$$
$$u_2 = x_1 + x_2 = u_3 - x_3$$
$$u_1 = x_1 = u_2 - x_2$$
$$F_3(u_3) = \max_{x_3} \left[-x_3^2 + F_2(u_2) \right]$$
$$F_2(u_2) = \max_{x_2} \left[x_2^2 + F_1(u_1) \right]$$
$$F_1(u_1) = \left[x_1^2 \right] = (u_2 - x_2)^2.$$

Substituting in $F_2(u_2) = \max_{x_2}\left[x_2^2 + (u_2 - x_2)^2 \right]$, by calculus, a function is a maximum if its partial differential equals zero. $F_2(u_2) = \left[x_2^2 + (u_2 - x_2)^2 \right]$ is maximum if

$$\frac{\partial F_2(u_2)}{\partial x_2} = [2x_2 - 2(u_2 - x_2)] = 0 \quad \text{or} \quad x_2 = 2u_2.$$

Hence $F_2(u_2) = 5u_2^2$. Now

$$F_3(u_3) = \max_{x_3} \left[-x_3^2 + F_2(x_2) \right]$$
$$= \max_{x_3} \left[-x_3^2 + 5u_2^2 \right]$$
$$= \min_{x_3} \left[-x_3^2 + 5(u_3 - x_3)^2 \right].$$

Again, by calculus, a function is maximum, if its partial differentiation equals zero.

$$F_3(u_3) = \left[-x_3^2 + 5(u_3 - u_3)^2 \right]$$

is maximum if

$$\frac{\partial F_3(u_3)}{\partial x_3} = [-2x_3 - 10(u_3 - x_3)] = 0 \quad \text{or} \quad x_3 = \frac{5}{4}u_3.$$

Hence

$$F_3(u_3) = \left(\frac{-5}{4}\right)u_3^2, \quad u_3 \geq 14.$$

Obviously $F_3(x_3)$ is the maximum for $u_3 = 14$ which means that the maximum value of $(x_1^2 + x_2^2 - x_3^2)$ is therefore (-245). Back substitution to calculate the x_1, x_2, and x_3 gives

$$x_3 = \left(\frac{5}{4}\right)u_s = \left(\frac{5}{4}\right)(14) = 17.5, \quad u_2 = u_3 - x_3 = 14 - 17.5 = -3.5$$

$$x_2 = 2u_2 = -7, \quad u_1 = u_2 - x_2 = 3.5 = x_1.$$

Then the final result is $f_{max} = (-245)$, $x_1 = (3.5)$, $x_2 = (-7)$, and $x_3 = (17.5)$.

Example 8.11.2

The network shown in Figure 8.8 illustrates a transmission model with node (1) representing the generation unit, and nodes (2)–(7) representing load centers. The values associated with branches are power losses.

1. Derive an optimal policy of supplying the load at node (8) from the generation node (1) with minimum losses.
2. Derive an optimal policy of supplying the load at node (8) from the generation node (1) with minimum losses at node (1), but also supplying a load at node (4).

Solution

1. This is a minimization problem,

$$F_i(x_j) = \min_{u_j} \lfloor u_j + F_{j-1}(x_{j-1}) \rfloor,$$

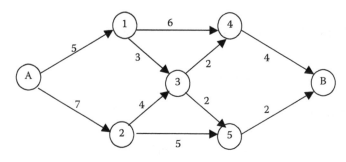

FIGURE 8.8
Transmission model for Example 8.11.2.

where, u_j is the connection from stage $(j-1)$ to stage j, based on the fact that $F_1(u_1) = u_1$. Then, $F_1(2) = 3$ and $F_1(3) = 5$.

$$F_2(x_2) = \min_{u_2} [u_2 + F_1(x_1)] = \min[F_2(4), F_2(5)]$$
$$F_2(4) = \min[10 + F_1(2), 11 + F_1(3)] = \min[10 + 3, 11 + 5] = 13$$
$$F_2(5) = \min[4 + f_1(3), 12 + F_1(2)] = \min[4 + 5, 12 + 3] = 9.$$

Then, $F_2(x_2) = \min[F_2(4), F_2(5)] = \min[13,9] = 9$. In a similar way,

$$F_3(x_3) = \min_{u_3} [u_3 + F_2(x_2)] = \min_{u_3} [F_3(6), F_3(7)].$$
$$F_3(6) = \min[15 + F_2(4), 10 + F_2(5)] = \min[15 + 13, 10 + 9] = 19$$
$$F_3(7) = \min[9 + F_2(4), 16 + F_2(5)] = \min[9 + 13, 16 + 9] = 22.$$

Then

$$F_3(x_3) = \min[F_3(6), F_3(7)] = \min[19, 22] = 19$$
$$F_4(x_4) = \min_{u_4} [u_4 + F_3(x_3)] = \min[9 + F_3(6), 8 + F_3(7)]$$
$$= \min[(9 + 19), (8 + 23)] = 29.$$

Then the minimum route is {node (1) – node (3) – node (4) – node (6) – node (8)}, and the value of the losses is (28).

2. We must pass by the node (4). Then the system is divided into two separate problems omitting node (5) as shown in Figure 8.9.

For the first part of the system:

$$F_1(u_1) = u_1.$$

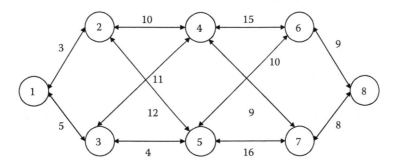

FIGURE 8.9
Reduced transmission model.

Then

$$F_1(2) = 3 \quad \text{and} \quad F_1(3) = 5$$

$$F_2(4) = \min_{u_2} [u_2 + F_1(x_1)] = \min[10 + F_1(2), \ 11 + F_1(3)]$$

$$= \min[13,16] = 13.$$

For the second part of the system:

$$F_3(x_3) = \min_{u_3} [u_3 + F_2(x_2)] = \min_{u_3} [F_3(6), \ F_3(7)]$$

$$= \min[(13 + 15), \ (13 + 9)] = 22$$

$$F_4(x_4) = \min_{u_4} [u_4 + F_3(x_3)] = \min[9 + F_3(6), \ 8 + F_3(7)]$$

$$= \min[(9 + 28), \ (8 + 22)] = 30.$$

Then the minimum route is {node (1) – node (2) – node (4) – node (7) – node (8)}, and the value of the losses is (30).

Example 8.11.3

A computer company has accepted a contract to supply 100 computers at the end of the first month and 150 at the end of the second month. The cost of manufacturing a computer is given by $\$(70x + 0.2x^2)$ where x is the number of manufactured units in that month. If the company manufactures more computers than needed in the first month, there is an inventory carrying charge of $80 for each unit carried over to the next month. Find the number of computers manufactured in each month to maintain the cost at a minimum level, assuming that the company has enough facilities to manufacture up to 250 computers per month.

Solution

The total cost is composed of the production cost and the inventory carrying costs. The constrained optimization problem can be stated as follows:

Minimize: $f(x_1,x_2) = (70x_1 + 0.2x_1^2) + (70x_2 + 0.2x_2^2) + 80(x_1 - 80)$

Subject to

$$x_1 \geq 100, \quad x_1 + x_2 = 250$$

and x_1, x_2 are positive integers,

where x_1, x_2 denote the number of computers manufactured in the first and second months, respectively. This problem can be considered as a two-stage decision problem as shown in Figure 8.10.

Assuming the optimal solution of the first month equals $x_1^* = 100 + u_2$, the objective of the first month will be

$$f_1^* = 70(100 + u_1) + 0.2(100 + u_1)^2 = 9000 + 110u_1 + 0.2u_1^2.$$

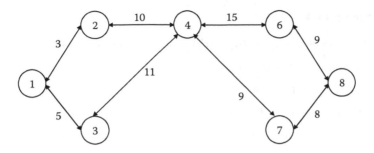

FIGURE 8.10
Two-stage decision problem.

The cost incurred in the second month is given by

$$f_2(x_2,u_1) = 70x_2 + 0.23x_2^2 + 10u_1.$$

The total cost will be

$$F(x_1,x_2,u_1) = 9000 + 110u_1 + 0.2u_1^2 + 70x_2 + 0.2x_2^2 + 10u_1$$
$$= 9000 + 120u_1 + 0.2u_1^2 + 70x_2 + 0.2x_2^2.$$

But the amount of inventory at the beginning of the second month plus the production in the second month must be equal to the supply in the second month. We have

$$x_2 + u_1 = 150 \Rightarrow u_1 = 150 - x_2.$$

Substituting for u_1 in the total cost function, we get

$$F = 9000 + 120(150 - x_2) + 0.2(150 - x_2)^2 + 70x_2 + 0.2x_2^2$$
$$= 31{,}500 - 110x_2 + 0.4x_2^2,$$

since this last equation is a function only in x_2. Then we can get the optimum value by setting

$$\frac{\partial F}{\partial x_2} = 0.0 \Rightarrow 110 = 0.8x_2 \Rightarrow x_2^* = 138.$$

Checking the second derivative,

$$\frac{\partial^2 F}{\partial x_2^2} = 0.8 > 0.$$

Then the second month's production of 80 corresponds to the minimum cost. The first month's production and the inventory will be $x_1^* = 112$, $u_1 = 12$ and the minimum total cost equals ($23937.6).

Example 8.11.4

$$\text{Max } P = (4d_1 - d_1^2) + (4d_2 - 2d_2^2) + (4d_3 - 3d_3^2) + (4d_4 - 4d_4^2)$$

s.t. $d_1 + d_2 + d_3 + d_4 = c$

$d_k \geq 0, \quad k = 1, \ldots, 4$

Set $u_4 = d_1 + d_2 + d_3 + d_4$

$u_3 = d_1 + d_2 + d_3 = u_4 - d_4$

$u_2 = d_1 + d_2 = u_3 - d_3$

$u_1 = d_1 = u_2 - d_2$

$$F_4(u_4) = \max_{d_4} \left[4d_4 - 4d_4^2 + F_3(u_3)\right]$$

$$F_3(u_3) = \max_{d_3} \left[4d_3 - 3d_3^2 + F_2(u_2)\right]$$

$$F_2(u_2) = \max_{d_2} \left[4d_2 - 2d_2^2 + F_1(u_1)\right]$$

$$F_1(u_1) = \left[4d_1 - d_1^2\right] = 4u_2 - 4d_2 - u_2^2 + 2u_2d_2 - d_2^2$$

Substituting $F_2(u_2) = \max_{d_2} \left[-3d_2^2 + 4u_2 - u_2^2 + 2u_2d_2\right]$

For $(\partial^2 F_2(u_2)/\partial u d_2)|_{\max} = 0 = -6d_2 + 2u_2$, We obtain $d_2 = (1/2)u_2$

Hence

$$F_2(u_2) = -3\left(\frac{1}{3}u_2\right)^2 + 4u_2 - u_2^2 + 2u_2\left(\frac{1}{3}u_2\right) = 4u_2 - \frac{2}{3}u_2^2$$

$$F_3(u_3) = \max_{d_3}\left[4d_3 - 3d_3^2 + 4u_2 - \frac{2}{3}u_2^2\right] = \max_{d_3}\left[-3\frac{2}{3}d_3^2 + 4u_3 - \frac{2}{3}u_3^2 + \frac{4}{3}u_3d_3\right]$$

For $\max(\partial^2 F_3/\partial d_3) = (22/3)d_3 + (4/3)u_3 = 0$, we obtain $d_3 = (2/11)u_3$

Hence

$$F_3(u_3) = -\frac{11}{3}\left(\frac{2}{11}u_3\right)^2 + 4u_3 - \frac{2}{3}u_3^2 + \frac{4}{3}u_3\left(\frac{2}{11}u_3\right) = 4u_3 - 0.303u_3^2$$

And

$$F_4(u_4) = \max_{d_4}\left[4d_4 - 4d_4^2 + 4u_3 - 0.303u_3^2\right]$$

$$= \max_{d_4}\left[-4.303d_4^2 + 4u_4 - 0.303u_4^2 + 0.606u_4d_4\right]$$

For $(\partial^2 F_4/\partial d_4) = -8.606d_4 + 0.606u_4 = 0$, we obtain $d_4 = 0.07u_4$

Hence $F_4(u_4) = -4.303(0.07u_4)^2 + 4u_4 - 0.303(u_4^2 + 6.06u_4(0.07u_4))$

$$= 0.1002u_4^2 + 4u_4$$

Let $u_4 = c$, maximum value of $P = 0.1002c^2 + 4c$, back substituting to calculate d_1, d_2, d_3, and d_4 gives,

$$d_4 = 0.07c, \quad u_3 = u_4 - d_4 = 0.93c$$

$$d_3 = \frac{2}{11} u_3 = 0.169c, \quad u_2 = u_3 - d_3 = 0.761c = 0.761c$$

$$d_2 = \frac{1}{3} u_2 = 0.254c, \quad u_1 = u_2 - d_2 = 0.507c$$

$$d_1 = u_1 = 0.507c$$

SOLUTION

$$P_{max} = 0.1002c^2 + 4c, \quad \begin{bmatrix} d_1 \\ d_2 \\ d_3 \\ d_4 \end{bmatrix} = \begin{bmatrix} 0.507c \\ 0.254c \\ 0.169c \\ 0.07c \end{bmatrix} = c \begin{bmatrix} 0.507 \\ 0.254 \\ 0.169 \\ 0.07 \end{bmatrix}.$$

Example 8.11.5

Solve Example 8.11.1 completely

Max $x_2^1 + x_2^2 - x_3^2$

s.t. $x_1 + x_2 + x_3 \geq 14$

Put $u_3 = x_1 + x_2 + x_3$

$u_2 = x_1 + x_2 = u_3 - x_3$

$u_1 = x_1 = u_2 - x_2$

$F_3(u_3) = \max_{x_3} \left[-x_3^2 + F_2(u_2) \right]$

$F_2(u_2) = \max_{x_2} \left[x_2^2 + F_1(u_1) \right]$

Let $x_1 = u_1 = u_2 - x_2$, $F_1(u_1) = x_1^2 = (u_2 - x_2)^2$
Substitute into $F_2(u_2)$ we get

$$F_2(u_2) = \max_{d_2} \left[x_2^2 + u_2^2 - 2u_2x_2 + x_2^2 \right]$$

At the maximum point $(\partial^2 F_2/\partial x^2) = 0$, we obtain $x_2 = 1/2u_2$
Hence

$$F_2(u_2) = \frac{1}{2} u_2^2$$

Now

$$F_3(u_3) = \max_{d_3} \left[-x_3^2 + \frac{1}{2} u_2^2 \right]$$

But

$$u_2 = u_3 - x_3, \quad u_2 = u_3 - x_3, \quad F_3(u_3) = \max_{x_3} \left[\frac{1}{2} x_3^2 + \frac{1}{2} u_3^2 - u_3 x_3 \right]$$

For $(\partial^2 F_2 / \partial x_2) = -x_3 - u_3 = 0$, we obtain $x_3 = -u_3$
  Hence, $F_3(u_3) = u_3^2$
  But $u_3 = x_1 + x_2 + x_3 \geq 14$, thus $u_{3\max} = 14$
  $\therefore F_3(u_3)_{\max} = 196 = $ max value of $(x_1^2 + x_2^2 - x_3^2)$
Back substituting to calculate x_1, x_2 and x_3 gives

$$x_3 = -u_3 = -14$$

$$u_2 = u_3 - x_3 = 28$$

$$x_2 = \frac{1}{2} u_2 = 14$$

$$u_1 = u_2 - x_2 = 14$$

$$x_1 = u_1 = 14$$

SOLUTION

  $f_{\max} = 196$, $x_1 = 14$, $x_2 = 14$, and $x_3 = -14$.

8.12   Conclusions

This chapter presented DP as an optimization approach able to transform a complex problem into a sequential set of simpler problems. Both discrete and continuous problems were considered. The formulation of a multistage decision process was explained step by step by discussing its representation as well as the types of multistage decision problems. Characteristics of the DP method were discussed in detail. The concept of suboptimization and the principle of optimality were used as an initial stage in presenting the formulation of DP as a consequence of the simple optimization problem. This chapter also discussed the different forms of DP approaches used in continuous and discrete optimization problems, in particular forward and backward DP. The derivation of the recursive formulae used for both approaches was presented. Computational procedures were shown through some illustrative examples that also explained the computational economy in DP. All the discussion was devoted to optimization problems with only one constraint; however, a discussion for systems with more than one constraint and the conversion of a final value problem into an initial value problem was presented in the last two sections of the chapter, with an illustrative example provided to support the argument. Finally, the chapter gave a set of unsolved problems for training and understanding the DP approach.

8.13 Problem Set

PROBLEM 8.13.1

Consider a transportation problem with m sources and n destinations. Let a_i be the amount available at source i, $i = 1, 2, \ldots, m$, and let b_j be the amount demanded at destination j, $j = 1, 2, \ldots, n$. Assuming that the cost of the transporting x_{ij} units from source i to destination j is $h_{ij}(x_{ij})$, formulate the problem as a DP model.

PROBLEM 8.13.2

Solve the following linear programming problem as a DP model.

Maximize $Z = 4x_1 + 4x_2$

Subject to

$2x_1 + 7x_2 \leq 21$

$7x_1 + 2x_2 \leq 21$

$x_i \geq 0$

and nonnegative for all *is*.

PROBLEM 8.13.3

Solve the following nonlinear programming problem as a DP model.

Maximize $Z = 7x_1^2 + 6x_1 + 5x_2^2$

Subject to

$x_1 + 2x_2 \leq 10$

$x_1 - 3x_2 \leq 9$

$x_1 \geq 0$

and nonnegative for all *is*.

TABLE 8.18

Data for Problem 8.13.4

Resource Unit	Plant 1	Plant 2	Plant 3
1	5	7	8
2	9	10	11
3	12	12	14
4	14	13	16

PROBLEM 8.13.4

Find the allocation for maximal profit when four resource units are available as shown in Table 8.18, where the numbers denote dollars ($) per resource unit.

PROBLEM 8.13.5

Maximize the hydroelectric power $P(X)$, $X = (x_1, x_2, x_3)^T$, produced by building dams on three different river basins, where

$$P(X) = f_1(x_1) + f_2(x_2) + f_3(x_3),$$

and $f_i(x_i)$ is the power generated from the ith basin by investing resource x_i. The total budgetary provision is $x_1 + x_2 + x_3 \leq 3$. The functions f_1, f_2, and f_3 are given in Table 8.19. An integer solution is required for this problem.

PROBLEM 8.13.6

Use Table 8.20 regarding buying and selling only in a deregulated power system with limited energy storage. A utility has a limited electrical energy storage parity of 1000 MW h and must use this to the greatest advantage. The price of the energy changes from month to month in the year. Assume that the energy storage system must be inspected (cleaned) once a year: it takes one month to complete and is scheduled for the first of July each year.

TABLE 8.19

Data for Problem 8.13.5

x_i	1	2	3	4
f_1	0	2	4	6
f_2	0	1	5	6
f_3	0	3	5	6

TABLE 8.20

Data for Problem 8.13.6

Months	Price ($)	MW h
January	18	20
February	18	19
March	18	16
April	15	17
May	14	16
June	15	16
July	17	16
August	17	17
September	17	18
October	17	19
November	17	19
December	18	19

PROBLEM 8.13.7

The network shown in Figure 8.11 represents a DC transmission system. In the network, the nodes represent substations. The values associated with the branches are voltage drops (pu). Derive an optimal policy of supplying a load at node (6) from the generation at node (1) with minimum voltage drop.

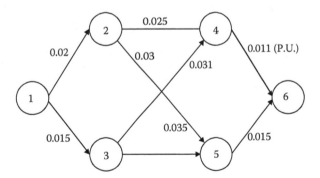

PROBLEM 8.13.8

(a) Define the dynamic program. Illustrate with all the notations and blocks change for multistage problem.

(b) Showing all steps, use DP method to solve the following problem.

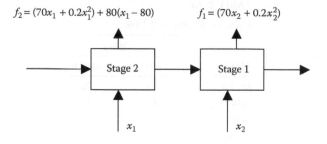

FIGURE 8.11
DC transmission system for Problem 8.13.7.

$$\text{Maximize } P = \left(4d_1 - d_1^2\right) + \left(4d_2 - 2d_2^2\right) + \left(4d_3 - 3d_3^2\right) + \left(4d_4 - 4d_4^2\right)$$

$$\text{s.t. } d_1 + d_2 + d_3 + d_4 = C$$

$$d_k \geq 0, \quad k = 1, \ldots, 4$$

PROBLEM 8.13.9

Use DP technique to find the maximum path through the following network.

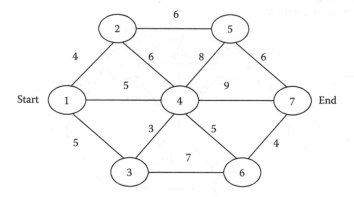

PROBLEM 8.13.10

Shortest-route problem: the network below has different routes from A to B by passing through a number of other points. The lengths of the individual routes are shown. Determine the minimum distance from A to B using DP approach. (Find the optimal solution with clear definition of the stages, states, and return function.)

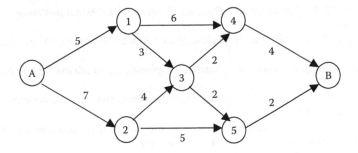

PROBLEM 8.13.11

Solve the following problem by using DP technique

$$\text{Minimize } \sum_{i=1}^{10} y_i^2$$

Subject to

$$\prod_{i=1}^{10} y_i = 8, \, y_i$$

PROBLEM 8.13.12

Solve the following linear programming problem using DP:

Maximize $z = 4x_1 + 10x_2$

Subject to

$2x_1 + 7x_2 \leq 18$

$7x_1 + 2x_2 \leq 14$

$x_1, x_2 \geq 0$

References

1. Bellman, R., *Dynamic Programming*, Princeton University Press, Princeton, NJ, 1957.
2. Bellman, R. and Dreyfus, S., *Applied Dynamic Programming*, Princeton University Press, Princeton, NJ, 1962.
3. Dano, S., *Nonlinear and Dynamic Programming*, Springer-Verlag, New York, 1975.

4. Hadley, G. F., *Nonlinear and Dynamic Programming*, Addison-Wesley, Reading, MA, 1964.
5. Howard, R. A., *Dynamic Programming and Markov Processes*, Wiley, New York, 1960.
6. Kaufman, A. and Cruon, R., *Dynamic Programming*, Academic Press, New York, 1967.
7. Larson, R. E., *State Increment Dynamic Programming*, American Elsevier, New York, 1968.
8. Markowitz, H. and Manne, A. S., On the solution of discrete programming problems, *Econometrica*, 25(1), 84–110, January 1957.
9. Neinhauser, G. L., *Introduction to Dynamic Programming*, Wiley, New York, 1966.
10. Sasieni, M. W., Yaspan, A., and Friedman, A., *Operations Research, Methods and Problems*, Wiley, New York, 1959.
11. Wagner, H. M., *Principles of Management Sciences*, Prentice-Hall, Englewood Cliffs, NJ, 1970.
12. Denardo, E. V., *Dynamic Programming Models and Applications*, Prentice Hall, Englewood Cliffs, NJ, 1982.
13. Dreyfus, S. E. and Law, A. M., *The Art and Theory of Dynamic Programming*, Academic Press, New York, 1977.
14. Dixon, P. and Norman, J. M., An Instructive Exercise in Dynamic Programming, *IIE Transactions*, 16(3), 1984, 292–294.

9

Lagrangian Relaxation

9.1 Introduction

In the last decade, Lagrangian relaxation has grown from a successful but largely theoretical concept to a tool that is the backbone of a number of large-scale applications. Although there have been several surveys of Lagrangian relaxation (e.g., Fisher [1] and Geofrion [2] and an excellent textbook treatment by Muckstadt and Koenig [6]), more extensive use of Lagrangian relaxation in practice has been inhibited by the lack of a "how to do it" exposition similar to the treatment usually extended to linear, dynamic, and integer programming in operations research texts. This chapter is designed to at least partially fill that void and should be of interest to both developers and users of Lagrangian relaxation algorithms.

In many large-scale optimization problems for planning in production, feasibility of the optimization problems present in linear programming (LP) allows the determination of optimal solution by first decomposing a problem into smaller subproblems and then solving the subproblems almost independently. The procedure would allow the procedure of many large-scale problems completely solvable is the decomposition. It exploits the sparsity of many large-scale problems. Several practical applications in various fields such as power systems, management sciences, and economics have taken advantage of Lagrangian relaxation in solving complex scheduling problems [3,6–9].

Many variants of Lagrange relaxation have been posed by researchers in power systems operation for unit commitment (UC). The technique has been developed to take advantage of iterative technique and eliminate certain coupling constraints by adding them to the master problem thereby enabling the independent and simultaneous solutions of many resulting subproblems. Again, in the power system area it is a preferred solution method for mostly the large set of constraints in a UC problem with hydrodispatch problem where start-up, shutdown, and ramp rates are involved. Other approaches such as resistive planning in optimal power flow (OPF) have shown some promise where security constraints of OPF for subproblems are used. This technique allows for speed which hardly offset constraints and easy procedure exists for relaxing constraints by adding them as part to objective function for optimization.

Lagrangian relaxation is based upon the observation that many difficult programming problems can be modeled as relatively easy problems complicated by a set of side constraints. To exploit this observation, we create a Lagrangian problem in which the complicating constraints are replaced with a penalty term in the objective function involving the amount of violation of the constraints and their dual variables. The Lagrangian problem is easy to solve and provides an upper bound (to a maximization problem) on the optimal value of the original problem. It can thus be used in place of LP relaxation to provide bounds in a branch-and-bound algorithm. The Lagrangian approach offers a number of important advantages over LP relaxations.

The Lagrangian relaxation concept is first formulated in general terms and its use is then demonstrated extensively on a numerical example.

9.2 Concepts

Consider an integer programming problem of the following form:

$$\left.\begin{array}{l} Z = \max{[cx]} \\ \text{Subject to} \\ Ax \le b \\ Dx \le e \\ x \ge 0 \text{ and integral} \end{array}\right\}, \tag{9.1}$$

where
   x is the $n \times 1$ vector and the elements of x are integers
   b is the $m \times 1$ vector
   e is of order $k \times 1$, and all other matrices have conformable dimension

We assume that the constraints of (9.1) have been partitioned into the two sets $Ax \le b$ and $Dx \le e$ so that (9.1) is relatively easy to solve if the constraint set $Ax \le b$ is removed. To create the Lagrangian problem, we first define an **m** vector of nonnegative multipliers u and add the nonnegative term $u(b - Ax)$ to the objective function of (9.1) to obtain Equation 9.2:

$$\left.\begin{array}{l} \text{Maximize } cx + u[b - Ax] \\ \text{Subject to} \\ Ax \le b \\ Dx \le e \\ x \ge 0 \text{ and integral} \end{array}\right\}. \tag{9.2}$$

It is clear that the optimal value of this problem for u fixed at a nonnegative value is an upper bound on Z because we have merely added a nonnegative

TABLE 9.1

Several Issues Related to a Lagrangian Relaxation-Based System

Question	Answer
(a) Which constraints should be relaxed?	The relaxation should make the problem significantly easier, but not too easy
(b) How to compute good multipliers u	There is a choice between a general-purpose procedure called the subgradient method and smarter methods that may be better but which are, however, highly problem specific
(c) How to deduce a good feasible solution to the original problem given a solution to the relaxed problem	The answer tends to be problem specific

term to the ejective function. At this point, we create the Lagrangian problem by removing the constraints $Ax \leq b$ to obtain

$$Z_D(u) = \text{Max}[ex] + u[b - Ax]$$

Subject to

$$Dx < e$$

$$x > 0.$$

Since removing the constraints $Ax \leq b$ cannot decrease the optimal value, $Z_D(u)$ is also an upper bound on Z. Moreover, by assumption the Lagrangian problem is relatively easy to solve. There are three major questions in designing a Lagrangian relaxation-based system, and some answers can (roughly speaking) be given. Table 9.1 summarizes these issues.

A numerical example is used to illustrate considerations (a), (b), and (c) as well as to compare Lagrangian relaxation to the use of LP to obtain bounds for use in a branch-and-bound algorithm.

9.3 Subgradient Method for Setting the Dual Variables

The subgradient method is demonstrated by an illustrative example. Consider the following problem.

$$\text{Maximize } Z = 16x_1 + 10x_2 + 4x_4 \tag{a}$$

Subject to

$$8x_1 + 2x_2 + x_3 + 4x_4 = 10 \tag{b}$$

$$x_1 + x_2 \leq 1, \tag{c}$$

$$x_3 + x_4 \leq 1. \tag{d}$$

The numerical example is used to develop and demonstrate a method for obtaining dual variable values that produce a tight bound. Ideally, u should solve the following dual problem,

$$Z_D = \text{Min } Z_d(u) \quad \text{for } u \geq 0.$$

Before presenting an algorithm for this problem, it is useful to develop some insight by trying different values for the single duel variable u in the example. We use

$$Z_D(u) = \max\left(16 - 8u\right)x_1 + (10 - 2u)x_2 + (0 - u)x_3 + (4 - 4u)x_4 + 10u. \quad \text{(e)}$$

Take $u = 0$
Here we have

$$Z_D(u) = \max\{16x_1 + 10x_2 + 0x_3 + 4x_4\}.$$

Note that x_1 has a coefficient larger than that of x_2, and x_4 has a coefficient larger than that of x_3.

$$x_1 = 1 \quad \text{and} \quad x_2 = 0 \quad \text{satisfy } x_1 + x_2 \leq 1$$
$$x_3 = 0 \quad \text{and} \quad x_4 = 0 \quad \text{satisfy } x_3 + x_4 \leq 1.$$

By substitution we get:

$$Z_D(0) = 20.$$

Also from (a):

$$Z = [16x_1 + 10x_2 + 4x_4] = 20.$$

We now test constraint (b):

$$8x_1 + 2x_2 + x_3 + 4x_4 = 12,$$

which is a violation. Therefore, this solution is not feasible.

It is useful to think of the single constraint (b) that we have dualized as a resource constraint with the right-hand side representing the available supply of some resource and the left-hand side the amount of the resource demanded in a particular solution. We can then interpret the dual variable u as a price charged for the resource. It turns out that if we can discover a price for which the supply and demand for the resource are equal, then this value will also give a tight upper bound. However, such a price might not exist.

With $u = 0$, we discover that the Lagrangian relaxation solution demand for the resource exceeds the available supply by two units, suggesting that we should use a larger value for u.

Take $u = 6$

Here we now have

$$Z_D(6) = \max\{-32x_1 - 2x_2 - 6x_3 - 20x_4 + 60\}.$$

All coefficients are negative, and maximization takes place with all variables set to 0.

$$x_1 = 0 \quad \text{and} \quad x_2 = 0 \quad \text{satisfy } x_1 + x_2 \leq 1$$
$$x_3 = 0 \quad \text{and} \quad x_4 = 0 \quad \text{satisfy } x_3 + x_4 \leq 1.$$

By substitution we get:

$$Z_D(6) = 60.$$

Also from (a):

$$Z = [16x_1 + 10x_2 + 4x_4] = 0.$$

We now test constraint (b)

$$8x_1 + 2x_2 + x_3 + 4x_4 = 0.$$

This solution is feasible. For $u = 6$ we discover that we have overcorrected in the sense that all variables are 0 in the Lagrangian solution and none of the resource is used. We next try a sequence of dual values in the interval between 0 and 6.

Take $u = 3$

Here we have:

$$Z_D(3) = \max\{-8x_1 + 4x_2 - 3x_3 - 8x_4\} + 30.$$

The variables with negative coefficients are set to 0.

$$x_1 = 0 \quad \text{and} \quad x_2 = 1 \quad \text{satisfy } x_1 + x_2 \leq 1$$
$$x_3 = 0 \quad \text{and} \quad x_4 = 0 \quad \text{satisfy } x_3 + x_4 \leq 1.$$

By substitution we get:

$$Z_D(3) = 34.$$

Also from (a):

$$Z = [16x_1 + 10x_2 + 4x_4] = 10.$$

We now text constraint (b):

$$8x_1 + 2x_2 + x_3 + 4x_4 = 0.$$

This solution is feasible.

Take $u = 2$
Here we have

$$Z_D(2) = \max\{6x_2 - 2x_3 - 4x_4\} + 20.$$

Thus:

$$Z = [16x_1 + 10x_2 + 4x_4] = 14.$$

$$x_1 = 0 \quad \text{and} \quad x_2 = 1 \quad \text{satisfy } x_1 + x_2 \le 1$$
$$x_3 = 0 \quad \text{and} \quad x_4 = 0 \quad \text{satisfy } x_3 + x_4 \le 1.$$

By substitution we get:

$$Z_D(2) = 26.$$

Also from (a):

$$Z = [16x_1 + 10x_2 + 4x_4] = 10.$$

We now test constraint (b):

$$8x_1 + 2x_2 + x_3 + 4x_4 = 2 \le 10.$$

This solution is feasible.

Take $u = 1$
Here we have

$$Z_D(1) = \max((16 - 8*1)x_1 + (10 - 2*1)x_2 + (0 - 1)x_3 + (4 - 4*1)x_4 + 10*1).$$

Thus

$$Z_D(1) = \max\{8x_1 + 8x_2 - x_3 - 0x_4\} + 10*1.$$

We definitely have $x_3 = 0$, since its coefficient is negative.

Take $x_1 = 1$

$$x_1 = 0 \quad \text{and} \quad x_2 = 1 \quad \text{satisfy } x_1 + x_2 \le 1$$
$$x_3 = 0 \quad \text{and} \quad x_4 = 0 \quad \text{satisfy } x_3 + x_4 \le 1.$$

By substitution we get:

$$Z_D(1) = 18.$$

Also from (a):

$$Z = [16x_1 + 10x_2 + 4x_4] = 16.$$

We now test constraint (b):

$$8x_1 + 2x_2 + x_3 + 4x_4 = 8 \leq 10.$$

This solution is feasible. We still have another option:

$$x_3 = 0 \quad \text{and} \quad x_4 = 1 \quad \text{satisfy } x_3 + x_4 \leq 1.$$

By substitution we get:

$$Z_D(1) = 18.$$

Also from (a):

$$Z - [16x_1 + 10x_2 + 4x_4] = 20.$$

We now test constraint (b):

$$8x_1 + 2x_2 + x_3 + 4x_4 = 12.$$

This solution is not feasible.

Take $x_1 = 0$

$$x_1 = 0 \quad \text{requires } x_2 = 1 \quad \text{to satisfy } x_1 + x_2 \leq 1$$
$$x_3 = 0 \quad \text{requires } x_4 = 0 \quad \text{to satisfy } x_3 + x_4 \leq 1.$$

By substitution we get:

$$Z_D(1) = 18.$$

Also from (a):

$$Z = [16x_1 + 10x_2 + 4x_4] = 10.$$

We now test constraint (b):

$$8x_1 + 2x_2 + x_3 + 4x_4 = 2 \leq 10.$$

This solution is feasible. There still exists another option:

$$x_3 = 0 \quad \text{and} \quad x_4 = 1 \quad \text{satisfy } x_3 + x_4 \leq 1.$$

By substitution we get:

$$Z_D(1) = 18.$$

Also from (a):

$$Z = [16x_1 + 10x_2 + 4x_4] = 14.$$

We now test constraint (b):

$$8x_1 + 2x_2 + x_3 + 4x_4 = 6 \leq 10.$$

This solution is feasible.

In the case of $u = 1$, we see that there are four alternative optimal Lagrangian solutions. Table 9.2 gives a list of seven values for u, together with the associated Lagrangian relaxation solution, the bound $Z_D(u)$, and Z for those Lagrangian solutions that are feasible for $u = 2$. For the values tested, the tightest bound of 18 was obtained with $u = 1$, but at the moment we lack any means for confirming that it is optimal.

It is possible to demonstrate that 18 is the optimal value for $Z_D(u)$ by observing that if we substitute any x into the objective function for the Lagrangian problem, we obtain a linear function in u. We use:

$$Z_D(u) = \max\,(16 - 8u)x_1 + (10 - 2u)x_2 + (0 - u)x_3 + (4 - 4u)x_4 + 10u.$$

TABLE 9.2

List of u Values with the Associated Lagrangian Relaxation Solution, the Bound $Z_D(u)$, and Z for those Lagrangian Solutions

u	x_1	x_2	x_3	x_4	$Z_D(u)$	Value of Lagrangian Solution Z (if Feasible)
0	1	0	0	1	20	—
6	0	0	0	0	60	0
3	0	1	0	0	34	10
2	0	1	0	0	26	10
1	1	0	0	0	18	16
—	1	0	0	1	18	—
—	0	1	0	0	18	10
—	0	1	0	1	18	14
1/2	1	0	0	1	19	—
3/4	1	0	0	1	18.5	—

Take $x_1 = x_2 = x_3 = x_4 = 0$

$$Z_D(u) = \max (16 - 8u)0 + (10 - 2u)0 + (0 - u)0 + (4 - 4u)0 + 10u$$
$$= 10u.$$

Take $x_1 = 1, x_2 = x_3 = x_4 = 0$

$$Z_D(u) = \max (16 - 8u)1 + (10 - 2u)0 + (0 - u)0 + (4 - 4u)0 + 10u$$
$$= 16 + 2u.$$

Take $x_2 = 1, x_1 = x_3 = x_4 = 0$

$$Z_D(u) = \max (16 - 8u)0 + (10 - 2u)1 + (0 - u)0 + (4 - 4u)0 + 10u$$
$$= 10 + 8u.$$

Take $x_2 = x_4 = 1, x_1 = x_3 = 0$

$$Z_D(u) = \max (16 - 8u)0 + (10 - 2u)1 + (0 - u)0 + (4 - 4u)0 + 10u$$
$$= 14 + 4u.$$

Take $x_1 = x_4 = 1, x_2 = x_3 = 0$

$$Z_D(u) = \max (16 - 8u)1 + (10 - 2u)0 + (0 - u)0 + (4 - 4u)1 + 10u$$
$$= 20 - 2u.$$

Figure 9.1 exhibits this family of linear functions for all Lagrangian relaxation solutions that are optimal for at least one value of u.

The fact that we must maximize the Lagrangian objective means that for any particular value of u, $Z_D(u)$ is equal to the largest of these linear functions. Thus, the $Z_D(u)$ function is given by the upper envelope of this family of linear equations, shown as a darkened piecewise linear function in Figure 9.1. From this figure it is easy to see that $u = 1$ minimizes $Z_D(u)$.

Figure 9.2 also provides motivation for a general algorithm for finding u. As shown, the $Z_D(u)$ function is convex and differentiate except at points where the Lagrangian problem has multiple optimal solutions. At differentiable points, the derivative of $Z_D(u)$ with respect to u is given by $8x_1 + 2x_2 + x_3 + 4x_4 - 10$, where x is an optimal solution to (LR$_u$).

These facts also hold in general with the gradient of the $Z_D(u)$ function at differentiable points given by $(\mathbf{Ax} - \mathbf{b})$. These observations suggest that it might be fruitful to apply a gradient method to the minimization of $Z_D(u)$ with some adaptation at the points where $Z_D(u)$ is nondifferentiable. This has been properly achieved in a procedure called the subgradient method. At points where $Z_D(u)$ is nondifferentiable, the subgradient method selects arbitrarily from the set of alternative optimal Lagrangian solutions and uses the vector $(\mathbf{Ax} - \mathbf{b})$ for the solution as though it were the gradient of $Z_D(u)$.

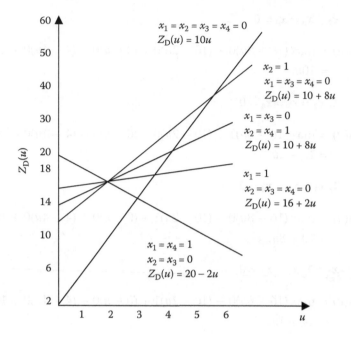

FIGURE 9.1
Family of linear equations for Lagrangian relaxation solution.

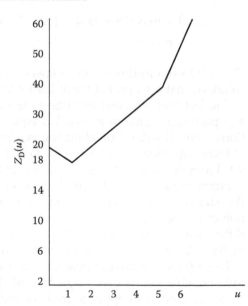

FIGURE 9.2
Composite behavior of $Z_D(u)$ featuring its convex nature and its nondifferentiability property.

The result is a procedure that determines a sequence of values for u by beginning at an initial point u^0 and applying the formula

$$u^{k+1} = \max\{0,\, u^k - t_k(b - Ax^k)\}.$$

In this formula, t_k is a scalar stepsize and x^k is an optimal solution to (LR_d^k), the Lagrangian problem with dual variables set to u^k. The nondifferentiability also requires some variation in the way the stepsize is normally set in a gradient method.

9.4 Setting t_k

To gain insight into a sensible procedure for setting t_k, we discuss the results of the subgradient method applied to the example with three different rules for t_k. For example, we have

$$u^{k+1} = \max\left\{0, u^k - t_k\left(10 - 8x_1^{(k)} - 2x_2^{(k)} - 4x_4^{(k)}\right)\right\} \qquad (9.3)$$

9.4.1 Case 1: Subgradient Method with $t_k = 1$ for All k

In this first case, t_k is fixed at 1 on all iterations. As a result the formula is

$$u^{k+1} = \max\left\{0, u^k - \left(10 - 8x_1^{(k)} - 2x_2^{(k)} - 4x_4^{(k)}\right)\right\}.$$

We start with

$$u^0 = 0.$$

Recall the table entry for $u = 0$.

u	x_1	x_2	x_3	x_4
0	1	0	0	1

$$u^1 = \max\{0, 0 - (10 - 8 \times 1 - 2 \times 0 - 4 \times 1)\}$$
$$= \max\{0, -(-2)\}$$
$$u^1 = 2.$$

Recall the table entry for $u = 2$.

u	x_1	x_2	x_3	x_4
2	0	1	0	0

$$u^2 = \max\{0, 2 - (10 - 8 \times 0 - 2 \times 1 - 4 \times 0)\}$$
$$= \max\{0, -6\}$$
$$u^2 = 0.$$

We now use values for $u=0$ as before,

$$u^3 = 0 - (-2) = 2,$$

and use values for $u=2$ as before,

$$u^4 = \max[0, 2 - 8] = 0.$$

We see that the subgradient method oscillates between the values $u=0$ and $u=2$.

9.4.2   Case 2: Subgradient Method with $t_k = 1, 0.5, 0.25, \ldots$

In this second case, t_k converges to 0 with each successive value equal to half the value of the previous iteration.

$$t_k = \left(\frac{1}{2}\right)^k.$$

Therefore,

$$u^{k+1} = \max\left\{0, u^k - \left[\frac{1}{2}\right]^k \left(10 - 8x_1^{(k)} - 2x_2^{(k)} - 4x_4^{(k)}\right)\right\}.$$

We start once again with

$$u^0 = 0.$$

Here we have $t_0 = 1$, and we get the same result as in the preceding case.

$$u^1 = 0 - (-2) = 2.$$

With $u=2$, we get

u	x_1	x_2	x_3	x_4
2	0	1	0	0

$$u^2 = \max\{0, u^{12} - [1/2]^1(10 - 8 \times 0 - 2 \times 1 - 4 \times 0)\}$$
$$= \max\{0, 2 - 0.5(10 - 2)\}$$
$$= \max\{0, 2 - 4\}$$
$$u^2 = 0.$$

Proceeding similarly, we get

$$u^3 = 0 - \left(\frac{1}{4}\right)(-2) = \frac{1}{2}$$

$$u^4 = \frac{1}{2} - \left(\frac{1}{8}\right)(-2) = \frac{3}{4}$$

$$u^5 = \frac{3}{4} - \left(\frac{1}{16}\right)(-2) = \frac{7}{8}$$

$$u^6 = \frac{7}{8} - \left(\frac{1}{32}\right)(-2) = \frac{15}{16}.$$

In this case, the subgradient method behaves nicely and converges to the optimal value of u.

9.4.3 Case 3: Subgradient Method with $t_k = 1, 1/3, 1/9, \ldots$

In this final case, t_k also converges to 0, but more quickly. Each successive value is equal to one-third the value of the previous iteration,

$$t_k = \left(\frac{1}{3}\right)^k.$$

Therefore,

$$u^{k+1} = \max\left\{0, u^k - \left[\frac{1}{3}\right]^k \left(10 - 8x_1^{(k)} - 2x_2^{(k)} - 4x_4^{(k)}\right)\right\}.$$

We start once again with

$$u^0 = 0$$

$$u^1 = 0 - (-2) = 2$$

$$u^2 = \max\left(0, 2 - \left(\frac{1}{3}\right)(8)\right) = 0$$

$$u^3 = 0 - \left(\frac{1}{9}\right)(-2) - \frac{2}{9}$$

$$u^4 = \frac{2}{9} - \left(\frac{1}{27}\right)(-2) = 0.296$$

$$u^5 = 0.296 - \left(\frac{1}{81}\right)(-2) = 0.321$$

$$u^6 = 0.321 - \left(\frac{1}{243}\right)(-2) = 0.329$$

$$u^7 = 0.329 - \left(\frac{1}{729}\right)(-2) = 0.332.$$

In this case, the subgradient method converges to $u = 1/3$, showing that if the stepsize converges to 0 too quickly, then the subgradient method will converge to a point other than the optimal solution.

From these examples, we suspect that the stepsize in the subgradient method should converge to 0 but not too quickly. These observations have been confirmed in a result (see Held et al. [5]) that states that if as $k \to \infty$,

$$t_k \to 0 \quad \text{and} \quad \sum_{i=1}^{k} t_i \to \infty$$

then $Z_D(u^k)$ converges to its optimal value Z_D. Note that Case 3 actually violates the second condition since $\sum_{i=1}^{k} t_i \to 2$, thus showing that these conditions are sufficient but not necessary.

A formula for t_k that has proven effective in practice is

$$t_k = \frac{\lambda_k \left[Z_D(u^k) - Z^* \right]}{\sum_{i=1}^{m} \left[b_i - \sum_{j=1}^{n} a_{ij} x_j^k \right]^2}. \tag{9.4}$$

In this formula, Z^* is the objective value of the best-known feasible solution to (p) and λ_k is a scalar chosen between 0 and 2. Frequently the sequence λ_k is determined by starting with $\lambda_k = 2$ and reducing λ_k by a factor of 2 whenever $Z_D(u^k)$ has failed to decrease in a specified number of iterations.

Justification for this formula, as well as many other interesting results on the subgradient method, is given in Held et al. [5]. The feasible value Z^* initially can be set to 0 and then updated using the solutions that are obtained on those iterations in which the Lagrangian problem solution turns out to be feasible in the original problem. Unless we obtain a u^k for which $Z_D(u^k) = Z^*$, there is no way of proving optimality in the subgradient method. To resolve this difficulty, the method is usually terminated upon reaching a specified iteration limit.

Other procedures that have been used for setting multipliers are called multiplier-adjustment methods. Multiplier-adjustment methods are heuristics for the dual problem that are developed for a specific application and exploit some special structures of the dual problem in that approach. The first highly successful example of a multiplier-adjustment method was Erlenkotter's [X] algorithm for the uncapacitated location problem.

By developing a multiplier-adjustment method specifically tailored for some problem class, one is usually able to improve on the subgradient method. However, because the subgradient method is easy to program and has performed robustly in a wide variety of applications, it is usually at least the initial choice for setting the multipliers in Lagrangian relaxation.

Returning to our example, we have obtained through the application of Lagrangian relaxation and the subgradient method a feasible solution with a value of 16 and an upper bound on the optimal value of 18. At this point, we could stop and be content with a feasible solution proven to be within about 12% of optimality, or we could complete the solution of the example to optimality using branch-and-bound, with bounds provided by our Lagrangian relaxation. In the next section, we show how such an approach compares with more traditional LP-based branch-and-bound algorithms.

9.5 Comparison with LP-Based Bounds

In this section, we compare Lagrangian relaxation with the upper bound obtained by relaxing the integrality requirement on x and solving the resulting linear program. Let Z_{LP} denote the optimal value of (p) with integrality on x relaxed. We start by comparing Z_{LP} for the example with the best upper bound of 18 obtained previously with Lagrangian relaxation. To facilitate this comparison, we first write out the standard LP dual of the example. Let u, v_1, and v_2 denote dual variables on constraints 9.5 through 9.8 and w_j a dual variable on the constraint $x_j \le 1$.

Then, the LP dual form is

$$\text{Min } 10u + v_1 + v_2 + w_1 + w_2 + w_3 + w_4 \tag{9.5}$$

$$8u + v_1 + w_1 \ge 16 \tag{9.6}$$

$$2u + v_1 + w_2 \ge 10 \tag{9.7}$$

$$u + v_2 + w_3 \ge 0 \tag{9.8}$$

$$4u + v_2 + w_4 \ge 4 \tag{9.9}$$

$$u, v_1, v_2, w_1, \ldots, w_4 \ge 0. \tag{9.10}$$

The optimal solution to the primal LP is $x_1 = 1, x_2 = 0, x_3 = 0, x_4 = 1/2$, and the optimal solution to the dual LP is $u = 1, v_1 = 8, v_2 = w_1 = \cdots = w_4 = 0$. In order to verify that each of these solutions is optimal, we simply substitute them in the primal and dual and observe that each is feasible and gives the same objective value 18.

This exercise has demonstrated two interesting facts: $Z_{LP} = 18$, the same upper bound we obtained with Lagrangian relaxation; and the LP dual variable value of $u = 1$ on constraint 9.7 is exactly the value that gave the minimum upper bound of 18 on the Lagrangian problem. These observations are part of a pattern that holds generally and is nicely summarized in a result from Geoffrion [2] which states that $Z_D \le Z_{LP}$ for any Lagrangian relaxation. This fact is established by the following sequence of relations between optimization problems.

$$Z_D = \text{Min}\{\text{Max}(\mathbf{cx} + u(\mathbf{b} - \mathbf{Ax}))\}$$
$$u \geq 0$$
$$\mathbf{Dx} \leq \mathbf{e}$$
$$\mathbf{x} \geq 0 \text{ and integral,}$$

by LP duality

$$\text{Max } \mathbf{cx}$$
$$\mathbf{Ax} \leq \mathbf{b}$$
$$\mathbf{Dx} \leq \mathbf{e}$$
$$\mathbf{x} \geq 0.$$

Besides showing that $Z_D \leq Z_{LP}$ the preceding logic indicates when $Z_D = Z_{LP}$ and when $Z_D < Z_{LP}$. The inequality in the sequence of relations connecting Z_D and Z_{LP} is between the Lagrangian problem and the Lagrangian problem with integrality relaxed. Hence, we can have $Z_D < Z_{LP}$ only if this inequality holds strictly or, conversely, $Z_D = Z_{LP}$ only if the Lagrangian problem is unaffected by removing the integrality requirement on x. In the Lagrangian problem for the original example, the optimal values of the variables will be an integer whether we require it or not. This implies that we must have $Z_D = Z_{LP}$, something that we have already observed numerically.

This result also shows that we can improve the upper bound by using a Lagrangian relaxation in which the variables are not naturally integral.

9.6 Improved Relaxation

An alternative relaxation for the example is given below.

$$Z_D(v_1, v_2) = \text{Max}(16 - v_1)x_1 + (10 - v_1)x_2 + (0 - v_2)x_3 + (4 - v_2)x_4 + v_1 + v_2$$
$$(9.10)$$

Subject to

$$8x_1 + 2x_2 + x_3 + 4x_4 \leq 10 \tag{9.11}$$

$$0 < x_j < 1, \quad j = 1, \ldots, 4 \tag{9.12}$$

$$x_j \text{ integer}, \quad j = 1, \ldots, 4. \tag{9.13}$$

In this relaxation, we have dualized constraints, Equations 9.12 and 9.13, and obtained a relaxation that is a knapsack problem. Although this problem is known to be difficult in the worst case, it can be solved practically using a variety of efficient knapsack algorithms such as dynamic programming.

TABLE 9.3

Subgradient Method Applied to Improved Relaxation

v_1	v_2	λ_k	x_1	x_2	x_3	x_4	$Z_D(v_1, v_2)$
Feasible with $Z = 4$							
0	0	1	1	1	0	0	26
13	0	1	0	0	0	1	17
Feasible with $Z = 16$							
0	0	1	1	1	0	0	26
11	0	1	1	0	0	0	16

Because the continuous and integer solutions to the knapsack problem can differ, the analytic result obtained in the previous section tells us that this relaxation may provide bounds that are better than LP.

This is confirmed empirically in Table 9.3, which shows the application of the subgradient method to this relaxation. We begin with both dual variables equal to 0 and in four iterations, we converge to a dual solution in which the upper bound of 16 is equal to the objective value of the feasible solution obtained when we solve the Lagrangian problem. Hence, Lagrangian relaxation has completely solved the original problem. In this example, we have set the stepsize using the formula given previously with $\lambda_k = 1$.

This example illustrates that with careful choice of which constraints to dualize, Lagrangian relaxation can provide results that are significantly superior to LP-based branch-and-bound. The choice of constraints is to some extent an art much like the formulation itself. Typically, one will construct several alternative relaxations and evaluate them, both empirically and analytically, using the result on the quality of bounds presented in the previous section. The alternative relaxations can be constructed in one of the two ways: begin with an integer programming formulation and select different constraints to dualize, or alternatively, begin with some easy to solve model such as the knapsack or shortest route problem which is close to, but not exactly the same as, the problem one wishes to solve. Then try to add a set of side constraints to represent those aspects of the problem of interest that are missing in the simpler model. A Lagrangian relaxation can be obtained by dualizing the side constraints that have been added.

9.7 Summary of Concepts

Up to this point, the concept of Lagrangian relaxation has been developed piecemeal on an example. We can now formulate and present a generic Lagrangian relaxation algorithm. Figure 9.3 shows a generic Lagrangian

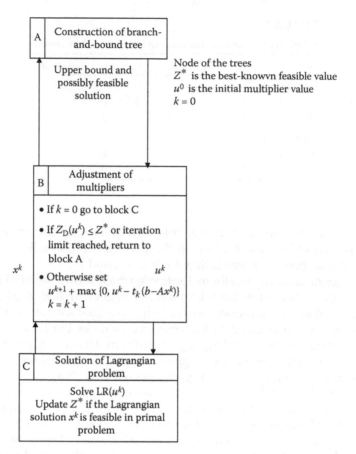

FIGURE 9.3
Generic Lagrangian relaxation algorithm.

relaxation algorithm consisting of several major steps. The first step is the standard branch-and-bound process in which a tree of solution alternatives is constructed with certain variables fixed to specified values at each node of the tree. These specified values are passed from block A to block B together with Z^*, the objective value of the currently best-known feasible solution, and starting multipliers u^0.

In blocks B and C, we iterate between adjusting the multipliers with the subgradient, updating the formula to obtain a new multiplier value u^k, and solving the Lagrangian problem to obtain a new Lagrangian solution x^k. The process continues until we either reach an iteration limit or discover an upper bound for this node that is less than or equal to the current best-known feasible value Z^*. At this point, we pass back to block A the best upper bound we have discovered together with any feasible solution that may have been obtained as a result of solving the Lagrangian problem. In Fisher's experience, it is rare in practice that the Lagrangian solution will be feasible in the original problem (p). However, it is not uncommon that the Lagrangian solution will be nearly feasible and can be made feasible with some minor modifications. A systematic procedure for doing this can be applied in block C and constitutes what might be called a "Lagrangian heuristic." Lagrangian heuristics have been vital to the computational success of many applications, such as those described in Fisher [1], and may well prove to be as important as the use of Lagrangians to obtain upper bounds.

It is not uncommon in large-scale applications to terminate the process depicted in block B before the branch-and-bound tree has been explored sufficiently to prove optimality. In this case, the Lagrangian algorithm is really a heuristic with some nice properties, such as an upper bound on the amount by which the heuristic solution deviates from optimality.

9.8   Past Applications

A brief description of several instances in which Lagrangian relaxation has been used in practice should give the flavor of the kinds of problems for which Lagrangian relaxation has been successful.

Bell et al. describe the successful application of the algorithm to Air Products and Chemicals, which has resulted in a reduction of distribution cost of about $2 million per year.

Fisher [1] discusses the application, in the Clinical Systems Division of Du Pont, of an algorithm for vehicle routing that is based on a Lagrangian relaxation algorithm for the generalized assignment problem.

Graves and Lamar [4] treat the problem of designing an assembly by choosing from available technology a group of resources to perform certain operations. The choices cover people, single-purpose machines, narrow-purpose pickplace robots, and general-purpose robots. Their work has been applied in a number of industries, including the design of robot assembly systems for the production of automobile alternators. Graves [3] has also discussed the use of Lagrangian relaxation to address production planning problems from an hierarchical perspective.

Mulvey [9] is concerned with condensing a large database by selecting a subset of representative elements. He has developed a Lagrangian relaxation-based clustering algorithm that determines a representative subset for which the loss in information is minimized in a well-defined sense. He has used this algorithm to reduce the 1977 U.S. Statistics of Income File for Individuals maintained by the Office of Tax Analysis from 155,212 records to 74,762 records.

The application described in Shepardson and Marsten [7] involves the scheduling of personnel who must work two duty periods, a morning shift and an afternoon shift. Their algorithm determines optimal schedules for each worker so as to minimize cost and satisfy staffing requirements. Helsinki City Transport has applied this algorithm to bus crew scheduling.

Van Roy and Gelders discuss the use of Lagrangian relaxation for a particular problem arising in distribution.

In each of the applications described above, development of the Lagrangian relaxation algorithm required a level of involvement on the part of skilled analysts that is similar to that required in the use of dynamic programming. Just as some insight into a problem is required before dynamic programming can be applied fruitfully, it is generally nontrivial to discover a Lagrangian relaxation that is computationally effective. Moreover, once this has been done, the various steps in the algorithm must be programmed more or less from scratch. Often this process can be made easier by the availability of an "off the shelf" algorithm for the Lagrangian problem if it is a well-known model, such as a network flow, shortest route, minimum spanning tree, or knapsack problem.

Despite the level of effort required in implementing Lagrangian relaxation, the concept is growing, in popularity because the ability it affords to exploit special problem structure often is the only hope for coping with large real problems.

For the future, it remains to be seen whether Lagrangian relaxation will continue to exist as a technique that requires a significant ad hoc development effort or whether the essential building blocks of Lagrangian relaxation will find their way into user-friendly mathematical programming codes such as LINDO or IFPS OPTIMUM.

9.9 Summary

9.9.1 Overview

The branch-and-bound technique for solving integer programming problems is a powerful solution technique despite the computational requirements involved.

Most computer codes based on the branch-and-bound technique differ from standard known procedures in the details of selecting the branching variable at a node and the sequence in which the subproblems are examined.

These rules are based on heuristics developed through the basic disadvantage of the branch-and-bound algorithm, which is the necessity of solving a complete linear program at each node. In large problems, this could be very time consuming, particularly when the only information needed at the node may be its optimum objective value. This point is clarified by realizing that once a good bound is obtained, many nodes can be discarded as their optimum objective values are then known.

The preceding point led to the development of a procedure whereby it may be unnecessary to solve all the subproblems of the branch-and-bound tree. The idea is to estimate an upper bound and assume a maximization problem on the optimum objective valued at each node. Should this upper bound become smaller than the objective associated with the best available integer solutions, the node is discarded.

9.9.2 Algorithm of Solution Using Lagrangian Relaxation Approach

Consider an integer programming problem of the following form:

$Z = \max c^T x$

Subject to

$Ax \leq b$

$Dx \leq e$

$x \leq 0$ and integer,

where x is an $n*1$, elements of x are integers, b is an $m*1$, c is a $k*1$, and all other matrices have comformable dimensions. The algorithm is summarized as follows:

Step 1. We assume that the constraints of the problems have been partitioned into two sets $Ax \leq b$ and $Dx \leq e$ so that the problem is relatively easy to solve if the constraint set $Ax \leq b$ is removed.

Step 2. To create the Lagrangian problem we first define an m vector of nonegative multipliers U and add the nonnegative term $U(b - Ax)$ to the objective function to form $Z_D(U)$.

$Z_D(U) = \max c^T x + U(b - Ax)$

Subject to

$Dx \leq e$

$x \geq 0$ and integral.

It is clear that the optimal value of this problem for U fixed at a nonnegative value is an upper bound on Z because we have merely added a nonnegative form to the objective function.

Step 3. The new objective $Z_D(U)$ is nondifferentiable at points of the Lagrangian problem solution. This led to the use of the subgradient method for minimization of $Z_D(U)$ with some adaptation at the points where $Z_D(U)$ is nondifferentiable.

Assuming we begin at a multiple value U^0, then

$$U^{k+1} = \max\{0, U^k - t_k(\mathbf{b} - \mathbf{A}\mathbf{x}^k)\},$$

t_k is a stepsize, and x^k is an optimal solution to $LR_v k$.

The stepsize, t_k, should converge to 0 but not too quickly; that is,

$$k \to \infty, \quad t_k \to 0, \quad \sum_{i=1}^{k} t_i \to \infty.$$

$Z_D(U^k)$ converges to its optimal value.

A formula for t_k is given by

$$t_k = \frac{\lambda_k[Z_D(U^k) - Z^*]}{\sum_{i=1}^{n}\left[b_i - \sum_{j=1}^{m} a_{ij} X_j^k\right]^2}. \tag{9.14}$$

Z^* is the objective value of the best-known feasible solution and λ_k is a scalar chosen between 0 and 2, and reducing λ_k by a factor of 2 whenever $Z_D(U^k)$ has failed to decrease in a specified number of iterations.

The feasible value Z^* initially can be set to (0) and then updated using the solutions obtained on those iterations in which the LR solution must be feasible in the original problem.

Step 4. Continue the procedure in step 3 until we either reach an iteration limit or discover an upper bound for this node that is less than or equal to the best-known feasible value Z^*.

9.9.3   Power System Application: Scheduling in Power Generation Systems

9.9.3.1   Model

A simplified version of the power scheduling problem is formulated as a mixed integer programming problem having a special structure that facilitates rapid computation.

First, reserve and demand constraints are included in the basic model. Then other constraints are included in the formulation without affecting the basic structure of the problem.

The integer variables in the model indicate whether a specified generating unit is operating during a period.

$x_{it} = 1 \to$ generating unit i is operating in period t

$x_{it} = 0 \to$ generating unit i is off

$i = 1, \ldots, I, \quad t = 1, 2, \ldots, T.$

The continuous variable y_{ikt} represents the proportion of the available capacity M_{ik} that is actually used at period t: $k = 1, \ldots, k_i$.

$k_i \rightarrow$ number of linear segments of the production cost curve

$yk_{it} \rightarrow 0$ only if $x_{it} = 1$; hence this constraint is introduced: $0 \leq y_{ik} \leq x_{it}$.

The total energy output from generator i at time t is given by

$$m_i x_{it} + \sum_{k=i}^{k_i} M_{ik} y_{ikt}, \qquad (9.15)$$

where
m_i is the minimum unit capacity
M_i is the maximum unit capacity

$$w_i = \begin{cases} 1 & \text{if the unit is started up at time } t \\ 0 & \text{otherwise} \end{cases}$$

$$z_{it} = \begin{cases} 1 & \text{if the unit is shut down in period } t \\ 0 & \text{otherwise} \end{cases}$$

c_i is the start-up cost of generator i
g_i is the operating cost of unit i at its minimum capacity for 1 h
h_t is the number of hours in period t

The objective function to be minimized for the basic T-period scheduling model is

$$\sum_{i=1}^{I} \left\{ \sum_{t=1}^{T} \left\{ c_i w_i + d_i z_{it} + h_i g_i x_{it} + \sum_{k=1}^{k_i} M_{ik} h_t g_{ik} y_{ikt} \right\} \right\}. \qquad (9.16)$$

Let D_t represent the demand level at time t. Then, the demand constraint is given by

$$m_i x_{it} + \sum_{k=1}^{k_i} M_{ik} y_{ikt} \geq D_t. \qquad (9.17)$$

Let the reserve R_t be the minimum quantity at time t. Then, the reserve constraint is given by

$$\sum_{i=1}^{I} M_i x_{it} \geq R_t. \qquad (9.18)$$

By imposing only the constraints

$$w_{it} \geq x_{it} - x_{i,t-1}, \quad w_{it} \geq 0$$
$$z_{it} \geq x_{i,t-1} - x_{it}, \quad z_{it} \geq 0$$
$$x_{it} = 0 \text{ or } 1.$$

then the objective function to be minimized w_{it}, z_{it} must equal either 0 or 1. The mixed integer model for the basic power scheduling problem is

$$\text{Minimize} \sum_{i=1}^{I} \left\{ \sum_{t=1}^{T} \left\{ c_i w_{it} + d_i z_{it} + h_t g_i x_{it} + \sum_{k=1}^{k_i} M_{ik} h_{ik} g_{ik} y_{ikt} \right\} \right\}. \quad (9.19)$$

Subject to

$$\sum_{i}^{k_i} k = 1 \, m_i x_{it} + \sum_{k=1}^{k_i} M_{ik} y_{ikt} \geq D_t \quad (9.20)$$

$$\sum_{i=1}^{I} M_i x_{it} \geq R_t \quad (9.21)$$

$$0 < y_{ikt} \leq x_{it} \quad (9.22)$$

$$w_{it} \geq x_{it} - x_{i,t-1} \quad (9.23)$$

$$z_{it} \geq x_{i,t-1} - x_{it} \quad (9.24)$$

$$w_{it} \geq 0 \quad (9.25)$$

$$z_{it} \geq 0 \quad (9.26)$$

$$0 \leq x_{it} \leq 1 \quad (9.27)$$

$$x_{it} \text{ integer.}$$

9.9.3.2 Relaxation and Decomposition of the Model

Lagrangian relaxation is used to decompose the problem into I single generator subproblems. The advantage of decomposing by generator is that the constraints and costs that depend on the state of the generator from period to period can easily be considered in the subproblems.

The solution of the relaxed problem provides a lower bound on the optimal solution of the original problem. The Lagrangian relaxation model is

$$\sum_{i=1}^{I} \left\{ \sum_{t=1}^{T} \left\{ c_i w_{it} + d_i Z_{it} + h_t g_i x_{it} + \sum_{k=1}^{k_i} M_{ik} h_{it} g_{ik} y_{ikt} \right\} \right\}$$
$$+ \sum_{t=1}^{T} v_t \left(D_t - \sum_{i=1}^{I} m_i x_{it} - \sum_{i=1}^{I} \sum_{k=1}^{k_i} m_{ik} y_{ikt} \right) + \sum_{t=1}^{T} u_t \left(k_t - \sum_{i=1}^{I} m_i x_{it} \right) \quad (9.28)$$

$$C_{set} \begin{cases} 0 \le y_{ikt} \le x_{it} \\ w_{it} \ge x_{it} - x_{i,t-1} \\ z_{it} \ge x_{i,t-1} - x_{it} \\ w_{it} \ge 0 \\ z_{it} \ge 0 \\ 0 \le x_{it} \le 1 \\ x_{it} \text{ integers} \end{cases} , \qquad (9.29)$$

where v_t and u_t are nonnegative real numbers (Lagrange multipliers).

The Lagrangian relaxation problem decomposes into I single generator subproblems of the form:

$$\text{Minimize } \sum_{t=1}^{T} \left\{ c_i w_{it} + d_i z_{it} + (h_t g_i - v_t M_i - v_t m_i) x_{it} \right.$$

$$\left. + \sum_{k=1}^{k_i} M_{ik}(h_t g_{ik} - v_t) y_{ikt} \right\} \qquad (9.30)$$

Subject to

C_{set} of constraints.

Figure 9.4 shows a graph that could be used for the solution of the subproblems in the basic model.

The upper state in each period represents the ON state and the lower state represents the OFF state for the generator. The transition arcs on the graph represent feasible decisions and the arc lengths are the costs associated with the decision. If the generator is OFF in period $t-1$ then $x_{i,t-1}=0$, and if it is ON in period t, $x_{it}=1$.

Consequently, $w_{it} \ge 1$, $z_{it} \ge 0$, $0 \le y_{ikt} \le 1$ to minimize the cost. Given the values of $x_{i,t-1}$ and x_{it} we set $w_{it}=1$, $z_{it}=0$, $y_{ikt}=0$ if $g_{ik}h_t - v_t \ge 0$ and $y_{ikt}=1$ if $g_{ik}H_t - v_t < 0$. The lengths of other arcs can be determined in a similar way.

A path through the graph specifies an operating schedule for the generator and the problem of finding the minimum cost schedule becomes a shortest path problem on a cyclic state graph.

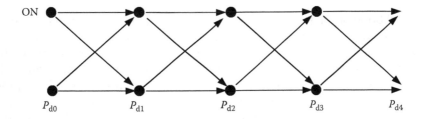

FIGURE 9.4
Basic model state graph.

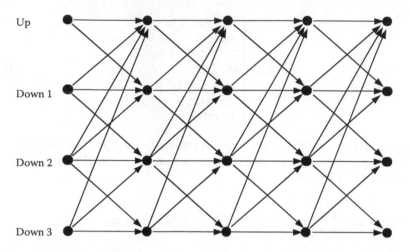

FIGURE 9.5
Graph with time-dependent start-up.

By expanding the state graph, it is also possible to represent certain aspects of the real problem that were omitted from the basic model. The state graph of the model is shown in Figure 9.5.

The state model in Figure 9.5 can accommodate many extensions such as

1. Time-dependent start-up cost
2. Minimum up and down constraints
3. Generator availability restriction

The time-dependent start-up costs are added to the model by varying the costs on the transition arcs that lead from a downstate to an upstate. The minimum up- and downtimes are enforced by eliminating some of the transition arcs in the state graph.

For example, if a generator must be OFF for at least three time periods, the transition arcs from the downstate 1 tend down two states to the up-states in the figure. Two would be eliminated.

9.9.3.3 Solution Technique

The backbone of the technique is a branch-and-bound procedure that builds an enumeration tree for the 0–1 variable x_{ij}. Each node in the tree is characterized by a set of x_{it} variables with fixed values. For the basic model the problem represented by a node in the enumeration tree has the form of the basic problem with the appropriate set of the x_{it} variables fixed. The Lagrangian relaxation of the problem at each node is solved to obtain a lower bound on the optimal solution for the problem at the node. To obtain the solution at a node a simple shortest path algorithm is used at the node to solve each of the I single generator subproblems using dynamic programming.

9.10 Illustrative Examples

Example 9.1

Consider the multidimensional problem

Maximize $z = 10x_1 + 5x_2 + 8x_3 + 7x_4$

Subject to

$6x_1 + 5x_2 + 4x_3 + 6x_4 \leq 40$

$3x_1 + x_2 \leq 15$

$x_1 + x_2 \leq 10$

$x_3 + 2x_4 \leq 10$

$x_j \leq 0.$

1. Develop the of the Lagrange relaxation formulation for this problem.
2. Formulate the maximization of the subproblem for $k=0$.

SOLUTION

Step 1. In matrix notation, the problem is

Maximize $Z = \mathbf{cx}$

Subject to

$\mathbf{Ax} \leq \mathbf{b}$

$\mathbf{Dx} \leq \mathbf{e},$

where $\mathbf{x} = [x_1, x_2, x_3, x_4]^t$ and $\mathbf{c} = [10, 5, 8, 7]^t$, with

$$\begin{bmatrix} 6 & 5 & 4 & 6 \\ 3 & 1 & 0 & 0 \\ 1 & 1 & 0 & 0 \\ 0 & 0 & 1 & 2 \end{bmatrix} \begin{bmatrix} x_1 \\ x_2 \\ x_3 \\ x_4 \end{bmatrix} \leq \begin{bmatrix} 40 \\ 15 \\ 10 \\ 10 \end{bmatrix}.$$

By partitioning this set of constraints, we obtain

$$\mathbf{Ax} \leq \mathbf{b}$$

$$\mathbf{Dx} \leq \mathbf{e},$$

where

$$A = [6 \quad 5 \quad 4 \quad 6], \quad b = [40]$$

$$D = \begin{bmatrix} 3 & 1 & 0 & 0 \\ 1 & 1 & 0 & 0 \\ 0 & 0 & 1 & 2 \end{bmatrix}, \quad \text{and} \quad c = \begin{bmatrix} 15 \\ 10 \\ 10 \end{bmatrix}.$$

Step 2. Lagrange is defined as

$$Z_D(u) = \text{Max}(cx) + u(b - Ax)$$

$$\therefore Z_D(u) = 10x_1 + 5x_2 + 8x_3 + 7x_4 + u(40 - 6x_1 - 5x_2 - 4x_3 - 6x_4)$$

Subject to

$$Dx \le e,$$

which is

$$3x_1 + x_2 \le 15$$

$$x_1 + x_2 \le 10$$

$$x_3 + 2x_4 \le 10.$$

Step 3. Assume $U^k = 1$ for $k = 0$. Then

Maximize

$$Z_D(u^k) = 10x_1 + 5x_2 + 8x_3 + 7x_4 + 40 - 6x_1 - 5x_2 - 4x_3 + 6x_4$$
$$= 4x_1 + 4x_3 + x_4 + 40,$$

such that $Dx \le e.$

9.11 Conclusions

This chapter discusses a Lagrangian problem in which the complicated constraints were replaced with a penalty term in the objective function involving the amount of violation constraints and their dual variables. The Lagrangian relaxation concept and setting method were first discussed in Sections 9.2 through 9.4. A comparison with an LP-based bound was presented in Section 9.5, followed by an improved relaxation concept. Practical applications, such as for power systems, were presented in Sections 9.7 to 9.9.

The Lagrange relaxation technique has been presented in this chapter. Simplex method relaxes or eliminates complex constraints by adding them to the master problem thereby enabling the independent solution to many subproblems.

The procedure for solving nonlinear programming optimization problems is shown in this chapter. The coupling constraints are handling them as additional constraints. It stems from Dantzig–Wolf description method for solving large linear problems. The algorithm involves the determination of λ value and controls the control and assorted state variables to be optimized. Application of Lagrangian relaxation (LR) to large-scale problems is generally applicable to security constrained economic dispatch OPF problem. Schemes to carry out implementation of Lagrange relaxation in UC is for scheduling for real-time application are identified in [10].

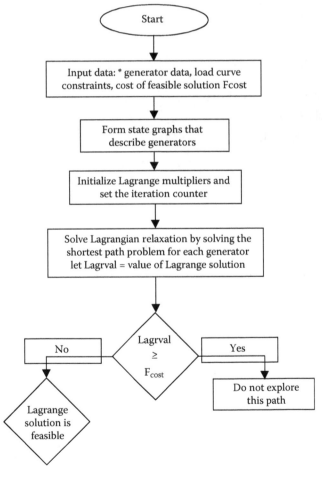

FIGURE 9.6
Algorithm for Lagrangian method.

9.12   Problem Set

PROBLEM 9.1

Maximize $z = 4x_1 + x_2 + 6x_3$

Subject to

$3x_1 + 2x_2 + 4x_3 \leq 17$

$x_1 \leq 2$

$x_3 \leq 2$

$x_1 \geq 1$

$x_2 \geq 1$

$x_3 \leq 1.$

PROBLEM 9.2

Maximize $z = 3x_1 + 5x_2$

Subject to

$3x_1 + 2x_2 \leq 18$

$x_1 \leq 4$

$x_j \leq 0.$

PROBLEM 9.3

Apply the principle to the following problem.

Maximize $z = 6x_1 + 7x_2 + 3x_3 + 5x_4 + x_5 + x_6$

Subject to

$x_1 + x_2 + x_3 + x_4 + x_5 + x_6 \leq 50$

$x_1 + x_2 \leq 10$

$x_2 \leq 8$

$5x_3 + x_4 \leq 12$

$x_5 + x_6 \leq 5$

$x_5 + 5x_6 \leq 50$

$x_j \leq 0.$

PROBLEM 9.4

Solve the following problem using the LR algorithm.

Maximize $z = 10x_1 + 2x_2 + 4x_3 + x_4$

Subject to

$x_1 + 4x_2 - x_3 \geq 8$

$2x_1 + x_2 + x_3 \geq 2$

$3x_1 + x_4 + x_5 \geq 4$

$x_1 + 2x_4 + x_5 \geq 10$

$x_j \leq 0.$

PROBLEM 9.5

Solve the following problem using the LR algorithm.

Maximize $P = 4x_1 + 5x_2 + 5x_3 + 4x_4$

Subject to

$2x_1 + 2x_2 + 3x_3 + 4x_4 \leq 7$

$x_1 - x_2 + x_3 - x_4 \leq 0$

$x_i \in \{0,1\}$

PROBLEM 9.6

Apply the LR algorithm to

Maximize $f = \left(0.3x_1^2 + 10\right)u_1 + \left(0.4x_2^2 + 14\right)u_2$

Subject to

$- x_1 u_1 - x_2 u_2 + 9.5 = 0$

$0 \leq x_1 \leq 0$

$0 \leq x_2 \leq 12$

$u_i \in \{0,1\}$

References

1. Fisher, M. L., The Lagrangian relaxation method for solving integer programming problems, *Management Science*, 27(1), 1981, 1–18.
2. Geoffrion, A. M., Lagrangian relaxation and its uses in integer programming, *Mathematical Programming Study*, 2, 1974, 82–114.

3. Graves, S. C., Using Lagrangian techniques to solve hierarchical production planning problems, *Management Science*, 28(3), 1982, 260–275.
4. Graves, S. C. and Lamar, B. W., An integer programming procedure for assembly system design problems, *Operations Research*, 31(3), 1983, 522–545.
5. Held, M. H., Wolfe, P., and Crowder, H. D., Validation of subgradient optimization, *Mathematical Programming*, 6(1), 1974, 62–88.
6. Muckstadt, J. A. and Koenig, S., An application of Lagrangian relaxation to scheduling in power generation systems, *Operations Research*, 25(3), 1977.
7. Shepardson, F. and Marsten, R. E., A Lagrangian relaxation algorithm for the two duty period scheduling problem, *Management Science*, 26, 1980, 274–281.
8. Shepardson, F. and Marsten, R. E., Solving a distribution problem with side constraints, *European Journal of Operations Research*, 6(1), 1981, 61–66.
9. Mulvey, J., Reducing the US Treasury's taxpayer data base by optimization, *Interfaces*, 10(5), 1980, 101–112.
10. Santos, A. and Costa, G. R. M., Optimal-power-flow solution by Newton's method applied to an augmented Lagrangian function, *IEE Proceedings—Generation, Transmission and Distribution*, 142(1), 1995, 33–36.

10

Decomposition Method

10.1 Introduction

In large-scale systems, a special class of linear programming (LP) problems is posed as multidimensional problems and is represented by the decomposition principle, a streamlined version of the LP simplex method. The decomposition principle has special characteristic features in that its formulation exploits certain matrices with distinct structures. These matrices, representing the formulated problems, are generally divided into two parts, namely, one with the easy constraints and the other with the complicated constraints. The partitioning is done such that the desired diagonal submatrices and identity matrices are obtained in the reformulation of the problem.

Now, in the decomposition principle [1,2,5,9], the method enables large-scale problems to be solved by exploiting these special structures. Therefore, we note that the decomposition method [4,6,7,8] can be used for any matrix **A** in the formulation:

Minimize

$Z = \mathbf{c}^T \mathbf{x}$

Subject to

$\mathbf{Ax} = \mathbf{b}$.

However, the method becomes more vivid when the matrix **A** has a certain structure, as explained in the following section.

10.2 Formulation of the Decomposition Problem

Consider the LP problem

Minimize $Z = \mathbf{c}^T \mathbf{x}$

Subject to

$$\mathbf{Ax} = \mathbf{b}$$

$$x_i \geq 0 \quad \text{for } \forall i \in [I, N],$$

where

     Z is the scalar objective function
     $\mathbf{c}^T$ is the coefficient vector of objective function
     $\mathbf{A}$ is the coefficient matrix of the equality constraints
     $\mathbf{b}$ is the vector of the inequality constraints
     $\mathbf{x}$ is the vector of unknown or decision variables

If matrix $\mathbf{A}$ has the special property form of a multidivisional problem, then by applying a revised simplex method, we start with

$$\mathbf{A} = \begin{bmatrix} A_1 & A_2 & \cdots & A_N \\ A_{N+1} & 0 & \cdots & 0 \\ 0 & A_{N+2} & \cdots & 0 \\ \vdots & \vdots & \vdots & \vdots \\ 0 & 0 & \cdots & A_{2N} \end{bmatrix} \quad \text{"complicated" constraints "easy" constraints}$$

It should be noted that some A_{N+1} blocks are empty arrays. Vector $\mathbf{b}$ is also partitioned accordingly into $N+1$ vectors such that

$$\mathbf{b} = [\mathbf{b}_0, \mathbf{b}_1, \ldots, \mathbf{b}_N]^T.$$

Similarly, vector $\mathbf{c}$ is also partitioned into N row vectors to obtain

$$\mathbf{c} = [\mathbf{c}_0, \mathbf{c}_1, \ldots, \mathbf{c}_N]^T.$$

Similarly, we have

$$\mathbf{x} = [\mathbf{x}_0, \mathbf{x}_1, \ldots, \mathbf{x}_N]^T.$$

Therefore, the problem takes the form:

     Minimize $Z = \displaystyle\sum_{j=1}^{N} c_j x_j$

     Subject to

$$\sum_{j=1}^{N} L_j x_j = b_0$$

$$A_j x_j = b_j$$

$$x_j \geq 0 \quad \forall j \in \{1, N\},$$

and the submatrices A_{N+j} correspond to L_j.

Notably, each of the constraints $A_j x_j = b_j$ defines the boundary of a convex polytope S_j thereby greatly reducing the computational effort. The set of points $\{X_j\}$ such that $x_j \geq 0$ and $A_{N+j}x_j < b_j$ constitutes a convex set with a finite number of extrema points. These points represent the corner-point feasible solution for the subproblem with these constraints. Then any solution x_j to the easy subproblem j that satisfies the constraints $A_j x_j = b_j$, where $x_j > 0$, can also be written as

$$x_j = \sum_{i \in S_j} \lambda_{ij} x_{ij},$$

where x_{ij} is assumed to be known and $\sum_{i \in S_j} \lambda_{ij} x_{ij} = 1$, with $\lambda_{ij} \leq 0$, $\forall j \in \{1, N\} \wedge i \in \{1, S_j\}$. It is further assumed that the polytope so formed by the subset of constraints $A_j x_j = b_j$ contains S_j vertices and is bounded. This is not the case for any x_j that is not a feasible solution of the subproblem.

Suppose x_{ij} is known, then let

$$L_{ij} \cong L_j x_{ij} \quad \text{and} \quad C_{ij} \cong C_j x_{ij},$$

such that the problem can be reformulated with fewer constraints as

$$\text{Minimize } Z = \sum_{j=1}^{N} \sum_{i=1}^{S_j} C_{ij} \lambda_{ij}$$

Subject to

$$\sum_{j=1}^{N} \sum_{i=1}^{S_j} \lambda_{ij} l_{ij} = b_0$$

with

$$\sum_{i=1}^{S_j} \lambda_{ij} = 1$$

$$\lambda_{ij} \geq 0 \quad \forall j \in \{1, N\} \quad \text{and} \quad \forall i \in \{1, S_j\}.$$

This formulation is a transformation from the partitioned problem to a revised problem that has reduced the number of rows from $m_0 + \sum_{j=1}^{N} m_j$ to $m_0 + N$ rows. However, it has greatly increased the number of variables from

$$\sum_{j=1}^{N} n_j \quad \text{to} \quad \sum_{j=1}^{N} S_j.$$

Fortunately, we do not have to consider all the x_{ij} variables if the revised simplex method is to be used.

10.3   Algorithm of the Decomposition Technique

This section introduces a typical decomposition algorithm. The solution steps are as follows:

> *Step 1*. Reformulate the LP problem into N LP subproblems and let $\mathbf{A}'$ represent the matrix of constraints and $\mathbf{c}'$ represent the vector of objective coefficients.
>
> *Step 2*. Initialization. Assume $\mathbf{x} = 0$ is a feasible solution to the original problem. Set $j = 1$ and $x_{jk}^* = 0$, where $j \in \{1, N\}$ and determine the basis matrix $\mathbf{B}$ and the vector of the basic variable coefficients $\mathbf{c_B}$ in the objective function.
>
> *Step 3*. Compute the vector $\mathbf{c_B^{-1} A'} - \mathbf{c}'$ and set ρ_{ij} to the minimum value (use the revised simplex method).
>
> *Step 4*. Compute the vector $(z_{jk} - c_{jk})$ for all $k = 1, 2, \ldots, n_j$ that corresponds to ρ_{ij} using;

$$(z_{jk} - c_{jk}) = \mathbf{c_B}(\mathbf{B}^{-1})_{1, m_0} A_j x_{jk}^* + \mathbf{c_B}(\mathbf{B}^{-1})_{1, m_0+j} - c_j x_{jk}^*$$

$$= \left(\mathbf{c_B}(\mathbf{B}^{-1})_{1, m_0} A_j - c_j\right) x_{jk}^* + \mathbf{c_B}(\mathbf{B}^{-1})_{1, m_0+j},$$

where

   m_0 is the number of elements of b_0
   x_{jk}^* is the corner-point feasible solution for the set of constraints
       given by $x_j \geq 0$ and $A_{N+j} x_j \leq b_j$
   $(\mathbf{B}^{-1})_{1, m_0}$ is the matrix of the first m_0 columns of B^{-1}
   $(\mathbf{B}^{-1})_{1, m_0}+j$ is the matrix of the ith column of B^{-1}

> *Step 5*. Use an LP approach to solve for the optimal W_j in the new problem that is given by

$$\text{Minimize } W_j = \left(\mathbf{c_B}(\mathbf{B}^{-1})_{1, m_0} A_j - c_j\right) x_j + \mathbf{c_B}(\mathbf{B}^{-1})_{1, m_0+j}$$

   Subject to

$$x_j \geq 0 \quad \text{and} \quad A_{N+j} x_j \leq b_j.$$

> *Step 6*. Obtain W_j^*, the optimal objective value of W_j, which is $W_j = \text{Min}(z_{jk} - c_{jk})$ for all values of k. The corresponding optimal solution is $x_{jk}^* = x_j$.
>
> *Step 7*. Determine the coefficient of the elements of $\mathbf{x_s}$ that are nonbasic variables as elements of $\mathbf{c_B}(\mathbf{B}^{-1})_{1, m_0}$.

Step 8. *Optimality test.* **IF** all coefficients of $c_B(B^{-1})_{1,m_0}$ are nonnegative, **THEN** the current solution is optimal. Go to step 10.

Otherwise find the minimum of the coefficients of $c_B(B^{-1})_{1,m_0}$ and select the corresponding entering basic variable.

IF the minimum of the coefficients of $c_B(B^{-1})_{1,m_0} = p_{jk}$, **THEN** identify the value of x_{jk}^* and the original constraints of p_{jk}.

Step 9. Repeat step 3 for all $j \in \{1, N\}$.

Step 10. Apply the revised simplex method and obtain the final optimal solution.

Step 11. Print/display final solution and **end**.

Notably, under the assumption that $x = 0$ is a feasible solution to the original problem, the initialization step utilizes the corresponding solution as the initial point or as the initial basic feasible solution. That is, we select x_s as the initial set of basic variables along with one variable p_{jk} for each of the subproblems j, where $j \in \{1, N\}$ such that $x_{jk}^* = 0$. Finally, successive iterations are performed until the optimal solution is found and the best value of p_{jk} is used to replace the value of x_j for the optimal solution to conform with that of the original problem.

10.4 Illustrative Example of the Decomposition Technique

Consider the problem

$$\text{Maximize } Z = 4x_i + 6x_2 + 8x_3 + 5x_4$$

Subject to

$$
\begin{array}{llllll}
x_1 & +3x_2 & +2x_3 & +4x_4 & \leq 20 \\
2x_1 & +3x_2 & +6x_3 & +4x_4 & \leq 25 \\
x_1 & +x_2 & & & \leq 5 \\
x_1 & +2x_2 & & & \leq 8 \\
& & 4x_3 & +3x_2 & \leq 12
\end{array}
$$

and

$$x_j \geq 0, \quad j \in (1, 2, \ldots, 4).$$

Solution
In the reformulated problem, the partitioned **A** matrix that reflects the easy and complicated constraints is

$$A = \begin{bmatrix} 1 & 3 & \vdots & 2 & 4 \\ 2 & 3 & \vdots & 6 & 4 \\ \cdots & \cdots & \vdots & \cdots & \cdots \\ 1 & 1 & \vdots & 0 & 0 \\ 1 & 2 & \vdots & 0 & 0 \\ \cdots & \cdots & \vdots & \cdots & \cdots \\ 0 & 0 & \vdots & 4 & 3 \end{bmatrix}.$$

Therefore, $N = 2$ and

$$A_1 = \begin{bmatrix} 1 & 3 \\ 2 & 3 \end{bmatrix}, \quad A_2 = \begin{bmatrix} 2 & 4 \\ 6 & 4 \end{bmatrix},$$

$$A_3 = \begin{bmatrix} 1 & 1 \\ 1 & 2 \end{bmatrix}, \quad A_4 = [4 \quad 3].$$

In addition,

$$c_1 = [4 \quad 6], \quad c_2 = [8 \quad 5],$$

$$X_1 = \begin{bmatrix} x_1 \\ x_2 \end{bmatrix}, \quad X_2 = \begin{bmatrix} x_3 \\ x_4 \end{bmatrix}, \quad b_0 = \begin{bmatrix} 20 \\ 25 \end{bmatrix}, \quad b_1 = \begin{bmatrix} 5 \\ 8 \end{bmatrix}, \quad b_2 = [12].$$

To prepare for demonstrating the solution to this problem, we first examine its two subproblems individually and then the reformulation of the overall problem.

Subproblem 1

Maximize $Z_1 = [4 \quad 6] \begin{bmatrix} x_1 \\ x_2 \end{bmatrix}$

Subject to

$$\begin{bmatrix} 1 & 1 \\ 1 & 2 \end{bmatrix} \begin{bmatrix} x_1 \\ x_2 \end{bmatrix} \leq \begin{bmatrix} 5 \\ 8 \end{bmatrix} \quad \text{and} \quad \begin{bmatrix} x_1 \\ x_2 \end{bmatrix} \geq \begin{bmatrix} 0 \\ 0 \end{bmatrix}.$$

It can be seen that this subproblem has four extreme points ($n_1 = 4$). One of these is the origin, considered the "first" of these extreme points, so

$$X_{11}^* = \begin{bmatrix} 0 \\ 0 \end{bmatrix}, \quad X_{12}^* = [50], \quad X_{13}^* = \begin{bmatrix} 2 \\ 3 \end{bmatrix}, \quad X_{14}^* = \begin{bmatrix} 0 \\ 4 \end{bmatrix},$$

where $\rho_{11}, \rho_{12}, \rho_{13},$ and ρ_{14} are the respective weights on these points.

Subproblem 2

Maximize $Z_2 = \begin{bmatrix} 8 & 5 \end{bmatrix} \begin{bmatrix} x_3 \\ x_4 \end{bmatrix}$

Subject to

$\begin{bmatrix} 4 & 3 \end{bmatrix} \begin{bmatrix} x_3 \\ x_4 \end{bmatrix} \leq [12]$ and $\begin{bmatrix} x_3 \\ x_4 \end{bmatrix} \geq \begin{bmatrix} 0 \\ 0 \end{bmatrix}$.

Its set of feasible solutions is

$$X_{21}^* = \begin{bmatrix} 0 \\ 0 \end{bmatrix}, \quad X_{22}^* = \begin{bmatrix} 3 \\ 0 \end{bmatrix}, \quad X_{23}^* = \begin{bmatrix} 0 \\ 4 \end{bmatrix},$$

where ρ_{21}, ρ_{22}, and ρ_{23} are the respective weights on these points.

By performing the $c_j x_{jx}^*$ vector multiplications and the $A_j x_{jk}^*$ matrix multiplications, the following reformulated version of the overall problem can be obtained.

Maximize $Z = 20\rho_{12} + 26\rho_{13} + 24\rho_{14} + 24\rho_{22} + 20\rho_{23}$

Subject to

$5\rho_{12} + 11\rho_{13} + 12\rho_{14} + 6\rho_{22} + 16\rho_{23} + x_{s1} = 20$

$10\rho_{12} + 13\rho_{13} + 12\rho_{14} + 18\rho_{22} + 16\rho_{23} + x_{s2} = 25$

$\rho_{11} + \rho_{12} + \rho_{13} + \rho_{14} = 1$

$\rho_{21} + \rho_{22} + \rho_{23} = 1$

and

$$\rho_{1k} \geq 0 \quad \text{for } k = 1,2,3,4$$
$$\rho_{2k} \geq 0 \quad \text{for } k = 1,2,3$$
$$x_{si} \geq 0 \quad \text{for } i = 1,2.$$

However, we should emphasize that the complete reformulation normally is not constructed explicitly; rather, just parts of it are generated as needed during the progress of the revised simplex method.

To begin solving this problem the initialization step selects x_{s1}, x_{s2}, ρ_{11}, and ρ_{12} to be the initial basic variables, so that

$$X_B = \begin{bmatrix} x_{s1} \\ x_{s2} \\ \rho_{11} \\ \rho_{21} \end{bmatrix}.$$

Therefore, since $A_1 x_{11}^* = 0$, $A_2 x_{21}^* = 0$, $c_1 x_{11}^* = 0$, and $c_2 x_{21}^* = 0$, then

$$\mathbf{B} = \begin{bmatrix} 1 & 0 & 0 & 0 \\ 0 & 1 & 0 & 0 \\ 0 & 0 & 1 & 0 \\ 0 & 0 & 0 & 1 \end{bmatrix} = \mathbf{B}^{-1}, \quad \mathbf{X_B} = \mathbf{b}' = \begin{bmatrix} 20 \\ 25 \\ 1 \\ 1 \end{bmatrix}, \quad \mathbf{c_B} = [0 \; 0 \; 0 \; 0],$$

for the initial basic solution.

To begin testing for optimality, let $j = 1$, and solve the LP problem.

Minimize $W_1 = (0 - c_1)x_1 + 0 = -4x_1 - 6x_2$

Subject to

$A_3 x_1 \le b_1$ and $x_1 \ge 0$.

The optimal solution of this problem is

$$x_1 = \begin{bmatrix} 2 \\ 3 \end{bmatrix} = x_{13}^*,$$

such that $W_1^* = -26$. Next, let $j = 2$ and solve the LP problem.

Minimize $W_2 = (0 - c_2)x_2 + 0 = -8x_3 - 5x_4$

Subject to

$A_4 x_2 \le b_2$ and $x_2 \ge 0$.

The solution of this problem is

$$x_2 = \begin{bmatrix} 3 \\ 0 \end{bmatrix} = x_{22}^*$$

such that $W_2^* = -24$. Finally, since none of the slack variables are nonbasic, no more coefficients need to be calculated. It can now be concluded that because both $W_1^* < 0$ and $W_2^* < 0$, the current basic solution is not optimal. Furthermore, since W_1^* is the smaller of these, ρ_{13} is the new entering basic variable.

For the revised simplex method to now determine the leaving basic variable, it is first necessary to calculate the column of $\mathbf{A}'$ giving the original coefficients of ρ_{13}. This column is

$$\mathbf{A}_k' = \begin{bmatrix} A_1 x_{13}^* \\ 1 \\ 0 \end{bmatrix} = \begin{bmatrix} 11 \\ 13 \\ 1 \\ 0 \end{bmatrix}.$$

Proceed in the usual way to calculate the current coefficient of ρ_{13} and the right-hand side column,

$$\mathbf{B}^{-1}\mathbf{A}'_k = \begin{bmatrix} 11 \\ 13 \\ 1 \\ 0 \end{bmatrix}, \quad \mathbf{B}^{-1}\mathbf{b}' = \begin{bmatrix} 20 \\ 25 \\ 1 \\ 1 \end{bmatrix}.$$

By considering only the strictly positive coefficients, the minimum ratio of the right-hand side to the coefficient is the $(1/1)$ in the third row, so that $r = 3$; that is, ρ_{11} is the new leaving basic variable. Thus, the new values of $\mathbf{X_B}$ and $\mathbf{c_B}$ are

$$\mathbf{X_B} = \begin{bmatrix} x_{s1} \\ x_{s2} \\ \rho_{13} \\ \rho_{21} \end{bmatrix}, \quad \mathbf{c_B} = \begin{bmatrix} 0 & 0 & 26 & 0 \end{bmatrix}.$$

By using a matrix inversion technique to find the value of B^{-1}, we obtain such that

$$\mathbf{B}^{-1}_{new} = \begin{bmatrix} 1 & 0 & -11 & 0 \\ 0 & 1 & -13 & 0 \\ 0 & 0 & 1 & 0 \\ 0 & 0 & 0 & 1 \end{bmatrix}.$$

The current basic feasible solution is now tested for using the optimality conditions of the revised simplex method. In this case, $W_1 = (0 - c_1)x_1 + 26 = -4x_1 - 6x_2 + 26$ and the minimum feasible solution of this problem is

$$x_1 = \begin{bmatrix} 2 \\ 3 \end{bmatrix} = x^*_{13}.$$

This yields $W^*_1 = 0.0$. Similarly, $W_2 = (0 - c_2)x_2 + 0 = -8x_3 - 5x_4$ such that the minimum solution of this problem is

$$x_2 = \begin{bmatrix} 3 \\ 0 \end{bmatrix} = x^*_{22}.$$

This yields $W^*_2 = -24$.

Finally, since none of the slack variables are nonbasic, no more coefficients need to be calculated. It can now be concluded that because both $W^*_2 < 0$, the current basic solution is not optimal, and ρ_{22} is the new basic variable. Proceeding with the revised simplex method.

$$\mathbf{A}'_k = \begin{bmatrix} A_2 x^*_{22} \\ 0 \\ 1 \end{bmatrix} = \begin{bmatrix} 6 \\ 18 \\ 0 \\ 1 \end{bmatrix}.$$

This implies that

$$\mathbf{B}^{-1}\mathbf{A}'_k = \begin{bmatrix} 6 \\ 18 \\ 0 \\ 1 \end{bmatrix}, \quad \mathbf{B}^{-1}\mathbf{b}' = \begin{bmatrix} 9 \\ 12 \\ 1 \\ 1 \end{bmatrix}.$$

Therefore, the minimum positive ratio is $(12/18)$ in the second row, so that $r = 2$; that is, x_{s2} is the new leaving basic variable.

The new inverse of the $\mathbf{B}$ matrix is now:

$$\mathbf{B}^{-1}_{new} = \begin{bmatrix} 1 & \left(\frac{-1}{3}\right) & \left(\frac{-20}{3}\right) & 0 \\ 0 & \left(\frac{1}{18}\right) & \left(\frac{-13}{18}\right) & 0 \\ 0 & 0 & 1 & 0 \\ 0 & \left(\frac{-1}{18}\right) & \left(\frac{13}{18}\right) & 1 \end{bmatrix}.$$

$$\mathbf{X}_B = \begin{bmatrix} x_{s1} \\ p_{22} \\ p_{13} \\ p_{21} \end{bmatrix}, \quad \mathbf{c}_B = [0 \quad 24 \quad 26 \quad 26 \quad 0].$$

Now test whether the new basic feasible solution is optimal.

$$W_1 = \left([0 \ 24 \ 26 \ 0] \begin{bmatrix} 1 & \left(\frac{-1}{3}\right) \\ 0 & \left(\frac{1}{18}\right) \\ 0 & 0 \\ 0 & \left(\frac{-1}{18}\right) \end{bmatrix} \begin{bmatrix} 1 & 3 \\ 2 & 3 \end{bmatrix} - [4 \ 6] \right) \begin{bmatrix} x_1 \\ x_2 \end{bmatrix} + [0 \ 24 \ 26 \ 0] \begin{bmatrix} \left(\frac{-20}{3}\right) \\ \left(\frac{-13}{18}\right) \\ 1 \\ \left(\frac{13}{18}\right) \end{bmatrix}$$

$$= \left(\begin{bmatrix} 0 & \left(\frac{4}{3}\right) \end{bmatrix} \begin{bmatrix} 1 & 3 \\ 2 & 3 \end{bmatrix} - [4 \ 6] \right) \begin{bmatrix} x_1 \\ x_2 \end{bmatrix} + \frac{26}{3}$$

$$= -\frac{4}{3} x_1 - 2x_2 + \frac{26}{3}.$$

Therefore, the feasible solution is

$$\mathbf{X}_1 = \begin{bmatrix} 2 \\ 3 \end{bmatrix} = x^*_{13}, \quad \text{with } W^*_1 = \left(\frac{2}{3}\right).$$

Similarly,

$$W_2 = \left(\begin{bmatrix} 0 & \left(\frac{3}{4}\right) \end{bmatrix} \begin{bmatrix} 2 & 4 \\ 6 & 4 \end{bmatrix} - [8 - 5] \right) \begin{bmatrix} x_3 \\ x_4 \end{bmatrix} + 0.0$$

$$= \frac{1}{3} x_4,$$

and the minimum solution now is

$$X_2 = \begin{bmatrix} 0 \\ 0 \end{bmatrix} = X_{21}^*$$

and the corresponding objective value is $W_2^* = 0.0$. Finally, $W_1^* \geq 0.0$ and $W_2^* \geq 0$, which means that the feasible solution is optimal. To identify this solution, set

$$X_B = \begin{bmatrix} x_{s1} \\ \rho_{22} \\ \rho_{13} \\ \rho_{21} \end{bmatrix} = B^{-1}b' = \begin{bmatrix} 1 & \left(\frac{-1}{3}\right) & \left(\frac{-20}{3}\right) & 0 \\ 0 & \left(\frac{1}{18}\right) & \left(\frac{-13}{18}\right) & 0 \\ 0 & 0 & 1 & 0 \\ 0 & \left(\frac{-1}{18}\right) & \left(\frac{13}{18}\right) & 1 \end{bmatrix} \begin{bmatrix} 20 \\ 25 \\ 1 \\ 1 \end{bmatrix} = \begin{bmatrix} 5 \\ \frac{2}{3} \\ 1 \\ \frac{1}{3} \end{bmatrix}$$

$$\therefore X_1 = \begin{bmatrix} x_1 \\ x_2 \end{bmatrix} = \sum_{k=1}^{4} \rho_{1k} x_{1k}^* = x_{12}^* = \begin{bmatrix} 2 \\ 3 \end{bmatrix}.$$

and

$$X_2 = \begin{bmatrix} x_3 \\ x_4 \end{bmatrix} = \sum_{k=1}^{3} \rho_{2k} x_{2k}^* = \frac{1}{3} \begin{bmatrix} 0 \\ 0 \end{bmatrix} + \frac{2}{3} \begin{bmatrix} 2 \\ 3 \end{bmatrix} = \begin{bmatrix} 2 \\ 0 \end{bmatrix}.$$

Thus, an optimal decision variable for the problem is $x_1 = 2$, $x_3 = 3$, $x_3 = 2$, and $x_4 = 0$. The corresponding value of the objective function is $Z = 42$.

10.5 Conclusions

This chapter discussed the decomposition method for a special class of LP multidimensional problems. This optimization method is applicable to a wide lass of problems such as in Refs. [3–5] with a special LP structures. Formulation of the decomposition problem was shown in Section 10.2. The algorithm of the decomposition technique and an illustration example of the decomposition method were given in Sections 10.3 and 10.4.

10.6   Problem Set

PROBLEM 10.1

Maximize $z = 4x_1 + x_2 + 6x_3$

Subject to

$3x_1 + 2x_2 + 4x_3 \leq 17$

$x_1 \leq 2$

$x_3 \leq 2$

$x_1 \geq 1$

$x_2 \geq 1$

$x_3 \geq 1.$

PROBLEM 10.2

Consider the multidivisional problem

Maximize $z = 10x_1 + 5x_2 + 8x_3 + 7x_4$

Subject to

$6x_1 + 5x_2 + 4x_3 + 6x_4 \leq 40$

$3x_1 + x_2 \leq 15$

$x_1 + x_2 \leq 10$

$x_3 + 2x_4 \leq 10$

$x_j \geq 0.$

Use the decomposition principle to solve this problem.

PROBLEM 10.3

Maximize $z = 3x_1 + 5x_2$

Subject to

$3x_1 + 2x_2 \leq 18$

$x_1 \leq 4$

$x_j \geq 0.$

PROBLEM 10.4

Apply the decomposition principle to the following problem.

Maximize $z = 6x_1 + 7x_2 + 3x_3 + 5x_4 + x_5 + x_6$

Subject to

$x_1 + x_2 + x_3 + x_4 + x_5 + x_6 \leq 50$

$x_1 + x_2 \leq 10$

$x_3 \leq 8$

$5x_3 + x_4 \leq 12$

$x_5 + x_6 \geq 5$

$x_5 + 5x_6 \leq 50$

$x_j \geq 0.$

PROBLEM 10.5

Indicate the necessary changes in the decomposition algorithm in order to apply it to minimization problems. Then solve the problem:

Maximize $z = 5x_1 + 3x_2 + 8x_3 - 5x_4$

Subject to

$x_1 + x_2 + x_3 + x_4 \geq 25$

$5x_1 + x_2 \leq 20$

$5x_1 - x_3 \geq 5$

$x_3 + x_4 = 20$

$x_j \geq 0.$

PROBLEM 10.6

Solve the following problem using the decomposition algorithm.

Maximize $z = 10x_1 + 2x_2 + 4x_3 + x_4$

Subject to

$x_1 + 4x_2 - x_3 \geq 8$

$2x_1 + x_2 + x_3 \geq 2$

$3x_1 + x_4 + x_5 \geq 4$

$x_1 + 2x_4 - x_5 \geq 10$

$x_j \geq 0.$

PROBLEM 10.7

Solve the following problem by the decomposition algorithm:

Maximize $P = 3x_1 + 4x_2 + x_3 + 2x_4$

Subject to

$$
\begin{aligned}
x_1 + x_2 + x_3 + x_4 &\leq 38 \\
5x_1 + x_2 &\leq 12 \\
x_3 + x_4 &\geq 6 \\
x_3 + 5x_4 &\leq 45
\end{aligned}
$$

$x_1, x_2, x_3, x_4 \geq 0$

PROBLEM 10.8

Solve the following problem by the decomposition algorithm:

Maximize $z = x_1 + 3x_2 + 5x_3 + 2x_4$

Subject to

$$
\begin{aligned}
2x_1 + x_2 &\leq 9 \\
5x_1 + 3x_2 + 4x_3 &\geq 10 \\
x_1 + 4x_2 &\leq 8 \\
x_3 - 5x_4 &\leq 4 \\
x_3 + x_4 &\leq 10
\end{aligned}
$$

$x_1, x_2, x_3, x_4 \geq 0$

References

1. Bell, E. J., Primal-dual decomposition programming, USGR & DR Order AD-625 365 from CFSTI, Operations Research Center, University of California, Berkeley, August 1965.
2. Birge, J. R., A Dantzig–Wolfe decomposition variant equivalent to basis factorization, *Mathematical Programming Study* 24, 1985, 43–64, R.W. Cottle (Ed.).
3. Glickman, T. and Sherali, H. D., Large scale network distribution of pooled empty freight cars over time, with limited substitution and equitable benefits, *Transportation Research, B (Methodology)*, 19(2), 1985, 85–94.
4. Himmelblau, D. M., *Decomposition of Large Scale Problems*, North-Holland, Amsterdam, 1973.
5. Hu, T. C., Decomposition algorithm for shortest paths in a network, *Operations Research*, 16(1), 1968, 91–102.
6. Lasdon, L. S., Duality and decomposition in mathematical programming, *IEEE Transactions on Systems Science and Cybernetics*, 4(2), 1968, 86–100.

7. Lemke, C. E., The dual method for solving the linear programming problem, *Naval Research Logistics Quarterly*, 1(1), 1954, 36–47.
8. Ritter, K., A decomposition method for linear programming problems with coupling constraints and variables, MRC Report No. 739, Mathematics Research Center, US Army, University of Wisconsin, MA, April 1967.
9. Dantzig, G. B., Orden, A., and Wolfe, P., Generalized simplex method for minimizing a linear form under linear inequality restraints, *Pacific Journal of Mathematics*, 5, 1955, 183–195.

11

State Estimation

11.1 Historical Perspective of State Estimation

Online state estimation is concerned with computing solutions of the basic load-flow problem every few minutes using online data telemetered periodically to the energy control center (ECC). As shown in Figure 11.1, this is one at present for the internal bulk transmission system of the utility concerned. Data exchanges with other neighboring utilities for the purpose of developing an external network equivalent model will be made easier if every utility is an online state estimator. An external equivalent representation will be necessary to perform online contingency analysis. Without an external equivalent model the uses of online state estimation will be limited to the monitoring of voltage levels, phase angles, line flows, and network topology. Another benefit is to use state estimator outputs short-term load forecasting.

In the basic load-flow problem the input/demand variables describe steady-state behavior of the system. In actual online systems one may measure these inputs and demands directly. The demands are injection quantities each of which is the sum of several power flow solutions. In principle, one can measure any meaningful set of system quantities that use those measurements as inputs to a system of equations whose solution yields values of state variables (bus voltage magnitudes and angles).

Of significance, in any online process, it is the fact that measurements will always have errors associated with them. Thus by measuring more quantities than the necessary minimum number, one can use the statistical theory of state estimation to filter out some of the error in the measurements. Hopefully, the solutions obtained will be more accurate than the measurements themselves. Although the increased accuracy of solutions is desirable, what is more significant is the ability to filter out the so-called bad data or highly erroneous measurements. Bad data will occur because of infrequent malfunctions in measuring instruments, possible communication errors, and other factors. Only when there are redundant measurements can one hope to develop a rational and automatic self-checking algorithm to insure the reliability of online load-flow solutions.

Because it addresses itself to the statistics of errors in measurements, online state estimation is by necessity a stochastic approach to the problem.

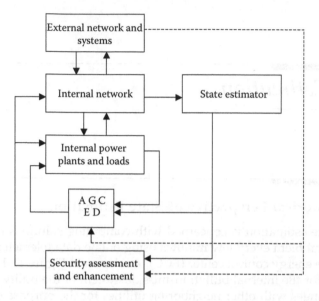

FIGURE 11.1
Simplified block diagram illustrating the role of online state estimation in an energy control center.

The obtained solutions are stochastic in nature with probabilistic character-istics. The importance of this stochastic approach lies not only in addressing issues of measurement errors but also in addressing other factors of uncer-tainty like those arising from modeling inaccuracies. Such inaccuracies result from errors in the values of line and transformer admittances due to initial approximations in obtaining admittance values, weather effects, and others. These are usually small but significant errors especially when their effects are compared to those produced by measurement errors. Large modeling errors occur, sometimes, because of incorrect topology determination by the online system. The major topics associated with online state estimation are

- Weighted least squares estimation (WLSE)
- Model error correction by means of parameter estimation
- Detection and identification of bad data
- Selection and location of measuring instruments

Vertically integrated utilities provide bundled services to customers aim-ing at high reliability with the lowest cost. In the traditional environment, utilities perform both power network and marketing functions. Although energy management systems (EMS) technology has been used to a certain extent, utilities were not pressed to utilize tools that demanded accurate real-time network models such as optimal power flows and available transfer capability determination. This is bound to change in the emerging competi-tive environment.

11.1.1 Conventional State Estimation

In conventional state estimation, network real-time modeling is decomposed into (1) the processing of logical data (the statuses of switching devices) and (2) the processing of analog data (e.g., power flow, power injection, and voltage magnitude measurements). During topology processing, the status of breakers/switches is processed using a bus section/switching-device network model. During observability analysis and state estimation, the network topology and parameters are considered as given, and analog data are processed using the bus *i* branch network model. In the conventional approach, logical data are checked by the topology processor and the analog data are checked by the state estimator.

11.1.2 Generalized State Estimation

In generalized state estimation there is no clear-cut distinction between the processing of logical and analog data since network topology processing may include local, substation level, state estimation, whereas when state estimation is performed for the whole network, parts of it can be modeled at the physical level (bus section/switching-device model). The term generalized is used to emphasize the fact that not only states, but also parts of the network topology, or even parameters, are estimated.

The explicit modeling of switches facilitates bad data analysis when topology errors are involved (incorrect status of switching devices). In this case, state estimation is performed on a model in which parts of the network can be represented at the physical level. This allows the inclusion of measurements made on zero impedance branches and switching devices. The conventional states of bus voltages and angles are augmented with new state variables. Observability analysis is extended to voltages at bus sections and flows in switching devices, and if their values can be computed from the available measurements they are considered to be observable.

For a zero impedance branch, or a closed switch, the following constraints, or pseudomeasurements, are included in state estimation:

$$V_k - V_m = 0 \quad \text{and} \quad \theta_{km} = \theta_k - \theta_m = 0.$$

In this case, P_{km} and Q_{km} are used as additional state variables. These variables are independent of the complex nodal voltages $V_k e^{j\theta_k}$ and $V_m e^{j\theta_m}$, since Ohm's law (in complex form) cannot be used to compute the branch current as a function of these voltages.

For open switches, the same additional state variables are included in state estimation. In the case of open switches the pseudomeasurements are as follows:

$$P_{km} = 0 \quad \text{and} \quad Q_{km} = 0.$$

No pseudomeasurements are added in the case of switches with unknown status. (There are situations in which the wrong status of a switching device can affect state estimation convergence. In these cases it may be preferable to treat such status as unknown and proceed with state estimation, which hopefully will include the estimation of the correct status.)

The ideas above can be extended to branches with unknown impedances. (The same comment regarding the impact of unknown status on state estimation convergence applies to branches with impedances with large errors.) Consider where the branch impedance z_{km} is unknown, whereas for simplicity all branches incident to k and m are assumed to have known impedances. As with zero impedance branches and with closed/open breakers, Ohm's law cannot be used to relate the state variables $V_k e^{j\theta_k}$ and $V_m e^{j\theta_m}$, associated with the terminal nodes k and m, with the branch complex power flows $P_{km} + jQ_{km}$ and $P_{mk} + jQ_{mk}$. These power flows can be used as additional states, although they are not independent, since they are linked by the constraint $I_{km} + I_{mk} = 0$, which can be expressed by the two following pseudomeasurements:

$$P_{km} V_m + (P_{mk} \cos \theta_{km} - Q_{mk} \sin \theta_{km}) V_k = 0$$
$$Q_{km} V_m + (P_{mk} \sin \theta_{km} + Q_{mk} \cos \theta_{km}) V_k = 0. \tag{11.1}$$

A power injection measurement at node k can be expressed as the summation of the flow state variables $P_{km} + Q_{km}$ and the flows in all other branches incident to k. Since only the flows in regular branches are functions of the nodal state variables, the unknown impedance will not form part of the measurement model. A similar analysis holds for power injection measurement at nodal and power flow measurements made in the unknown impedance branch. Once the network state is estimated, the value of the unknown parameter can be computed from the estimates.

More complex network elements such as a transmission line π equivalent model require the consideration of additional constraints (pseudomeasurements) in addition to the inclusion of flow state variables. Consider, for example, the equivalent π model, where the series branch impedance is to be estimated. In this case, power flows P'_{km}, Q'_{km}, P'_{mk}, and Q'_{mk} are considered to be additional states. The terminal power flows P_{km}, Q_{km}, P_{mk}, and Q_{mk} are then expressed in terms of the new state variables rather than as a function of the terminal bus voltages; as a consequence, the series branch impedance will not appear in the measurement model, and this can be written as

$$P_{km} = P_{kk} + P'_{km} \quad \text{and} \quad Q_{km} = Q_{kk} + Q'_{km} \tag{11.2}$$

$$P_{mk} = P_{mm} + P'_{mk} \quad \text{and} \quad Q_{mk} = Q_{mm} + Q'_{mk}. \tag{11.3}$$

Notice that in Equations 11.2 and 11.3, the power flows P_{kk}, Q_{kk}, P_{mm}, and Q_{mm}, are written in terms of the shunt parameters as usual. The bus injection measurements at buses k and m are expressed in terms of the terminal power flows as described earlier.

These added states are not entirely independent, so it is necessary to include the following relationship in the model:

$$P'_{km} V_m + (P'_{mk} \cos \theta_{km} - Q'_{mk} \sin \theta_{km}) V_k = 0$$
$$Q'_{km} V_m + (P'_{mk} \sin \theta_{km} + Q'_{mk} \cos \theta_{km}) V_k = 0. \tag{11.4}$$

Now consider the situation where a π equivalent model in which the shunt elements are made dormant. The state variables and measurement model are the same as those in the previous model. The constraints linking the state variables in the case of a balanced π model is $y^{sh}_{kk} = y^{sh}_{mm}$.

Expressing the shunt admittances in terms of the corresponding active and reactive power flows yields the two following pseudomeasurements:

$$P_{kk} V^2_m - P_{mm} V^2_k = 0$$
$$Q_{kk} V^2_m - Q_{mm} V^2_k = 0. \tag{11.5}$$

11.2 Simple Mathematical Background

11.2.1 Definition of Static State Estimation

In a typical static state estimation application many quantities of interest are measured and telemetered periodically every few seconds to the ECC. Such quantities include measurements of items such as

- Real and reactive line power flows
- Bus voltage magnitudes at generation and load busses
- Real and reactive generation at generation busses
- Real and reactive bus loads at load busses
- Transformer tap settings

In addition, on–off status quantities like breaker status are also telemetered to establish the exact network configuration. We assume that the given measurements are more than needed, i.e., there is a subset of measurements that will provide a load-flow solution. Each measured quantity can be expressed as follows:

$$z_i = z^t_i + v_i, \tag{11.6}$$

where
  z_i is the measured value
  v_i is the measurement error
  z_i^t is the true (but unknown) value of the measured quantity

The error v_i is obviously not known. What is known is a statistical measure associated with v_i. This measure can be adequately estimated from the calibration curves of the measuring instrument involved. The measure is given in terms of standard deviation error. Statistically, we can say that

$$E(v_i) = 0 \quad \text{and} \quad E(v_i^2) = \sigma_i^2. \tag{11.7}$$

This means that the error v_i, on the average, is zero, and that its standard deviation is σ_i. In actual applications σ_i may depend on the actual magnitude of the quantity measured, i.e., $\sigma_i = \sigma_i (z_i)$. We assume that σ_i is a constant for a given meter. A final statistical assumption is that of independence of errors coming from two different instruments, i.e.,

$$E(v_i v_j) = 0, \quad i \neq j. \tag{11.8}$$

In principle, the true but unknown value z_i^t is related to the true but unknown state vector $\mathbf{x}$ and the network admittance parameter vector $\mathbf{p}$ by the relation:

$$z_i^t = h_i(\mathbf{x}, \mathbf{p}), \tag{11.9}$$

where
  $\mathbf{x}$ is the vector of all complex bus voltages
  $\mathbf{p}$ is the vector of given series and shunt admittances

For example, we can write

$$\left.\begin{aligned}
T_{ij} &= T_{ij}^t + v_{T_{ij}} \\
T_{ij}^t &= V_i^2(g_{ij} + g_{sij}) - V_i V_j(g_{ij}\cos(\delta_i - \delta_j) + b_{ij}\sin(\delta_i - \delta_j)) \\
U_{ij} &= U_{ij}^t + v_{U_{ij}} \\
U_{ij}^t &= -V_i^2(b_{ij} + b_{sij}) - V_i V_j(g_{ij}\sin(\delta_i - \delta_j) - b_{ij}\cos(\delta_i - \delta_j))
\end{aligned}\right\} \tag{11.10–11.13}$$

In general, we write

$$z = h(\mathbf{x}, \mathbf{p}) + \mathbf{v}, \tag{11.14}$$

where

$$\mathbf{z} = \begin{bmatrix} z_1 \\ \vdots \\ z_m \end{bmatrix} = \text{vector of measured values}$$

$$h(\mathbf{x}) = \begin{bmatrix} h_1(\mathbf{x}) \\ \vdots \\ h_m(\mathbf{x}) \end{bmatrix}$$

(11.15)

$$x = \begin{bmatrix} x_1 \\ \vdots \\ x_n \end{bmatrix} = \begin{bmatrix} \delta_2 \\ \vdots \\ \delta_{n_b} \\ Vd_1 \\ \vdots \\ V_{n_b} \end{bmatrix}$$

$$E(\mathbf{v}) = 0$$
$$E(\mathbf{v}\mathbf{v}^{\mathrm{T}}) = \mathbf{R} = \text{covariance of } \mathbf{v}.$$

(11.16)

$\mathbf{R}$ is a diagonal matrix of all measurement variances, i.e.,

$$\mathbf{R} = \begin{bmatrix} \sigma_1^2 & 0 & \cdots & 0 \\ 0 & \sigma_2^2 & \cdots & 0 \\ \vdots & \vdots & \ddots & 0 \\ 0 & 0 & \cdots & \sigma_m^2 \end{bmatrix}.$$

(11.17)

It is necessary to have $m \geq n$ in such a way that a subset of n measurements can yield a solution of all state variables. This is known as the observability criterion. Normally m is 2–3 times the value of n allowing for a considerable amount of redundancy in the measurement information.

11.3 State Estimation Techniques

11.3.1 Method

Over the last few decades, many state estimation methods have been proposed and some of them were successfully applied to electric power industry [1]. In the classical normal method for state estimation called the normal equations (NE), a slow convergence or even nonconvergence can occur when dealing with large coefficient matrices. When the following factors are present, using the NE method may result in ill conditioning:

- Disparity in weighting factors
- Large number of injection measurements
- Connection of short and long transmission lines

Other methods called the orthogonal transformation method and the hybrid method directly perform the QR decomposition of the Jacobian matrix. The QR decomposition, also known as the QR factorization of a matrix, is a decomposition of the matrix into an orthogonal and triangular matrix. It is a method often used to solve the least squares problem. As a result, these are more stable and preferable to the NE method. There exist, however, some disadvantages. Namely, with regard to the orthogonal transformation method, Q needs to be stored which would require costly memory to store the nonspare and high dimension matrix. With the hybrid method is comparatively less stable than the orthogonal transformation method.

Some authors have used interior point methods for weighted least absolute value (WLAV) state estimation [2], while there are other methods such as auto tuning of measurement weights [3]. We will discuss the method and formulation of the least square state estimation [4].

11.3.1.1 Least Squares Estimation (LSE)

In least square calculations, we are trying to minimize the sum of measurements residuals

$$J(x) = \sum_{i=1}^{Nm} \frac{(Z_i - f_i(x))^2}{\sigma^2},$$ (11.18)

where
f_i is the function that is used to calculate the value using measurements by the ith measurement
σ^2 is the variance for the residual
$J(x)$ is the measurement residual
N_m is the number of independent measurements
Z_i^{meas} is the ith measured quantity

Assuming a given vector x of n random variables $x_1, x_2, \ldots, x_n$, and another vector y are related by Equation 11.19

$$y = Hx + r,$$ (11.19)

where
   H is the known matrix of dimension $m \times n$
   r is the zero mean random variable of same dimension as y
   x is the variable to be estimated
   y is the variable whose numerical values are available

In this least squares method, we have to obtain the best possible value of the vector x from the given values of the vector y. We start by taking the expectation of Equation 11.19. We then obtain

$$\bar{y} = H\bar{x}, \tag{11.20}$$

where $\bar{x}$ and $\bar{y}$ are the expected values of x and y, respectively. As we develop this method, we need to assume that $\hat{x}$ represents the desired estimate of x so that $\hat{c}$ is given by

$$\hat{y} = H\hat{x}. \tag{11.21}$$

The error can then be given by

$$\tilde{y} = y - \hat{y}. \tag{11.22}$$

The estimate $\hat{x}$ is the LSE if it is calculated by minimizing the estimation index J using

$$J = \tilde{y}^T \tilde{y}. \tag{11.23}$$

We then obtain the following expression using Equations 11.21 and 11.22

$$J = y^T y - y^T H\hat{x} - \hat{x}^T H^T y + \hat{x}^T H^T H\hat{x}.$$

We are then faced with a minimization problem to be solved such that

Min $J = f_{(\hat{x})}$ \hspace{3cm} (11.24)

s.t. $\operatorname*{grad}_{\hat{x}} J = 0.$

Solution to Equation 11.24 leads to

$$H^T H\hat{x} - H^T y = 0. \tag{11.25}$$

Equation 11.25 is called the NE and can be solved explicitly for the LSE of vector $\hat{x}$ as follows,

$$\hat{x} = (H^T H)^{-1} H^T y. \tag{11.26}$$

Example 11.3.1

Assume $H = \begin{bmatrix} 1 & 1 \\ 0 & 1 \\ 1 & 0 \end{bmatrix}$, estimate two random variables x_1 and x_2 by using the data for a three-dimensional vector y.

The matrix $H^T H$ is given by

$$H^T H = \begin{bmatrix} 2 & 1 \\ 1 & 2 \end{bmatrix} \text{ and its inverse } (H^T H)^{-1} = \begin{bmatrix} 2/3 & -1/3 \\ -1/3 & 2/3 \end{bmatrix}.$$

The vector $H^T y$ can now be formed and along with the inverse of $H^T H$, the estimate of x can be obtained using

$$\hat{x} = \begin{bmatrix} (2/3)y_1 - (1/3)(y_2 - y_3) \\ -(1/3)y_1 + (2/3)y_2 + (1/3)y_3 \end{bmatrix}.$$

11.3.1.2 Weighted Least Square Estimation

The weighted least squares state estimator [4] is to find the best state vector x, which minimizes the performance index of m measurements. We first obtain the ordinary least square estimate equation, listed in Equation 11.26 by minimizing the index function that puts equal weight to the errors of estimation of all the components of vector y. The estimation index is defined by

$$J = \tilde{y}' W \tilde{y}, \tag{11.27}$$

where W is the real symmetric weighting matrix of dimension $m \times m$.

The method of LSE is extended to the weighted form of and the NE is derived in the following form

$$H^T W H \hat{x} - H^T W y = 0. \tag{11.28}$$

The desired weighted least square (WLS) estimate is then found using

$$\hat{x} = (H^T W H)^{-1} H^T W y. \tag{11.29}$$

Equation 11.29 can then be rewritten as

$$\hat{x} = ky, \tag{11.30}$$

where

$$k = (H^T W H)^{-1} H^T W. \tag{11.31}$$

The following relationship is then obtained.

$$\hat{x} = KHx + kr$$
$$= (H^T WH)^{-1}(H^T WH)x + kr$$

or

$$\hat{x} = x + kr \tag{11.32}$$

and

$$E\{\hat{x}\} = E\{x\}. \tag{11.33}$$

Note that in Equation 11.33, called an unbiased estimate when satisfied, the error r is statistically independent of the columns in H and that the vector r has a zero mean. Hence the estimation error can be inferred to be zero on average. We can then write

$$\tilde{x} = kr \tag{11.34}$$

and the covariance of the estimation error is given as

$$P_x = KRK^T, \tag{11.35}$$

where R is the covariance of the error vector v.

Since the covariance P_x is a measure of the accuracy of the estimation, a smaller trace of this matrix would result in a more suitable estimate. To select the best choice of the weighting matrix, we need to set $W = R^{-1}$. We can now solve for the optimum value of the error covariance matrix using

$$P_x = (H^T R^{-1} H)^{-1}. \tag{11.36}$$

Also, K depends on H given by the relationship.

$$K = (H^T WH)^{-1} H^T W \tag{11.37}$$

where

$$W = R^{-1} \tag{11.38}$$

Example 11.3.2

Assume that we want to obtain the WLSE of the variable x from Example 11.3.1, by choosing the following weighting matrix,

$$W = \begin{bmatrix} 0.2 & & \\ & 1 & \\ & & 0.2 \end{bmatrix}.$$

The matrix

$$H^T W H = \begin{bmatrix} 0.4 & 0.2 \\ 0.2 & 1.2 \end{bmatrix}$$

and the matrix

$$H^T W = \begin{bmatrix} 0.2 & 0 & 0.2 \\ 0.2 & 1 & 0 \end{bmatrix}.$$

The WLEs estimate of the vector x is obtained using Equation 11.29. We therefore get

$$\hat{x} = \begin{bmatrix} (5/11)y_1 & -(5/11)y_2 & (6/11)y_3 \\ (1/11)y_1 & (10/11)y_2 & -(1/11)y_3 \end{bmatrix}.$$

Comparing the result of Example 11.3.1, we can clearly notice the effect of introducing the weighting on the estimate. Note that in the matrix H, the data for y_2, is given more weight and hence considered more valuable and in turn makes the components of x more dependant on y_2.

The matrix k can also be found using Equation 11.31.

$$K = \begin{bmatrix} (5/11) & -(5/11) & (6/11) \\ (1/11) & (10/11) & -(1/11) \end{bmatrix}.$$

Assuming that the covariance of the measurement error $R = I$, then using Equation 11.36 we can obtain P_x such that

$$P_x = (1/121) \begin{bmatrix} 86 & -51 \\ -51 & 102 \end{bmatrix}.$$

The choice of W previously yields unacceptably large estimation error variances. We therefore choose the weighting matrix to be equal to the identity matrix, $W = I$. The matrix K can now be solved as

$$K = \begin{bmatrix} 1/3 & -1/3 & 2/3 \\ 1/3 & 2/3 & -1/3 \end{bmatrix}.$$

Now the error covariance can be given by

$$P_x = (1/9) \begin{bmatrix} 6 & -3 \\ -3 & 3 \end{bmatrix}.$$

As expected, the error variances are now much smaller.

11.4 Applications to Power Network

11.4.1 State Estimation in Power Systems

The main goal of the power systems state estimator is to find a robust estimate for the unknown complex voltage at every bus in the modeled network [5]. Since inexact measurements—such as those from a SCADA system—are used to calculate the complex voltages, the estimate will also be inexact. This introduces the problem of how to devise the best estimate for the voltages given the available measurements. Of the many criteria used to develop a robust state estimator, the following three are regarded as the most common [6]:

1. *Maximum likelihood*: Maximizes the probability that the estimated state variable is near the true value.
2. *Weighted least squares (WLS)*: Minimizes the sum of the squared weighted residuals between the estimated and the actual measurements.
3. *Minimum variance*: Minimizes the expected value of the sum of the squared residuals between components of the estimated state variable and the true state variable.

The state estimation problem in power systems is derived from the relationship of the available system measurements to the system unknowns [7], given as

$$z_i = h_i(x) + \eta_i \quad i = 1, \ldots, q, \tag{11.39}$$

where

z_i is the available system measurement (q such measurements)
x is the system variable (voltage magnitudes $|V_i|$, and phases θ_i)
η_i is the measurement error typically modeled as a random variables with a zero mean Gaussian distribution
$h_i(x)$ is the nonlinear algebraic function relating the system variables to the system measurements, for various measurements

Random samples of the system measurements are taken and the objective is to find the best estimate of the system variables based on these random samples. These estimates are referred to as static estimates since the system is at equilibrium, and the methods discussed are referred to as static state estimators. Various applications of dynamic state estimators in power systems exist and focus on either applying Kalman filtering techniques [8,9] to a measurement model similar to Equation 11.39 or applying observers to track the dynamics of a power system [10–13]. In the observer-based methods, the model of the system includes the dynamic states of the system associated with the performance of the system generators, which is neglected in Refs. [8,9]. In particular, for the approach in Ref. [13], the observer-based state estimator is constructed using a DAE model of the power system.

There are multiple classes of estimators for static estimation, and each estimator is based on a unique cost function which is minimized in obtaining the system variables. A sample of commonly used objective functions ($C(x)$)

TABLE 11.1

Sample Object Functions and Estimators for Solving the State
Estimation Problem

WLSs	$C(x) = \frac{1}{2} \sum_i \frac{(z_i - h_i(x))^2}{\sigma_i^2}$	σ_i^2: Variance of error
Nonquadratic estimators	$C(x) = \sum_i \rho \frac{z_i - h_i(x)}{\sigma_i}$	ρ: Function that varies with residual
WLAV	$C(x) = \sum_i \frac{\lvert z_i - h_i(x) \rvert}{\sigma_i}$	
Least median of squares	$C(x) = \text{median}\{(z_i - h_i(x))^2\}$	

is provided in Table 11.1. For each of these methods there are multiple
approaches used to improve the solution accuracy/efficiency of the problem.
Detailed information about the various state estimation solvers in power systems can be found in Refs. [6,14], and a summary of commonly used methods
and their properties is given in Ref. [15]. Of particular interest is the commonly
adopted WLS estimator, which is described in the following section.

11.4.1.1 WLSs Estimator

Because of redundant measurements, the solution $\hat{x}$ of x is obtained by
minimizing the WLSs performance index J given by

$$J = [\mathbf{z} - h(\mathbf{x}, \mathbf{p})]^T \mathbf{R}^{-1} [\mathbf{z} - h(\mathbf{x}, \mathbf{p})] \tag{11.40}$$

with respect to $\mathbf{x}$. The vector $\mathbf{p}$ is assumed to be known exactly. Hence, we
can drop $\mathbf{p}$ from Equation 11.40. The concise state estimation problem statement becomes

$$\text{Given } z = h(x) + v. \tag{11.41}$$

Such that $E(v) = 0$, $E(w^T) = R$. Compute the best estimate $\hat{x}$ of x which
minimizes.

$$J = [\mathbf{z} - h(\mathbf{x})]^T \mathbf{R}^{-1} [\mathbf{z} - h(\mathbf{x})] \tag{11.42}$$

with respect to $\mathbf{x}$

At the minimum of J we should expect that

$$\left. \frac{\partial J}{\partial \mathbf{x}} \right|_{\hat{x}} = 0, \tag{11.43}$$

where $\hat{x}$ is the state vector at the minimum of J and is referred to as the
best estimate of $\mathbf{x}$. Given the above definition of J, we assert that the zero

gradient condition just stated will yield the following n-dimensional vector equation

$$0 = \mathbf{H}^T(\hat{\mathbf{x}})\mathbf{R}^{-1}(\mathbf{z} - h(\hat{\mathbf{x}})). \tag{11.44}$$

The necessary conditions for solution are a set of nonlinear algebraic equations requiring an iterative solution method. As in the Newton–Raphson method one would linearize the system equations around a nominal value of the state vector $\mathbf{x}$. Let $\hat{x}^0$ be such a nominal solution, we write

$$
\begin{aligned}
0 &= \mathbf{H}^T(\hat{\mathbf{x}})\mathbf{R}^{-1}[\mathbf{z} - h(\hat{\mathbf{x}})] \\
&= [\mathbf{H}^T(\hat{x}^0) + \Delta\mathbf{H}^T(\hat{\mathbf{x}})]\mathbf{R}^{-1}[\mathbf{z} - h(\hat{x}^0) - \mathbf{H}(\hat{x}^0)(\hat{\mathbf{x}} - \hat{x}^0)] \\
&\quad + \text{higher-order terms} \\
&\approx \mathbf{H}^T(\hat{x}^0)\mathbf{R}^{-1}(\mathbf{z} - h(\hat{x}^0)) - \mathbf{H}^T(\hat{x}^0)\mathbf{R}^{-1}\mathbf{H}(\hat{x}^0)(\hat{\mathbf{x}} - \hat{x}^0). \tag{11.45}
\end{aligned}
$$

In the linearization the term $\mathbf{HR}^{-1}(\mathbf{z} - h(\mathbf{x}))$ was not included. This is since in the vicinity of the actual solution, the $(\mathbf{z} - h(\mathbf{x}))$ vector is small. Hence the entire term is close to being a higher-order term. The above derivation is called a quasilinearization.

In the iterative scheme the next guess $\hat{x}^{-1}$ is defined as the solution of

$$0 = \mathbf{H}^T(\hat{x}^0)\mathbf{R}^{-1}(\mathbf{z} - h(\hat{x}^0)) - \mathbf{H}^T(\hat{x}^0)\mathbf{R}^{-1}\mathbf{H}(\hat{x}^0)(\hat{\mathbf{x}} - \hat{x}^0) \tag{11.46}$$

hence we write

$$\hat{x}^1 = \hat{x}^0 + (\mathbf{H}_0^T\mathbf{R}^{-1}\mathbf{H}_0)^{-1}\mathbf{H}_0^T\mathbf{R}^{-1}(\mathbf{z} - h(\hat{x}^0)), \tag{11.47}$$

and in general we obtain

$$\hat{x}^{k+1} = \hat{x}^k + (\mathbf{H}_k^T\mathbf{R}^{-1}\mathbf{H}_k)^{-1}\mathbf{H}_k^T\mathbf{R}^{-1}(\mathbf{z} - h(\hat{x}^k)), \tag{11.48}$$

where $k = 0, 1, \ldots,$ and

$$\mathbf{H}_k = \mathbf{H}(\hat{x}^k). \tag{11.49}$$

Sparse matrix techniques are directly applicable to the basic WLS algorithm derived here. Some general comments are given as follows:

1. Since the matrix $\mathbf{H}^T\mathbf{R}^{-1}\mathbf{H}$ (known as the information matrix) is symmetrical, one can store only the upper triangle of that matrix. For a symmetrical matrix one can easily show that the upper and lower triangular factors are related by the relation

$$L_{ji} = U_{ij}L_{ii}, \quad j > I \tag{11.50}$$

Therefore only the lower triangular or the upper triangular factors plus the diagonal of the lower triangular factors need to be stored.

2. In computing the vector $\mathbf{H}^T\mathbf{R}^{-1}(\mathbf{z} - h(\mathbf{x}))$, we use

$$\mathbf{H}^T\mathbf{R}^{-1}(\mathbf{z} - h(\mathbf{x})) = \sum_{i=1}^{m} \left(\frac{\partial h_i}{\partial \mathbf{x}}\right) \frac{(\mathbf{z} - h_i(\mathbf{x}))}{\sigma_i^2}. \tag{11.51}$$

Since $h_i(\mathbf{x})$ depends on only a few components of the $\mathbf{x}$ vector, the vector $\partial h_i/\partial x$ is highly sparse. In this identity one computes the contribution of each measurement to the overall expression separate and adds it to previous contributions.

3. We evaluate the information matrix by means of the identity:

$$\mathbf{H}^T\mathbf{R}^{-1}\mathbf{H} = \sum_{i=1}^{m} \left(\frac{\partial h_i}{\partial \mathbf{x}}\right)\left(\frac{\partial h_i}{\partial \mathbf{x}}\right)^T \frac{1}{\sigma_i^2}. \tag{11.52}$$

The contribution of each measurement to the information matrix is computed separately and added to previous contributions. We need to be careful when using compact storage arrays for off-diagonal terms.

4. Sparsity of the information matrix degrades because of injection measurements. In load-flow analysis, the number of off-diagonal terms in the Jacobian due to an injection measurement at a bus with b neighbors is $2b$. However, in the information matrix this number is $b(2b-1)$. The sparsity of the load-flow and state estimation cases are identical when the state estimator is based on line flow and bus voltage measurements only.

5. Basic WLS algorithm is fast in terms of convergence. Normally 3–4 iterations are sufficient as in the Newton–Raphson load-flow case.

11.4.2 Statistical Properties of State Estimator Outputs

The expected value of the optimal estimate vector $\hat{x}$ can be shown to be given by

$$E(\hat{\mathbf{x}}) = \mathbf{x}. \tag{11.53}$$

Moreover, the covariance matrix of $\hat{x}$ is given by

$$\mathrm{cov}(\hat{\mathbf{x}}) = [\mathbf{H}^T\mathbf{R}^{-1}\mathbf{H}]^{-1}. \tag{11.54}$$

We are also interested in the expected values of the index (J/m) corresponding to the fit of estimates of measured quantities to the measurements themselves. This can be shown to be given by

$$E(J/m) = \frac{m - n}{n}.$$ (11.55)

For $m - n$ (no redundancy) $E(J/m) = 0$ and the estimates fit the measurements perfectly. For $m \to \infty$ (infinite redundancy), $E(J/m) \to 1$ and the estimates approach the true value. In addition, the index J^t/m corresponds to the fit of the estimates to the true values of the noisy measurements. This can be shown to be given by

$$E(J^t/m) = n/m.$$ (11.56)

For $m = n$, we get $E(J^t/m) = 1$, and for $m \to \infty$, $E(J^t/m)\ m \to 0$

In general the index J is chi-square distributed with $m - n$ degrees of freedom.

11.4.2.1 Decoupled WLS and DC Models

In transmission systems, the sensitivity of real power to changes in bus voltage magnitudes and the sensitivity of reactive power to changes in bus phase angle changes are low. This characteristic leads to the decoupled measurement model where the measurement vector is partitioned as

$$z = \begin{bmatrix} z_{\text{real}} \\ z_{\text{reactive}} \end{bmatrix} = \begin{bmatrix} h_P(x) \\ h_Q(x) \end{bmatrix} + \begin{bmatrix} e_P \\ e_Q \end{bmatrix} \quad x = [\theta_i, V_i]$$ (11.57)

for the case of real (P) and reactive (Q) measurements in the system (either line flows or injections). The corresponding measurement Jacobian and gain matrices become

$$J = \begin{bmatrix} J_{P\theta} & 0 \\ 0 & J_{QV} \end{bmatrix} \quad G = \begin{bmatrix} G_{P\theta} & 0 \\ 0 & G_{QV} \end{bmatrix}$$ (11.58)

leading to the system estimates being obtained by

$$\Delta\theta = G_{P\theta}^{-1} J_{P\theta}^{T} R_{\text{real}}^{-1}(z_{\text{real}} - h_P(x))$$
$$\Delta V = G_{QV}^{-1} J_{QV}^{T} R_{\text{reactive}}^{-1}(z_{\text{reactive}} - h_Q(x)).$$ (11.59)

In addition, if all bus voltages are assumed known and set to 1 pu and all branch resistances are neglected, the DC measurement model is obtained, where the system estimates are obtained using

$$\theta = G_{P\theta}^{-1} J_{P\theta}^{T} R_{\text{real}}^{-1} z_{\text{real}}^{-1}, \qquad (11.60)$$

which represents the first θ iterations of the decoupled model in Equation 11.60.

11.4.2.2 Including Equality Constraints

Typically, the measurement vector z will be composed of both metered measurements and pseudomeasurements, which are known values, such as zero injections. These measurements are added to enhance the accuracy and sensitivity of the system estimates; however, they can cause ill-conditioned measurement Jacobians. One approach is to remove these measurements from the measurement vector, and instead, solve the state estimation problem by minimizing the cost function $C(x)$, subject to the added constraints:

$$\tilde{c}(x) = 0, \qquad (11.61)$$

where $c(x)$ is obtained from the pseudomeasurements and Equation 11.61 represents these measurements in the form of solution constraints. More about the specific properties of this approach can be found in Refs. [6,14].

On–off status quantities, like breaker status, are also telemetered in order to establish the exact network configuration. At the moment, we shall assume that the given measurements are more than needed, i.e., there is a subset of measurements that will provide a load-flow solution.

A careful selection of five of those measurements will be sufficient for solving the problem. For example, we can choose the following five quantities:

$$V_1, T_{12}, U_{12}, T_{23}, U_{23},$$

where T_{ij} and U_{ij} are the respective real and reactive power flows on line $i-j$. An unsatisfactory selection of measurement quantities may consist of

$$V_1, T_{12}, U_{12}, T_{21}, U_{21}.$$

This is so because from these measurements no information is obtainable on the voltage and angle of bus 3. Each measurement quantity, e.g., T_{ij}, can be expressed as follows:

$$z_i = z_i^t + v_i, \qquad (11.62)$$

where
   z_i is the measured value
   v_i is the measurement error
   z_i^t is the true (but unknown) value of the measured quantity

The error v_i is obviously not known. What is known is a statistical measure associated with v_i. This measure can be adequately estimated from the calibration current (analog-to-digital) conversion. The measure is given in terms of standard deviation of error. Statistically, we can say that

$$E[v_i] = 0$$
$$E[v_i^2] = \sigma_i^2 \quad . \tag{11.63}$$

This means that the error v_i is zero, on the average, and that its standard deviation is σ_i. In actual applications σ_i may depend on the actual magnitude of the quantity measured, i.e., $\sigma_i = \sigma_i(z_i)$. For example, the measurement error when the meter registers 1000 MW may be larger than the error when it reads 100 MW. For the moment we shall assume that σ_i is a constant for the given meter. A final statistical assumption is that of independence of errors coming from two different instruments, i.e.,

$$E[v_i v_j] = 0, \quad i \neq j. \tag{11.64}$$

In principle, the true but unknown value z_i^t is related to the true but unknown state vector $\mathbf{x}$, and the network admittance parameter vector $\mathbf{p}$ in the relation:

$$T_{ij} = T_{ij}^t + v_{T_{ij}}$$
$$T_{ij}^t = V_i^2(g_{ij} + g_{sij}) - V_i V_j(g_{ij} \cos(\delta_i - \delta_j) + b_{ij} \sin(\delta_i - \delta_j)) \tag{11.65}$$
$$U_{ij} = U_{ij}^t + v_{u_{ij}}$$
$$U_{ij}^t = -V_i^2(b_{ij} + b_{sij}) - V_i V_j(g_{ij} \sin(\delta_i - \delta_j) - b_{ij} \cos(\delta_i - \delta_j)). \tag{11.66}$$

In general we can write

$$z = h(x,p) + v, \tag{11.67}$$

where

$$z = \begin{bmatrix} z_1 \\ \vdots \\ z_m \end{bmatrix} = \text{Vector of measured valued.}$$

$$h(x) = \begin{bmatrix} h_1(x) \\ \vdots \\ h_m(x) \end{bmatrix} \tag{11.68}$$

$$x = \begin{bmatrix} x_1 \\ \vdots \\ x_n \end{bmatrix} = \begin{bmatrix} \delta_1 \\ \vdots \\ \delta_{n_b} \\ V_1 \\ \vdots \\ V_{n_b} \end{bmatrix} \tag{11.69}$$

$$E[v] = 0$$
$$= \text{covariance of } v. \tag{11.70}$$
$$E[vv^T] = R$$

R is a diagonal matrix of all measurement variances, i.e.,

$$R = \begin{bmatrix} \sigma_1^2 & 0 & \cdots & 0 \\ 0 & \sigma_2^2 & \cdots & 0 \\ \vdots & \vdots & \ddots & 0 \\ 0 & 0 & \cdots & \sigma_m^2 \end{bmatrix}. \tag{11.71}$$

It is necessary to have $m \geq n$ in such a way that a subset of n measurements can yield a solution of all state variables. This is known as the observability criterion. Normally m is 2–3 times the value of n allowing for a considerable amount of redundancy in the measurement information.

Because of the presence of redundant measurements, the solution $\hat{x}$ of x is obtained by minimizing the WLSs performance index J given by

$$J = [z - h(x,p)]^T R^{-1} [z - h(x,p)] \tag{11.72}$$

with respect to x. At this stage the vector p is assumed to be known exactly. Hence we can drop p from the above expressions for the sake of simplicity until we discuss the problem of parameter uncertainty. The concise state estimation problem statement becomes:

$$\text{Given } z = h(x) + v \tag{11.73}$$

such that $E(v) = 0$, $E(vv^T) = R$. Compute the best estimate $\hat{x}$ of x which minimizes

$$J = [z - h(x,p)]^T R^{-1} [z - h(x,p)] \tag{11.74}$$

with respect to x.

11.4.2.3 Necessary Solution Conditions

As stated, the state estimation problem consists of an unconstrained minimization of J given in Equation 11.74 with respect to x. At the minimum J we should expect that

$$\frac{\partial J}{\partial x}\bigg|\hat{x} = 0, \tag{11.75}$$

where $\hat{x}$ is the state vector at the minimum of J and is referred to as the best estimate of x. Given the above definition of J, one concludes that the zero gradient condition just stated will yield the following n-dimensional vector equation:

$$0 = H^T(\hat{x})R^{-1}(z - h(\hat{x})). \tag{11.76}$$

11.4.3 Model Parameter Identification—Sources of Inaccuracy

Calculated values of transmission line and transformer admittances normally contain various inaccuracies. In the transmission line case errors will arise as a result of factors such as

- Mathematical approximations used in calculations, e.g., truncation of Taylor series expansion.
- Simplified modeling assumptions, e.g., flat earth, completely transposed lines, no mutual effects relative to lines in same right-of-way, etc.
- Occasional gross human errors due to manual data handling at the initial input phase.
- Whether effects which modify conductor temperatures, causing different levels of sagging. This in turn can modify both line resistances and inductances.

Studies have concluded that errors of the order of 5% of normal values are quite possible. Obviously, in cases of human input data errors, the resulting parameter errors can be much larger.

11.4.4 State Estimation in Deregulated Environment

Vertically integrated utilities provide bundled services to customers aiming at high reliability with the lowest cost. In the traditional environment utilities perform both power network and marketing functions. Although EMS technology has been used to a certain extent, utilities were not pressed to utilize tools that demanded accurate real-time network models such as optimal power flows and available transfer capability determination [1]. This is bound to change in the emerging competitive environment.

In the new environment, the pattern of power flows in the network is less predictable than it is in the vertically integrated systems, in view of the new possibilities associated with open access and the operation of the transmission network under energy market rules. Although reliability remains a central issue, the need for the real-time network models becomes more important than before due to new energy market-related functions which will have to be added to new and existing EMS. These models are based on the results yielded by state estimation and are used in network applications such as optimal power flow, available transfer capability, voltage, and transient stability. The new role of state estimation and other advanced analytical functions in competitive energy markets was discussed by Singh et al. [2]. Hence, the implementation of real-time network analysis functions is crucial for the proper independent system operation (ISO). Based on these network models, operators will be able to justify technical and economical decisions, such as congestion management and the procurement for adequate ancillary services, and to uncover potential operational problems related to voltage and transient stability [1].

Reviews of the state of the art in state estimation algorithms were presented in Refs. [4–6]. Comparative studies of numerically robust estimators for power networks can be found in Ref. [7]. A review of the state of the art in bad data analysis was provided in Ref. [8]. A comprehensive bibliography on state estimation up to 1989 can be found in Ref. [9]. Generalized state estimation, which includes the estimation of states, parameters, and topology, is discussed in Refs. [10,11]. A review of external system modeling was presented in Ref. [12], and more recently, the state of the art on this subject was reviewed by the *IEEE Task Force on External Network Modeling* [13].

11.4.4.1   Network Real-Time Modeling

Network real-time models are built from a combination of snapshots of real-time measurements and static network data. Real-time measurements consist of analog measurements and statuses of switching devices, whereas static network data correspond to the parameters and basic substation configurations. The real-time model is a mathematical representation of the current conditions in a power network extracted at intervals from state estimation results. Ideally, state estimation should run at the scanning rate (say, at every two seconds). Due to computational limitations, however, most practical estimators run every few minutes or when a major change occurs.

11.4.4.2   Impact of the Changing Marketplace

The recent creation of ISO, with the need to control the grid, has increased the size of the networks that have to be modeled by state estimation. This trend compounds with two other phenomena: the spatial expansion stemming from the merger of several companies and the growing requirements for representing the network at lower voltage levels. As a result, supervised networks with tens of thousands of buses are becoming more and more common. The difficulties are not limited to network sizes, however.

Perhaps the most important issue in the new competitive environment is the way poorly estimated network models will affect the determination of prices of electricity. Metering quality and redundancy levels can vary significantly in a large network, measurement redundancy being more deficient at lower voltage levels. High redundancy levels will be necessary for adequate model building, and even more so for topology estimation, i.e., when parts of the network are represented at the physical level. It is envisaged that the range of measurement redundancy should evolve from today's 1.7–2.2 to 2.5–3.0. In addition to this improvement in the redundancy levels, the location of new meter should also take into consideration the need for estimating topology (statuses of switching devices).

Time skew among the measurements is normally present in most existing EMS systems, but its effect on state estimation is hardly noticeable in smaller networks. With larger networks, however, large pockets of data, sometimes involving an entire company's network, can be skewed by significant amounts from the rest of the data. This can affect both state estimation convergence and bad data processing: part of the skewed data will appear in the boundary nodes as multiple bad data (they contain errors that are conforming). Even though the dynamics of the system could be taken into consideration to correct the effect of time skew without much additional effort (e.g., using time-tagged measurements), when the protection system is activated or when the system is ramping up or down at a rapid pace in the area affected by the time skew, it is unlikely to exist a satisfactory solution to this problem in view of the scan rates that are currently used by the industry (1–2 s).

11.5 Illustrative Examples

Example 11.5.1

Establish the necessary minimality conditions associated with the set equations:

$$
\left.\begin{aligned}
1.1 &= z_1 = x_1 x_2 + v_1 \\
2.0 &= z_2 = x_1 + x_2^2 + v_2 \\
1.9 &= z_3 + x_1^2 + x_2 + v_3
\end{aligned}\right\}. \tag{11.77–11.79}
$$

with

$$
E[v_i^2] = 0.01, \quad i = 1, 2, 3. \tag{11.80}
$$

SOLUTION

The Jacobian matrix is given by

$$H(x) = \begin{bmatrix} \dfrac{\partial h_1}{\partial x_1} & \dfrac{\partial h_1}{\partial x_2} \\[2mm] \dfrac{\partial h_2}{\partial x_1} & \dfrac{\partial h_2}{\partial x_2} \\[2mm] \dfrac{\partial h_3}{\partial x_1} & \dfrac{\partial h_3}{\partial x_2} \end{bmatrix}. \tag{11.81}$$

For $h_1 = x_1 x_2$, $h_2 = x_1 + x_2^2$, $h_3 = x_1^2 + x_2$, one obtains

$$H(x) = \begin{bmatrix} x_2 & x_1 \\ 1 & 2x_2 \\ 2x_1 & 1 \end{bmatrix}. \tag{11.82}$$

The error covariance matrix **R** is given by

$$\begin{bmatrix} 0 \\ 0 \end{bmatrix} = H^T(\hat{x}) R^{-1} [z - h(\hat{x})]$$

$$= \begin{bmatrix} \hat{x}_2 & 1 & 2\hat{x}_1 \\ \hat{x}_1 & 2\hat{x}_2 & 1 \end{bmatrix} \begin{bmatrix} 100 & 0 & 0 \\ 0 & 100 & 0 \\ 0 & 0 & 100 \end{bmatrix} \begin{bmatrix} 1.1 - \hat{x}_1\hat{x}_2 \\ 2.0 - \hat{x}_1 - \hat{x}_2^2 \\ 1.9 - \hat{x}_1^2 - \hat{x}_2 \end{bmatrix}$$

$$= 100 \begin{bmatrix} \hat{x}_2(1.1 - \hat{x}_1\hat{x}_2) + \left(2 - \hat{x}_1 - \hat{x}_2^2\right) + 2\hat{x}_1\left(1.9 - \hat{x}_1^2 - \hat{x}_2\right) \\ \hat{x}_1(1.1 - \hat{x}_1\hat{x}_2) + 2\hat{x}_2\left(2 - \hat{x}_1 - \hat{x}_2^2\right) + \left(1.9 - \hat{x}_1^2 - \hat{x}_2\right) \end{bmatrix}. \tag{11.83}$$

Since the necessary conditions are a set of nonlinear algebraic equations, an iterative solution would be most logical. Using the Newton–Raphson method, the system equations can be linearized around a normal value of the state vector **x**. Let $\hat{x}^0$ be such a nominal solution. As a result we can write

$$0 = H^T(\hat{x}) R^{-1} [z - h(\hat{x})] \tag{11.84}$$

$$= [H^T(\hat{x}^0) + \Delta H^T(\hat{x})] R^{-1} [z - h(\hat{x}^0) - H(\hat{x}^0)(\hat{x} - \hat{x}^0)] + \text{Higher-order terms.}$$

$$\approx H^T(\hat{x}^0) R^{-1} (z - h(\hat{x}^0)) - H^T(\hat{x}^0) R^{-1} H(\hat{x}^0)(\hat{x} - \hat{x}^0) \tag{11.85}$$

Note that in the above linearization, the term $\Delta H R^{-1}(z - h(x))$ was not included. That is because around the actual solution, the vector $(z - h(x))$ is small. Therefore the whole term is close to being a high-order term. The derivation is hence called a quasilinearization.

In the iterative scheme the next guess $\hat{x}^1$ is defined as the solution of

$$0 = H^T(\hat{x}^0) R^{-1} (z - h(\hat{x}^0)) - H^T(\hat{x}^0) R^{-1} H(\hat{x}^0)(\hat{x} - \hat{x}^0) \tag{11.86}$$

hence we can write

$$\hat{x}^1 = \hat{x}^0 + \left(H_0^T R^{-1} H_0\right)^{-1} H_0^T R^{-1} (z - h(\hat{x}^0)) \tag{11.87}$$

and in general we obtain

$$\hat{x}^{k+1} = \hat{x}^k + \left(H_k^T R^{-1} H_k\right)^{-1} H_k^T R^{-1}(z - h(\hat{x}^k)), \tag{11.88}$$

where

$$k = 0, 1, \ldots \quad \text{and} \quad H_k = H(\hat{x}^k) \tag{11.89}$$

Example 11.5.2

Given the linear set of algebraic equations:

$$z = Hx + v, \tag{11.90}$$

where H is $n \times m$ matrix, find the best estimate $\hat{x}$ is the WLSs sense.

SOLUTION

Since $h(x) = Hx$, the Jacobian matrix simply is H. The minimality condition is given by

$$0 = H^T R^{-1}(z - H\hat{x})$$
$$= H^T R^{-1} z - H^T R^{-1} Hx. \tag{11.91}$$

Hence we obtain

$$\hat{x} = (H^T R^{-1} H)^{-1} H^T R^{-1} z. \tag{11.92}$$

11.6 Conclusion

Online state estimation in an ECC processes incoming raw data and generates a statistically reliable solution of the load-flow problem. It uses measurement system redundancy to detect and identify the so-called bad data. The redundancy also helps in smoothing out normal measure errors to produce accurate estimates of bus voltages and angles, as well as line flows and unmeasured loads.

It is demonstrated that network modeling errors can degrade the quality of state estimator outputs. A parameter estimator, which is an extension of the state estimator, can be used to provide better estimates of transmission line and transformer admittances. The state/parameter estimator combination may be used in conjunction with a robust bad data detection scheme to identify both, bad measurements and network configuration errors.

Once tools of state/parameter estimation and bad data identification have been developed, one can simulate the effectiveness of various measurement system configurations. The objective here involves studying trading among the goals of: increased reliability in bad data detection; improved ability to detect and correct model errors; improved accuracy of voltage and power flow estimates; and reduced measurement system cost.

State estimation is a key function in determining real-time models for interconnected networks as seen from EMS. In this environment, a real-time model is extracted at intervals from snapshots of real-time measurements. It is generally agreed that the emerging energy markets will demand network models more accurate and reliable than ever. This can only be achieved with state estimators that can reliably deal with both state, topology (status), and parameter estimation.

11.7   Problem Set

PROBLEM 11.7.1

For the following set of equations:

$$z_1 = 2.2 = x_1 + x_2 + v_1 \qquad (11.93)$$
$$z_2 = 2.9 = x_1 + 2x_2 + v_2 \qquad (11.94)$$
$$z_3 = 4.9 = 4x_1 + 2x_2 + v_3 \qquad (11.95)$$
$$z_4 = 7.2 = 3x_1 + 6x_2 + v_4 \qquad (11.96)$$

assume that $E[v_i] = 0$, $i = 1, \ldots, 4$, and

$$E[vv^T] = \begin{bmatrix} 0.01 & 0 & 0 & 0 \\ 0 & 0.01 & 0 & 0 \\ 0 & 0 & 0.01 & 0 \\ 0 & 0 & 0 & 0.01 \end{bmatrix} \qquad (11.97)$$

a. Find the best estimates $\hat{x}_1$ and $\hat{x}_2$.
b. What is the covariance matrix of

$$\hat{x} = \begin{bmatrix} \hat{x}_1 \\ \hat{x}_2 \end{bmatrix} \qquad (11.98)$$

Denote that by **P**. Prove your results.
c. Suppose that a new measurement:

$$z_5 = 7.0 = x_1 + 6x_2 + v_5 \qquad (11.99)$$

is made, with $E[v_5] = 0$ and $\text{var}[v_5] = 0.04$. Show that the best estimate of the state using this extra measurement can be expressed as a function of the old estimate of the state obtained in part (a) the covariance $\mathbf{P}$ obtained in part (b), z_5, the vector $\begin{bmatrix} 1 \\ 6 \end{bmatrix}$, and $\text{var}[v_5] = 0.04$. Evaluate the new state estimate. (Hint: y may employ the matrix inversion lemma).

PROBLEM 11.7.2

Given that

$$Z = Hx + v \qquad (11.100)$$

such that the $\text{cov}[v] = R$. Let $\mathbf{P}$ be the covariance matrix of $\hat{x}$. For the new scalar measurement:

$$z' = h^T x + v', \qquad (11.101)$$

where $E\lfloor (v')^2 \rfloor = \rho^2$ show that the best estimate of the state vector including the new measurement is given by

$$\hat{x}' = \hat{x} + \frac{Ph}{\rho^2 + h^T Ph}[z' - h^T \hat{x}] \qquad (11.102)$$

PROBLEM 11.7.3

For the three-bus system shown in Figure 11.2 the following measurements from a snapshot are obtained:

$$z_1 = V_1 = 1.01$$
$$z_2 = V_2 = 1.0$$
$$z_3 = V_3 = 1.0$$
$$z_4 = T_{12} = 1.0$$
$$z_5 = T_{13} = 1.97$$

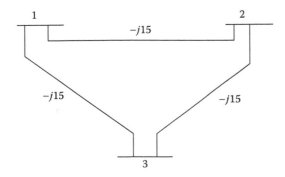

FIGURE 11.2
Network for Problem 11.7.3.

$$z_6 = P_1 = 2.59$$
$$z_7 = T_{21} = -0.95$$
$$z_8 = T_{23} = 1.0$$
$$z_9 = P_2 = 0.5$$
$$z_{10} = T_{31} = -1.5$$
$$z_{11} = T_{32} = -1.05$$
$$z_{12} = P_3 = -3.0,$$

where
P_i is the real power injection at bus i
T_{ij} is the real flow from bus i to bus j

It is known that in the absence of bad data the variance of any measurement in the above set is $\sigma_i^2 = 10^{-4}$.

a. Identify the bad measurement by means of preestimation analysis.
b. With the bad measurement eliminated compute the first iteration of the WLSs estimate using the initial guess of 1 pu voltage magnitudes and zero phase angles.

PROBLEM 11.7.4

The measurement vector z is related to the state vector x by

$$Z = Hx + v, \tag{11.103}$$

where $\dim[z] = m$, $\dim[x] = n < m$, $\text{Rank}[H] = n$, $E[v] = 0$, $\text{cov}[v] = R$.

d. Let $\hat{x}$ be the best estimate of the state vector in the weighted squares sense. Determine the covariance matrix of the vector

$$\hat{y} = F\hat{x}, \tag{11.104}$$

where F is an $r \times n$ constant matrix. Call this covariance matrix Q.
b. Let $F = H$, i.e.,

$$\hat{y} = H\hat{x}. \tag{11.105}$$

Determine the trace of the matrix QR^{-1}.

PROBLEM 11.7.5

For the set of measurements and associated equations:

$$z_1 = 3.4 = x_1 + 2x_2 + v_1$$
$$z_2 = -2 = x_1 - 2x_2 + v_2 \tag{11.106}$$
$$z_3 = 0 = -x_1 + 2x_2 + v_3$$

assume that the error covariance matrix is given by

$$R = 10^{-2} \begin{bmatrix} 1 & 0 & 0 \\ 0 & 3 & 0 \\ 0 & 0 & 1 \end{bmatrix}. \tag{11.107}$$

a. Determine the best estimate

$$\hat{x} = \begin{bmatrix} \hat{x}_1 \\ \hat{x}_2 \end{bmatrix}. \tag{11.108}$$

b. What is the covariance matrix of $\hat{x}$?
c. Assume now that measurement z_3 is exact, i.e., $\sigma_3 = 0$. It implies that

$$R = 10^{-2} \begin{bmatrix} 1 & 0 & 0 \\ 0 & 3 & 0 \\ 0 & 0 & 0 \end{bmatrix}. \tag{11.109}$$

Can you determine $\hat{x}$ with this information? How?

PROBLEM 11.7.6

For the sample system shown in Figure 11.3, assume that the three meters have the following characteristics shown in Table 11.2.

Calculate the best estimate for the phase angles δ_1 and δ_2 given the following measurements in Table 11.3.

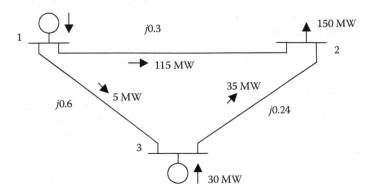

FIGURE 11.3
Figure for Problem 11.7.6.

TABLE 11.2

Meter Measurements for Problem 11.7.5

Meter	Full Scale (MW)	Accuracy (MW)	σ (pu)
M12	150	±6	0.03
M13	150	±5	0.02
M32	150	±0.4	0.004

TABLE 11.3

Meter Measurements for Problem 11.7.5

Meter	Measured Value (MW)
M12	115.0
M13	4.0
M32	35.5

References

1. Zhengchun, D., Zhenyong, N., and Wanliang, F., Block QR decomposition based Power System State Estimation Algorithm, *ScienceDirect*, July 11, 2005.
2. Singh, H. and Alvarado, F. L., Weighted least absolute value state estimation using interior point methods, *IEEE Transactions on Power Systems*, 9(3), August 1994.
3. Zhong, S. and Abur, A., Auto tuning of measurement weights in WLS state estimation, *IEEE Transactions on Power Systems*, 19(4), November 2004.
4. Kothari, D. P. and Nagrath, I. J., *Modern Power Systems Analysis*, McGraw-Hill, New York, 2008.
5. Jeffers, R. F., Techniques for wide area state estimation in power systems, Master of Science in Electrical Engineering Thesis, Virginia Polytechnic Institute and State University, Virginia, November 29, 2006.
6. Abur, A. and Exposito, A. G., *Power System State Estimation Theory and Implementation*, Marcel Dekker, New York, 2004, Chapter 1–2, pp. 1–36.
7. Dafis, C. J., An observability formulation for nonlinear power systems modeled as differential algebraic systems, Doctor of Philosophy thesis, Drexel University, Philadelphia, February 2005.
8. Rousseaux, P., Cutsem, T. V., and Dy Liacco, T. E., Whither dynamic state estimation, *Electrical Power and Energy Systems*, 12(2), 1990, 104–116.
9. Handschin, E., Real-time data processing using state estimation in electrical power systems, *Real-Time Control of Electrical Power Systems*, 1, 1972, 29–57.
10. Ueda, R., Takata, H., Yoshimura, S., and Takata, S., Estimation of transient state of multi-machine power system by extended linear observer, *IEEE Transactions on Power Apparatus and Systems*, PAS-96(2), 1977, 681–687.
11. Takata, H., Ueda, R., Fujita, T., and Takata, S., An iterative sequential observer for estimating the transient states of a power system, *IEEE Transactions on Power Apparatus and Systems*, PAS-96(2), 1977, 673–679.

12. Takata, H., Ueda, R., Fujita, T., and Takata, S., On the estimation of transient state of power system by discrete nonlinear observer, *IEEE Transactions on Power Apparatus and Systems*, PAS-94(6), 1975, 2135–2140.
13. Scholtz, E., Sonthikorn, P., Verghese, G. C., and Lesieutre, B. C., Observers for dynamic state estimation of swing motions in power systems, *Proceedings of the Power Systems Computation Conference (PSCC)*, Sevilla, Spain, 2002.
14. Monticelli, A., *State Estimation in Electric Power Systems: A Generalized Approach*, Kluwer Academic, Boston/Dordrecht/London, 1999.
15. Wu, F. F., Power system state estimation: A survey, *Electrical Power and Energy Systems*, 12(2), 1990, 80–87.

12

Optimal Power Flow

12.1 Introduction

The idea of optimal power flow (OPF) was introduced in the early 1960s as an extension of conventional economic dispatch to determine the optimal settings for control variables while respecting various constraints. The term is used as a generic name for a large series of related network optimization problems [4,6,8,12,14].

The development of OPF in the last two decades has tracked progress closely in numerical optimization techniques and advances in computer technology. Current commercial OPF programs are able to solve very large and complex power systems optimization problems in a relatively short time. Many different solution approaches have been proposed to solve OPF problems.

For OPF studies, the power system network is typically modeled at the main transmission level, including generating units. The model may also include other auxiliary generating units and representation of internal or external parts of the system are used in deciding the optimum state of the system.

In a conventional power flow, the values of the control variables are prespecified. In an OPF, the values of some or all of the control variables need to be found so as to optimize (minimize or maximize) a predefined objective. The OPF calculation has many applications in power systems, real-time control, operational planning, and planning [2,21,28,30,34,35]. OPF is available in most of today's energy management systems (EMSs).

OPF continues to gain importance due to the increase in power system size and complex interconnections [4,8,18]. For example, OPF must supply deregulation transactions or provide information on what reinforcement is required. The trade-offs between reinforcements and control options are decided by carrying out OPF studies. It is clarified when a control option maximizes utilization of an existing asset (e.g., generation or transmission), or when a control option is a cheaper alternative to installing new facilities. Issues of priority of transmission access and VAr pricing or auxiliary costing to afford fair price and purchases can be done by OPF [10,14,24].

The general OPF problem is posed as minimizing the general objective function $F(x,u)$ while satisfying the constraints $g(x,u) = 0$ and $h(x,u) \leq 0$, where $g(x,u)$ represents nonlinear equality constraints (power flow equations) and $h(x,u)$ is nonlinear inequality constraints on the vectors x and u. The vector x contains the dependent variables including bus voltage magnitudes and phase angles and the MVAr output of generators designed for bus voltage control. The vector x also includes fixed parameters such as reference bus angles; noncontrolled generator MW, MVAr, and outputs; noncontrolled load on fixed voltage; line parameters; and so on. The vector u consists of control variables involving:

Active and reactive power generation

Phase-shifter angles

Net interchange

Load MW and MVAr (load shedding)

DC transmission line flows

Control voltage settings

LTC transformer tap settings

Line switching

Table 12.1 shows a selection of objectives and constraints commonly found in OPF formulation. The time constants of the control process are relatively long, allowing the OPF implementation to achieve optimality adequately. The quality of the solution depends on the accuracy of the model studied. It is also important that the proper problem definition with clearly stated objectives be given at the onset. No two-power system companies have the same type of devices and operating requirements. The model form presented here allows OPF development to easily customize its solution to different cases under study [16,22,25–27,31,32].

To account for optimal dispatch of units, minimum loss, minimum voltage deviation, and security of the power system, the term OPF has been conceived. It is based on a concept largely based on static optimization method for minimizing a scalar optimization function (e.g., cost). It was first introduced in the 1960s by Tinney and Dommel [6]. It employs first-order gradient algorithm for minimization objective function subject to equality and inequality constraints. The solution methods were computationally intensive and did not gain attention compared to tradition power flow. The next generation OPF has been greater as power systems operation or planning need to know the limit, the cost of power, incentive for adding units, and building transmission systems serving a particular load entity. The OPF remains a major tool for

1. Scheduling power system control
2. Unit commitment, pricing of electricity market
3. Reliability or congestion regardless of its use

TABLE 12.1

Some Objectives and Constraints Commonly Found in OPF

Objectives	Constraints
Active power objectives	Limits on control variables
• Economic dispatch (minimum cost losses, MW generation or transmission losses)	• Generator output in MW
	• Transformer taps limits
	• Shunt capacitor range
• Environmental dispatch	Operating limits on line and transformer flows
• Maximum power transfer	
Reactive power objectives	• MVA, Amps, MW, and MVAr
• MW and MVAr loss minimization	• MW and MVAr interchanges
General goals	• MW and MVAr reserve margins (fixed/dynamic)
• Minimum deviation from a target schedule	• Voltage, angle (magnitude, difference)
• Minimum control shifts to alleviate violations	Control parameters
• Least absolute shift approximation of control shift	• Use of engineering rules to offer more controls for handling violation
	• Control effectiveness (more control with sufficient effects)
	• Limit priorities engineering preferable operating limit enforcement (cost benefit)
	• Control rates change and trajectories
	• Voltage stability
	Local and nonoptimized controls
	• Generator voltage, generator real power, transformer output voltage, MVAr, and shunt/SVC controls
	Equipment ganging and sharing
	• Tap changing
	• Generator MVAr sharing
	• Control ordering

It is based on standard requirements or specifications. Objectives must be modeled and its practicality with possible solutions. The constraints also have to be of appropriate design to satisfy upper and lower bounds and value to have a feasible solution. The constraints are equality and inequality. The constraints, when fixed an upper or a lower limit, are referred to as binding or active constraints. Second, when the constraints are scheduled within the limits they are called inactive.

Finding the active inequality constraints is the most difficult part of solving the OPF problems. No direct methods exist for solving OPF without using intermediate optimization method to identify current active sets. The challenges of OPF are many and they include:

1. Methodology. Some OPF solutions can be static. They are typically based on optimization methods such as linear programming (LP) and nonlinear programming (NLP) discussed earlier.

2. Issues of speed and accuracy have been proved to be important appropriate solution method. It depends on the model to achieve the accuracy and speed.

3. Knowledge of OPF usage setup time requires training.

4. Data required for OPF implementation problems converge error in data and inadequate models.

5. Inclusion of cost-benefit analysis approaches that factor in the impact of both technical and nontechnical operation constraints [17].

12.2 OPF—Fuel Cost Minimization

Fuel cost minimization is primarily an operational planning problem. It is also a useful tool in planning functions. This is usually referred to as economic dispatch, the aim of which is to obtain the active power generation of the units committed for operation, such that the total fuel cost is minimized while satisfying operational feasibility constraints [19,24].

12.2.1 Modeling Issues

Fuel cost minimization requires knowledge of the fuel cost curves for each of the generating units. An accurate representation of the cost curves may require a piecewise polynomial form, or can be approximated in several ways, with common ones being:

1. Piecewise linear
2. Quadratic
3. Cubic
4. Piecewise quadratic

A linear approximation is not commonly used while the piecewise linear form is used in many production-grade LP applications. A quadratic approximation is used in most NLP applications. Control variables are usually the independent variables in an OPF, including:

1. Active power generation
2. Generator bus voltages
3. Transformer tap ratios

4. Phase-shifter angles
5. Values of switchable shunt capacitors and inductors

The use of all of the above as control variables should give the best (least expensive) solution. For a regular OPF, the usual constraints are

1. Network power balance equations at each node
2. Bounds on all variables
3. Line-flow constraints
4. Others such as transformer tap ratios of parallel transformers

However, this may not be the most desired solution depending on certain other factors such as additional constraints. The following assumptions are made in modeling the objectives and constraints:

1. Fuel cost curves are smooth and quadratic in nature.
2. Only active power generations are controlled for cost minimization. Transformer tap ratios, generation voltages, shunt capacitors, and inductor positions are held at their nominal set values throughout the optimization.
3. Current flows are controlled approximately using voltage and phase angle restriction across the lines.
4. Contingency constraints are neglected.

12.2.2 Mathematical Description of the Objective Functions and Constraints for Cost Minimization

The objective function is given by the following fuel cost model:

$$F(P_g) = \sum_{i=1}^{N_g} \left(\alpha_i + \beta_i P_{g_i} + \gamma_i P_{g_i}^2 \right), \tag{12.1}$$

subject to equality constraints representing the active and reactive electric network balance,

$$P_i - P_{g_i} + P_{d_i} = 0, \quad i = 1, \ldots, N_b, \tag{12.2}$$

$$Q_i - \sum_{i \in \text{gen/synch}} Q_{g_i} + Q_{d_i} = 0, \quad i = 1, \ldots, N_b, \tag{12.3}$$

where

$$P_i = V_i \sum_{j=1}^{N_b} V_j Y_{ij} \cos(\theta_i - \theta_j - \psi_{ij}), \quad i = 1, \ldots, N_b, \tag{12.4}$$

$$Q_i = V_i \sum_{j=1}^{N_b} V_j Y_{ij} \sin (\theta_i - \theta_j - \psi_{ij}), \quad i = 1, \dots, N_b, \qquad (12.5)$$

together with the inequality constraints,

$$V_{i\min} \le V_i \le V_{i\max}, \quad i = 1, \dots, N_b, \qquad (12.6)$$

$$P_{g_i\min} \le P_{g_i} \le P_{g_i\max}, \quad i = 1, \dots, N_g, \qquad (12.7)$$

$$Q_{g_i\min} \le Q_{g_i} \le G_{g_i\max}, \quad i = 1, \dots, N_{g_q}, \qquad (12.8)$$

$$-k_{V_i}I_{l\max} \le V_i - V_j \le k_V I_{l\max}, \quad l = 1, \dots, N_l, \qquad (12.9)$$

i, j defined by l,

$$-k_{\theta_i}I_{l\max} \le \theta_i - \theta_j \le k_{\theta_i}I_{l\max}, \quad l = 1, \dots, N_l, \qquad (12.10)$$

i, j defined by l.
   The controls are the active generator power outputs.

$F(P_g)$ is the total fuel cost, as a function of P_g

P_{g_i} is the active power generation at unit i

$\alpha_i, \beta_i,$ and γ_i are the fuel cost parameters of unit i

N_g is the number of dispatchable generation units

N_{g_q} is the number of PV buses, including generators and synchronous condensers

N_b is the total number of buses

N_l is the total number of lines

V_i and V_j are the voltage magnitude at buses i and j

θ_i and θ_j are the phase angles at buses i and j

P_i is the net active power injections at node i

Q_i is the net reactive power injection at node i

Y_{ij} is the magnitude of the complex admittance matrix element at the ith row and jth column

ψ_{ij} is the phase angle of the complex admittance matrix element at position i, j

$I_{l\max}$ is the maximum allowable current flow in branch l

$V_{i\min}$ and $V_{i\max}$ are the lower and upper bound on the voltage magnitude at bus i

k_{v_i} and k_{θ_i} are the conversion factors to convert the maximum allowable current flow to an appropriate maximum allowable voltage and phase angle difference across the ends of the line l

$Q_{g_{i\min}}$ and $Q_{g_{i\max}}$ are the lower and upper bounds on the reactive generation at bus i

12.3 OPF—Active Power Loss Minimization

Active power loss minimization (referred to as loss minimization) is usually required when cost minimization is the main goal with control variables being active generator power outputs. When all control variables are utilized in a cost minimization (such as is reasonable when contingency constraints are included), a subsequent loss minimization will not yield further improvements. When cost minimization is performed using only the active power generations as control variables, a subsequent loss minimization computation using a different set of control variables can be useful in obtaining a better voltage profile and lower current flow along the lines. This will involve less risk of low-voltage insecurities during contingencies as well as a lower risk of current flow constraint violations during contingencies. The primary application of loss minimization is in operations, similar to cost minimization. In planning, loss minimization can be a useful tool in conjunction with a planning objective, providing more secure optimal solutions for planning purposes. This is especially useful in studies that neglect contingency constraints.

Loss minimization can be graphically represented as shown in Figure 12.1, which demonstrates that the process attempts to minimize the square of the distance between two voltage vectors connected across a transmission line. We see in the figure that in loss minimization, both magnitude and phase angle of voltage vectors across each line are minimized.

There are two basic approaches to loss minimization, namely, the slack bus approach and the summation of losses on individual lines. The slack bus approach is by far the least complicated approach, where the slack bus generation is minimized. The objective function is linear in this case and can be handled by any LP or NLP method. The disadvantage of this approach is that it can only minimize the total active power loss of the system. It is sometimes desirable to minimize the losses in a specific area only, and the above approach may not be applicable to this type of situation.

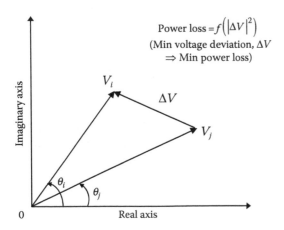

FIGURE 12.1
Graphical representation of loss minimization.

The second approach does not have the disadvantage mentioned earlier, but is more involved computationally. The objective function turns out to be more complicated when expressing voltages in polar form. When using the rectangular form, the objective function is simplified to a quadratic form. The polar form can be used in NLP methods. A quadratic form is preferred and the rectangular form of voltage representation is utilized. Due to the need for optimizing certain geographic areas, the summation approach is implemented.

12.3.1 Modeling Issues for Loss Minimization

In loss minimization, the usual control variables are

1. Generator bus voltage magnitudes
2. Transformer tap ratios
3. Switchable shunt capacitors and inductors
4. Phase-shifter angles

Out of these, a great deal of control can be achieved by using generator bus voltages and transformer tap ratios as control variables. Phase-shifter angles are normally used to alleviate line overloads. Since loss minimization indirectly takes care of line flows via the objective, line overloads are expected to be at a minimum. Active power generations are usually not employed as control variables in order to minimize changes to the economic dispatch solution for an integrated implementation.

In the formulation of loss minimization, generator voltages and transformer tap ratios are used as control variables. Transformer tap ratios are treated as continuous variables during the optimization, after which they are adjusted to the nearest physical tap position and reiterated holding the taps at the adjusted values. This approximation is justified based on the small step size usually found in transformers.

The constraints for loss minimization, as well as for other objectives described, are similar to those discussed earlier for cost minimization. The following assumptions are made in the formulation of the loss objective:

1. Loss minimization is done following a cost minimization, and thus the active power generations excluding the slack bus generation are held at their optimal values.
2. Generator bus voltages and transformer tap ratios are used as control variables. Shunt reactances and phase-shifter angles where available are held at nominal values.
3. Transformer tap ratios are treated as continuous variables during the optimization, after which they are adjusted to the nearest physical tap position and reiterated.

4. Current flows are controlled approximately, using restrictions on the real and imaginary components of the complex voltage across the lines.
5. Contingency constraints are neglected.

12.3.2 Mathematical Description of the Objective Functions and Constraints for Loss Minimization

The objective functions to be minimized are given by the sum of line losses

$$P_L = \sum_{k=1}^{N_l} P_{l_k}. \tag{12.11}$$

Individual line losses P_{l_k} can be expressed in terms of voltages and phase angles as

$$P_{l_k} = g_k \left[V_i^2 + V_j^2 - 2 V_i V_j \cos(\theta_i - \theta_j) \right], \quad k = 1, \ldots, N_l. \tag{12.12}$$

This expression involves transcendental functions. Transforming the above to equivalent rectangular form, we have

$$P_{l_k} = g_k \left[e_i^2 + f_i^2 + \left(e_j^2 + f_j^2 \right) - 2(e_i e_j + f_i f_j) \right], \quad k = 1, \ldots, N_l, \tag{12.13}$$

which simplifies to

$$P_{l_k} = g_k \left[(e_i - e_j)^2 + (f_i - f_j)^2 \right], \quad k = 1, \ldots, N. \tag{12.14}$$

The objective function can now be written as

$$p_{l_k} = \sum_{k=1}^{N} g_k \left[(e_j - e_j)^2 + (f_i - f_j)^2 \right]. \tag{12.15}$$

This is a quadratic form and is suitable for implementation using the quadratic interior point method, where

P_L is the total active power loss
P_{l_k} is the active power loss in branch k
g_k is the series conductance of line k
e_i and f_i are the real and imaginary components of the complex voltage at node i
e_j and f_j are same as e_i, f_i for node j

G_{ij} and B_{ij} are the real and imaginary components of the complex admittance matrix elements

$\tau_{k_{p1}}$ and $\tau_{k_{p2}}$ are the transformer tap ratios for transformers in parallel

N_{p_i} is the number of such transformers in the ith set of parallel transformers

$\tau_{i,\min}$ and $\tau_{i,\max}$ are the lower and upper bounds on the transformer tap ratio at the ith transformer

k_{e_i} and k_{f_i} are the equivalent of k_{v_i} and k_{θ_i} in rectangular form

$K_{e_{i\min}}$ and $K_{f_{i\min}}$ are the conversion factors, to convert the lower voltage bound to an equivalent rectangular form

$K_{e_{i\max}}$ and $K_{f_{i\max}}$ are the conversion factors, to convert the upper voltage bound to an equivalent rectangular form

$\tilde{g}_{l_i}$ is the shunt conductance of line l on side i (i.e., half total)

g_{l_i} is the series conductance of line (connected to node i)

τ_m is the transformer tap ratio of mth transformer

$\tilde{b}_{l_i}$ and b_{l_i} are the equivalent susceptance of $\tilde{g}_{l_i}$ and g_{l_i}

$\tilde{b}_i$ is the shunt reactance at node i

l_i is the lines connected to node i

N_{l_i} is the total number of l_i

g_m and b_m are the series conductance and susceptance of transformer m

N_τ is the total number of transformers

The constraints are equivalent to those specified in Section 12.2 for cost minimization, with voltage and phase angle expressed in rectangular form. The equality constraints are given by

$$P_i - \underset{i\in\text{gen}}{P_{g_i}} + P_{d_i} = 0, \quad i-1,\ldots, N_b, \tag{12.16}$$

$$Q_i - \underset{i\in\text{gen/synch}}{Q_{g_i}} + Q_{d_i} = 0, \quad i = 1,\ldots, N_b, \tag{12.17}$$

where

$$P_i = e_i\left[\sum_{j=1}^{N_b}(G_{ij}e_j - B_{ij}f_j)_i\right] + f_i\left[\sum_{j=1}^{N_b}(G_{ij}f_j - B_{ij}e_j)\right], \quad i = 1,\ldots, N_b, \tag{12.18}$$

$$Q_i = f\left[\sum_{j=1}^{n}(G_{ij}e_j - B_{ij}f_j)\right] - e_i\left[\sum_{j=1}^{n}(G_{ij}f_j - B_{ij}e_j)\right], \tag{12.19}$$

and also,

$$\tau_{k_{p1}} - \tau_{k_{p2}} = 0, \quad k = 1, \ldots, N_{pi}, \quad k_{p2} = 1, \ldots, N_{pi}, \quad k_{p1} \neq k_{p2}, \qquad (12.20)$$

$i \in$ sets of parallel transformers, for parallel transformers. The inequality constraints are given by

$$V_{i\min} \leq V_i \leq V_{i\max}, \quad i = 1, \ldots, N_b, \qquad (12.21)$$

$$P_{g_{i\min}} \leq P_{g_i} \leq P_{g_{i\max}}, \quad i \notin \text{slackbus}, \qquad (12.22)$$

$$Q_{g_{i\min}} \leq Q_{g_i} \leq Q_{g_{i\max}}, \quad i = 1, \ldots, N_{g_\eta}, \qquad (12.23)$$

$$\tau_{i\min} \leq \tau_i \leq \tau_{i\max}, \quad i = 1, \ldots, N_\tau, \qquad (12.24)$$

$$-k_{e_i} I_{l\max} \leq e_i - e_j \leq k_{e_l} I_{l\max}, \quad l = 1, \ldots, N_l, \qquad (12.25)$$

i, j defined by l.

$$-k_{f_1} I_{l\max} \leq f_i - f_j \leq k_{f_t} I_{t\max}, \quad l = 1, \ldots, N_l, \qquad (12.26)$$

i, j defined by l.

Note that Equation 12.21 is not linear in the rectangular formulation. It may be linearized or an approximate linear form may be used as

$$K_{e_{i\min}} V_{i\min} \leq e_i \leq k_{e_{i\max}} V_{i\max}, \quad i = 1, \ldots, N_b, \qquad (12.27)$$

$$K_{f_{i\min}} V_{i\min} \leq f_i \leq k_{f_{i\max}} V_{i\max}, \quad i = 1, \ldots, N_b. \qquad (12.28)$$

For the exact form V is given by

$$V_i^2 = e_i^2 + f_i^2, \quad i = 1, \ldots, N_b. \qquad (12.29)$$

The transformer tap ratio τ controls the optimization via the admittance matrix. The relationship is as follows:

$$G_{ii} = \sum_{i_i \in N_{l_i}} \left(\underset{l_i \in \text{all lines}}{\tilde{g}_{l_i}} + \underset{\substack{i_l m \in t/f \text{ with tapside} i}}{\frac{g_{l_i}}{\tau_m^2}} + \underset{\substack{i_j \in \text{lines or} t/f \text{ with tapside} \neq i}}{g_{l_i}} \right), \quad i = 1, \ldots, N_b. \qquad (12.30)$$

12.4 OPF—VAr Planning

The application of VAr planning as the name implies is in power system planning. It is aimed at minimizing the installation cost of additional reactive support necessary to maintain the system in a secure manner. The planning priority is to minimize cost and also to minimize future operations costs. This is necessary since cost of equipment and apparatus can be prohibitive in achieving an overall cost-effective planning scenario.

VAr planning involves identification of accurate VAr sites and measurable quantities of reactive sources to achieve system security. The analysis involves modeling to account for the discrete nature of the reactive power. This must generally be done using curve fitting or planning experience before it can simulate an optimum decision process.

The results obtained from VAr planning allow indirect sites, sizing, and costing of components. The extensive calculations, required for the siting and sizing elements, are normally done using OPF concepts. The traditional computations necessary for this purpose involve mathematical programming techniques—LP and NLP.

In the VAr planning process, we conduct voltage optimization which, being similar to loss minimization, is handled in much the same way. Loss minimization can be classified as a vector form of voltage optimization. In voltage optimization, the aim is to maintain the system voltage magnitudes as close as possible to a nominal voltage, such as one per unit. This is pictorially represented in Figure 12.2.

At the optimal solution, the voltage vectors stay within a narrow band close to nominal voltage values as shown by the shaded area in the figure. Hence, we see that the endpoints of two voltage vectors are brought closer to one another in a radial sense only, as opposed to loss minimization, where the endpoints are brought closer together in a vectorial sense. The application is mainly for operations with potential for use in planning when combined with a suitable planning objective. We present here a model, its formulation, and the associated algorithms for VAr planning.

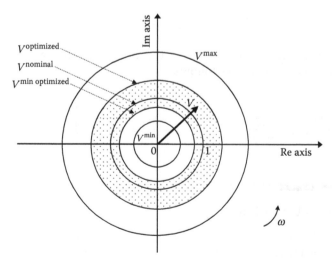

FIGURE 12.2
Description of voltage optimization.

12.4.1 Modeling Issues for VAr Planning Type I Problem

VAr installation costs are usually modeled as linear functions. The inductive and capacitive components of the VArs may be combined and modeled as one piecewise linear function as shown in Figure 12.3. With minimum and maximum values given, the existing VAr at a given site can be

1. All capacitive
2. All inductive
3. Both inductive and capacitive
4. None

The capacitive and inductive VArs can also be modeled as separate variables as shown in Figure 12.4. This is useful for developing models for capacitive and inductive charges needed in the VAr optimization process.

The modeling of capacitive and inductive VArs as separate variables is useful when including contingency constraints in the formulation. For

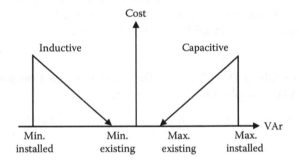

FIGURE 12.3
Cost curve for VAr support.

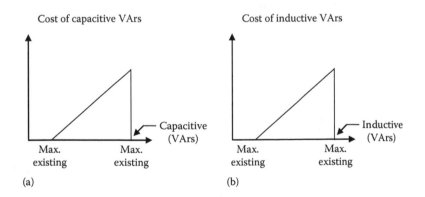

FIGURE 12.4
Model of discrete representations of capacitive and inductive variables. (a) Cost curve for capacitive VAr and (b) cost curve for inductive VAr.

example, a contingency involving a loss of load may require inductive compensation at a site while a certain line causing low-voltage situations may require capacitive compensation at the same site. This is not facilitated in the representation of Figure 12.3.

Combining planning and optimal objectives in VAr planning results in the two-objective optimization problem. The units of the planning objective may be in dollars. The units of the operations objective (such as fuel cost) may be in dollars per hour. Hence, if the objective is not scaled with respect to the other, the result will be meaningless. Some existing practical approaches are

1. Convert the planning costs to a comparable operations cost using life cycle costing. This is commonly used by the industry.
2. Convert the operations cost to a long-term cost comparable to the life of the installed equipment.

Voltage optimization can be performed by minimizing the sum of absolute voltage deviations from a nominal voltage, or by minimizing the sum of squares of the voltage deviations from a nominal voltage. The former is a piecewise linear objective, while the latter is a quadratic form, ideally suited for implementation using a quadratic OPF method such as the quadratic interior point.

12.4.2 Mathematical Description of the Objective and Constraints for Type I Problem for VAr Planning

The objective function and constraints for voltage optimization are described in mathematical form following a brief description of the notation not already discussed.

1. Mathematical notation

   $F(v)$ = Objective function to be minimized

   $V_{i,\text{nom}}$ = Nominal voltage (could be one per unit)
2. Mathematical description (voltage optimization)

$$F(v) = \sum_{i=1}^{N_b} (V_i - V_{i,\text{nom}})^2. \tag{12.31}$$

The objective function to be minimized is given by subject to the equality constraints given by Equations 12.31 and 12.32, repeated below with slight modifications.

$$P_i - \underset{i \in \text{gen}}{P_{g_i}} + P_{d_i} = 0, \quad i - 1, \ldots, N_b, \tag{12.32}$$

$$Q_i - \underset{i \in \text{gen/synch}}{Q_{g_i}} + Q_{d_i} = 0, \quad i - 1, \ldots, N_b, \tag{12.33}$$

where

$$P_i = V_i \sum_{j=1}^{N_b} V_j \left[G_{ij} \cos(\theta_i - \theta_j) + B_{ij} \sin(\theta_i - \theta_j) \right], \quad (12.34)$$

$$Q_i = V_i \sum_{j=1}^{N_b} V_j \left[G_{ij} \sin(\theta_i - \theta_j) + B_{ij} \cos(\theta_i - \theta_j) \right], \quad (12.35)$$

and also,

$$\tau_{k_{p1}} - \tau_{k_{p2}} = 0, \quad k_{p1} = 1, \dots, N_{pi}, \quad k_{p2} = 1, \dots, N_{pi}, \quad k_{p1} \neq k_{p2}, \quad (12.36)$$

$i \in$ sets of parallel t/f,

$$V_{i\min} \leq V_i \leq V_{i\max}, \quad i = 1, \dots, N_b, \quad (12.37)$$

$$P_{g_i\min} \leq P_{g_i} \leq P_{g_i\max}, \quad i \notin \text{slackbuses}, \quad (12.38)$$

$$Q_{g_i\min} \leq Q_{g_{i,k}} \leq Q_{g_i\max}, \quad i = 1, \dots, N_q, \quad (12.39)$$

$$\tau_{i\min} \leq \tau_i \leq \tau_{i\max}, \quad i = 1, \dots, N_\tau, \quad (12.40)$$

$$-k_{v1}I_{l\max} \leq V_i - V_j \leq k_{v1}I_{l\max}, \quad l = 1, \dots, N_l, \quad i, j \in l, \quad (12.41)$$

$$-k_\theta I_{l_1\max} \leq \theta_i - \theta_j \leq k_\theta I_{l\max}, \quad l = 1, \dots, N_l, \quad i, j \in l. \quad (12.42)$$

12.4.3 Type II Problem for VAr Planning

12.4.3.1 Control Variables

In designing the VAr/OPF problem, we start with the definition of the control variables. The control variables to be used depend on the objective specified.
For objective (1), the control variables are

1. Generator bus voltages
2. Transformer tap ratios
3. Shunt capacitors (existing and additional)
4. Shunt inductors (existing and additional)

For objective (2), the control variables are

1. Active power generations
2. Generator bus voltages
3. Transformer tap ratios

4. Shunt capacitors (existing and additional)

5. Shunt inductors (existing and additional)

For objective (2), generator bus voltages and transformer tap ratios will be controlled only when contingency constraints are employed.

12.4.3.2 Constraints

The constraints for VAr planning usually include contingency constraints, since security during contingencies is a primary objective in VAr planning. In some applications, contingencies are considered on a case-by-case basis. The advantage of this approach is the reduction of the problem to several smaller subproblems. The disadvantage is the increased problem size, especially for large systems, as the total amount of VAr support deemed necessary by the common set contained in the individual solutions will not necessarily be the optimal VAr support required. In order to guarantee a true optimal solution, it is necessary to consider the contingency cases jointly and solve them as one composite problem.

12.4.3.3 Assumptions

A major concern in VAr planning is the nature of the variables being optimized. The shunt inductances and capacitances come in discrete form and the inclusion of integer variables in the optimization require special mixed-integer programming techniques that do not perform well in non-linear power system applications. An approximation commonly adopted is to assume the variables as continuous during the optimization and clamp them to the nearest physical value at the optimal point. The solution may require reiteration with clamped quantities for a more accurate solution.

An alternative method is to move the variable closer to a physically available value during optimization using penalty functions. The penalty parameter is controlled during optimization in such a way as to avoid forcing the variable to a physical value far from the optimal point assuming continuity. The disadvantage of this method is that the penalty function changes the function's convexity and convergence can occur at a local rather than a global minimum. The first approximation is used in this study. It is assumed that the voltage at generator buses and transformer tap ratios do not change during a contingency.

12.4.4 Mathematical Description of the Objective and Constraints for Type II Problem for VAr Planning

The objective function and constraints for VAr planning are described in mathematical form, following a brief description of notation not already described.

12.4.4.1 Mathematical Notation

$F(q)$	VAr objective function to be minimized
$q_{c(i,k)}^{\text{total}}$	total capacitive VAr required at bus i during contingency k
$q_{r(i,k)}^{\text{total}}$	total inductive VAr required at bus i during contingency k
q_{ci}^{existing}	existing capacitive VAr sites
q_{ri}^{existing}	existing inductive VAr sites
N_c	number of capacitive VAr sites
N_r	number of inductive VAr sites
N_k	number of contingency cases
S_{ci}	unit cost of capacitive VAr (\$)
S_{ri}	unit cost of inductive VAr (\$)
$Q_{g_{i,k}}$	reactive power generation at bus i during contingency k
$V_{i,k}$	voltage at PQ buses during contingency k
$Q_{i,k}$	phase angle at all but the slack bus during contingency k
$q_{ci,k}$	value of optimal $q_{ci,k}$ averaged over all k

12.4.4.2 Mathematical Description of VAr Planning

The objective function to be minimized for VAr planning can be mathematically expressed as

$$F(q) = \sum_{i=1}^{N_c} S_{ci} \left[\max_{\text{all} k} \left\{ q_{c(i,k)}^{\text{total}} - q_{ci}^{\text{existing}} \right\} \right] + \sum_{i=l}^{N_r} S_{ri} \left[\max_{\text{all} k} \left\{ q_{r(i,k)}^{\text{total}} - q_{ri}^{\text{existing}} \right\} \right], \quad (12.43)$$

where a negative value for the expressions inside braces is treated as zero. This formulation, although providing a true mathematical description of the problem (based on the assumptions), has the disadvantage of including discrete variables in the optimization, even when q_c and q_r are assumed continuous. An alternate solution is to use the average over all k instead of the maximum as

$$F(q) = \frac{t}{N_k} \left[\left[\sum_{k=1}^{N_k} \left\{ q_{c(i,k)}^{\text{total}} - q_{ri}^{\text{existing}} \right\} \right] + \sum_{i=l}^{N_r} S_{ri} \sum_{k=l}^{N_k} \left\{ q_{r(i,k)}^{\text{total}} - q_{ri}^{\text{existing}} \right\} \right]. \quad (12.44)$$

The scalar quantity $1/N_k$ can be removed from the objective. The disadvantage of this method is shown in Figure 12.5, where we see that the average VAr requirement over all k is kept low, but the maximum is very much larger. The amount of VAr to be installed at bus i is determined by the maximum. Another possible alternative is to minimize the squared average deviations over all values of k. This will minimize large individual deviations

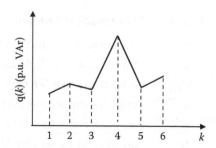

FIGURE 12.5
Optimal solution for site 1.

from the average. An objection might arise here that the objective is no longer cost, but cost squared. The answer to this is that, unless Equation 12.43 is minimized as specified, any other alternative is an approximation, and any approximation that gives the least expensive solution should be the choice for implementation. The objective function for the latter is given by

$$F(q) = \sum_{i=1}^{N_c} S_{ci}^2 \sum_{k=1}^{N_k} \left\{ q_{c(i,k)}^{total} - q_{ci}^{existing} \right\}^2 + \sum_{i=1}^{N_r} S_{ri}^2 \sum_{k=1}^{N_k} \left\{ q_{r(i,k)}^{total} - q_{ri}^{existing} \right\}^2. \quad (12.45)$$

Equation 12.44 is linear while Equation 12.45 is quadratic.

$$P_{i,k} - \underset{i \in \text{gen}}{P_{g_i}} + P_{d_i} = 0, \quad i = 1, \ldots, N_b, \quad k = 0, \ldots, N_k. \quad (12.46)$$

The equality constraints are given by

$$Q_{i,k} - \underset{i \in \text{gen/synch}}{Q_{g_i}} + Q_{d_i} + \underset{i \in \text{VAr}}{q_{i,k}} = 0, \quad i = 1, \ldots, N_b, \quad k = 0, \ldots, N_k, \quad (12.47)$$

$$P_{i,k} = V_{i,k} \sum_{j=1}^{N_b} V_{j,k} \left[G_{i,k} \cos(\theta_{i,k} - \theta_{j,k}) + B_{ij} \sin(\theta_{i,k} - \theta_{j,k}) \right], \quad (12.48)$$

$$Q_{i,k} = V_{i,k} \sum_{j=1}^{N_b} V_{j,k} \left[G_{i,j} \sin(\theta_{i,k} - \theta_{j,k}) + B_{ij} \cos(\theta_{i,k} - \theta_{j,k}) \right], \quad (12.49)$$

where (Note that $k = 0$ specifies the intact system. In cases where k is not specified, $k = 0$ is assumed.)

$$\tau_{k_{p1}} - \tau_{k_{p2}} = 0, \quad k_{p1} = 1, \ldots, N_{pi}, \quad k_{p2} = 1, \ldots, N_{pi}, \quad (12.50)$$

$i \notin$ sets of parallel t/f, with the inclusion of parallel transformers.
The inequality constraints are given by

$$V_{i\min} \le V_i \le V_{i\max},$$

$$V_{i\min} \le V_i, \quad 0 \le V_{i\max}, \quad i = 1, \ldots, N_b, \quad i \in \text{pv buses,}$$

$$V_{i\,\text{min}\,c} \leq V_{i,k} \leq V_{i\,\text{max}\,c}, \quad i = 1, \ldots, N_b, \quad i \in \text{PQ buses},$$

$$P_{g_{i\,\text{min}}} \leq P_{g_i} \leq P_{g_{i\,\text{max}}}, \quad i \in \text{PQ buses}, \quad k = 1, \ldots, N_k,$$

$$Q_{g_{i\,\text{min}}} \leq Q_{g_{i,k}} \leq Q_{g_{i\,\text{max}}}, \quad i \in \text{pv buses}, \quad k = 1, \ldots, N_k,$$

$$\tau_{i\,\text{min}} \leq \tau_i \leq \tau_{i\,\text{max}}, \quad i = 1, \ldots, N_k,$$

$$-k_{v1}I_{l\,\text{max}} \leq V_{i,k} - V_{j,k} \leq k_{v1}I_{l\,\text{max}}, \quad I = 1, \ldots, N_l, \quad i,j \in l, \quad k = 1, \ldots, N_k,$$

$$-k_{\theta}I_{l\,\text{max}} \leq \theta_{i,k} - \theta_{j,k} \leq k_{\theta 1}I_{l\,\text{max}}.$$

Note that generator bus voltages, transformer tap ratios, and active power generations are unchanged during contingencies. The transformer tap ratios control the optimization via the admittance matrix as before. We also have

$$0 \leq q_{ci,k} \leq q_{ci\,\text{max}}, \quad i = 1, \ldots, N_c, \quad k = 1, \ldots, N_k. \tag{12.51}$$

$$0 \leq q_{ri,k} \leq q_{ri,k,\,\text{max}}, \quad i = 1, \ldots, N_r, \quad k = 1, \ldots, N_k, \tag{12.52}$$

A minimum VAr installation is desired to avoid installation of very small quantities at sites. This can be achieved by clamping the optimal VAr to zero or to the minimum VAr (whichever is closer), in the event that the optimal VAr is lower than the minimum allowed. After clamping, the solution may require reiteration for a more accurate solution by using

$$q_{ci,\,\text{min}} \leq q_{ci,k} \leq q_{ci,\,\text{max}}, \tag{12.53}$$

$$q_{ri,\,\text{min}} \leq q_{ri,k} \leq q_{ri,\,\text{max}}. \tag{12.54}$$

The control variables used in VAr planning additionally include

$$V_i, \quad i = 1, \ldots, N_{g_q},$$

$$P_{g_i}, \quad i = 1, \ldots, N_g.$$

The second objective function is given as

$$F(q,P_g) = F^*(q) + \sum_{i=1}^{N_g} \left(\alpha_i + \beta_i P_{g_i} + \gamma_i P_{g_i}^2 \right), \tag{12.55}$$

where $F^*(q)$ is a modification of $F(q)$ described earlier. The modifications consist of changes to S_{ci} and S_{ri} to convert the planning cost to an equivalent operations cost. Assuming that money is borrowed at interest r, and the life of the equipment is n years, we can write:

$$S_{ci}\ (\$/\text{VAr}) = S_{ci} \frac{(1+r)^{n_{ci}}}{n_{ci} \times 365 \times 24}\ (\$/\text{VAr})/\text{h}, \tag{12.56}$$

$$S_{ri} \ (\$/\text{VAr}) = S_{ri} \frac{(1+r)^{n_{ri}}}{n_{ri} \times 365 \times 24} \ (\$/\text{VAr})/\text{h}, \qquad (12.57)$$

$$P_{g_{i\,\min}} \leq P_{g_i} \leq P_{g_{i\,\max}}, \quad i \in \text{gen.} \qquad (12.58)$$

12.5   OPF—Adding Environmental Constraints

The Clean Air Act Amendments (CAAAs) of 1990 require the power industry to reduce its SO_2 emissions level by 10 million tons per year from the 1980 level, and its NO_x levels by about 2 million tons per year. The SO_2 provisions of the Act are to be implemented in phases: Phase I, which began in 1995, required 262 generating units from 110 power plants to limit their SO_2 emissions to 2.5 lb/MBtu, and Phase II, beginning in 2000, requires all units to emit under 1.2 lb/MBtu. To prevent utilities from shifting emissions from Phase I to Phase II units, an underutilization (or burn) provision mandates a minimum generation level at the Phase I units. If this provision is not met, the utility concerned has to either surrender allowances proportionally, or designate one or more Phase II units as compensating units subject to the same restrictions as Phase I units. Energy conservation measures and unexpected demand reductions are taken into account for the underutilization constraint [7,13,20].

12.5.1   Modeling Issues for Environmental Constraint

There are various ways to model environmental constraints. The model adopted here assumes that both SO_2 and NO_x emissions can be expressed as separable quadratic functions of the real power output of the individual generating units. More specifically, the same heat-rate functions are used for calculating the fuel and each of the emission types.

For configuration I, the SO_2 emission constraints can be expressed as

$$S_i(U_i) \leq ES_{\max}, \qquad (12.59)$$

where

$$S = \sum_{j \in \Phi} \alpha_j H_j(P_j) \qquad (12.60)$$

and

$ES_{\max}$ is the SO_2 upper limit for the power system being analyzed
Φ is the set of all committed Phase I units, for the ith configuration
$\alpha_j =$ Appropriate conversion coefficient for the jth unit

$H_j(P_j)$ is the heat rate for unit j, expressed as a quadratic form $a_j P_j^2 + b_j P_j + c_j$

P_j is the real power output of unit j for the ith configuration

The NO_x can be similarly expressed as

$$N_i(U_i) \leq EN_{max}, \tag{12.61}$$

where

$$N = \sum_{j \in \Phi} \beta_j H_j(P_j). \tag{12.62}$$

The underutilization (burn) constraints are of the form:

$$N = \sum_{j \in \Phi} \Gamma_j H_j(P_j) \geq B_{min}, \tag{12.63}$$

where B_{min} is the required minimum generation at the Phase I units.

An important feature of the Act is a provision that allows utilities to trade and bank emission allowances (granted to all Phase I units) with the provision that the national upper limit of 10 million tons be met. An annual auction is held to promote trading of allowances. One way to model allowance trading is to add an extra penalty term to the objective function reflecting the current market price of allowances, and relaxing the maximum number of allowances that can be purchased.

Since the emission constraints are all specific to a given configuration, they can be written in the generic form:

$$E_i(Z_i) \leq 0. \tag{12.64}$$

12.6 Commonly Used Optimization Technique in Linear Programming (LP)

The following requirements need to be met by any contemplated solution techniques for the OPF problem [12,14,18,21].

Reliability. The performance of OPF calculations must be reliable for application in real time. They must converge to realistic answers and if not, then adequate justifications must be provided. The more operationally stressed the power system is, the more mathematically difficult the OPF problem is to solve. The acceptance of the OPF industry is based on its reliable performance at all times. Failing to do so, OPF will not gain acceptance.

Speed. The OPF calculations involve computation of nonlinear objective functions and nonlinear constraints with tens of thousands of variables. This therefore requires solution methods that converge fast.

Flexibility. The OPF solution methods simulate real-life power system operation and control situations, and new requirements are continually being defined for calculations. Therefore, robust and flexible OPF algorithms must accommodate and adapt to a wide range of objectives and constraint models.

Maintainability. Due to new knowledge of system models, and perceived priorities of objectives and constraints, an OPF algorithm must include a rule-based scheme and easy to maintain features for real-time application.

12.6.1   LP

The LP-based algorithm solves OPF problems as a succession of linear approximations:

Minimize $F(x^0 + \Delta x u^0 + \Delta u)$ (12.65)

Subject to

$g'(x^0 + \Delta x u^0 + \Delta u) = 0$ (12.66a)

$h'(x^0 + \Delta x u^0 + \Delta u) \leq 0,$ (12.66b)

where
  x^0, u^0 are the initial values of x and u
  Δx, Δu are the shift about this initial point
  g', h' are the linear approximations to the original are the nonlinear constraints

The basic steps required in the LP-based OPF algorithm are as follows:

Step 1. Solve the power flow problem for nominal operating conditions.

Step 2. Linearize the OPF problem (express it in terms of changes about the current exact system operating point) by

1. Treating the limits of the monitored constraints as changes with respect to the values of these quantities, accurately calculated from the power flow.

2. Treating the incremental control variables Δu as changes about the current control variable values (affected by shifting the cost curves).

Step 3. Linearize the incremental network model by

1. Constructing and factoring the network admittance matrix (unless it has not changed since last time performed).
2. Expressing the incremental limits obtained in Step 2.2 in terms of the incremental control variables $\Delta\mathbf{u}$.

Step 4. Solve the linearly constrained OPF problem by a special dual, piecewise linear relaxation LP algorithm computing the incremental control variables.

Step 5. Update the control variables $u = u + \Delta\mathbf{u}$ and solve the exact nonlinear power flow problem.

Step 6. If the changes in the control variables in Step 4 are below user-defined tolerances, the solution has not been reached. If not, go to Step 4 and continue the cycle.

Notably, Step 4 is considered the key step since it determines the computational efficiency of the algorithm. The algorithm solves the network and test operating limits in sparse form while performing minimization in the non-sparse part. For Steps 1 and 5, solving the exact nonlinear power flow problem $g(\mathbf{x},\mathbf{u}) = 0$ is required to provide an accurate operating $\mathbf{x}^0$. This offers either a starting point for the optimization process or a new operating point following the rescheduling of control variables. The power flow solution may be performed using either the Newton–Raphson (N–R) power flow or the fast decoupled power flow (FDPF) technique.

As shown in Equation 12.65, the optimization problem that is solved at each iteration is a linear approximation of the actual optimization problem. Steps 2 and 3 in the LP-based OPF algorithm correspond to forming the linear network model and expressing it in terms of changes about the operating point.

Linearized network constraints models may be derived using either a Jacobian-based coupled formulation given by

$$\begin{bmatrix} \Delta P \\ \Delta Q \end{bmatrix} = J \begin{bmatrix} \Delta\delta \\ \Delta V \end{bmatrix} \quad \text{or} \quad \Delta u_{PQ} = J\Delta x, \tag{12.67}$$

or a decoupled formulation based on the modified FDPF equations expressed by

$$B'\Delta\delta = \Delta P, \tag{12.68}$$

$$B''\Delta V = \Delta Q. \tag{12.69}$$

The latter is used in most applications of LP-based OPF. The linear coupled and decoupled network models are considered separately.

12.6.1.1   Definition of LP Problem Structure

Let m represent the number of constraints and n represent the number of decision variables. If $m > n$, the problem is under-solved. If $m = n$, the linearized problem has a unique solution and cannot be optimized. For the case when $n > m$, values may arbitrarily be chosen for $n - m$ of the variables, while values for the remaining variables are then uniquely defined.

The m variables are termed the basis variables. The remaining $n - m$ variables are termed nonbasis variables. During the course of the solution, variables are exchanged between the basis and nonbasis sets. At any given time, however, exactly m variables must reside in the basis for a problem with m equality constraints. The objective is to choose values for $n - m$ nonbasis variables that minimize Δz. The values for the remaining m variables are determined by the solution.

The LP algorithm changes only a simple variable in the nonbasis set at a time, and this variable is given the subscript i. The remaining $n - m - 1$ nonbasis variables, denoted by j, remain constant. Hence, Δz equations can now be reformulated in terms of the basis set of variables b and a given nonbasis variable i:

$$\Delta z = C_b \Delta x_b + C_i \Delta x_i, \tag{12.70}$$

$$B\Delta x_b + p_i \Delta x_i = 0, \tag{12.71}$$

$$\Delta x_j = 0, \tag{12.72}$$

where the subscript "b" represents the subset of variables termed the basis variables. The variables in this subset change during the solution procedure. A variable in x_b may be from x_c or x_s. The number of variables in x_b is equal to the number of constraints.

C_b is the subset of incremental costs associated with x_b

x_i is the single nonbasis variable chosen to enter the basis

p_i is the column of matrix A associated with x_i that enters the basis

C_i is the incremental cost associated with x_i

Similar to the vector $\mathbf{x}$, $\mathbf{c_i}$ can include elements from both $\mathbf{c_c}$ and $\mathbf{c_s}$. Therefore

$$\{x_c, x_s\} = \{x_b, x_j, x_i\},$$

$$\{c_c, c_s\} = \{x_b, x_j, x_i\}.$$

Initially, $x_b = x_s$.

12.6.1.2   LP Iteration

The LP solution procedure is iterative (see Figure 12.6). Each iteration involves selecting a variable to enter the basis in exchange for a variable

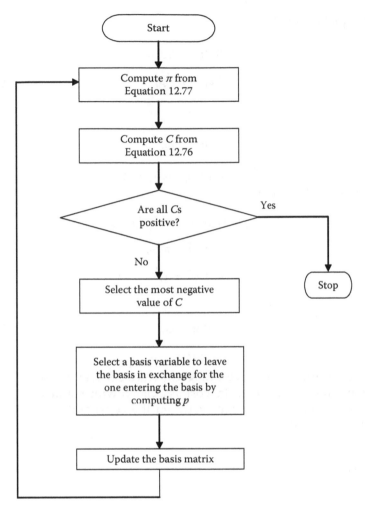

FIGURE 12.6
Algorithmic steps for implementing the LP algorithm.

leaving the basis. The variable basis is the one that achieves the greatest reduction in Δz per unit of movement. The variable leaving the basis is the one that first reaches a limit or breakpoint. The basis matrix is then updated to reflect the exchange in variables along with any changes in sensitivities.

12.6.1.3 Selection of Variable to Enter Basis

To select a variable i to enter the basis, it is necessary to derive an equivalent cost and sensitivity that represent not only the movement of i itself, but also the simultaneous movement of all the basis variables.

From Equation 12.71,

$$\Delta x_b = -B^{-1}p_i - \Delta x_b = -p_i \Delta x_i, \tag{12.73}$$

$$p_i = B^{-1}p_i \tag{12.74a}$$

or

$$B_{p_i} = p_i. \tag{12.74b}$$

Substituting Equation 12.74b into Equation 12.70 gives the equivalent cost sensitivity c_i, with respect to variable i:

$$\Delta z = (c_i - c_b B^{-1})\Delta x_i = (c_i - \pi' p_i)\Delta x_i = c_i \Delta X_i, \tag{12.75}$$

$$c_i = c_j - \pi' p_i, \tag{12.76}$$

$$\pi = (B - 1)'c_b \tag{12.77a}$$

or

$$B'\pi = c_b. \tag{12.77b}$$

where
   p_i is the negative of the vector of sensitivities between x_i and x_b
   c_i is the composite incremental cost associated with x_i and x_b
   π is the vector of sensitivities between the constraint limits and the objective function

12.6.2   LP Applications in OPF

Example 12.1

The LP-based OPF problem formulation requires the objective function to be expressed as a set of separable, convex, and continuous cost curves:

$$F = \sum c_t(u_t).$$

In the LP formulation, the active power and the reactive power problems are solved separately. In the active power problems the only controls are active power controls, and correspondingly in the reactive power problems the only controls are reactive power controls.

The most common OPF objective function is the active power production cost. The cost curve of a generator is obtained from the corresponding incremental heat rate (IHR) curve. For each segment of the piecewise linear IHR curve, the corresponding quadratic cost curve coefficients are found. This gives a piecewise quadratic cost curve for the unit. A piecewise linear cost curve is obtained by evaluating the piecewise quadratic cost curve at various MW values and creating linear segments between these points. This piecewise linear cost curve is what the optimization process uses during processing. The cost curve for interchange with an external company represents the actual cost of exchanging power with that company. It is obtained from the transaction data by arranging the available blocks of power according to increasing cost.

No direct economic cost is associated with phase-shifters or load shedding. Thus, artificial cost curves have to be assigned to these controls. The cost curves may be thought of as penalty functions; that is, there is a cost penalty to be paid for moving these controls away from their initial value. The basic shape of the cost curve for a phase-shifter has been represented by

$$c_i(u_i) = k_i\left(u_i - u_i^0\right)^2, \tag{12.78}$$

where
  u_i is the variable (the phase-shifter angle)
  u_i^0 is the initial value of the control variable
  k_i is the cost curve weighting factor

Example 12.2

The active power minimum control-shift minimization objective aims to limit the rescheduling of active power controls to the minimum amount necessary to relieve all constraint violations. If the initial power flow solution does not involve constraint violations, then no rescheduling is required. This problem is similar in many ways to the cost optimization problem. The control variables that may be used and the constraints that are observed are identical to those in the cost optimization problem. The difference is in the cost representing the generator MW outputs and the interchange control variables. These cost curves are now defined as piecewise linear approximations to the quadratic penalty terms. None of the control variables have an actual economic cost. They are all artificial costs.

The minimum number of active controls minimization objective uses a linear V-shaped curve for each control, with zero value of cost at the target initial control value. In practice, the most sensitive controls are moved one at a time within their full available control range in order to eliminate constraint violations. The result is a minimum number of the controls being rescheduled.

Example 12.3

The minimum control shift and minimum number of control objectives can be used for reactive power optimization. The cost curves for the reactive power minimum-shift minimization objective are obtained from penalty functions. The minimum number of reactive controls minimization objective is the same as for the active power counterpart except that the V-shaped cost curves are used as the reactive power controls.

Example 12.4

In the active power loss minimization problem, the active power generation profile is held fixed and the reactive power profile is varied in order to achieve the minimum loss solution. In addition, phase-shifters can be used to change the MW flow pattern in such a way as to reduce losses. The objective function can be written as

$$F = \sum RI^2. \tag{12.79}$$

This objective function is nonseparable. Since the LP program requires a linearized formulation, the approach then is to minimize the changes in system power losses. The change in system power losses ΔP_L is related to the control variable changes by

$$\Delta P_L = \left(\frac{\partial P_L}{\partial \phi}\right)^T \Delta\phi + \left(\frac{\partial P_L}{\partial V_c}\right)^T \Delta V_c + \left(\frac{\partial P_L}{\partial t}\right)^T \Delta t + \left(\frac{\partial P_L}{\partial b}\right)^T \Delta b. \tag{12.80}$$

The sensitivities in this equation are obtained from a loss penalty factor calculation, using a transpose solution with the Jacobian matrix factors of the coupled model.

In this coupled formulation all control variables are represented explicitly and the corresponding rows and columns are prebordered to the Jacobian matrix. This linear approximation is valid over a small region, which is established by imposing limits on the changes in control variables from their current values. This separable linearized objective is subject to the usual linearized network constraints.

The linearized region must be sufficiently small relative to the local curvature of the nonlinear transmission loss hypersurface in order to achieve appreciable loss reduction. However, if the region is too small, the solution will require an excessive number of iterations. To cope with this problem a heuristic approach for contracting or expanding the linearized region can be invoked. Another difficulty is that the number of binding constraints may be significantly larger than for any other previously mentioned OPF problem, resulting in prolonging the solution time. Although not best suited for loss minimization, a well-tuned LP algorithm can successfully solve the problem in a reasonable time.

12.6.3   Interior Point

Since we discussed interior point methods in detail earlier, we restrict ourselves here to some observations pertinent to the preceding discussion of other general NLP methods.

The current interest in interior point algorithms was sparked by Karmarkar's projective scaling algorithm for LP, which is based on two key ideas:

Steepest descent direction is much more effective in improving the iterate X^k if the iterate is at the center of the polytope formed by the linear constraints rather than if it were closer to the boundary.

Transformation of the decision space can be found that places the iterate at the center of the polytope without altering the problem.

Under certain conditions, one can show that the projective scaling algorithm is equivalent to logarithmic barrier methods, which have a long history in LP and NLP. This led to the development of Mehrotra's primal-dual predictor–corrector method, an effective interior point approach. The main ideas in all these barrier methods are

Convert functional inequalities to equalities and bound constraints using slack variables.

Replace bound constraints by adding them as additional terms in the objective function using logarithmic barriers.

Use Lagrange multipliers to add the equalities to the objective and thus transform the problem into an unconstrained optimization problem.

Use Newton's method to solve the first-order conditions for the stationary points of the unconstrained problem.

Interior point methods can be applied to OPFs [15,33] by using a successive LP technique, and employing an interior point method for solving the linear programs. The other way is to directly apply interior point methods to the NLP formulation using the relation to barrier methods as outlined above.

12.6.3.1 OPF Formulation (Method II)

Objective function (loss minimization)

Minimize P_L

Subject to

$$P_{Gi} - P_{di} - P_i(V,\theta,T) = 0, \quad i = 1, 2, \ldots, N_{bus}, \quad i \neq slack$$

$$Q_{Gi} - Q_{di} + q_{ci} - q_{ri} - T_i(V,\theta,T) = 0, \quad i \in generator$$

$$P_{slack} = F_i(V,\theta,T) = 0, \quad i = slack$$

$$Q_{di} - q_{ci} + q_{ri} + T_i(V,\theta,T) = 0, \quad i = 1, 2, \ldots, N_{bus}, \quad i \neq generator$$

$$-I_{L_{max}}^2 \leq \frac{V_i^2 + V_j^2 - 2V_iV_j \cos(\theta_i - \theta_j)}{Z_L^2} \Bigg/ \leq I_{L_{max}}^2$$

$$Q_{Gi\,min} \leq Q_{Gi} \leq Q_{Gi\,max}$$

$$V_{i_{min}} \leq V_{di} \leq V_{i_{max}}$$

$$T_{i_{min}} \leq T_i \leq T_{i_{max}}$$

$$V_{Gi\,min} \leq V_{Gi} \leq V_{Gi\,max}.$$

For this type of OPF (min loss), control variable

$$u = \begin{bmatrix} V_G \\ T \\ P_L \end{bmatrix}$$

and dependent variable

$$x = \begin{bmatrix} V_D \\ Q_G \\ \theta \end{bmatrix}.$$

Write the OPF problem into the following mathematical programming problem.

$$\text{Min } F = \frac{1}{2} U^T G U + R^T x + C$$

Subject to

$$h(u,x) = 0$$

$$g(u,x) \leq 0$$

$$x_{min} \leq x \leq x_{max}$$

$$u_{min} \leq u \leq u_{max}.$$

Solve the OPF problem using quadratic programming (QP) and/or the quadratic interior point method (see Figure 12.7).

Thus it is necessary to linearize the nonlinear constraints around the base load flow solution for small disturbances. The dependent variable of load flow X can also be eliminated using the implicit functions of the control variable u.

First linearize the equality constraint $h(u,x)$,

$$\Delta x = -h_x^{-T} \left(h_u^T \Delta u + h \right), \tag{a}$$

and linearize the nonequality constraint $g(u,x)$,

$$g_u^T \Delta u + g_x^T \Delta x + g \leq 0. \tag{b}$$

Combine (a) and (b),

$$\left(g_u^T - g_x^T h_x^{-T} h_u^T \right) \Delta u + \left(g - h_x^{-T} h \right) \leq 0.$$

Thus, the quadratic form of the OPF problem is

$$F = \frac{1}{2} \Delta U^T G \Delta U + R^T \Delta U + C$$

Subject to

$$\left(g_u^T - g_x^T h_x^{-T} h_u^T\right)\Delta u + \left(g - h_x^{-T}h\right) \leq 0$$

$$\Delta u_{\min} \leq \Delta u \leq \Delta u_{\max}$$

$$\Delta x_{\min} \leq -h_x^{-T}h_u^T \Delta u \leq \Delta x_{\max}$$

$$\Delta x_{\min} = x_{\min} - x_{\text{base}} + h_x^{-T}h$$

$$\Delta x_{\max} = x_{\max} - x_{\text{base}} + h_x^{-T}h$$

$$\Delta u_{\min} = u_{\min} - u_{\text{base}}$$

$$\Delta u_{\max} = u_{\max} - u_{\text{base}}.$$

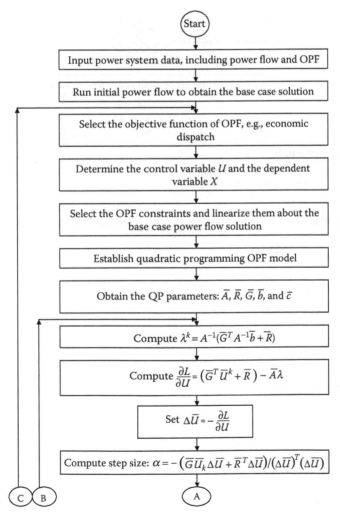

FIGURE 12.7
OPF implementation flowchart by quadratic interior point method.

(*continued*)

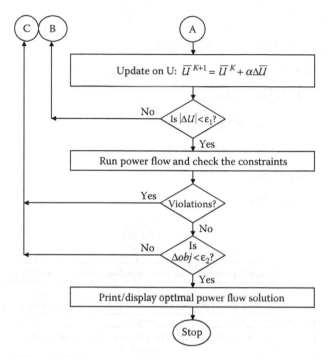

FIGURE 12.7 (continued)

The quadratic form of the OPF can be expressed as follows:

Minimize $F = \dfrac{1}{2} U^T G U + R^T U + C$

Subject to

$b^{\min} \leq AU \leq b^{\max}$

or

Minimize $F = \dfrac{1}{2} \bar{U}^T \bar{G} \bar{U} + \bar{R}^T \bar{U} + \bar{C}$

$\bar{A} \bar{U} = \bar{b}$

$\bar{R} = [R,0]^T, \quad \bar{U} = [U, S_1, S_2], \quad \bar{b} = [b^{\min}, b^{\max}]$

$$\bar{A} = \begin{bmatrix} A & -I & 0 \\ A & 0 & I \end{bmatrix}, \quad \bar{G} = \begin{bmatrix} G & 0 & 0 \\ 0 & 0 & 0 \\ 0 & 0 & 0 \end{bmatrix} \bar{C} = C,$$

where I is an identity matrix.

Solve the OPF problem

Minimize $F = \frac{1}{2}\bar{U}^T\bar{G}\bar{U} + \bar{R}^T\bar{U} + \bar{C}$

Subject to

$\bar{A}\bar{U} = \bar{b}$

using the interior point method and starting at initial feasible point U^0 at $k=0$.

Step 1. $D^k = \text{diag}\lfloor \bar{U}_1^k, \ldots, \bar{U}_n^k \rfloor$.

Step 2. $B^k = \bar{A}D^k$.

Step 3. $dp^k = \{(B^k)^T(B^k(B^k)^T)^{-1}B^k - 1\}D^k(\bar{G}\bar{u}^k + \bar{R})$.

Step 4.

$$\beta_a = \begin{cases} \dfrac{-1}{r}, & r > 0 \\ 10^6, & r \geq 0 \end{cases},$$

where $r = \min\{dp_j^k, j \in (s_1, s_2)\}$.

Step 5.

$$\beta_2 = \begin{cases} (dp^k)^T(dp^k)/T, & T > 0 \\ 10^6, & T \leq 0 \end{cases},$$

where $T = (D^k dp^k)^T \bar{G}(D^k dp^k)$.

Step 6.

$$U^{k+1} = U^k + \beta D^k dp^k,$$

where $\beta = \min\{\beta_1, \beta_2\}$.

Set $k := k+1$, and go to Step 2. End when $dp < \varepsilon$.

12.7 Commonly Used Optimization Techniques in Nonlinear Programming

12.7.1 NLP

Consider an objective function $f(X)$; the negative gradient of $f(X)$, $-\nabla f(X)$, is a direction vector that points toward decreasing values of $f(X)$. This direction is a descent direction for $f(X)$. Disregard the constraints for the moment. That is, assume that the problem is unconstrained. Then, the optimal solution can be obtained using the following algorithm:

Step 1. Assume an initial guess X^0.

Step 2. Find a descent direction D^k.

Step 3. Find a *step length* a^k to be taken.

Step 4. Set $X^{k+1} = X^k + a^k D^k$.

Step 5. If $\|X^{k+1} - X^k\| \leq \varepsilon$, stop. X^{k+1} is declared to be the solution where ε is a tolerance parameter.

Step 6. Increment k. Go to Step 2.

12.7.1.1 Finding the Descent Direction

There are several ways to obtain D^k with the simplest being to set D^k equal to $-\nabla f(X^k)$. This is the steepest descent direction. A more efficient approach is the Newton method which obtains D^k by solving the following system of equations:

$$\nabla^2 f(X^k)D^k = -\nabla f(X^k), \tag{12.81}$$

where $\nabla^2 f(X^k)D^k$ is the Hessian matrix evaluated at X^k. In Newton's approach, there are two alternatives to evaluate expressions for the Hessian. One is to approximate it using the finite difference method. Another involves using automatic differentiation software.

The exact computation of the Hessian can be time-consuming or difficult. Also, the Hessian may not be positive-definite as required by Newton's method. Quasi-Newton methods build up an approximation [9] to the Hessian at a given point X^k using the gradient information from previous iterations. D^k is obtained by solving:

$$B^k D^k = -\nabla f(X^k), \tag{12.82}$$

where B^k is the approximate Hessian for the point X^k.

12.7.1.2 Finding the Step Length

The step length α^k is required to be positive and such that $f(X^{k+1}) < F(X^k)$. The value of α^k can be obtained by solving the following one-dimensional optimization problem.

$$\underset{\alpha}{\text{Min }} f(X^k + \alpha D^k). \tag{12.83}$$

The problem is typically solved by a fast procedure (such as quadratic or cubic interpolation) which is very approximate, since a precise solution is not needed.

The above discussion has used a type of minimization technique called the line-search method. A less common technique referred to as the trust-region approach calculates the next iterate using $X^{k+1} = X^k + s^k$, where s^k is chosen so as to obtain sufficient decrease in $f(X)$. Trust-region methods are useful when the Hessian is indefinite.

12.7.1.3 Treatment of the Constraints

Now we consider how the equality and inequality constraints can be satisfied while minimizing the objective. The Lagrangian function plays a central role in constrained optimization.

$$L(X,\lambda) = f(X) - \sum_{i=1}^{a} \lambda_i g_i + \sum_{j=1}^{b} \lambda_{j+a} h_i, \qquad (12.84)$$

where
  $\lambda \in \Re^{a+b}$ is the vector of Lagrange multipliers
  g_i and h_i are the elements of the constraint vectors

The Lagrangian multipliers measure the sensitivity of the objective function to the corresponding constraints. Estimating the proper value of these multipliers is an important issue in constrained optimization.

Another important (and in many ways, an alternative) function is the augmented Lagrangian function. Before discussing this function, it is useful to consider an equivalent formulation of the optimization problem. Each functional inequality can be replaced by an equality and a bound constraint. As a simple example of how this can be done, consider an inequality $h_1(z_1) \leq 0$. This can be written as $h_1(z_1) + z_2 = 0$, and $z_2 \geq 0$, where z_2 is referred to as a slack variable. Thus, a number of new slack variables are introduced corresponding to the number of functional inequalities resulting in the following formulation.

$$\text{Minimize}_x \; f(X) \qquad (12.85)$$

Subject to

$$C(X) = 0 \qquad (12.86)$$

$$X^{\min} \leq X \leq X^{\max}, \qquad (12.87)$$

where
  $C: \Re^{m+n+b} \to \Re^{a+b}$ is the vector function of equalities
  $X \in \Re^{m+n+b}$ is the vector of decision variables obtained by augmenting the original vector with the slack variables

The augmented Lagrangian function can now be defined as

$$L_A(X, \lambda, \rho) = f(X) - \lambda^T C(X) + \frac{1}{2}(X)^T C(X),$$ (12.88)

where ρ is some positive penalty constant.

Choosing a proper value for the penalty parameter ρ becomes an important issue. If ρ is too large, the efficiency of the solution approach will be impaired. A large ρ is also likely to lead to ill-conditioning of the Hessian of the augmented Lagrangian function and, consequently, cause difficulties for methods that rely on such a Hessian or a suitable approximation. If ρ is too small, the solution approach may not be able to converge to a solution that satisfies Equation 12.86. Depending on the solution approach to be discussed, either the original formulation or the formulation given in Equations 12.85 through 12.87 is used.

A concept that is used in many approaches is that of active or binding constraints. An inequality constraint is said to be active or binding if it is strictly satisfied. Consider the inequality constraints. We can define a set such that $H_a(X) = 0$ is the active subset of the set of inequalities. Note that the composition of this subset varies with the iterate X^k. Thus $H_a(Z^k)$ is the subset of active inequalities corresponding to X^k. By definition, all equality constraints are always active. So, the overall set of active constraints $A(X^k)$ is the union of the sets $G(X^k)$ and $H_a(X^k)$.

12.7.2 Sequential Quadratic Programming (SQP)

This algorithm is an extension of the quasi-Newton method for constrained optimization. The method solves the original problem by repeatedly solving a QP approximation. A QP problem is a special case of an NLP problem wherein the objective function is quadratic and the constraints are linear. Both the quadratic approximation of the objective and the linear approximation of the constraints are based on Taylor series expansion of the nonlinear functions around the current iterate X^k.

The objective function $f(X)$ is replaced by a quadratic approximation; thus:

$$q^k(D) = \nabla f(X^k)D + \frac{1}{2}D^T \nabla_{zz}^2 L(X^k, \lambda^k)D.$$ (12.89)

The step D^k is calculated by solving the following QP subproblem.

Minimize $q^k(D)$ (12.90)
$\quad\;\; D$

Subject to

$G(X^k) + J(X^k)D = 0$ (12.91)

$H(X^k) + I(X^k)D \leq 0,$ (12.92)

where J and I are the Jacobian matrices corresponding to the constraint vectors G and H, respectively.

The Hessian of the Lagrangian $\nabla^2_{zz} L(X^k, \lambda^k)$ that appears in the objective function, Equation 12.90, is computed using a quasi-Newton approximation. Once D^k is computed by solving Equations 12.90 through 12.92, X is updated using

$$X^{k+1} = X^k + \alpha^k D^k, \tag{12.93}$$

where α^k is the step length.

Finding α^k is more complicated in the constrained case. This is because α^k must be chosen to minimize constraint violations in addition to minimizing the objective in the chosen direction D^k. These two criteria are often conflicting and thus a merit function is employed to reflect the relative importance of these two aims. There are several ways to choose a merit function with one choice being:

$$P_1(X, v) = f(X) + \sum_{j=1}^{a} v_i |g_i| + \sum_{j=1}^{b} v_{a+j} \max[h_j, 0], \tag{12.94}$$

where
  $v \in \Re^{a+b}$ is the vector of positive penalty parameters
  g_i and h_i are elements of the constraint vectors $G(X)$ and $H(X)$, respectively

For the merit function $P_1(X)$ as defined in Equation 12.94, the choice of v is defined by the following criterion,

$$v_i \geq |\lambda_i|, \quad i = 1, 2, \ldots, a, a+1, \ldots, b,$$

where the λ_i are Lagrange multipliers from the solution of the QP subproblem of Equations 12.90 through 12.92 that defines D^k. Furthermore, the step length α^k is chosen so as to approximately minimize the function given by

$$P_1(X^k + \alpha D^k, v).$$

A different merit function that can be used is known as the augmented Lagrangian merit function:

$$L_A(X, \lambda, v) = f(X) - \sum_{i=1}^{a} \lambda_k g_i + \frac{1}{2} \sum_{j=1}^{b} v_i g_i^2 + \sum_{j=a}^{b} \Phi_{j-a}\left(X, v_j, \lambda_j^2\right), \tag{12.95}$$

where

$$\phi_{j-a}(X, v_j, \lambda_j) = \frac{1}{v_j}\left[\max\left(0, (\lambda_j + v_j h_{j-a})^2\right) - \lambda_j^2\right],$$

and g_i and h_i are elements of the constraint functions $G(X)$ and $H(X)$, respectively, v is the vector of positive penalty parameters, and λ'_i are Lagrange multipliers from the solution of the QP subproblem given by Equations 12.90 through 12.92 that defines D^k.

If Equation 12.90 is used as the merit function, the step length α^k is chosen to approximately minimize

$$L_A\left(X^k + \alpha D^k, \lambda^k + \alpha\left(\lambda^{k+1} - \lambda_k\right), v\right),$$

where

D^k is the solution of the QP subproblem given by Equations 12.90 through 12.92

λ^{k+1} is the associated Lagrange multiplier

12.7.3 Augmented Lagrangian Methods

These methods are based on successive minimization of the augmented Lagrangian function in Equation 12.88 corresponding to the NLP formulation. Therefore these methods solve the following subproblem successively.

$$\underset{z}{\text{Minimize}} \; L_A(X, \lambda^k, \rho^k) \tag{12.96}$$

Subject to

$$X^{\min} \leq X \leq X^{\max}, \tag{12.97}$$

where

L_A is the augmented Lagrangian function

λ^k is the vector of the Lagrangian multipliers that is updated every iteration

ρ^k is the positive penalty parameter that is updated heuristically

By solving Equations 12.96 and 12.97, we get X^{k+1}. Then λ^{k+1} is obtained using

$$\lambda^{k+1} = \lambda^k + \rho CCX^k. \tag{12.98}$$

This method is relatively unexplored for the OPF problem.

12.7.4 Generalized Reduced Gradients

The general reduced gradients (GRG) class uses the equality constraints to eliminate a subset of decision variables to obtain a simpler problem. We partition the decision vector X into two vectors X_B and X_N, X_B is the vector of basic variables that we want to eliminate using the equality constraints. X_N is the vector of the remaining variables, called nonbasic variables. Then,

$$X_B = W(X_N), \tag{12.99}$$

where $W(\cdot)$ is chosen such that

$$C(W(X_N), X_N) = 0. \tag{12.100}$$

The mapping $W(\cdot)$ as in Equation 12.99 is usually defined by the implicit relation in Equation 12.100. Updates of X_N are obtained by solving Equation 12.100 using an appropriate procedure such as Newton's method. For example, a Newton update of X_B is of the form:

$$X_B^{k+1} = X_b^k - \left(\frac{\partial}{\partial X_B} C(X_B, Z_N)^{-1}\right)\Bigg|_{X_B^k, X_N^k} C(X_B^k, X_N^k). \tag{12.101}$$

The problem can be formulated as the reduced problem:

$$\underset{Z_n}{\text{Minimize}}\ f(W(X_N), X_N) \tag{12.102a}$$

Subject to

$$X_N^{min} \leq X_N \leq X_N^{max}. \tag{12.102b}$$

The vector X_N can in turn be partitioned into two sets: X_F and X_S. The fixed variables X_F are held at either their lower or upper bounds during the current iteration. The superbasic variables X_S are free to move within their bounds. Thus, the reduced problem given by Equation 12.102a and b is solved through successive minimization of the following subproblem.

$$\underset{X_s}{\text{Minimize}}\ f(W(X_S, X_F), X_S X_F) \tag{12.103a}$$

Subject to

$$X_S^{min} \leq X_S \leq X_S^{max} \tag{12.103b}$$

$$(X_F - X_F^{min})(X_F^{max} - X_F) = 0. \tag{12.103c}$$

Since the constraints in Equation 12.103b and c are simple bound constraints, the subproblem can be solved by using the negative gradient of the objective function in Equation 12.103 as the descent direction. This gradient, $-\nabla_{X_S} f(W(X_S, X_F), X_S X_F)$, is referred to as the reduced gradient since it involves only a subset of the original decision variables. If a superbasic variable violates one of its bounds, it is converted into a fixed variable held at that bound. Note that to solve Equation 12.103a through c, some GRG methods use the reduced Hessian objective function to obtain a search direction. Instead of computing the reduced Hessian directly, quasi-Newton schemes are employed in the space of the superbasic variables.

Each solution of Equation 12.103a through c is called a minor iteration. At the end of each minor iteration, the basic variable vector X_B is updated using Equation 12.101. At this point, a check is made to see whether certain elements can be moved from the fixed variable set X_F into the superbasic set X_S. The composition of all three vectors X_B, X_F, and X_S is usually altered at the end of each minor iteration. In other words, the decision vector X is repartitioned between the minor iterations.

The GRG scheme was applied to the OPF problem with the main motivation being the existence of the concept of state and control variables, with the load flow equations providing a natural basis for the elimination of the state variables. The availability of good load flow packages provides the needed sensitivity information for obtaining a reduced problem in the space of the control variables with the load flow equations and the associated state variables eliminated.

12.7.4.1 OPF Formulation Using QP Reduced Gradient Method

The objective function is cost minimization.

$$\text{Minimize } F(P_G) = \sum_i \left(\alpha_i P_{Gi}^2 + \beta_i P_{Gi} + \gamma_i \right)$$

Subject to

$$P_{Gi} - P_{di} - F_i(V,\theta,T) = 0, \quad i = 1, 2, \dots, N_{bus}, \quad i \neq \text{slack}$$

$$Q_{Gi} - Q_{di} + q_{ci} - q_{ri} - T_i(V,\theta,T) = 0, \quad i \in \text{generator}$$

$$P_{slack} = F_i(V,\theta,T) = 0, \quad i = \text{slack}$$

$$Q_{di} - q_{ci} + q_{ri} + T_i(V,\theta,T) = 0, \quad i = 1, 2, \dots, N_{bus}, \quad i \neq \text{generator}$$

$$-I_{L_{max}}^2 \leq \left[V_i^2 + V_j^2 - 2V_i V_j \cos(\theta_i - \theta_j) \right] \Big/ Z_L^2 \leq I_{L_{max}}^2$$

$$P_{Gi\,min} \leq P_{Gi} \leq P_{Gi\,max}$$

$$Q_{Gi\,min} \leq Q_{Gi} \leq Q_{Gi\,max}$$

$$V_{i\,min} \leq V_i \leq V_{i\,max}$$

$$T_{i\,min} \leq T_i \leq T_{i\,max}$$

$$V_{Gi\,min} \leq V_{Gi} \leq V_{Gi\,max}.$$

For this type of OPF (min cost), control variable

$$u = \begin{bmatrix} V_G \\ T \\ P_G \end{bmatrix},$$

and dependent variable

$$x = \begin{bmatrix} V_D \\ Q_G \\ \theta \end{bmatrix}.$$

Writing the OPF problem into mathematical programming problems, we have

Minimize $F = \dfrac{1}{2}U^T GU + R^T x + C$

Subject to

$h(u,x) = 0$

$g(u,x) \leq 0$

$x_{min} \leq x \leq x_{max}$

$u_{min} \leq u \leq u_{max}.$

Solve the OPF problem using QP (see Figure 12.8).

It is necessary to linearize the nonlinear constraints around the base load flow solution for small disturbances. The dependent variable of load flow X can also be eliminated using the implicit function of the control variable U.

First, linearize the equality constraint $h(u,x)$,

$$\Delta x = -h_x^{-T}\left(h_u^T \Delta U + h\right). \tag{a}$$

Now, linearize the inequality constraint $g(u,x)$,

$$g_u^T \Delta U + g_x^T \Delta x + g \leq 0. \tag{b}$$

Combining (a) and (b), we get

$$\left(g_u^T - g_x^T h_x^{-T} h_u^T\right)\Delta U + \left(g - h_x^{-T} h\right) \leq 0.$$

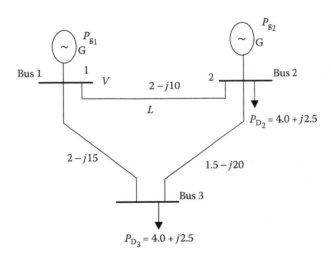

FIGURE 12.8
OPF implementation flowchart by QP method.

Thus, the quadratic form of the OPF problem is

$$\text{Minimize } F = \frac{1}{2}\Delta U^T G \Delta U + R^T \Delta U + C$$

Subject to

$$\left(g_u^T - g_x^T h_x^{-T} h_u^T\right)\Delta u + \left(g - h_x^{-T} h\right) \leq 0$$

$$\Delta u_{\min} \leq \Delta u \leq \Delta u_{\max}$$

$$\Delta x_{\min} \leq -h_x^{-T} h_u^T \Delta u \leq \Delta X_{\max}$$

$$\Delta x_{\min} = x_{\min} - x_{\text{base}} + h_x^{-T} h$$

$$\Delta x_{\max} = x_{\max} - x_{\text{base}} + h_x^{-T} h$$

$$\Delta u_{\min} = u_{\min} - u_{\text{base}}$$

$$\Delta u_{\max} = u_{\max} - u_{\text{base}}.$$

The quadratic form of OPF can be expressed as follows:

$$\text{Minimize } F = \frac{1}{2}U^T G U + R^T U + C$$

Subject to

$$b^{\min} \leq AU \leq b^{\max}$$

or

$$\text{Minimize } F = \frac{1}{2}\bar{U}^T \overline{GU} + \bar{R}^T \bar{U} + \bar{C}$$

$$\overline{AU} = \bar{b}$$

$$\bar{R} = [R,0]^T, \quad \bar{U} = [U,S_1,S_2], \quad \bar{b} = [b^{\min},b^{\max}]$$

$$\bar{A} = \begin{bmatrix} A & -I & 0 \\ A & 0 & I \end{bmatrix}, \quad \bar{G} = \begin{bmatrix} G & 0 & 0 \\ 0 & 0 & 0 \\ 0 & 0 & 0 \end{bmatrix}, \quad \bar{C} = C,$$

where I is an identity matrix.
   Solve the OPF Problem

$$\text{Minimize } F = \frac{1}{2}\bar{U}^T \overline{GU} + \bar{R}^T \bar{U} + \bar{C}$$

Subject to

$$\overline{AU} = \bar{b}.$$

Use QP with the reduced gradient method, and assume initial values of U^0 at $k=0$.

Step 1. Compute $\lambda^k = A^{-1} (\bar{G}^T A^{-1} \bar{b} + \bar{R})$.

Step 2. Compute $\partial L/\partial U = (\bar{G}^T \bar{U}_k \bar{R}) - \bar{A}\lambda$.

Step 3. Let $\Delta \bar{U} = -(\partial L/\partial U)$.

Step 4. If $|\Delta \bar{U}| < \varepsilon$, stop; obtain optimal solution. Otherwise go to next step.

Step 5. Compute optimal step size α, $\alpha = -(\overline{GU}_k \Delta \bar{U} + \bar{R}^T \Delta \bar{U})/(\Delta \bar{U})^T (\Delta \bar{U})$.

Step 6. Update $\bar{U}^{K+1} = \bar{U}^K + \alpha \Delta \bar{U}$.

Step 7. Set $K = K+1$ and go to Step 2.

12.7.5 Projected Augmented Lagrangian

The projected augmented Lagrangian method successively solves subproblems of the form:

$$\underset{x}{\text{Minimize }} L_A^\sim(X, X^k, \lambda^k, \rho) \tag{12.104a}$$

Subject to

$$C^k(X, X^k) = 0 \tag{12.104b}$$

$$X^{min} \leq X \leq X^{max}, \tag{12.104c}$$

where

$$L_A^\sim(X, X^k, \lambda^k, \rho) = f(X^k) - \lambda^{k^T}(C - C^k) + \frac{1}{2}\rho(C - C^k)^T(C - C^k)$$

$$C^k(X, X^k) = C(X^k) + J(X^k)[X - X^k],$$

where
  X^k is the solution obtained from the kth iteration
  $J(X^k)$ is the Jacobian matrix of $C(X)$ evaluated at X^k
  λ^k is the vector of Lagrange multipliers corresponding to X^k
  ρ is the penalty parameter, adjusted heuristically

The procedure used to solve each subproblem of Equation 12.104a through c is similar to the GRG scheme. The variables are partitioned into basic, superbasic, and nonbasic variables. The nonbasic variables are held at one of their bounds. The basic variables are eliminated using the linearized constraints, Equation 12.104b, and the reduced problem is solved to obtain a new value for the superbasic variables. If a superbasic variable reaches one of its bounds, it is converted to a nonbasic variable.

The gradient of the augmented Lagrangian is thus projected into the space of the active constraints. The active constraints can be written in the form:

$$\begin{bmatrix} B & S & N \\ & & I \end{bmatrix} \begin{bmatrix} X_b \\ X_S \\ X_N \end{bmatrix} = \begin{bmatrix} b \\ b_N \end{bmatrix},$$

where

$$A = \begin{bmatrix} B & S & N \\ & & I \end{bmatrix}$$

is the active constraint matrix. Consider the operator

$$W = \begin{bmatrix} -B^{-1}S \\ I \\ 0 \end{bmatrix},$$

which has the property that $AW = 0$. That is, this operator W affects a transformation into the null space of the active constraints. Let $-\nabla L_A^{\sim}$ be the negative gradient of the objective function in Equation 12.104a. Then the vector $-W^T \nabla L_A^{\sim}$ is the negative reduced gradient of the objective function, and points in the direction that lowers the objective without violating the active constraints. Let $\nabla^2 L_A^{\sim}$ be the Hessian of the objective function. Then the matrix $W^T (\nabla^2 L_A^{\sim})$ is the reduced Hessian. If the reduced Hessian is positive-definitive and if the reduced gradient is nonzero, then a feasible descent direction D^k can be obtained using

$$W^T (\nabla^2 L_A^{\sim}) W D_S^k = -W^T \nabla L_A^{\sim} \quad \text{and} \quad D^k = W D_S^k.$$

A popular implementation uses a quasi-Newton approach to finding the search direction D_S^k by replacing the Hessian with a positive-definitive approximation. Given D_S^k a new estimate of X can be obtained, where α^k is obtained, using a line-search such that it lowers the value of the objective in Equation 12.104a.

12.7.6   Discussion on Nonlinear OPF Algorithms

The GRG method was one of the first to be used in OPF packages. Its main attraction is its ability to use standard load flow methods to eliminate the power flow equalities and obtain a reduced problem that is easier to solve. The SQP method is better able to handle nonlinear objectives and constraints present in the OPF. However, SQP is currently not competitive for large-scale systems. The same is true of the projected augmented Lagrangian approach, which does not seem to have a future for OPF applications. The interior point method is presently the most favored method for OPFs, because of the robustness of the underlying approach.

12.7.6.1 Decomposition Strategies

We present some decomposition strategies that can be instrumental in saving extended OPF formulations such as security-constrained OPF [11,23]. All decomposition strategies aim to solve NH subproblems independently. First, we discuss adding security constraints to the OPF formulation.

12.7.6.2 Adding Security Constraints

The traditional notion of security has relied almost exclusively on preventive control. That is, the requirement has been that the current operating point be feasible in the event of the occurrence of a given subset of the set of all possible contingencies. In other words, the base-case control variables are adjusted to satisfy postcontingency constraints that are added to the original formulation:

$$\text{Minimize}_{U_0,Z_j} f(Z_0) \tag{12.105a}$$

Subject to

$$G_i(U_0,Z) = 0, \quad i = 0, 1, 2, \ldots, N \tag{12.105b}$$

$$H_i(U_0,Z_i) \leq 0, \quad i = 0, 1, 2, \ldots, N, \tag{12.105c}$$

where
$i = 0$ is the base-case
$i > 0$ is the ith postcontingency configuration
N is the total number of contingencies considered (i.e., those selected by a security assessment procedure)
$U_i \in \Re^m$ is the vector of control variables for configuration i
$Z_i \in \Re^n$ is the vector of state variables for configuration i
$X_i \in \Re^{m+n} = [U_i \ X_i]^T$ is the decision vector for the ith configuration
$f: \Re^{m+n} \to \Re^1$ is the base-case objective function representing operating costs
$G_i: \Re^{m+n} \to \Re^a$ is the vector function representing the load flow constraints for the ith configuration
$H_i: \Re^{m+n} \to \Re^b$ is the vector function representing operating constraints for the ith configuration

The formulation shown in Equation 12.105a through c is very conservative in that it allows no room for postcontingency corrective actions. It places much more emphasis on maximizing security than on minimizing operating cost. In today's competitive environment, such a formulation is not easily justifiable, given that there is a small but nonzero correction time (about 15–30 min) available for implementing postcontingency changes to the control variables. Hence, it is preferable to use a corrective control formulation as follows:

$$\underset{X_0, X_i}{\text{Minimize}}\ f(X_0) \tag{12.106a}$$

Subject to

$$G_i(X_i) = 0, \quad i = 0, 1, 2, \ldots, N \tag{12.106b}$$

$$H_i(X_i) \leq 0, \quad i = 0, 1, 2, \ldots, N \tag{12.106c}$$

$$\phi_i(U_i - U_0) \leq \Theta_i, \quad i = 0, 1, 2, \ldots, N, \tag{12.106d}$$

where
   $\phi_i(\cdot)$ is the distance metric (say Euclidean norm)
   Θ_i is the vector of upper bounds reflecting ramp-rate limits

The last set of constraints, Equation 12.106d, called the coupling constraints, reflects the fact that the rate of change in the control variables of the base-case (like the real power output of generators) is constrained by upper bounds (which are, typically, system specific). Note that without the coupling constraints, the constraints, Equation 12.106b and c, are separable into $N+1$ disjoint configuration-specific sets, which indicates the decomposability of the above formulation into $N+1$ subproblems.

   The decomposition strategies are based on the corrective control scheme of Equation 12.106a through c, and differ mainly in the manner in which they handle the coupling constraints of Equation 12.106d that impede independent solutions of the subproblems: decomposition strategies are indispensable in handling security and environmental constraints.

12.8   Illustrative Examples

Illustrative Example 12.1

A loss minimization problem for the given three-bus power system is shown in Figure 12.9.
   The generator voltages' magnitude and cost functions are given as

$$V_1^0 = 1.0 \text{ pu},$$

$$V_2^0 = 1.0 \text{ pu},$$

$$F(P_{G_1}) = 4P_{G_1} + 0.05P_{G_1}^2, \quad 0 \leq P_{G_1} \leq 4.0 \text{ pu},$$

$$F_2(P_{G_2}) = 3P_{G_2} + 0.07P_{G_2}^2, \quad 0 \leq P_{Gi} \leq 3.0 \text{ pu},$$

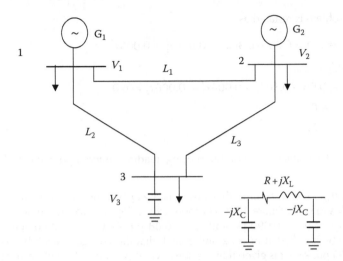

Transmission line: 300 km, 200 kV Line
R = 35 ohms,
ωL = 150 ohms
$\omega C = 5.0 \times 10^{-4}$ mhos

FIGURE 12.9
Single-line diagram for Illustrative Example 12.1.

and the loss formulation is given by

$$P_L = 0.06 - 0.30P_{G_1} + 0.004P_{G_1}^2 + 0.006P_{G_2}^2.$$

Step 1. Express the loss minimization problem
Objective:

Minimize $P_L = 0.06 - 0.30P_{G_1} + 0.004P_{G_1}^2 + 0.006P_{G_2}^2$

Subject to

$P_{G_1} + P_{G_2} + 0.06 - 0.30P_{G_1} + 0.004P_{G_1}^2 + 0.006P_{G_2}^2 = 6.0$ pu

$0 \leq P_{G_1} \leq 4.0$ pu

$0 \leq P_{G_2} \leq 3.0$ pu.

Step 2. Convert the loss minimization problem into a general mathematical expression. Let

$$\begin{aligned} \text{vector } x &= [x_1, x_2]^T \\ &= [P_{G_1}, P_{G_2}]. \end{aligned}$$

Then, the problem is restated as

Minimize $f(x) = 0.06 - 0.30x_1 + 0.004x_1^2 + 0.006x_2^2$

Subject to

$x_1 + x_2 + 0.06 - 0.30x_1 + 0.004x_1^2 + 0.006x_2^2 = 6.0$

$0 \le x_1 \le 4.0$

$0 \le x_2 \le 3.0.$

Step 3. Solve this mathematical problem using quadratic interior point method.

Case 1

VAr planning problem. Assume that bus 2 is a PQ bus for the given three-bus network shown in Figure 12.8. Let there exist a low voltage in the power system such that $|V_2| = 0.090$ pu and $|V_3| = 0.88$ pu. Assume shunt compensators are to be installed at buses 2 and 3 such that the voltage at each bus is raised to 0.95 pu. And it is given that the sensitivities ΔV_2 and ΔV_3 with respect to Δq_3 are

$$\Delta V_2 = 0.02\Delta q_2 + 0.010\Delta q_3,$$
$$\Delta V_3 = 0.07\Delta q_2 + 0.045\Delta q_3.$$

Step 1. Express the VAr planning problem

Minimize $Q = (\Delta q_2 + \Delta q_3)$

Subject to

$\Delta V_2 = 0.02\Delta q_2 + 0.010\Delta q_3 \ge 0.95 - 0.90$

$\Delta V_3 = 0.07\Delta q_2 + 0.045\Delta q_3 \ge 0.95 - 0.88.$

Step 2. Convert the VAr planning problem into a general mathematical expression. Let

$$\text{vector } x = [x_1, x_2]^T$$
$$= [q_2, q_3].$$

Then the problem is restated as

Minimize $f(x) = x_1 + x_2$

Subject to

$0.02x_1 + 0.010x_2 \ge 0.05$

$0.07x_1 + 0.045x_2 \ge 0.07$

$x_1 \ge 0 \quad \text{and} \quad x_2 \ge 0.$

Step 3. Solve this mathematical problem using an LP method.

Case 2

Voltage optimization using quadratic interior point method. The objective function to be minimized is

$$F(V) = \sum_{i=1}^{N_b} (V_i - V_i^{nom})^2$$

Subject to

$$P_i - P_{g_i} + P_{d_i} = 0, \quad \forall i \in \{1, N_b\}$$

$$Q_i - Q_{g_i} + Q_{d_i} = 0, \quad \forall i \in \{1, N_b\}$$

where, from the standard power flow equations expressed in rectangular form,

$$P_i = V_i \sum_{i=1}^{N_b} V_j[G_{ij} \cos(\theta_i - \theta_j) + B_{ij} \sin(\theta_i - \theta_j)],$$

$$Q_i = V_i \sum_{i=1}^{N_b} V_j[G_{ij} \sin(\theta_i - \theta_j) + B_{ij} \cos(\theta_i - \theta_j)] \quad \text{and} \quad V_i^{min} \le V_i \le V_i^{max}.$$

Thus, using the three-bus test system, the optimization problem becomes:

Minimize $F(V) = (V_1 - V_1^0)^2 + (V_2 - V_2^0)^2 + (V_3 - V_3^0)^2$

Subject to

$P_1 = P_{G_1}, \quad Q_1 = Q_{g_1}$

$P_2 = P_{g_2} - 2.0, \quad Q_2 = Q_{g_2} - 1.0$

$P_3 = 0 - 4.0, \quad Q_3 = 0 - 2.5.$

Note that P_{g_1} and P_{g_2} are obtained from the minimal cost calculation, and the data Q_{g_1} and Q_{g_2} are specified. Therefore, $Q_{g_1} = 1.2$ pu and $Q_{g_2} = 2.7$ pu.

Illustrative Example 12.2

A power system is shown in Figure 12.10. All three transmission lines are assumed identical and can be electrically described by π representation. System data are shown in Table 12.2.

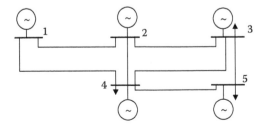

FIGURE 12.10
Power system diagram for Illustrative Example 12.2.

TABLE 12.2

System Data for Figure 12.10

Bus	Real P, P_G, MW	Reactive Q, Q_G, MVAr	Load, P_D	Load, Q_D	Voltage (kV)
1	Optimal	Unspecified	140	45	220
2	Optimal	Unspecified	50	25	220
3	0	Unspecified	140	50	220

The generator cost functions are

$$f_1 = 0.015P_1^2 + 2.0P_1,$$

$$f_2 = 0.010P_2^2 + 3.0P_2.$$

Find the optimum generation schedule (real power).

SOLUTION

Given

$$f_1 = 0.015P_1^2 + 2.0P_1,$$

$$f_2 = 0.0010P_2^2 + 3.0P_2,$$

The total system load is

$$P_D = P_{D_1} + P_{D_2} + P_{D_3}$$
$$= 140 + 50 + 140 \text{ MW} = 330 \text{ MW}.$$

Calculating the optimal generation schedule and forming the Lagrangian, we obtain

$$L(P_1, P_2, \lambda) = f_1 + f_2 + \lambda\left(P_D - \sum_{i=1}^{N_G} P_i\right),$$

$$\therefore L = 0.015P_1^2 + 2.0P_1 + 0.01P_2^2 + 3P_2 + \lambda(330 - P_1 - P_2).$$

The transmission losses are neglected.

By applying the optimality condition,

$$\frac{\partial L}{\partial P_1} = 0, \quad \frac{\partial L}{\partial P_2} = 0, \quad \frac{\partial L}{\partial \lambda} = 0,$$

$$\frac{\partial L}{\partial P_1} = 0.03P_1 + 2 - \lambda = 0,$$

$$\frac{\partial L}{\partial P_2} = 0.02P_2 + 3 - \lambda = 0,$$

$$\frac{\partial L}{\partial \lambda} = 330 - P_1 + P_2 = 0.$$

Hence, we obtain

$$\lambda = 0.03 P_1 + 2 = 0.02 P_2 + 3$$
$$\therefore\ 0.03 P_1 - 0.02 P_2 - 1 = 0.$$

But

$$P_2 = 330 - P_1$$
$$\therefore\ 0.03 P_1 - 0.02(330 - P_1) - 1 = 0$$
$$0.03 P_1 - 6.6 + 0.02 P_1 - 1 = 0$$
$$0.05 P_1 - 7.6 = 0$$
$$P_1 = 152\ \text{MW}.$$

Hence, $P_2 = 330 - 152 = 178$ MW.

Illustrative Example 12.3

A five-bus system is shown in Table 12.3. With P in MW, the cost functions in dollars per hour are as follows:

$$F_1 = 0.0060 P_{g1}^2 + 2.0 P_{g1} + 140,$$
$$F_2 = 0.0075 P_{g2}^2 + 1.5 P_{g2} + 120,$$
$$F_3 = 0.0070 P_{g3}^2 + 1.8 P_{g3} + 80.$$

Assuming that the voltage limits at all buses vary between 0.95 and 1.05 pu, and all generators are rated at 200 MW (see Table 12.4), then use the OPF program to obtain:

TABLE 12.3

Five-Bus System Impedance and Line Charging Data

Bus Code		Line	1/2 Line Charging	Line
From Bus i	To Bus j	Impedance (pu)	Susceptance (pu)	Limits (MW)
1	2	$0.02 + j0.06$	$j0.030$	30
1	3	$0.08 + j0.24$	$j0.025$	40
2	3	$0.06 + j0.18$	$j0.020$	50
2	4	$0.06 + j0.18$	$j0.020$	80
2	5	$0.04 + j0.12$	$j0.015$	40
3	4	$0.01 + j0.03$	$j0.010$	180
4	5	$0.08 + j0.24$	$j0.025$	120

TABLE 12.4

Initial Generation Schedule

Bus i	Bus Voltage, $V_i = \lvert V_i \rvert, V\angle\theta_i$		Power Generation		Load Level	
	Magnitude (pu)	Angle (°)	P_{gi} (MW)	Q_{gi} (MVAr)	P_{load} (MW)	Q_{load} (MVAr)
1	1.060	0.0	98.4	—	0	0
2	1.056	−2.27	40.0	23.2	20	10
3	1.044	−3.69	30.0	30.0	45	15
4	1.041	−4.16	0.0	10.0	40	5
5	1.030	−5.35	0.0	0.0	60	10

1. Absolute minimum cost of this system and the real and reactive generation schedule
2. Loss minimum of this system, the reactive power of generation, and the optimal voltage profile

SOLUTION

From the OPF program, we obtain:

1. Absolute minimum cost $= 2.7403$

   $P_{g1} = 97.48$ MW, $P_{g2} = 40.00$ MW, and $P_{g3} = 30.00$ MW

   $Q_{g1} = -17.86$ MVAr, $Q_{g2} = -0.260$ MVAr, and $Q_{g3} = 33.94$ MVAr

2. Loss minimum $= 0.024763$, $\lvert V_4 \rvert = 1.04535$ pu, $\lvert V_5 \rvert = 1.02052$ pu, $Q_{g1} = -18.87$ MVAr, and $Q_{g2} = 1.38$ MVAr

12.9 Conclusions

In this chapter, we discussed general LP and NLP approaches to the OPF problem [1,3,5,29]. We also presented an extended formulation of the problem to accommodate constraints pertaining to system security aspects. We discussed decomposition strategies that can be used to solve the extended OPF problem.

The OPF problem is in general nonconvex. This implies that multiple minima may exist that can differ substantially. Very little work has been done toward exploring this particular aspect of the problem. Furthermore, we have only considered a smooth formulation with continuous controls. However, many effective control actions are in fact discrete. Examples include capacitor switching (for voltage violations) and line switching (for line overload violations). Also, the generator cost curves are in reality fairly discontinuous although they are often modeled as smooth polynomials. Handling these discontinuities and nonconvexities is a challenge for existing OPF methods.

12.10 Problem Set

PROBLEM 12.1

Two generators supply the total load 800 MW; the generator cost functions and limits are given as follows:

$$f_1(P_{G_1}) = 850 + 50P_{G_1} + 0.01P_{G_1}^2,$$

$$50 \leq P_{G_1} \leq 300 \text{ MW},$$

$$f_2(P_{G_2}) = 2450 + 48P_{G_2} + 0.003P_{G_2}^2,$$

$$50 \leq P_{G_1} \leq 650 \text{ MW}.$$

Find the optimum schedule using NLP:

1. Neglecting generation limits
2. Considering generation limits

PROBLEM 12.2

A power system with two plants has the transmission loss equation:

$$P_L = 0.3 \times 10^{-3}P_1^2 + 0.5 \times 10^{-3}P_2^2.$$

The fuel-cost functions are

$$f_1 = 8.5P_1 + 0.00045P_1^2,$$

$$f_1 = 8.2P_2 + 0.0012P_2^2,$$

and system load is 600 MW. Use the N–R method to find the optimal generation schedule.

PROBLEM 12.3

A power system with two plants has a power demand of 1000 MW. The loss equation is given by

$$P_L = 4.5 \times 10^{-3}P_1 + 2.0 \times 10^{-3}P_2.$$

The fuel cost functions are

$$f_1 = .00214P_1 + 7.74, \quad 100 \le P_1 \le 600 \text{ MW},$$

$$f_2 = .00144P_2 + 7.72, \quad 100 \le P_2 \le 600 \text{ MW}.$$

Use LP to solve the optimal schedule:

1. Neglecting generation limits
2. Considering generation limits

PROBLEM 12.4

A subtransmission system has three buses which undergo low voltage; that is, $V_1 = 0.91$, $V_2 = 0.89$, and $V_3 = 0.90$. It is necessary to install shunt capacitors at buses 2 and 8, so that these three buses' voltage can be raised to 0.95 pu. The sensitivities ΔV_1, ΔV_2, and ΔV_3, with respect to Δq_2 and Δq_8, are given as

$$\Delta V_1 = 0.01\Delta q_2 + 0.03\Delta q_8,$$

$$\Delta V_2 = 0.07\Delta q_2 + 0.045\Delta q_8,$$

$$\Delta V_3 = 0.02\Delta q_2 + 0.01\Delta q_8.$$

Find the optimal VAr planning for this subsystem.

PROBLEM 12.5

Consider the following quadratic equation for the voltage deviation optimization problem.

Minimize $f(V) = \left(V_1 - V_1^0\right)^2 + \left(V_2 - V_2^0\right)^2$

Subject to

$V_1 + 2V_2 \le 4$

$V_1 + V_2 \ge 2$

$V_1, V_2 \ge 0.$

If $V_1^0 = V_2^0 = 1.0$ pu, solve the QP problem.

PROBLEM 12.6

A system has three plants. The system loss equation is given by

$$P_L = B_{11}P_1^2 + B_{22}P_2^2 + B_{33}P_3^2,$$

$$(B_{11}, B_{22}, B_{33})^T = (1.6 \times 10^{-4}, 1.2 \times 10^{-4}, 2.2 \times 10^{-4})^T.$$

The system load is 300 MW. Loss minimization is selected as the objective function. Solve this optimal problem using any NLP technique.

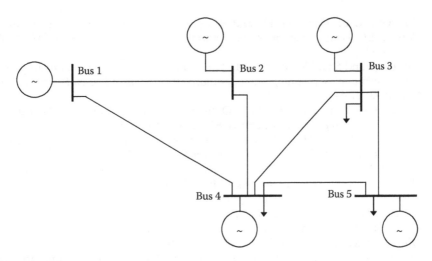

FIGURE 12.11
Figure for Problem 12.7.

PROBLEM 12.7

With P in MW, the cost equations, in dollars per hour, of a large power system are as seen in Figure 12.11.

The five generator data are as follows:

$$F_1 = 0.0015P_1^2 + 1.80P_1 + 40,$$

$$F_2 = 0.0030P_2^2 + 1.70P_2 + 60,$$

$$F_3 = 0.0012P_3^2 + 2.10P_3 + 100,$$

$$F_4 = 0.0080P_4^2 + 2.00P_4 + 25,$$

$$F_5 = 0.0010P_5^2 + 1.80P_5 + 120.$$

The total system real power load is 730 MW. Obtain the absolute minimum cost of this system using the NLP method.

References

1. Alsac, O., Bright, J., Prais, M., and Scott, B., Further developments in LP-based optimal power flow, *IEEE Transactions on Power Systems*, 5(3), 1990, 697–711.
2. Bacher, R. and Van Meeteren, H. P., Real time optimal power flow in automatic generation control, *IEEE Transactions on Power Systems*, PWRS-3, 1988, 1518–1529.
3. Burchett, R. C., Happ, H. H., and Vierath, D. R., Quadratically convergent optimal power flow, *IEEE Transactions on Power Apparatus and Systems*, PAS-103, 1984, 3267–3275.

4. Burchett, R. C., Happ, H. H., and Wirgau, K. A., Large scale optimal power flow, *IEEE Transactions on Power Apparatus and Systems*, PAS-101(10), 1982, 3722–3732.

5. Delson, J. and Shahidehpour, S. M., Linear programming applications to power system economics, planning and operations, *IEEE Transactions on Power Systems*, 7(3), 1992, 1155–1163.

6. Dommel, H. W. and Tinney, W. F., Optimal power flow solutions, *IEEE Transactions on Power Apparatus and Systems*, 87, 1968, 1886–1878.

7. El-Keib, A. A., Ma, H., and Hart, J. L., Economic dispatch in view of the Clean Air Act of 1990, *IEEE Transactions on Power Systems*, 9(2), 1994, 972–978.

8. Galiana, F. D. and Hunneault, M., A survey of the optimal power flow literature, *IEEE Transactions on Power Systems*, 6, 1991, 762–770.

9. Giras, T. C. and Talukdar, S. N., Quasi-Newton method for optimal power flows, *Electrical Power and Energy Systems*, 3(2), 1981, 59–64.

10. Granville, S., Optimal reactive dispatch through interior point method, *IEEE Transactions on Power Systems*, PWRS-9, 1994, 136–146.

11. Huang, W. and Hobbs, B. F., Optimal SO2 compliance planning using probabilistic production costing and generalized benders decomposition, *IEEE Transactions on Power Systems*, 6(2), 1991, 174–180.

12. Huneault, M. and Galiana, F. D., A survey of the optimal power flow literature, *IEEE Transactions on Power Systems*, 6(2), 1991, 762–770.

13. Lamont, J. W. and Obessis, E. V., Emission dispatch models and algorithms for the 1990s, *Presented at the 1994 IEEE/PES Summer Meeting*, 94 SM 526-4 PWRS, San Francisco, CA, July 24–28, 1994.

14. Momoh, J. A. and El-Hawary, M. E., A review of selected optimal power flow literature to 1993, Part I & II: Nonlinear and quadratic programming approaches, *IEEE Transactions on Power Systems (PES)*, 1999.

15. Momoh, J. A. and Zhu, J. Z., Improved interior point method to OPF problems, *IEEE Transactions on Power Systems (PESO)*, 14(3), 1999, 1114–1130.

16. Monticelli, S., Pereira, M. V. F., and Granville, S., Security-constrained optimal power flow with post-contingency corrective rescheduling, *IEEE Transactions on Power Systems*, PWRS-2(1), 1987, 175–182.

17. Papalexopoulos, A., Hao, S., Liu, E., Alaywan, Z., and Kato, K., Cost/benefits analysis of an optimal power flow: The PG&E experience, *IEEE Transactions on Power Systems*, 9(2), 1994, 796–804.

18. Stolt, B., Alsac, O., and Marinho, J. L., The optimal power flow problem, In *Electric Power Problems: The Mathematical Challenge*, A. M. Erisman, K. W. Neves, and M. H. Dwarakanath (Eds.), SIAM, Philadelphia, 1980, pp. 327–351.

19. Talaq, J. H., El-Hawary, F., and El-Hawary, M. E., A summary of environmental/ economic dispatch algorithms, *IEEE Transactions on Power Systems*, 9(3), 1994, 1508–1516.

20. Talaq, J. H., El-Hawary, F., and El-Hawary, M. E., Minimum emissions power flow, *IEEE Transactions on Power Systems*, 9(1), 1994, 429–442.

21. Talukdar, S. N. and Giras, T. C., A fast and robust variable metric method for optimal power flows, *IEEE Transactions on Power Apparatus and Systems*, PAS-101(2), 1982, 415–420.

22. Talukdar, S. N. and Ramesh, V. C., A multi-agent technique for contingency-constrained optimal power flows, *IEEE Transactions on Power Systems*, 9(2), 1994, 855–861.

23. Talukdar, S. N., Giras, T. C., and Kalyan, V. K., Decompositions for optimal power flow, *IEEE Transactions on Power Apparatus and Systems*, PAS-102(12), 1983, 3877–3884.
24. Wood, A. J. and Wollenberg, B. F., *Power Generation, Operation and Control*, Wiley, New York, 1984.
25. Wu, F., Gross, F., Luini, J. F., and Look, P. M., A two-state approach to solving large-scale optimal power flows, *Proceedings of PICA Conference*, 1979, pp. 126–136.
26. Handschin, E., Neise, F., Neumann, H., and Schultz, R., Optimal operation of dispersed generation under uncertainty using mathematical programming, *International Journal for Power and Energy Systems*, 28(9), 2006, 618–626.
27. Gan, D., Thomas, R. J., and Zimmerman, R. D., Stability-constrained optimal power flow, *IEEE Transactions on Power Systems*, 15(2), 2000, 535–540.
28. Fouad, A. A. and Tong, J., Stability constrained optimal rescheduling of generation, *IEEE Transactions on Power Systems*, 8(1), 1993, 105–112.
29. Alsac, O., Bright, J., Prais, M., and Stott, B. Further developments in LP-based optimal power flow, *IEEE Transactions on Power Systems*, 5(3), 1990, 697–711.
30. Chen, L., Tada, Y., Okamoto, H., Tanabe, R., and Ono, A., Optimal operation solutions of power systems with transient stability constraints, *IEEE Transactions on Circuits and Systems*, 48(3), Mar. 2001, 327–339.
31. Yuan, Y., Kubokawa, J., and Sasaki, H., A solution of optimal power flow with multicontingency transient stability constraints, *IEEE Transactions on Power Systems*, 18(3), 2003, 1094–1102.
32. Sun, Y., Yang, X., and Wang, H. F., Approach for optimal power flow with transient stability constraints, *IEE Proceedings—Generation, Transmission and Distribution*, 151(1), 2004, 8–18.
33. Wu, Y., Debs, A. S., and Marsten, R. E., A nonlinear direct predictor-corrector primal-dual interior point algorithm for optimal power flows, *IEEE Transactions on Power Systems*, 9(2), 1994, 876–893.
34. Aganagic, M. and Mokhtari, S., Security constrained economic dispatch using nonlinear Dantzig–Wolfe decomposition, *IEEE Transactions on Power Systems*, 12(1), 1997, 105–112.
35. Zhang, X., Song, Y. H., Lu, Q., and Mei, S., Dynamic available transfer capability (ATC) evaluation by dynamic constrained optimization, *IEEE Transactions on Power Systems*, 19(2), 2004, 1240–1242.

13

Pricing

13.1 Introduction

This chapter summarizes introductory aspects of price theory concepts commonly used in describing economic theory of markets. The ideas and formulations presented are generalized for modeling transactions in the equilibrium states of elastic and nonelastic market models. Special attention was given to marginal pricing and spot pricing methods commonly used in bilateral trading and settlement of transactions among market players.

Noteworthy, several economic foundations of microeconomics and price theory have been extended to many real-time applications including power system valuation strategies. In recent years, the advent of several deregulated markets following the unbundling of power system companies and ancillary services have led to a renewed effort by engineers and economists to improved power delivery efficiency with market-driven incentives. This has led to key areas of research and technology development such as

- Development of pricing mechanisms for issues such as settling power transactions, managing energy resources, and providing incentives or penalties in congestion management.
- Increased wheeling of power and centralized monitoring and control of power markets that is done by the transmission system operators or independent system operators.
- Development of standard market designs (SMDs) that incorporate power pricing mechanisms for managing generation, transmission, and distribution entities.

Specifically, the foundation concepts of marginal price have been used to improve market efficiency for the electric power system in areas such as

- Economic dispatch of generation that places a price cap on the lease expensive generator source subject to the system and network constraints.
- Optimal locations of generation markets in competitive environment and energy pricing of generation sales taking into consideration of the load and generation sites.

441

- Clearing prices that determines the feasibility of power delivery based on the generation bids and the load schedules for optimal power system supply–demand side management.

Economic theory indicates that when commodity prices are equal to marginal costs, the resulting levels of production and consumption will be most efficient, and that marginal prices are induced through competition. It involves bilateral trading where two parties (buyer and seller) agree on trading arrangements and transaction prices via negotiations. Common types include (1) customized long-term contracts, (2) over-the-counter (OTC), and (3) electronic trading exchanges.

13.2 Marginal Pricing

Marginal cost pricing is the principle that the market will, over time, cause goods to be sold at their marginal cost of production. Whether goods are in fact sold at their marginal cost will depend on competition and other factors, as well as the time frame considered. In the most general criticism of the theory of marginal cost pricing, economists note that monopoly power may allow a producer to maintain prices above the marginal cost; more specifically, if a good has low elasticity of demand (consumers are insensitive to changes in price) and supply of the product is limited (or can be limited), prices may be considerably higher than marginal cost. Since this description applies to most products with established brands, marginal pricing may be relatively rare; an example would be in markets for commodities [3,10,14,15,24].

In economics and finance, marginal cost is defined as the change in total cost that arises when the quantity produced changes by one unit or increment. Mathematically, the marginal cost, MC function, is expressed as the derivative of the total cost, TC function, with respect to quantity, q. Figure 13.1 shows a typical marginal cost curve (there is an analogous property of supply: the supply curve is the inverse function of marginal cost).

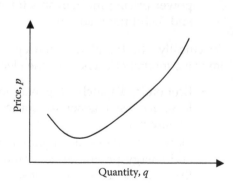

FIGURE 13.1
Typical marginal cost curve.

The marginal cost is given by

$$MC = \frac{\partial}{\partial q} = \frac{\partial(FC + f(q))}{\partial q} = \frac{\partial(f(q))}{\partial q}, \quad (13.1)$$

where the total cost function is expressed as $TC = FC + f(q)$ with $f(q)$ as the total production cost function and FC is the total fixed costs.

Generally, the marginal cost at each level of production includes any additional costs required to produce the next unit. At each level of production and time period being considered, marginal costs include all costs which vary with the level of production, and other costs are considered fixed costs. Also, it is a general principle that a producer should always produce (and sell) the last unit if the marginal cost is less than the market price. As the market price will be dictated by supply and demand, it leads to the conclusion that marginal cost equals marginal revenue (MR). This general principle of marginal cost and marginal cost pricing is important in economic definitions of efficiency.

Furthermore, for a piecewise linear market model, the average cost (AC) is therefore given as $AC = [FC + f(q)]q^{-1}$. This result has important implications, since it indicates that the producer who has already incurred the fixed costs should choose to produce (sell) even if the market price is less than AC. This is because the MR (the income received from selling the marginal unit) is greater than or equal to marginal cost [14].

Also, economists aptly named goods whose demand does not increase with income inferior goods give that individuals generally substitute to better quality and more expensive goods as their incomes rise relative to their rate of expenditures. Goods are typically normal (or superior), or inferior. When demand for a good increase with income levels, the good is termed normal and the converse applies to define inferior goods. Another factor that influences demand is the price of related goods. Such goods are examples of complements.

13.3 Marginal Costing

There are generally two types:

- Short-run marginal costs take into account capital equipment and overhead of the producer, any change in its production involving only changes in the inputs of labor, materials, and energy.
- On the other hand, long-run marginal costs allow all inputs, including capital items (plant, equipment, and buildings) to vary. A long-run cost function describes the cost of production as a function of

output assuming that all inputs are obtained at current prices, that current technology is employed, and everything is being built new from scratch.

Other definitions of costs include fixed costs that do not vary with output; variable cost or operating costs vary directly with the level of output; social costs of production incurred by society; and average total, fixed, and variable costs are the total, fixed, and variable costs divided by the quantity of output.

　Marginal decision-making therefore considers the incremental cost of a good with respect to an additional unit of its demand where the next unit of the good is the marginal unit. The practical usefulness of this concept is in making rational choice to maximize a goal function or benefit among competing alternatives. Computation of marginal benefits and cost, a natural extension of cost-benefit analysis with sensitivity studies, provides us with valuable conclusions. For instance, one could continue to increase an activity level in a viable way provided that the marginal benefit is greater than the marginal cost, and stops when an equilibrium between the two marginalized quantities is reached. Further insights into marginalism, a significant development in economics beyond the 1870s, helped define the concepts of marginal thinking commonly used today.

13.4　Marginal Revenue

MR is the additional revenue generated by an additional unit of production and is described as the change in revenue to the number of units sold. In exact terms, it is equal to the change in total revenue over the change in quantity for unit increment in the quantity and is given by

$$\text{MR} = \frac{\partial \text{TR}}{\partial q} = \frac{\partial p}{\partial q} \cdot q + \frac{\partial q}{\partial q} \cdot p \equiv p + q\frac{\partial p}{\partial q}. \tag{13.2}$$

Market conditions:

- For a company facing perfectly competitive markets, price does not change with quantity sold $(\partial p/\partial q)=0$ and the MR is equal to price $(\text{MR}=p)$.
- For a monopoly, the price received will decline with quantity sold such that $(\partial p/\partial q)<0$ and this implies that the MR is less than price, i.e., $\text{MR}<p$. This means that the profit-maximizing quantity, for which MR is equal to marginal cost, will be lower for a monopoly than for a competitive firm, while the profit-maximizing price will be higher.

- When MR is positive, $(\partial p/\partial q) > 0$ then the price elasticity of demand is said to be elastic, and when it is negative, $(\partial p/\partial q) < 0$, the price elasticity of demand is said to be inelastic. When $MR = 0$, the price elasticity of demand $(\partial p/\partial q) = -1$.

13.5 Pricing Policies for Regulated Systems and Markets

In the deregulated electric power industry as we know it today, electricity is essentially a commodity where goods and services are bought and sold by quantity and indistinguishable in quality or characteristics from one batch to another. And, regulation of the power industry is expected to benefit from price stability in which case both buyers and sellers know the price of power in advance and can depend on prices to be stable. This is because prices are not expected to vary quickly or unpredictably. However, as the competitive power market increase, so will the prices fluctuations that are functions of the market conditions that could also be influenced by a number of stochastic disturbances. This price volatility, and unpredictability of price ahead of time, makes planning energy resources more of a challenge for users.

Several well-established system operators are now being implemented to provide transparent trading mechanisms in trading floors for day-ahead and spot pricing of electricity. These trading environments include spot transactions, forward contracts, a future market, and price hedging. There are some salient points to note about some of these frameworks for transacting electricity sales.

Spot transactions are mainly conducted by telephone or computer network between two parties. It is an OTC market as opposed to an exchange. Spot markets do not necessarily have trading floors. A key advantage of the OTC market is that the terms of a contract do not have to have the specifications required by an exchange, but this approach usually lack transparency.

Additionally, forward and future markets were developed to reduce the risk of high price volatility that is often generated by spot trading. In future markets (accelerated since the mid-1980s), hedging against the risk of price fluctuations is achieved. A future contract is an agreement between two parties to buy or sell an asset at a certain future time for a certain price. On the other hand, a spot contract is an agreement to buy or sell an asset within a day or hour.

Pricing policies for real-time pricing are either fixed or variable. For fixed price policies, a market situation is analyzed for profit and an appropriate price is fixed amount until the situation changes, at which point they go through the process again. On the other hand, variable pricing is characterized by negotiations. Variants of variable pricing include price shading, auctions, and, due to advances in technology, real-time pricing mechanisms.

A special variant of real-time pricing is online auctions. All market participants can view the price changes soon after they occur. Traditional auctions are inefficient because they require bidders to be physically present. By solving this problem, online auctions reduce the transaction costs for bidders, increase the number of bidders, and increase the average bid price. In addition to these examples of variable pricing in the short term, there are also long-term pricing practices that could be used as variable pricing.

Under many deregulated marketplace structures, particularly those that in a nondiscriminant auction, all bidders are paid the market clearing price. Otherwise, they are paid only what they bid. A nondiscriminant auction method has been selected for many pools and power exchanges because overall, it provides lower and more stable costs.

13.6 Pricing Methods

One of the challenges ahead is the development of rules that allow the shared use of transmission system by utilities and third-party generation. Besides ensuring the technical quality of the transmission service (voltage control, static and dynamic security constraints, etc.), these rules should satisfy other criteria, including no cross-subsidies, transparency of cost allocation procedure, ease of regulation, ensure an adequate remuneration of present and future, transmission investments, economic signals for dimensioning, and continuity of the charge. Some methods have been proposed and may be classified as one of the following paradigms: embedded cost, incremental cost, and composite embedded–incremental cost. Belonging to the set of incremental methods, the short-term marginal cost has been the most popular due to its economic basis, that is, it can provide the economic signals for operation and dimension. Some limitations have been observed in its application to power transmission systems such as: not recovering all the transmission costs, the charges obtained may be highly volatile, and charges for the transmission system are based on generation costs rather than its own cost, the transmission system is usually not in the optimal condition, and so forth. On the other hand, the embedded cost methods provide, in general, an adequate remuneration of transmission systems and are easy to implement. These methods have been criticized due to their economic grounds. Several researchers have sought methods to hybridized costing methods that best captures the system performance and more accurately reflect the economic implication of marginal pricing and sensitivity of the utility or objective functions relative to changes in resources [6,18,19].

In these methods all system costs (existing transmission system, operation, and expansion) are allocated among the system users in proportion to their extent of use of the transmission resources. Some load-flow based methods are summarized briefly.

13.6.1 Megawatt-Mile (MWM) Method

The MWM method first calculates the flow on each circuit caused by the generation/load pattern of each agent based on a power flow model. Costs are then allocated in proportion to the ratio of power flow and circuit capacity:

$$R(u) = \sum_{\text{all } k} C_k \frac{|f_k(u)|}{\bar{f}_k},$$

where
  $R(u)$ is the allocated cost to agent u
  C_k is the cost of circuit k
  $f_k(u)$ is the k-circuit flow caused by agent u
  $\bar{f}_k$ is the k-circuit capacity

$$\text{Total cost} = \sum_{\text{all } k} C_k.$$

This method has been criticized as having no obvious grounding on economic theory. Its simplifying assumptions can help optimal transmission planning from a static analysis. As the total circuit power flows are usually smaller than the circuit capacities, this allocation rule does not recover all embedded costs. This means that the MWM scheme is only charging for a base-case network, but not for the transmission reserve.

13.6.2 Modulus Method (MM) or Usage Method

A simple way to ensure recovery of all embedded costs in the MWM method while retaining its advantages is to replace the circuit capacities by the sum of absolute power flows caused by all agents:

$$R(u) = \sum_{\text{all } k} C_k \frac{|f_k(u)|}{\sum_{\text{all } s} |f_k(s)|}.$$

For the transmission expansion interpretation, this MM assumes that all agents have to pay for the actual capacity use and for the additional reserve. This reserve may be due to the need of system meeting reliability, stability and security criteria, or due to system readjustments (i.e., due to planning errors caused by inherent uncertainties). However, there are no incentives to the agent that alleviates the circuit load, improving the system performance, and/or postponing transmission investments.

13.6.3   Zero Counterflow Method (ZCM)

Here, there is no charge for the agent whose power flow is in the opposite direction of the net flow. Only the agents that use the circuit in the same direction of the net flow pay in proportion to their flow.

$$R(u) = \begin{cases} \sum_{\text{all } k} C_k \dfrac{|f_k(u)|}{\sum_{\text{all } s=\Omega_{k+}} f_k(s)}, & \text{for } f_k(u) > 0, \\ 0, & \text{for } f_k(u) \leq 0 \end{cases}$$

where Ω_{k+} is the set of participants with positive flows on circuit k.

This method assumes that the net flow reduction is beneficial, even if there is already an excess installed capacity. Moreover, for a light loaded circuit, there is a discontinuity on the charges when the net flow changes the direction.

13.6.4   Dominant Flow Method (DFM)

This method is a hybrid of the previous two methods to overcome their inherent drawbacks. The scheme is to divide the circuit cost allocation $R(u)$ into two components, $R_1(u)$ and $R_2(u)$:

a. $R_1(u)$ is related to the circuit capacity that is actually being used, called base capacity. This capacity fraction corresponds to the circuit net flow and the associated cost is borne only by those participants with a positive flow w.r.t. the net flow. The allocation criterion of this portion is found by changing the k-circuit total cost C_k to C_{Bk} (cost of base capacity) where:

$$C_{Bk} = C_k \times \frac{f_k}{\bar{f}_k}.$$

b. $R_2(u)$ is related to the difference $\bar{f}_k - f_k$, called additional capacity. This capacity corresponds to the circuit reserve and as all participants take advantage of the reliability and security associated the corresponding fraction of total cost is borne by all participants, according to the C_k to C_{Ak} (cost of additional capacity).

The total allocation $R(u)$ is then given by the sum $R_1(u) + R_2(u)$. The MWM has a constant average charge irrespective of what is happening with the circuit and no incentive to the counter flow agent is provided. Also, it fails to collect the exact embedded cost. In the MM the average charge decreases as the transaction amount increases. This seems corrected from the agent viewpoint but it does not make sense when transmission system expansion and operation are considered. The ZCM gives an incentive to agent A (charge equal zero) when its flow is against to the net flow. When the net flow changes the direction the agent A bares the total cost. One can note the discontinuity when x is equal to the circuit capacity. This can lead to a great variation of the charge for a slight variation of x. In the DFM the agent A has incentive only when x is close to zero, i.e., when agent A actually alleviates the circuit load. When x approximates to I and becomes close to the agent B flow, the incentive decreases.

13.6.5   Alternative Pricing Methods

Pricing methods termed rolled-in methods (not based on load flow analysis) include Postage Stamp and Contract Path [23,25]. They ignore actual system operation and are likely to send incorrect economic signals to transmission customers.

Some of the alternative pricing methods also include:

1. Embedded costs. This has all transmission costs rolled in. Allocation of costs is done by coincident peak demand. It is based on the original cost less depreciation: it can lead to increased costs for exiting customers; it does not provide incentive for new lines; it does not account for reduction in other savings; and it provides incentive to build generation at wrong locations.

2. Short-run marginal cost. This includes the cost of incremental losses. It also includes incremental O & M and administrative costs. It includes other fuel costs for reduction in cost-saving transfers and may include congestion costs. It provides no incentive for the wheeling utility.

3. Long-run marginal cost. This includes carrying charges for facilities expansion. It includes same types of cost as short-run marginal cost. It gives less weight to near-term costs and provides little or not incentive for the wheeling utility. It is difficult to determine.

4. Value based on market-based pricing. In the U.S. under FERC, this was examined in the western systems power pool (WSPP) experiment—bids and offers. It may include open bidding for transmission rights.

13.7   Economic Basis of Shadow Prices in Linear Programming (LP)

The shadow price dictates the equilibrium between the changes in objective value per unit increase in the resource. These price indices are therefore functions of the constraints and not the decision variables.

WORKING DEFINITION

Shadow price of a resource refers to the change in the objective value of the optimal solution of a constrained optimization problem obtained by relaxing the constraint by one unit. More explicitly, it is the value of the Lagrange multiplier at the optimal solution, which means that it represents a small change in the objective value that results from small perturbations in the constraint.

This is consistent with the principle that at the optimal solution, the gradient of the objective function is a linear combination of the constraint function gradients with the weights equivalent to Lagrange multipliers. Each constraint in an optimization problem has a shadow price or dual variable.

In the case of LP problems, the primal solution dictates what actions should be taken in order to achieve the desired goal while the optimal dual variables provides worth or value to an additional unit of a resource. The dual solution of an LP problem is therefore termed shadow price (and its meaning is identical to that of Lagrange multipliers in the nonlinear programming [NLP] case). The primal LP takes the form $\text{Max} \, c^T x = \sum_{j=1}^{n} c_j x_j$ s.t. $Ax = \sum_{j=1}^{n} a_{ij} x_j \leq b_i$, $x_j \geq 0$ where $i \in \{1,m\}$ and $j \in \{1,n\}$ for n resources (or dual constraints) and m products (or dual variables). The dual form is therefore $\text{Min} \, b^T \lambda = \sum_{i=1}^{m} b_i \lambda_i$ s.t. $A^T \lambda = \sum_{i=1}^{m} a_{ij} \lambda_i \geq c_j$ and $\lambda_i \geq 0$. Note a_{ij} is the amount of resource i required to make a unit of product j and c_j is the monetary value per unit of product j.

THEOREM   Conditions for LP Convergence:

If the primal LP problem has at least one nondegenerate basic and optimal solution, then there is a positive epsilon, ε with the property $\delta_j \leq \varepsilon$ in the neighborhood of the optimal solution for all $i \in \{1,m\}$, then the problem $\text{Max} \, Z = \sum_{j=1}^{n} c_j x_j$ s.t. $\sum_{j=1}^{n} a_{ij} x_j \leq b_i + \delta_i$ and $x_j \geq 0$ has optimal value $Z^* + \sum_{i=1}^{m} \lambda_i^* \delta_i$ where λ_i^* is the set of optimal solutions for dual problem (shadow prices).

Here, λ_i^* give the maximum amount or willingness to pay above a marginal clearing price or trading price for each extra unit of resource i. Its relationship to the objective function and the constraint is readily established. From the dual LP, recall that B is the optimal basis for the primal problem and C_B is the final basic cost vector, then $Z^* = C_B B^{-1} b = \lambda^* b$. That is $\partial Z^* = \lambda^{*T}(\partial b) \Rightarrow \lambda^* = [\partial Z^* / \partial b]^T$. Alternatively, let $x^* = [x_B^*, 0]^T = [B^{-1} b, 0]^T$ where x^* is a nondegenerate optimal basic feasible solution to the standard form of the LP problem. Assuming that $x_B^* > 0$, then $b \leftarrow b + \Delta b$ does not cause the optimal basis B to change. Hence, in the optimal basic feasible solution, $\hat{x}_B^* = B^{-1}(b + \Delta b)$. The corresponding change in the objective is $\Delta Z^* = C_B^T B^{-1} \Delta b = \lambda^{*T} \Delta b$. Therefore, the change in the optimal value is given by $\lambda^{*T} = (\Delta Z^*/\Delta b)$ and, as Δb tend to zero in the limit, we obtain the same result: $\lambda^* = [\partial Z^*/\partial b]^T$. This solution is the vector of simpler multipliers for the primal problem or optimal dual variables.

13.7.1 Special Case of LP Problems with Two-Sided Bounded Variables

The formation of Karush-Kuhn Tucker (KKT) application will be most useful for solving locational marginal price (LMP) multipliers of a DC optimal power flow (OPF). The primal problem takes the form:

$$\text{Max}\, c^T x \quad \text{s.t. } Ax \le b \quad \text{with } x^{\min} \le x \le x^{\max}.$$

This can be written as

$$\text{Max}\, c^T x \quad \text{subject to } Ax \le b, \quad +x \le x^{\max}, \quad -x \le -x^{\min}.$$

In matrix notation, we have

$$\text{Max}\, c^T \begin{bmatrix} x \\ 0 \\ 0 \end{bmatrix} \quad \text{s.t. } \begin{bmatrix} A \\ +I \\ -I \end{bmatrix} x \le \begin{bmatrix} b \\ +x^{\max} \\ -x^{\min} \end{bmatrix}.$$

The dual problem is therefore:

$$\text{Min}\, [b \quad x^{\max} \quad x^{\min}] y \equiv b^T y + x^{\max} y + x^{\min} y$$
$$\text{s.t. } [A + I \quad -I] y = c, \quad y \ge 0.$$

This takes the form Min $c_{\text{new}}^T y$ s.t. $A_{\text{new}}^T y \ge c$ with $y \ge 0$. KKT conditions can now be applied as before.

13.7.2 Further Interpretation of Dual Shadow Prices Variables

The dual solutions, referred to as shadow prices, indicate the value of relaxing primal constraints by unit change and therefore represent a price tag or marginal cost for the specified additional unit the resource.

Notably, when a resource constraint is binding (the upper boundary or ceiling of the constraints is reached), no excess capacity exists, and the resource shadow price is positive. However, if the resource constraint is not binding (i.e., the constraint parameter is within its operating limits), excess capacity exists, and the resource shadow price is zero.

13.8   LMP

LMPs are the costs of supplying the next increment of load or demand as a specific node in the network taking into account the generation transactions (bids), load transactions (demands), network losses, transmission security constraints, and other operational constraints. In general, LMPs are obtained from an OPF formulated as a security constrained least-cost dispatch problem. The nodal prices differ from location to location due to the presence of network real power losses as well as congestion across critical interfaces. In cases where congestion plays a significant part in the marginal prices, the differences in LMPs will be used to settle congestion charges [2,23,25,27].

13.8.1   Components of LMP

The components of LMPs are due to energy, network losses, and congestion. From the formulation of the security-constrained OPF (SCOPF) problem, the marginal loss factor is given by

$$L_i = \frac{\partial P_{\text{Loss}}}{\partial P_i} = \lim_{\Delta P_i \Rightarrow 0} \left[\frac{\Delta P_{\text{Loss}}}{\Delta P_i} \right]. \tag{13.3}$$

Solution to the unconstrained OPF (economic dispatch without losses) yields the system shadow price reference, λ^{ref}. This problem can be linearized and solved using LP, or the exact solution can be obtained using a Newtonian optimization technique.

The marginal cost of losses in a power network is

$$\lambda_i^{\text{loss}} = \left(\frac{\partial P_{\text{Loss}}}{\partial P_i} \right) \left(\frac{\partial C}{\partial P_{gi}} \right)^{\text{T}}, \tag{13.4}$$

where
P_{gi} is the generation power
P_i is the injected power at the ith bus
P_{Loss} is the total system loss

Furthermore, the marginal cost of enforcing the binding constraints is given by

$$\lambda_i^{\text{congestion}} = -\sum_{j=1}^{NC} \mu_j \beta_{ji} \quad \text{or}$$

$$\lambda_i^{\text{congestion}} \equiv -\sum_{j=1}^{NC} \left\{ \left(\frac{\partial C}{\partial h_j^{\max}} \right) \left(\frac{\Delta P_j}{\Delta P_i} \right) \right\} \Bigg|_{\Delta P_{\text{ref}} \leftarrow \Delta P_{\text{ref}} - 1\,\text{MW}}, \tag{13.5}$$

where

μ$_j$ is the shadow price of the binding constraints

β_{ji} is the sensitivity of power flow constraints j relative to power injection at bus i

P_j is the power flow on the jth constraint

ΔP_i is the change in real power injection at bus i

ΔP_{ref} is the change in real power at the reference bus

C is the cost function in the SCOPF

The slack bus is the reference node used when computing this sensitivity factor. β_{ji} is also a shift factor or sensitivity of the real load at bus i on constraint j. Finally, compute the value of the LMP at each node i in the network by summing the energy, congestion, and loss components. Mathematically, LMP, λ_i that represents the sum of nodal price at the reference or slack bus, the marginal cost accounting for transmission loss distribution, and the marginal cost of transmission congestion relative to the power injections at each node is given by

$$\lambda_i = \lambda^{\text{ref}} + \lambda_i^{\text{loss}} + \lambda_i^{\text{congestion}} = \lambda^{\text{ref}} - L_i\lambda^{\text{ref}} - \left(\sum_{j=1}^{NC} \mu_j\beta_{ji}\right). \tag{13.6}$$

Further analyses will include computing settlements of congestion costs, transmission spot pricing, average LMPs, payments/dividends to ISO, and payments by load sets. Approach to compute the components of LMP is as follows.

Computing the reference shadow price (λ^{energy})

This is obtained by solving the unconstrained economic dispatch problem using the Lagrangian method.

$$L(P_g, \lambda) = C_T(P_g) - \lambda\left(\sum_{i\in NG} P_{gi} - \sum_{i\in ND} P_{Dk}\right). \tag{13.7}$$

By applying KKT condition

$$\frac{\partial L}{\partial P_{gi}} = 0 \quad \text{and} \quad \frac{\partial L}{\partial \lambda} = 0. \tag{13.8}$$

Solution to these equations yields the energy component of LMP.

Computing the marginal cost of losses (λ^{Loss})
Using loss penalty factors, this is given by

$$
\begin{aligned}
\lambda_i^{\text{Loss}} &= -L_i \lambda_{\text{ref}} \\
&= \left(\frac{\partial P_{\text{Loss}}}{\partial P_{gi}}\right)\lambda_{\text{ref}} = \left(\frac{\partial P_{\text{Loss}}}{\partial P_{gi}}\right)\left(\frac{\partial C_T}{\partial P_{gi}}\right)^{\text{T}}\Bigg|_{\text{losses}=0},
\end{aligned}
\tag{13.9}
$$

where B-coefficients are used to compute

$$
P_{\text{Loss}} = \sum \frac{B_{ii}P_{gi}}{|V_i|^2 \cos^2 \theta_i}, \quad \text{for AC system}
\tag{13.10}
$$

$$
P_{\text{Loss}} \cong \sum B_{ii}P_{gi}, \quad \text{for DC system.}
\tag{13.11}
$$

Computing the marginal cost of congestion ($\lambda^{\text{congestion}}$)
By definition, this is the cost associated with enforcing an active constraint. A constraint is said to be active if its limit is reached or exceeded.

$$
\lambda_i^{\text{congestion}} = -\sum_{j=1}^{NC} \mu_j \beta_{ji}
\tag{13.12}
$$

with

$$
\mu_j = \frac{\partial C_T}{\partial h_j^{\text{max}}}
\tag{13.13}
$$

$$
\beta_{ji} = \left(\frac{\Delta \vec{P_j}}{\Delta P_i}\right)\Bigg|_{\Delta P_{\text{ref}} \leftarrow \Delta P_{\text{ref}} - 1\,\text{MW}},
\tag{13.14}
$$

where
   μ_j is the shadow price of the binding constraint
   h_j^{max} is the ceiling of the constraint at the upper limit
   $\Delta \underline{P_i}$ is the change in injection at bus i on the line L_{ij} of the active constraint
   ΔP_j is the change in power flow on constraint j; $\Delta P_j = \Delta P_j = P_{ij}^{\text{old}} - P_{ij}^{\text{new}}$
   If $h_j^{\text{min}} < h_j < h_j^{\text{max}}$, then $\mu_j = 0$. Otherwise, $\mu_j \neq 0$ relative to the jth security or congestion constraint

The sensitivity term β_{ji} is the incremental power flow on the jth constraint when an additional unit of power (1 MW) is injected at the ith bus and the equivalent amount removed from the reference bus. Therefore, β_{ji} is computed as the sensitivity of the power flow constraint w.r.t. the injection at the ith bus subject to a reduction of power at the slack bus. Therefore, for the set of NC binding constraints under congestion conditions, the marginal cost of congestion is determined from

$$\lambda_i^{\text{congestion}} = -\sum_{j=1}^{NC} \left(\frac{\partial C_T}{\partial h_j^{\max}}\right)\left(\frac{\Delta \vec{P_j}}{\Delta P_i}\right)\Bigg|_{\Delta P_{\text{ref}} \leftarrow \Delta P_{\text{ref}} - 1\,\text{MW}} . \tag{13.15}$$

13.8.2 LMP in Energy Markets

13.8.2.1 Formulation for NLP Approximations

The determination of LMP or spot prices is obtained from OPF solutions. The general formulation can be summarized minimizing a welfare cost function subject to power balance, network, network security, and power market constraints. This typical formulation can be written as [1]

Minimize $\quad C_S^T P_S - C_D^T P_D$ \hfill (13.16)

Subject to

$f(x, u, P_S, P_D) = 0$ (power balance) \hfill (13.17)

$Q_G^{\min} \leq Q_G \leq Q_G^{\max}$ (gen Q-limits) \hfill (13.18)

$V^{\min} \leq V \leq V^{\max}$ (bus voltage limits) \hfill (13.19)

$|P_{ij}(x)| \leq P_{ij}^{\max}$ (thermal limits) \hfill (13.20)

$\lambda_c \leq \lambda_{c_o}$ (stability loading at critical points) \hfill (13.21)

$0 \leq P_S \leq P_S^{\max}$ (supply bids) \hfill (13.22)

$0 \leq P_D \leq P_D^{\max}$ (demand bids). \hfill (13.23)

This formulation is solved using LP or NLP techniques that are based in specialized techniques in classical optimization. These include interior point (IP) technology, Lagrangian or Newtonian approach, and Barrier penalty functions [2]. In the more general case, following this formulation above, we form the Lagrangian given in the generalized form:

$$L(x,u,\lambda) = f(x,u) + \sum_{i \in p}\left(\lambda_i^T g_i(x,u)\right) + \sum_{j \in m}\left(\lambda_j^T \left|h_j(x,u) - h_j^{\max}(x,u)\right|\right)$$

where λ_i and λ_j are Lagrange multipliers of the equality and inequality constraints in a typical OPF. Here, the Kuhn–Tucker (K–T) necessary optimality conditions, $L_x = 0$ and $L_\lambda = 0$, hold true. This approach, applied to the formulation above, gives rise to

$$\frac{\partial L}{\partial P_{S_i}} = C_{S_i} - \lambda_{S_i} + \mu P_{S_i}^{\max} \tag{13.24}$$

$$\frac{\partial L}{\partial D_{S_i}} = -C_{D_i} - \lambda_{D_i} + \mu P_{D_i}^{\max}, \tag{13.25}$$

where parameter μ is now a Barrier penalty function [1] and the aggregate LMPs for all ith nodes in the system is $\text{LMP}_i = \lambda_i$ for $\forall i \in N_{\text{buses}}$. These values represent the shadow price or marginal costs for each market participant located at the ith node in the power system. The computation of this LMP_i requires deterministic economic data, such as bid and cost schedules and load forecasts, and the conventional data used in a typical SCOPF.

By using the criteria for LMP in a more generalized form, this gives rise to lambda computation that can be decomposed into the components of energy, congestion, and losses. This summation represents the nodal price at the reference or slack bus, the marginal cost that accounts for the distribution of transmission losses, and the marginal cost of transmission congestion relative to the power injections. As shown in the previous chapters, LMP is given as

$$\lambda_i = \lambda^{\text{ref}} + \lambda_i^{\text{loss}} + \lambda_i^{\text{congestion}} = \lambda^{\text{ref}} - L_i\lambda^{\text{ref}} - \left(\sum_{j=1}^{NC} \mu_j\beta_{ji}\right). \tag{13.26}$$

$$L_i = \partial P_{\text{Loss}}/\partial P_i, \tag{13.27}$$

where
  μ_j is the shadow price of the binding constraint j in the binding set $j \in NC$
  β_{ji} is the shift factor or sensitivity of the real load at bus i on constraint j
  L_i is the loss factor
  P_i is the generator injection at the ith node

13.8.2.2 Formulation for LP-Based OPF

In the energy market, the market players (producers, consumers, and bilateral trade agents) submit their bids for energy production or consumption to the ISO, who is responsible for scheduling the system resources for each hour of the following day (day-ahead market). A SCOPF problem using linearized DC power flow equations [4] is formulated for ISOs to maximize the sum of the net consumer and producer surplus subject to transmission capacity and security constraints.

$$\max \sum_{i=1}^{d} E_i(D_i) - \sum_{i=1}^{g} C_i(P_i) \qquad (13.28)$$

Subject to

$$\mathbf{B} \cdot \theta = \mathbf{P} - \mathbf{D} \qquad (13.29)$$

$$\theta_{\text{ref}} = 0 \qquad (13.30)$$

$$\left| \frac{1}{x_{ij}} (\theta_i - \theta_j) \right| \leq F_{ij}^{\text{max}} \quad \text{for all lines } ij \qquad (13.31)$$

$$P_i^{\text{min}} \leq P_i \leq P_i^{\text{max}} \quad \text{for all units } i \qquad (13.32)$$

$$0 \leq D_i \leq D_i^{\text{max}} \quad \text{for all demands } i \qquad (13.33)$$

$$\left| \frac{1}{x_{ij}} (\theta_i - \theta_j) + \text{LODF}_{ij,mn} \frac{1}{x_{mn}} (\theta_m - \theta_n) \right| \leq F_{ij}^{\text{max},c}$$

$$\text{for all monitored lines } ij \text{ and contingent lines } mn, \qquad (13.34)$$

where
  Equation 13.29 represents the system DC power flow
  Equation 13.30 defines the slack bus voltage phase angle since det($\mathbf{B}$) = 0
     (the slack bus is not omitted in Equation 13.29)
  Equation 13.31 represents the transmission line power flow limits
  Equation 13.32 represents the unit active power output limits
  Equation 13.33 represents the bus active power demand limits
  Equation 13.34 is the short-term emergency rating of line ij relative to
     the normal rating, $F_{ij}^{\text{max},c}$
  LODFs are the line outage distribution factors

This problem is an LP optimization problem, which is solved for the unknown vectors $\mathbf{P}$, $\mathbf{D}$, and θ.

13.8.3 Computational Steps for LMP Using DC OPF

In this formulation, a single-area network with one slack bus is assumed. A DC OPF formulation based on sensitivity analysis of the constraints is summarized here. The optimization problem can be stated as

$$\text{Min}(c_{\text{P}}^{\text{T}} P - c_{\text{D}}^{\text{T}} P_{\text{D}}) = \sum_{i=1}^{NG} \sum_{k=1}^{ND} \left(c_{\text{P}i}^{\text{T}} P_i - c_{\text{D}k}^{\text{T}} P_{\text{D}k} \right)$$

(production − cost welfare function)

Subject to

1. Equality constraint of power balance $(P_D - P_g = 0)$

$$[-A_U \quad A_B]\begin{bmatrix} P \\ P_{flow} \end{bmatrix} + [A_L][P_D] = 0,$$

where A_U, A_B, and A_L are the bus-unit, bus-branch incidence, and bus-load incidence matrices.

2. Inequality constraint of line flow thermal limits $(P_{ij}^{min} \leq P_{ij} \leq P_{ij}^{max})$

$$\left[+\left(P_{ij} - P_{ij}^o\right)\right] = \left[+\sum_{k=1}^{N}\left\{S_{j-j,k}^{coeff}\left(P_k - P_k^o\right)\right\}\right] \leq \left[+P_{ij}^{max}\right]$$

$$\left[-\left(P_{ij} - P_{ij}^o\right)\right] = \left[-\sum_{k=1}^{N}\left\{S_{j-j,k}^{coeff}\left(P_k - P_k^o\right)\right\}\right] \leq \left[-P_{ij}^{min}\right],$$

where sensitivity coefficient matrix is obtained from the Kron reduction of admittance bus matrix for the system with bus 1 as the system reference such that

$$S^{coeff} = [\text{Diag}(Y)][A_B^T]\begin{bmatrix} 0 & 0 \\ 0 & [Y_{Bus}^r]^{-1} \end{bmatrix}.$$

3. Generation limits of energy bid limits

$$P^{min} \leq P \leq P^{max}.$$

4. Demand limits or bid limits for the loads

$$P_D^{min} \leq P_D \leq P_D^{max}.$$

The presence of distributed generators (DG) is incorporated in the load model, which assumes a negative injection to reduce the generating needs of utility-owned sources, increase the energy reserves, and reduce local area congestion and voltage problems.

The steps to compute the LMPs at each bus in the network are summarized in the following steps:

Step 1. Obtain network, generators, and load data and operating constraints, including those of any DG sources; obtain the load and generation profiles and demand curves, market data (supply and demand bids and limits), and time horizon of study.

Step 2. Initialize study time and other iteration control.

Step 3. Establish the power flow feasibility by solving the DC power flow equation summarized by $P_{ij} = (x_{ij})^{-1} (\theta_i - \theta_j)$ or $[P] = [Y_{bus}][\theta]$ such that $[\theta] = [Y_{bus}]^{-1}[P]$. Or, in matrix notation:

$$[P_{flow}] = [\text{Diag}(Y_{bus})][A_B^T][\theta] \quad \text{where } Y_{diag} \triangleq \text{diag}(x_{ij})^{-1}.$$

Step 4. Develop the DC OPF formulation according to the objective and constraints given above and derive A_U, A_B, and A_L which are the bus-unit, bus-branch incidence, and bus-load incidence matrices derived from the current network topology.

Step 5. Set up equality constraints, $A_B P_{flows} - A_U P + A_L P_D = 0$ such that

$$[-A_U \quad A_B] \begin{bmatrix} P \\ P_{flows} \end{bmatrix} + [A_L][P_D] = 0$$

or if the demand is unknown. For known demand,

$$\begin{bmatrix} -A_U & 0 \\ 0 & A_B \end{bmatrix} \begin{bmatrix} P \\ P_{flows} \end{bmatrix} = -[A_L][P_D]$$

and the associated Lagrange vector is $[\lambda_{energy} \quad \lambda_{flows}]^T$.

Step 6. Compute the sensitivity coefficient matrix from the Kron reduction of admittance bus matrix for the system with bus 1 as the system reference.

Step 7. Set up real power flow constraints for the network using:

$$\left[+\left(P_{ij} - P_{ij}^o\right)\right] = \left[+\sum_{k=1}^{N}\left\{S_{j-j,k}^{coeff}\left(P_k - P_k^o\right)\right\}\right] \leq \left[+P_{ij}^{max}\right]$$

$$\left[-\left(P_{ij} - P_{ij}^o\right)\right] = \left[-\sum_{k=1}^{N}\left\{S_{j-j,k}^{coeff}\left(P_k - P_k^o\right)\right\}\right] \leq \left[-P_{ij}^{min}\right].$$

Step 8. Set up overall inequality constraints for line flows, $|P_{ij}| \leq P_{ij}^{max}$, and power injections at the nodes, $0 \leq P_i \leq P_i^{max}$. Summarize as

$$\begin{bmatrix} P^{min} \\ P_{flow}^{min} \\ P_D^{min} \end{bmatrix} \leq \begin{bmatrix} P \\ P_{flow} \\ P_D \end{bmatrix} \leq \begin{bmatrix} P^{max} \\ P_{flow}^{max} \\ P_D^{max} \end{bmatrix}$$

or

$$\begin{bmatrix} P^{\min} \\ P^{\min}_{\text{flow}} \end{bmatrix} \leq \begin{bmatrix} P \\ P_{\text{flow}} \end{bmatrix} \leq \begin{bmatrix} P^{\max} \\ P^{\max}_{\text{flow}} \end{bmatrix} \quad \text{(for unknown load demands only).}$$

Step 9. Decompose the inequality constraints for the purposes of the dual LP formulation:

$$\begin{bmatrix} P \\ P_{\text{flow}} \\ P_{\text{D}} \end{bmatrix} \leq \begin{bmatrix} P^{\max} \\ P^{\max}_{\text{flow}} \\ P^{\max}_{\text{D}} \end{bmatrix} \quad \text{and} \quad \begin{bmatrix} -P \\ -P_{\text{flow}} \\ -P_{\text{D}} \end{bmatrix} \leq \begin{bmatrix} -P^{\min} \\ -P^{\min}_{\text{flow}} \\ -P^{\min}_{\text{D}} \end{bmatrix}$$

$$\begin{bmatrix} P \\ P_{\text{flow}} \end{bmatrix} \leq \begin{bmatrix} P^{\max} \\ P^{\max}_{\text{flow}} \end{bmatrix} \quad \text{and} \quad \begin{bmatrix} -P \\ -P_{\text{flow}} \end{bmatrix} \leq \begin{bmatrix} -P^{\min} \\ -P^{\min}_{\text{flow}} \end{bmatrix}$$

(for known load demands only).

Step 10. Recast the matrices for the objective function and constraints into the primal LP formulation, as given below and cast in state variable form:

$$\underset{x \in R^n}{\text{Min}} \, c^{\text{T}} x \text{ subject to } Ax = b, \quad x^{\min} \leq x \leq x^{\max}.$$

Step 11. Develop the LP dual problem formulation from the step above to the form:

$$\underset{\lambda \in R^m}{\text{Max}} \, [b^{\text{T}} \quad x^{\max} \quad x^{\min}] \, \lambda \text{ subject to } [A^{\text{T}} \quad I \quad -I]\lambda \geq c, \quad \lambda \geq 0,$$

where $\lambda = [\lambda_i | i \in \{1,m\}]^{\text{T}}$ is the vector of dual variables.

Step 12. Use a linear programming (LP) technique to compute $\lambda = [\lambda_i | i \in \{1,m\}]^{\text{T}}$. From the solution extract the LMPs in this partitioned vector. Here,

$$\lambda^{\text{T}} = \left[\lambda_{\text{LMP}_i} \vdots \lambda_{\text{gen\_max}} \vdots \lambda_{\text{gen\_min}} \vdots \lambda_{\text{flows\_max}} \vdots \lambda_{\text{flows\_min}} \right]^{\text{T}}.$$

Step 13. Compute $f^{\min}_{\text{Primal}} \equiv f^{\max}_{\text{Dual}}$.

Step 14. Transpose the algebraic solution to power system quantities.

Step 15. For other time stages, update input data and repeat from step 4. Otherwise, continue.

Step 16. Save and/or display all results: power flows and LMP of energy and congestion at all buses. Compute net settlement based on transactions and marginal energy prices between supply and demand nodded.

End.

13.8.4 Transmission Congestion Charges (TCCs)

The nodal prices (LMPs) are the Lagrange multipliers of the nodal power balance constraints [2,7,13,27]. Each unit is paid for every MW of production the nodal price of the bus that it is connected to. And each consumer pays for every MW of consumption the nodal price of the bus he/she is connected to. In case of congestion, the nodal prices differentiate from one another and the ISO collects a sum corresponding to TCCs. If a market player X performs a fixed power transaction of P_{km} MW from bus k to bus m during an H-hour period, then he should pay congestion TCCs to the ISO equal to

$$\mathrm{TCC}_x = \sum_{h=1}^{H} \left[(\mathrm{LMP}_m^h - \mathrm{LMP}_k^h) P_{km} \right].$$

If during a particular hour h, $\mathrm{LMP}_m^h - \mathrm{LMP}_k^h$, then the transaction performed by player X helps relieve congestion on the network and the ISO pays the player X for his/her valuable contribution during that hour.

Mathematical considerations used for electricity spot pricing involve:

1. System operator bases the dispatch on the bids of the market participants (i.e., on ex ante prices).
2. Presumption: observed power system dispatch is the optimal dispatch based on the bids.
3. A formal, but simplified, dispatch formulation is formed and then linearized around the observed solution.
4. Dual mathematical programming techniques are used to calculate prices which are consistent with both the dispatch and the bids.
5. These ex post prices provide incentives for long-term decision making.
6. A fundamental change in approach. Rather than determining the optimal operation, the model calculates prices to provide signals for decision makers to make long-run decisions.

The system representation is formulated as the primary problem with LP. The objective is to minimize generation fuel costs such that the constraints of energy balance, voltage magnitude at each node, and line flow at each transmission line are satisfied. The NLP problem is linearized around the dispatch in the form $\max c^T x$ s.t. $Ax = b$. The dual problem can be built from the primal problem as $\min b^T y$ s.t. $A^T y = c$. Also, the complementary slackness conditions make the dual objective function redundant where $(Ax - b)^T y = 0$ and $(A^T y - c)^T x = 0$. The dual constraints now define the prices of the primal constraints.

Key points to note:

1. Active and reactive power prices at the swing bus are the shadow prices of the energy conservation constraints. If there are no constraints on the transmission of power, there is only one marginal bus, which drives the prices at other buses.
2. Losses relative to the swing bus determine the prices at other buses.
3. If a transmission line is congested, there exists an additional marginal bus. This will have an impact on the price of the transmission line, which is the price difference between the two marginal buses.
4. Additional constraints such as contingency requirements or thermal power flow limits in SCOPF or contingency-constrained OPF (CCOPF) can be added to the primal problem.

The mechanism for the determination of electricity spot prices provides good incentives for market competition and coordination. Such prices can be found based on duality theory by solving a mathematical program.

13.9 Alternative OPF Formulation for Pricing using Duality in LP

OPF formulation is used for pricing with the assumption for short-term problems. Although the dispatch adjustment is continuous, the dispatch is treated as constant over a short period, say half an hour, for pricing purposes. The available generator capacities determined by the unit commitment problem are dispatched to determine the optimal operating level of each generator (and potentially load) and the transmission line power flows, while satisfying a range of system constraints. The swing bus is denoted by the index s, or when included in a set by the suffix S. We use PX to denote the set of all buses other than the swing bus, and PXS to be the set of all buses. Thus, for example, although individual buses are indicated by subscripts, P^{PXS} represents the vector of active power net injections for all buses in the system.

The OPF objective function has the form:

$$\text{Min Cost}\left(\mathbf{P}_G^{PXS}, \mathbf{Q}_G^{PXS}\right). \tag{13.35}$$

This equation states that the aim of the dispatcher is to minimize the total cost of generating active power P^{PXS} and reactive power Q^{PXS}, at each bus where $\text{Cost}\left(\mathbf{P}_G^{PXS}, \mathbf{Q}_G^{PXS}\right)$ describes the total fuel cost of generation.

This objective is assumed to be convex, but may be nondifferentiable at some points. In reality, these functions may be nonconvex, though separable with respect to each generator's output. In practice, however, a piecewise quadratic approximation is generally used, with smoothed transition across the nondifferentiable points.

In minimizing the objective function, it is necessary to ensure that active and reactive power are conserved, which is a fundamental physical constraint.

$$\sum_{i \in \text{PXS}} (P_{Gi} - P_{Di}) - L_P\left(\mathbf{P}_G^{\text{PX}} - \mathbf{P}_D^{\text{PX}}, \mathbf{Q}_G^{\text{PX}} - \mathbf{Q}_D^{\text{PX}}\right) = 0 \qquad (13.36)$$

$$\sum_{i \in \text{PXS}} (Q_{Gi} - Q_{Di}) - L_Q\left(\mathbf{P}_G^{\text{PX}} - \mathbf{P}_D^{\text{PX}}, \mathbf{Q}_G^{\text{PX}} - \mathbf{Q}_D^{\text{PX}}\right) = 0. \qquad (13.37)$$

The feasible range of the dispatch variables includes:

$$V_n^{\min} \leq V_n \leq V_n^{\max} \quad \forall n \in \text{PX} \qquad (13.38)$$

$$S_k^{\min} \leq S_k \leq S_k^{\max} \quad \forall k \in K \qquad (13.39)$$

$$P_{Gi}^{\min} \leq P_{Gi} \leq P_{Gi}^{\max} \quad \forall i \in \text{PXS} \qquad (13.40)$$

$$Q_{Gi}^{\min} \leq Q_{Gi} \leq Q_{Gi}^{\max} \quad \forall i \in \text{PXS} \qquad (13.41)$$

$$P_{Di} = P_{Di}^{\text{set}} \quad \forall i \in \text{PXS} \qquad (13.42)$$

$$Q_{Di} = Q_{Di}^{\text{set}} \quad \forall i \in \text{PXS}, \qquad (13.43)$$

where Equation 13.39 is the power flow limits on transmission lines. The demands are set externally to the dispatch problem by Equations 13.42 and 13.43. For an optimal observed setting of these values, the shadow prices on these bounds should equal the prices determined by a welfare maximizing objective when demand is not fixed.

13.9.1 Linearization of the OPF

Equations 13.36 through 13.43 describe a simple representation of an OPF problem. The linearization of this primal formulation about an observed dispatch results in prices which correspond to the actual dispatch, rather than to what was expected prior to the dispatch. The linearization is performed using a standard first-order Taylor's expansion of the nonlinear terms as

$$f(x) \approx f(x^*) + \frac{\partial f}{\partial x}(x - x^*). \qquad (13.44)$$

Here x^* denotes the observed value of a variable x. We assume x^* to be optimal. Due to the possibility of discontinuous first derivatives of the objective function, we must distinguish between increasing active or reactive power generation beyond their observed values, or reducing them below their observed values. Thus, we define:

$$P_{Gi} = P_{Gi}^* + P_{Gi}^+ - P_{Gi}^-. \tag{13.45}$$

$$Q_{Gi} = Q_{Gi}^* + Q_{Gi}^+ - Q_{Gi}^-. \tag{13.46}$$

$$P_{Gi}^+, P_{Gi}^-, Q_{Gi}^+, Q_{Gi}^- \geq 0. \tag{13.47}$$

Here P_{Gi}^+, an increase in active power generation, has a marginal fuel cost of c_{Pi}^+, which exceeds c_{Pi}^-, the marginal reduction in fuel cost associated with P_{Gi}^-, a decrease in active power generation. Analogous definitions, though in terms of reactive power generation, apply for c_{Qi}^+, c_{Qi}^-, Q_{Gi}^+, and Q_{Gi}^-.

Using Equations 13.45 through 13.47, the canonical form of the linearized OPF is that described by Equations 13.48 through 13.60. The shadow price associated with each constraint is given on the right.

$$\text{Min} \sum_{i \in PXS} \left(c_{Pi}^+ P_{Gi}^+ - c_{Pi}^- P_{Gi}^- \right) + \sum_{i \in PXS} \left(c_{Qi}^+ Q_{Gi}^+ - c_{Qi}^- Q_{Gi}^- \right). \tag{13.48}$$

Subject to

$$-\sum_{i \in PX} \frac{\partial P_s}{\partial P_i} \left(P_{Gi}^+ - P_{Gi}^- - P_{Di} \right) - \sum_{i \in PX} \frac{\partial P_s}{\partial Q_i} \left(Q_{Gi}^+ - Q_{Gi}^- - Q_{Di} \right)$$
$$+ \left(P_{Gs}^+ - P_{Gs}^- - P_{Ds} \right) = A_1^* : \lambda_P \tag{13.49}$$

$$-\sum_{i \in PX} \frac{\partial Q_s}{\partial P_i} \left(P_{Gi}^+ - P_{Gi}^- - P_{Di} \right) - \sum_{i \in PX} \frac{\partial Q_s}{\partial Q_i} \left(Q_{Gi}^+ - Q_{Gi}^- - Q_{Di} \right)$$
$$+ \left(Q_{Gs}^+ - Q_{Gs}^- - Q_{Ds} \right) = A_2^* : \lambda_Q \tag{13.50}$$

$$-V_n \geq -V_n^{\max} : v_{V_n}^+ \quad \forall n \in PX \tag{13.51}$$

$$V_n \geq V_n^{\min} : v_{V_n}^- \quad \forall n \in PX \tag{13.52}$$

$$-S_k \geq -S_k^{\max} : v_{S_k}^+ \quad \forall k \in K \tag{13.53}$$

$$S_k \geq S_k^{\min} : v_{S_k}^- \quad \forall k \in K \tag{13.54}$$

$$-P_{Gi}^+ \geq -P_{G_i}^{\max} + P_{Gi}^* : v_{P_i}^+ \quad \forall i \in PXS \tag{13.55}$$

$$-P_{Gi}^- \geq P_{G_i}^{\min} - P_{Gi}^* : v_{P_i}^- \quad \forall i \in PXS \tag{13.56}$$

$$-Q_{Gi}^+ \geq -Q_{G_i}^{\max} + Q_{Gi}^* : v_{Q_n}^+ \quad \forall i \in PXS \tag{13.57}$$

$$-Q_{Gi}^- \geq Q_{G_i}^{\min} - Q_{Gi}^* : v_{Q_i}^- \quad \forall i \in PXS \tag{13.58}$$

$$P_{Di} = P_{Di}^* : \beta_{P_i} \quad \forall i \in \text{PXS} \tag{13.59}$$

$$Q_{Di} = Q_{Di}^* : \beta_{Q_i} \quad \forall i \in \text{PXS}. \tag{13.60}$$

In each of Equations 13.55 through 13.58, one of the "change in generation" terms of Equations 13.45 and 13.46 is a slack variable, and has been dropped from the equation. Constants in the objective function have been ignored while those in the constraints Equations 13.49 and 13.50 have been moved to the right-hand side. Functional forms for the constants in the constraints are given as

$$A_1^* = P_s^* - \sum_{i \in \text{PX}} \frac{\partial P_s}{\partial P_i} P_{Di}^* - \sum_{i \in \text{PX}} \frac{\partial P_s}{\partial Q_i} Q_{Di}^* \tag{13.61}$$

$$A_2^* = Q_s^* - \sum_{i \in \text{PX}} \frac{\partial Q_s}{\partial P_i} P_{Di}^* - \sum_{i \in \text{PX}} \frac{\partial Q_s}{\partial Q_i} Q_{Di}^*. \tag{13.62}$$

13.9.2 LP Dual Construct

The dual mathematical programming problem corresponding to the linearized OPF can be formed in the usual manner, producing Equations 13.63 through 13.71. The primal variables corresponding to each dual constraint are shown on the right.

$$\text{Min } \lambda_P A_1^* + \lambda_Q A_2^* + \sum_{n \in \text{PX}} \left(-v_{V_n}^+ V_n^{\max} + v_{V_n}^- V_n^{\min} \right)$$

$$+ \sum_{k \in K} \left(-v_{S_k}^+ S_k^{\max} + v_{S_k}^- S_k^{\min} \right) + \sum_{i \in \text{PXS}} \beta_{P_i} P_{Di}^* + \sum_{i \in \text{PXS}} \beta_{Q_i} Q_{Di}^*$$

$$+ \sum_{i \in \text{PXS}} \left(v_{P_i}^+ \left(-P_{G_i}^{\max} + P_{G_i}^* \right) + v_{P_i}^- \left(P_{G_i}^{\min} - P_{G_i}^* \right) \right)$$

$$+ \sum_{i \in \text{PXS}} \left(v_{Q_i}^+ \left(-Q_{G_i}^{\max} + Q_{G_i}^* \right) + v_{Q_i}^- \left(Q_{G_i}^{\min} - Q_{G_i}^* \right) \right). \tag{13.63}$$

Subject to

Pricing relationships for the OPF demand settings:

$$\lambda_P \frac{\partial P_s}{\partial P_i} + \lambda_Q \frac{\partial Q_s}{\partial P_i} + \beta_{P_i} = 0 : P_{Di} \quad \forall i \in \text{PXS} \tag{13.64}$$

$$\lambda_P \frac{\partial P_s}{\partial Q_i} + \lambda_Q \frac{\partial Q_s}{\partial Q_i} + \beta_{Q_i} = 0 : Q_{Di} \quad \forall i \in \text{PXS} \tag{13.65}$$

Floor and ceiling constraints set by generator costs:

$$-\lambda_P \frac{\partial P_s}{\partial P_i} - \lambda_Q \frac{\partial Q_s}{\partial P_i} - v_{P_i}^+ \le c_{P_i}^+ : P_{Gi}^+ \quad \forall i \in \text{PXS} \tag{13.66}$$

$$\lambda_P \frac{\partial P_s}{\partial P_i} + \lambda_Q \frac{\partial Q_s}{\partial P_i} - v_{P_i}^- \le -c_{P_i}^+ : P_{Gi}^- \quad \forall i \in \text{PXS} \tag{13.67}$$

$$-\lambda_P \frac{\partial P_s}{\partial Q_i} - \lambda_Q \frac{\partial Q_s}{\partial Q_i} - v_{Q_i}^+ \le c_{Q_i}^+ : Q_{Gi}^- \quad \forall i \in \text{PXS} \tag{13.68}$$

$$\lambda_P \frac{\partial P_s}{\partial Q_i} + \lambda_Q \frac{\partial Q_s}{\partial Q_i} - v_{Q_i}^- \le -c_{Q_i}^+ : Q_{Gi}^- \quad \forall i \in \text{PXS}. \tag{13.69}$$

Pricing relationships for the OPF transmission line constraints:

$$-v_{S_k}^+ + v_{S_k}^- = 0 : S_k \quad \forall k \in K. \tag{13.70}$$

Pricing relationships for the OPF voltage constraints:

$$-v_{V_n}^+ + v_{V_n}^- = 0 : V_n \quad \forall n \in \text{PX}. \tag{13.71}$$

The simplified form of the dual is depicted as following assuming only active power is traded on a spot basis.

$$\text{Min} \sum_{i \in \text{PXS}} \beta_{P_i} \left(P_{Gi}^* - P_{Di}^* \right). \tag{13.72}$$

The constraints of Equations 13.64 and 13.65 can be reexpressed as

$$\beta_{P_i} = -\beta_{P_s} \frac{\partial P_s}{\partial P_i} - \beta_{Q_s} \frac{\partial Q_s}{\partial P_i} \quad \forall i \in \text{PX} \tag{13.73}$$

$$\beta_{Q_i} = -\beta_{P_s} \frac{\partial P_s}{\partial Q_i} - \beta_{Q_s} \frac{\partial Q_s}{\partial Q_i} \quad \forall i \in \text{PX}. \tag{13.74}$$

Note that the terms in Equations 13.73 and 13.74 involve data readily available once the dispatch is known. Here $(\partial P_s/\partial P_i)$, $(\partial Q_s/\partial P)$, $(\partial P_s/\partial Q_i)$, and $(\partial Q_s/\partial Q_i)$ describe the marginal generation and losses attributed to changes in the power flows. To evaluate these derivatives, we must apply a Jacobian coordinate transformation as the fundamental electrical equations only allow us to determine derivatives with respect to voltage magnitude and phase angle. In practice, this transformation amounts to solving a set of sparse linear equations, a problem which is straightforward to set up and solve.

Substituting the active power prices for all the nonswing buses from Equation 13.64 into Equations 13.66 and 13.67, and the reactive power prices for all the same buses from Equation 13.65 into Equations 13.68 and 13.69, produces Equations 13.75 and 13.76.

$$c_{P_i}^- - \left\langle v_{P_i}^- \right\rangle \leq \beta_{Pi} \leq c_{P_i}^+ + \left\langle v_{P_i}^+ \right\rangle \quad \forall i \in \text{PXS} \tag{13.75}$$

$$c_{Q_i}^- - \left\langle v_{Q_i}^- \right\rangle \leq \beta_{Qi} \leq c_{Q_i}^+ + \left\langle v_{Q_i}^+ \right\rangle \quad \forall i \in \text{PXS} \tag{13.76}$$

$$v_{P_i}^+, v_{P_i}^-, v_{Q_i}^+, v_{Q_i}^- \geq 0. \tag{13.77}$$

$\langle z \rangle$ denotes that z only appears in the dual formulation if the primal constraint to which it corresponds is binding.

References

1. F. C. Schweppe, M. C. Caramanis, R. D. Tabors, and R. E. Bohn, *Spot Pricing of Electricity*, Kluwer, Norwell, MA, 1988.
2. W. W Hogan, *Concave Networks for Electric Power Transmission: Technical Reference*, Energy and Environmental Policy Center, John F. Kennedy School of Government, Harvard University, Cambridge, 1991.
3. E. G. Read and B. J. Ring, Dispatch based pricing: Theory and application. In A.J. Turner (Ed.), *Dispatch Based Pricing*, Trans Power (NZ) Ltd., Wellington, 1995.
4. A. J. Wood and B. F. Wollenberg, *Power Generation Control, and Operation*, Wiley, New York, 1984.
5. R. Bacher, Power system models, objectives and constraints in optimal power flow calculations. In *Optimization in Planning and Operation of Electric Power Systems*, K. Frauendorfer, H. Glavitsch, and R. Bacher (Eds.), Physica Verlag, Springer, 1992, pp. 159–198.
6. R. Baldick, R. Grant, and E. Kahn, Theory and application of linear supply function equilibrium in electricity markets, *Journal of Regulatory Economics*, 25(2), 143–167, 2004.
7. R. D. Christie, B. F. Wollenberg, and I. Wangensteen, Transmission management in the deregulated environment, *Proceedings of the IEEE*, 88(2), 170–195, 2000.
8. R. Ethier, R. Zimmerman, T. Mount, R. Thomas, and W. Schulze, Auction design for competitive electricity markets, *HICSS Conference*, Maui, Hawaii, January 7–10, 1997.
9. I. Palacios-Huerta and T. J. Santos, A theory of markets, institutions, and endogenous preferences, *Journal of Public Economics*, 88, 601–627, 2004.
10. P. Holmberg, Comparing supply function equilibria of pay-as-bid and uniform-price auctions, Working Paper 2005:17, 2005.
11. C. J. Day, B. F. Hobbs, and J. Pang, Oligopolistic competition in power networks: A conjectured supply function approach, *IEEE Transactions on Power Systems*, 17 (3), 597–607, 2002.
12. J. Conejo, E. Castillo, R. Mínguez, and F. Milano, Locational marginal price sensitivities, *Transactions on Power Systems*, 20(4), 2026–2033, 2005.
13. M. Quelhas, E. Gil, and J. D. McCalley, Nodal prices in an integrated energy system, *International Journal of Critical Infrastructures*, 2(1), 50–69, 2006.

14. R. P. McAfee, *Introduction to Economic Analysis*, California Institute of Technology, Pasadena, California, July 24, 2006.
15. T. Orfanogianni and G. Gross, A general formulation for LMP evaluation, *IEEE Transactions on Power Systems*, 22, (3), 1163–1173, 2007.
16. L. Chen, H. Suzuki, T. Wachi, and Y. Shimura, Components of nodal prices for electric power systems, *IEEE Transactions on Power Systems*, 17(1), 41–49, 2002.
17. M. Rivier and J. I. Perez-Arriaga, Computation and decomposition of spot prices for transmission pricing, in *Proceedings of the 11th PSC Conference*, 1993.
18. B. Allaz and J. Vila, Cournot competition, forwards markets and efficiency, *Journal of Economic Theory*, 59, 1–16, 1993.
19. T. Wu, Z. Alaywan, and A. D. Papalexopoulos, Locational marginal price calculations using the distributed-slack power-flow formulation, *IEEE Transactions on Power Systems*, 20(2), 1188–1190, 2005.
20. X. Cheng and T. J. Overbye, An energy reference bus independent LMP decomposition algorithm, *IEEE Transactions on Power Systems*, 21(3), 1041–1049, 2006.
21. E. Litvinov, T. Zheng, G. Rosenwald, and P. Shamsollahi, Marginal loss modeling in LMP calculation, *IEEE Transactions on Power Systems*, 19(2), 880–888, 2004.
22. F. Li and R. Bo, DC OPF-based LMP simulation—Algorithm, comparison with AC OPF, and sensitivity, *IEEE Transactions on Power Systems*, 22(4), 1475–1485, 2007.
23. T. Overbye, X. Cheng, and Y. Sun, A comparison of the AC and DC power flow models for LMP calculations, in *Proceedings of the 37th Hawaii International Conference on System Sciences, IEEE*, Hawaii, 2004.
24. F. P. Sioshansi and W. Pfaffenberger, *Electricity Market Reform: An International Perspective*, Elsevier, Amsterdam, 2006.
25. A. A. El. Keib and X. Ma, Calculating short run marginal cost of active and reactive power production, *IEEE Transactions on Power System*, 12(2), 559–565, 1997.
26. J. Y. Choi, S. H. Rim, and J. K. Park, Optimal real time pricing of real and reactive powers, *IEEE Transactions on Power Systems*, 13(4), 1514–1520, 1998.
27. S. S. Oren, Economic inefficiency of passive transmission rights in congested electricity systems with competitive generation, *The Energy Journal*, 18(4), 63–83, 1997.

14

Unit Commitment

14.1 Introduction

Unit commitment is an operation scheduling function, which is sometimes called predispatch. In the overall hierarchy of generation resources management, the unit commitment function fits between the economic dispatch and the maintenance and production scheduling. In terms of timescales involved, unit commitment scheduling covers the scope of hourly power system operation decisions with a 1 day to 1 week horizon [1,2,4].

Unit commitment schedules the on-and-off times of the generating units, and calculates the minimum cost hourly generation schedule while ensuring that start-up and shutdown rates, minimum up and minimum down times are considered. The function sometimes includes deciding the practicality of interregional power exchanges, and meeting daily or weekly quotas for consumption of fixed-batch energies, such as nuclear, restricted natural gas contracts, and other fuels that may be in short supply [1,2,4].

The unit commitment decisions are coupled or iteratively solved in conjunction with coordinating the use of hydro including pumped storage capabilities and ensuring system reliability using probabilistic measures. The function may also include labor constraints due to crew policy and costs, which is the normal times that a full operating crew will be available without committing overtime costs. A foremost consideration is to adequately adopt environmental controls, such as fuel switching.

Systems with hydrostorage capability either with pumped hydrostations or with reservoirs on rivers usually require 1 week horizon times. On the other hand, a system with no "memory" devices and few dynamic components can use much shorter horizon times.

Most unit commitment programs operate discretely in time, at 1 h intervals. Systems with short horizon times can successfully deal with time increments as small as a few minutes. There is sometimes no clear distinction between the minute-by-minute dispatch techniques and some of the unit commitment programs with small-time increments.

Unit commitment has grown in importance recently, not only to promote system economy but also for the following reasons:

1. Start-up, shutdown, and the dynamic considerations in restarting the modern generating facilities are much more complex and costlier than they were for smaller, older units.

2. Growth in system size to the point where even small percentage gains have become economically very important.

3. Increase in variation between the peak and off-peak power demands.

4. System planning requires automated computerized schedulers to simulate the effect of unit selection methods on the choice of new generation.

5. Scheduling problem has grown out of the effective reach of the "earlier" techniques because of the large variety in efficiencies and types of power sources. The generation resource mix includes fossil-fueled units, peaking units, such as combustion turbines, stored and run-of-river hydro, pumped storage hydro, nuclear units, purchases and sales over tie-lines, and partial entitlements to units.

The application of computer-based unit commitment programs in electric utilities has been slow due to the following reasons:

1. Unit commitment programs are not readily transferred between systems. The problem is so large and complex that only the most important features can be included, and these vary a great deal among systems, thus requiring tailor-made applications.

2. There are political problems, constraints, and peculiarities of systems that are not easily amenable to mathematical solutions and may be very hard to model in the first place.

3. Operating situation changes so quickly and there are so much objective and subjective information about the system that the input requirements of sophisticated, computerized schedulers are discouraging.

4. As in other computer application areas, developing fully workable systems has been difficult, as has been the building of operator's confidence.

The unit commitment schedule is obtained considering many factors including

- Unit operating constraints and costs
- Generation and reserve constraints
- Plant start-up constraints
- Network constraints

The optimal operation and planning of power systems involve consideration of economy operation, system security, and start-up or downtime issues. The optimal scheduling units of operation with constraints to meet a given target or goal are generally referred to as the dynamic programming. This has been discussed as a general terminology under optimal power flow (OPF) which deals with security and reliability with technical and nontechnical constraints. The schedule of different resource intended or forced find units are done based on company standards. Research to include units with transmission controls is done above.

Several methods [16–32] have been proposed in unit commitment problem based on Lagrange relaxation method:

1. Table lookup approximation
2. Dynamic programming
3. Evolutionary programming

Genetic algorithms (GAs) are also being used in several applications of unit commitment. We will present details of GA in Chapter 15 that will involve its problem formulation, assumptions, algorithm, and test cases.

14.2 Formulation of Unit Commitment

14.2.1 Reserve Constraints

There are various classifications for reserve and these include units on spinning reserve and units on cold reserve under the conditions of banked boiler or cold start. The first constraint that must be met is that the net generation must be greater than or equal to the sum of total system demand and required system reserve. That is

$$\sum_{i=1}^{N} P_{g_i}(t) \geq (\text{Net demand} + \text{reserve}). \tag{14.1}$$

In case the units should maintain a given amount of reserve, its upper bounds must be modified accordingly. Therefore, we have

$$P_{g_i}^{\text{max}} = P_{g_i}^{\text{capacity}} - P_{g_i}^{\text{reserve}} \tag{14.2}$$

$$\text{Demand} + \text{losses} \leq \sum_{i=1}^{N} P_{g_i} - \sum_{i=1}^{N} P_{g_i}^{\text{reserve}} \tag{14.3}$$

$$C_{\text{cold}} = C_O(1 - e^{\alpha t}) + C_L, \tag{14.4}$$

where

C_{cold} is the cost to start an off-line boiler
α is the unit's thermal time constant
t is the time (s)
C_L is the labor cost up to the units
C_O is the cost to start-up a cold boiler

$$C_{\text{banked}} = C_B t + C_L, \tag{14.5}$$

where

C_B is the cost to start-up a banked boiler
t is the time (s)

14.2.2 Modeling in Unit Commitment

Nomenclature

F	Total operation cost on the power system
$F_i(t)$	Energy output of the ith unit at hour t
$F_i(E_i(t))$	Fuel cost of the ith unit at hour t when the generated power is equivalent to $E_i(t)$
N	Total number of units in the power system
T	Total time under which unit commitment is performed
$P_{g_i}(t)$	Power output of the ith unit at hour t
$\tilde{P}_{g_i}(t)$	Constrained generating capability of the ith unit at hour t
$P_{g_i}^{\max}$	Maximum power output of the ith unit
$P_{g_i}^{\min}$	Minimum power output of the ith unit
$S_i(t)$	Start-up cost of the ith unit at hour t
$f_i(t)$	Ramping cost of the ith unit at hour t
$P_D(t)$	Net system power demand at hour t
$P_R(t)$	Net system spinning-reserve hour t
$\lambda(t)$	Lagrangian multiplier for the system power balance constraint at hour t
$\mu(t)$	Lagrangian multiplier for the system spinning-reserve constraint at hour t
$u_i(t)$	Commitment state of the ith unit at hour t

$$F = \sum_{t=1}^{T} \sum_{i=1}^{N} [u_i(t)F_i(E_i(t)) + S_i(t) + f_i(t)]. \tag{14.6}$$

The constraint model for the unit commitment optimization problem is as follows:

1. System energy balance

$$\sum_{i=1}^{N} \frac{1}{2} [u_i(t)P_{g_i}(t) + u_i(t-1)P_{g_i}(t-1)] = P_D(t) \quad \text{for } t \in \{1,T\}. \tag{14.7}$$

2. System spinning-reserve requirements

$$\sum_{i=1}^{N} u_i(t)\tilde{P}_{g_i}(t) \geq P_D(t) + P_R(t) \quad \text{for } t \in \{1,T\}.$$ (14.8)

3. Unit generation limits

$$P_{g_i}^{\max} \leq P_{g_i}(t) \leq \tilde{P}_{g_i}(t) \quad \text{for } t \in \{1,T\} \quad \text{and} \quad i \in \{1,N\}.$$ (14.9)

4. Energy and power exchange

$$E_i(t) = \frac{1}{2}\left[P_{g_i}(t) + P_{g_i}(t-1)\right] \quad \text{for } t \in \{1,N\}.$$ (14.10)

14.2.3 Lagrangian Function for Unit Commitment

The Lagrangian function for unit commitment is expressed as

$$L\{P_{g_i}(t), u_i(t), \lambda(t), \mu(t)\}$$

$$= \sum_{t=1}^{T} \sum_{i=1}^{N} \left[F_i(E_i(t))u_i(t) + S_i(t) + f_i(t)\right]$$

$$- \sum_{t=1}^{T} \lambda(t)\left\{\sum_{i=1}^{N} \frac{1}{2}\left[u_i(t)P_{g_i}(t) + u_i(t-1)P_{g_i}(t-1)\right] - P_D(t)\right\}$$

$$- \sum_{t=1}^{T} \mu(t)\left\{\sum_{i=1}^{N} u_i(t)\tilde{P}_{g_i}(t) - P_D(t) - P_R(t)\right\}$$ (14.11)

$$\text{Min } L\{P_{g_i}(t), u_i(t), \tilde{\lambda}(t), \tilde{\mu}(t)\}$$

$$= \sum_{t=1}^{T} \sum_{i=1}^{N} \left[F_i(E_i(t))u_i(t) + S_i(t) + f_i(t) - \tilde{\lambda}'(0)P_{g_i}(0)\right]$$

$$- \sum_{t=1}^{T} \sum_{i=1}^{N} \left[\tilde{\lambda}'(t)u_i(t)P_{g_i}(t) + \tilde{u}_i(t)\tilde{P}_{g_i}(t)u_i(t)\right]$$

$$- \sum_{t=1}^{T} \left[\tilde{\lambda}'(t)P_D(t) + \tilde{\mu}(t)\{P_D(t) + P_R(t)\}\right],$$ (14.12)

where

$$\tilde{\lambda}'(0) = 0.5\tilde{\lambda}(1)$$ (14.13)

$$\tilde{\lambda}'(1) = 0.5\left(\tilde{\lambda}(1) + \tilde{\lambda}(2)\right)$$ (14.14)

$$\tilde{\lambda}'(T-1) = 0.5\left(\tilde{\lambda}(T-1) + \tilde{\lambda}(T)\right) \tag{14.15}$$

$$\tilde{\lambda}'(T) = 0.5\tilde{\lambda}(T). \tag{14.16}$$

Here we note that $\tilde{\lambda}'(0)P_{g_i}(0)$, for $i \in \{1,N\}$ are given by the initial conditions and thus can be ignored in searching for the optimal unit commitment scheme.

$$\text{Min } L\{P_{g_i}(t), u_i(t), \tilde{\lambda}(t), \tilde{\mu}(t)\}$$

$$= \sum_{t=1}^{T} \left[F_i(E_i(t))u_i(t) - \tilde{\lambda}'(t)P_{g_i}(t) - \tilde{\mu}'(t)P_{g_i}(t)\right]u_i(t)$$

$$- \sum_{t=1}^{T} [S_i(t) + f_i(t)]. \tag{14.17}$$

14.3 Optimization Methods

14.3.1 Priority List Unit Commitment Schemes

Economic scheduling techniques used until 1958 required that each unit operate at the same incremental operating cost. The unit commitment strategy was to drop a unit entirely from the system whenever the incremental cost calculation left it operating at below 10%–25% of its rated maximum capacity. The reasoning is that below some point the fixed operating costs make the unit too expensive to operate.

The other earlier techniques have been derived mainly from a method introduced by Baldwin et al. (1960), where for the first time the start-up costs and the minimum downtime requirements were considered. These methods are commonly called priority list or merit-order procedures, and they are in common use up to now.

The unit commitment techniques developed in the 1950s extended the previous incremental cost techniques to include minimum downtime and start-up costs. They built on strict priority-of-shutdown rules for different seasons, that is, for different daily load shapes. In a number of simulations, with the load decreasing, the least efficient unit among those that were on was tested for possible shutdown as follows:

1. Is it possible to restart this unit by the time the load reached its present level again?
2. Does the restart cost exceed the potential operating savings?

These priority lists were developed well ahead of time, and in actual operation the units were dropped from the system if they were

1. Next in line on the priority list.
2. There was not less than a predetermined critical interval before system loads would rise again to the same level (not necessarily the minimum downtime).

A number of refinements have been made on this original priority list method. One improvement was the addition of checks of the system spinning reserve at each trial scheduling (Kerr et al., 1966). Pseudoincremental costs were also introduced and iteratively adjusted to encourage the consumption of the appropriate quotas of fixed-batch energy supplies.

In the priority list unit commitment programs developed later, including those currently in use, the actual priority lists are, in effect, developed especially for the situation-at-hand in an iterative procedure that is generally deeply imbedded in complex heuristic algorithms. Additional features incorporated in these different heuristic techniques are energy interchange modeling, different start-up and shutdown orderings, unit response rates, minimum up and down times, transmission penalty factors, local area "must run" considerations, as well as others. These features are not particularly difficult to model, but the demanding task is to take all the custom-selected features for a particular system and construct a manageable scheduling program.

Although the concept of the priority list [8,12] was introduced only as a cursory first attempt at scheduling, it has remained one of the primary methods for using an approximation to reduce the dimensionality and complexity of the most sophisticated scheduling mechanisms. This method appears many times in the dynamic techniques and in the actual industrial applications.

14.3.2 Priority Criteria

The core of the priority list-based unit commitment scheduling program mechanism for ordering the units of the system according to a certain economic criterion so that the least expensive units is placed at the top of the list, and the proceeding to the more expensive ones. A number of variants have been used in the literature.

14.3.2.1 Type I: Fuel Cost-Based Lists

Static priority lists are obtained based on the average fuel cost of each unit operating at a certain fixed fraction of their maximum output. Thus the criterion is

$$M_i = \frac{F_i H_i}{P_{g_i}}\bigg|_{x P_{g_i}^{\max}}, \tag{14.18}$$

where
M_i is the priority index for the ith unit based on its average fuel cost
H_i is the heat rate curve of the ith unit (MBtu/h)

F_i is the fuel cost of the ith unit ($/MBtu)
$P_{g_i}^{max}$ is the maximum power of the ith unit (MW)
x is the fixed fraction of the maximum output of the ith unit

In many instances, the full load values are used. The full load average cost is simply the net heat rate at full load times the fuel cost per Btu. An alternative static ranking procedure called the "equal Lambda list" is based on assuming that units are operating at the same incremental cost and finding the cost of fuel per unit output of the unit

$$M_i = \frac{F_i H_i}{P_{g_i}}\bigg|_{\frac{\partial F_i}{\partial P_{g_i}}=\lambda}, \tag{14.19}$$

14.3.2.2 Type II: Incremental Fuel Cost-Based List

In Pang and Chen (1976), the criterion for placing individual units at various priority levels is the average incremental fuel cost

$$M_i = F_i \left(\frac{\partial H_i}{\partial P_{g_i}}\right)\bigg|_{avg}. \tag{14.20}$$

This is generally equivalent to using a one constant step unit incremental fuel cost, or unit full load fuel cost.

14.3.2.3 Type III: Incremental Fuel Cost with Start-Up Cost-Based List

In this case, the incremental fuel cost at the unit's point of maximum efficiency plus the ratio of the unit start-up cost to the expected energy produced by the unit producing maximum efficiency output for the minimum expected run time of the unit before cycling in hours is used as the priority order criterion

$$M_i = F_i \left(\frac{\partial H_i}{\partial P_{g_i}}\right)\bigg|_{\eta_{max}} + \left(\frac{S_i}{T_i^{min} P_{g_i}}\right)\bigg|_{\eta_{max}}, \tag{14.21}$$

where
H_i is the heat rate curve of the ith unit (MBtu/h)
F_i is the fuel cost of the ith unit ($/MBtu)
S_i is the start-up fuel cost of the ith unit
T_i^{min} is the minimum expected runtime of the ith unit before cycling (h)
$F_i(\partial H_i/\partial P_{g_i})|_{\eta_{max}}$ is the incremental fuel cost of the ith unit at maximum efficiency
$(S_i/(T_i^{min} P_{g_i}))|_{\eta_{max}}$ is the start-up cost component of the ith unit

The output power is computed at the maximum efficiency, η_{max}.

14.3.2.4 Type IV: Dynamic Priority Lists

In this approach, the ordering is based on economic dispatch results rather than the raw cost data. To obtain a dynamic priority list, one begins with load level equal to a preselected percentage of the total capacity of units to be ranked and obtains an economic dispatch solution including losses. Based on the optimum generation level for each unit, a measure of a unit's fuel cost is obtained as

$$\text{Index of fuel cost} = L_i \left[\frac{F_i(P_{g_i})}{P_{g_i}} \right], \qquad (14.22)$$

where L_i is the loss penalty factor of the ith unit. The highest cost unit is determined along with the optimal cost. The highest cost unit is removed and is placed in the ranking as the most expensive. The system load is reduced by the capacity of the unit removed and a second dispatch is calculated, and the highest cost unit for this load is removed, and included in the list as the second most expensive. The process is continued until the system load is at a level approximating base load. The remaining units are ranked using a static priority list [14,15].

14.3.3 Simple Merit-Order Scheme

Most priority list-based unit commitment schemes embody the following logic:

1. During each hour where the load is decreasing, determine whether shutting down the next unit on the priority list will result in sufficient generation to meet the load plus spinning-reserve requirements. If not, the unit commitment is not changed for the hour considered.

2. If the answer to (1) is yes, determine the number of hours, h before the unit will be needed again when the load increases to its present level. Determine whether h is greater than the minimum shutdown time for the unit. If not, the unit commitment is not changed for the hour considered.

3. If the answer to (2) is yes. Calculate the following:

    3.1 Sum of hourly production costs for the next h hours with the candidate unit up.

    3.2 Sum of hourly production costs for the next h hours with the candidate unit down plus the start-up cost for either cooling the unit or banking it.

4. Notably, if costs in 3.2 are less than those of 3.1, then the unit is shutdown, otherwise, the unit is kept on.

14.4   Illustrative Example

14.4.1   Lagrangian Relaxation Approach to Unit Commitment

The Lagrangian relaxation methodology has been demonstrated to have the capacity to handle systems consisting of hundreds of generating units effectively. The approach is claimed to be more efficient than other methods in solving large-scale problems and can handle various constraints more easily. The Lagrangian relaxation approaches acknowledge that the unit commitment problem consists of three ingredients.

1. Cost function, which is the sum of terms each of which, involves a single unit. The unit commitment problem seeks to minimize:

$$F = \sum_{t=1}^{T} \sum_{i=1}^{N} F_i(P_i(t), S_i(t), u_i(t)), \qquad (14.23)$$

where the individual terms are given by

$$F_i(P_i(t), S_i(t), u_i(t)) = C_i(P_{g_i}(T)) + u_i(t)[1 - u_i(t-1)]S_i(x_i(t)). \quad (14.24)$$

2. Set of coupling constraints (the generation and reserve requirements) involving all the units, one for each hour in the optimization period.

$$\sum_{i=1}^{N} R_{i,n}(P_i(t), u_i(t)) \geq P_{R,n}(t) \qquad (14.25)$$

for all $t \in \{1,T\}$ and requirements n.

The first requirement is the power balance constraint:

$$\sum_{t=1}^{N_{th}} R_{i,n}(P_i(t), u_i(t)), \qquad (14.26)$$

where $D(t)$ is the generation requirement at time t. As a result, in Equation 14.8, with $n=1$, we have

$$R_{i,1}(P_{g_i}(t), u_i(t)) = u_i(t)P_{g_i}(t) \qquad (14.27)$$

$$P_{R,2}(t) = P_D(t). \qquad (14.28)$$

The second requirement is the spinning-reserve requirement written as

$$\sum_{i=1}^{N} P_{g_i}^{max} u_i(t) \geq P_D(t) + P_R(t). \qquad (14.29)$$

In Equation 14.12, $P_R(t)$ is the MW spinning-reserve requirement during hour t. Therefore in Equation 14.8 for $n = 2$, we have

$$R_{i,2}\left(P_{g_i}(t),\ u_i(t)\right) = u_i(t)P_{g_i}^{max} \tag{14.30}$$

$$P_{R,2}(t) = P_D(t) + P_R(t). \tag{14.31}$$

3. Set of constraints involving a single unit:

$$L_i(P_{g_i}, u_i, t) \leq 0 \quad \text{for all } i \in \{1, N\}, \tag{14.32}$$

where

$$P_{g_i}(t) \in \{P_{g_i}(k) | k = 1, 2, \ldots, T\}$$
$$u_i(t) \in \{u_i(k) | k = 1, 2, \ldots, T\}.$$

The constraints of Equation 14.15 involve minimum up and down times, as well as unit loading constraints. In the nonlinear programming language, the problem posed is referred to as the primal problem.

The Lagrangian relaxation approach is based on Everett's work that showed that an approximate solution to the primal problem can be obtained by joining the coupling constraints to the cost function using the Lagrange multipliers λ_n to form the Lagrangian function.

$$L(P_{g_i}, u_i, t) = \sum_{t=1}^{T} \sum_{i=1}^{N} \left\{ L_i\left(P_{g_i}(t),\ u_i(t)\right) \right\}$$
$$- \sum_{t=1}^{T} \sum_{i=1}^{N} \left\{ \lambda_n(t) R_{i,n}\left(P_{g_i}(t),\ u_i(t)\right) \right\}. \tag{14.33}$$

The multipliers associated with the nth requirement for time t is denoted by $\lambda_n(t)$. In expanded form the Lagrangian function is

$$L(P_{g_i}, u_i, t) = \sum_{t=1}^{T} \sum_{i=1}^{N} \left\{ C_i\left(P_{g_i}(t)u_i(t)\right) + u_i(t)[1 - u_i(t-1)]S_i(x_i(t)) \right\}$$
$$+ \sum_{t=1}^{T} \left\{ \lambda_1(t) \left[P_D(t) - \sum_{i=1}^{N} P_{g_i}(t) \right] \right\}$$
$$+ \sum_{t=1}^{T} \left\{ \lambda_2(t) \left[P_D(t) + P_R(t) - \sum_{i=1}^{N} u_i(t)P_{g_i}^{max} \right] \right\}, \tag{14.34}$$

where
λ_1 is the Lagrangian multiplier without spinning-reserve constraints
λ_2 is the Lagrangian multiplier with spinning-reserve constraints

The resulting relaxed problem is to minimize the Lagrangian function, subject to $\lambda_1(t) \geq 0$ and $\lambda_2(t) \geq 0$. In addition, $P_{g_i}(t)$, $u_i(t)$, and $x_i(t)$ should satisfy Equation 14.15, which is given by

$$L(\lambda_1,\lambda_2,t) = \text{Min} \sum_{t=1}^{T} \sum_{i=1}^{N} \left\{ C_i\left(P_{g_i}(t)u_i(t)\right) + u_i(t)[1 - u_i(t-1)]S_i(x_i(t)) \right\}$$

$$+ \sum_{t=1}^{T} \left\{ \lambda_1(t) \left[P_D(t) - \sum_{i=1}^{N} P_{g_i}(t) \right] \right\}$$

$$+ \sum_{t=1}^{T} \left\{ \lambda_2(t) \left[P_D(t) + P_R(t) - \sum_{i=1}^{N} u_i(t)P_{g_i}^{\max} \right] \right\}. \tag{14.35}$$

The dual problem is

$$\text{Maximize } L_{\text{dual}}\left(\tilde{\lambda}(t)_1, \tilde{\lambda}_1(t)\right). \tag{14.36}$$

The formulation involves maximization of a minimum where the solution of the dual problem is an iterative process. For a fixed $\tilde{\lambda}_1$ and $\tilde{\lambda}_2$, Maximize $L_{\text{dual}}(\tilde{\lambda}(t)_1, \tilde{\lambda}_1(t))$ is determined by minimizing the right-hand side of Equation 14.15. The global procedure is represented by the update of the multipliers, and its objective is to maximize $L_{\text{dual}}(\tilde{\lambda}(t)_1, \tilde{\lambda}_1(t))$. Subsequent to finding the dual optimal a search for a feasible suboptimal solution is conducted. The flowchart of the Lagrangian relaxation method is shown in Figure 14.1.

Determining $L_{\text{dual}}(\tilde{\lambda}(t)_1, \tilde{\lambda}_1(t))$ is much simpler than the solution of the primal problem because of the following reasons:

1. Cost function can be written as a sum of terms each involving only one unit.
2. Coupling constraints between the units have already been relaxed.
3. Since each of the constraints involves one unit, the operation of each unit can be considered independently.

14.4.2 Single Unit Relaxed Problem

The minimization of the right-hand side of Equation 14.13 can be separated into subproblems, each of which deals with one generating unit only. Based on Equation 14.16, the single unit relaxed problem is stated as

$$\text{Minimize } L\left(P_{g_i}(t), u_i(t)\right) = \sum_{t=1}^{T} \left\{ F_i\left(P_{g_i}(t), u_i(t)\right) - \sum_{n=1}^{K} \left[\lambda_n(t)R_{i,n}\left(P_{g_i}(t), u_i(t)\right)\right] \right\}. \tag{14.37}$$

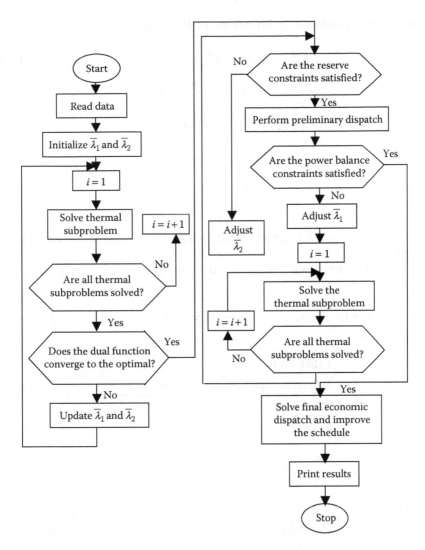

FIGURE 14.1
Flowchart of the Lagrangian relaxation algorithm for unit commitment.

Alternatively, based on Equation 14.14, the problem is restated as

$$
\text{Minimize } L\left(P_{g_i}(t), u_i(t)\right) = \sum_{t=1}^{T} \left\{ C_i\left(P_{g_i}(t)u_i(t)\right) + u_i(t)[1 - u_i(t-1)]S_i(x_i(t)) \right.
$$

$$
\left. + \lambda_1(t)P_{g_i}(t) + \lambda_2(t)P_{g_i}^{\max}(t) \right\} \tag{14.38}
$$

subject to the constraints of Equations 14.10, written as

$$P_{g_i}^{\min} \leq P_{g_i}(t) \leq P_{g_i}^{\max} \quad \text{if } u_i(t) = 0 \tag{14.39}$$

$$P_{g_i}(t) = 0 \quad \text{if } -x_i^{\text{down}} \leq x_i(t) \leq -1 \tag{14.40}$$

and the minimum up and down time constraints

$$u_i(t) = 1 \quad \text{if } 1 \leq x_i(t) \leq x_i^{\text{up}} \tag{14.41}$$

$$u_i(t) = 0 \quad \text{if } -x_i^{\text{down}} \leq x_i(t) \leq -1, \tag{14.42}$$

where

$x_i(t)$ is the cumulative uptime if $x_i(t) > 0$ and the cumulative downtime
   $x_i(t) > 0$
x_i^{up}, x_i^{down} is the minimum up and down times of the ith unit, respectively

Furthermore, $x_i(t)$ can be related to $u_i(t)$ by the following difference equations in Table 14.1.

TABLE 14.1

Difference Equations That Relate $x_i(t)$ to $u_i(t)$

Value of $x_i(t)$	Condition
$x_i(t) + 1$	$x_i(t) \geq 1$ and $u_i(t) = 1$
1	$x_i(t) \leq -1$ and $u_i(t) = 1$
$x_i(t) - 1$	$x_i(t) \leq -1$ and $u_i(t) = 0$
-1	$x_i(t) \geq 1$ and $u_i(t) = 0$

This problem can be solved easily by dynamic programming or any other method. The state variables are just $x_i(t)$. The number of required up states is x_i^{up} and the number of required down states is

$$\text{Max}(x_i^{\text{down}}, x_i^{\text{cool}}), \tag{14.43}$$

where x_i^{cool} is the time required for a unit to cool down completely so that the start-up cost is independent of downtime for downtimes greater than x_i^{cool}.

The operating limits and the minimum running and shutdown time constraints are treated implicitly by this method.

The dual problem is then decoupled into small subproblems, which are solved separately with the remaining constraints. Meanwhile, the dual function is maximized with respect to the Lagrangian multipliers, usually by a series of iterations based on the subgradient method.

14.4.3 Lagrangian Relaxation Procedure

Implementing the Lagrangian relaxation method involves the following key steps:

1. Finding the multipliers $\lambda_n(t)$ to obtain the solution to the relaxed problem near the optimum.
2. Estimating how close to the optimum is the solution obtained
3. Obtaining the actual solution to the relaxed problem

To start, Everett [7] shows that if the relaxed problem is solved with any sets of multipliers $\lambda_n(t)$ and the resulting value of the right-hand side of Equation 14.8 is $R_n^*(t)$. That is

$$\sum_{i=1}^{N} R_{i,n}\left(P_{g_i}^*(t),\ u_i^*(t)\right) = R_n^*(t), \tag{14.44}$$

where $P_{g_i}^*(t)$, $u_i^*(t)$ is the optimal solution.

This implies that $P_{g_i}^*(t)$ and $u_i^*(t)$ yields the optimum solution of the original problem with $P_{R,n}(t)$ replaced by $R_n^*(t)$.

The optimization requirement is met if $\lambda_n(t)$ can be found such that the resulting $R_n^*(t)$ are equal to $P_{R,n}(t)$. Unfortunately, this cannot always be done. As a result there will be a difference between the cost obtained by solving the relaxed problem or dual and the optimum cost for the original problem. This difference is referred to as the duality gap, and can be explained graphically as shown in Figure 14.2. The lower curve is a plot of $L_{\text{dual}}(\tilde{\lambda}_1,\tilde{\lambda}_2)$, which has been defined by Equation 14.15 as

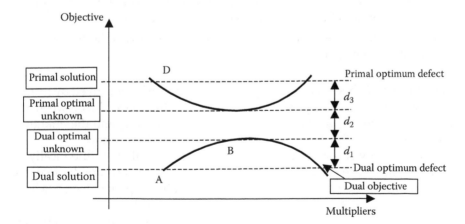

FIGURE 14.2
Duality gap of a relaxed problem.

$$L_{\text{dual}}(\tilde{\lambda}_1\tilde{\lambda}_2) = \text{Min} \sum_{t=1}^{T} \sum_{i=1}^{N} \left[C_i\left(P_{g_i}(t)u_i(t)\right) + u_i(t)[1 - u_i(t-1)]S_i(x_i(t)) \right]$$

$$+ \sum_{t=1}^{T} \left\{ \lambda_1(t) \left[P_D(t) - \sum_{i=1}^{N} P_{g_i}(t) \right] \right\}$$

$$+ \sum_{t=1}^{T} \left\{ \lambda_2(t) \left[P_D(t) + P_R(t) - \sum_{i=1}^{N} u_i(t)P_{g_i}^{\text{max}} \right] \right\}. \tag{14.45}$$

The minimization is with respect to P_i and u_i, and the plot corresponds to various values of the multipliers $\tilde{\lambda}_1$ and $\tilde{\lambda}_2$. The following points are identified in the graph:

- Point A is a known solution to the dual optimum (through the iterations).
- Point B is the unknown optimal solution to the dual problem.

The difference d_1, between the value of $L_{\text{dual}}(\tilde{\lambda}_1, \tilde{\lambda}_2)$ at Point A and the dual optimum at Point B, is a defect that can be improved upon by optimizing the dual problem. The upper curve corresponds to the objective of the primal problem defined by Equation 14.6 as

$$F = \sum_{t=1}^{T} \sum_{i=1}^{N} F_i\left(P_{g_i}(t), u_i(t)\right). \tag{14.46}$$

The points identified on the curve are

- Point C is the unknown optimal solution to the primal problem, and corresponds to the minimum cost for a feasible solution.
- Point D corresponds to the value of primal cost corresponding to the dual solution of Point A.

The difference d_3, between C and D, is a defect that can be improved by further optimizing the dual or primal problems. The difference d_2, between the unknown optimum value of the primal problem (Point C) and the unknown optimum value of the dual problem (Point B), is the duality gap.

Duality theory shows that for nonconvex problems there will typically be a duality gap. Since the commitment decision variables $x_i(t)$ are discrete, the unit commitment problem is nonconvex. The duality gap, however, has been shown to go to zero as the problem size gets bigger [2].

Duality theory also generates guidelines on how to update the multipliers $\lambda_n(t)$ so that the solution to the relaxed problem is near the optimum solution. Let $L_{\mathrm{dual}}(\lambda_n(t))$ be the value of the Lagrangian at the solution to the relaxed problem, then good values of the multipliers $\lambda_n(t)$ can be obtained by maximizing $L_{\mathrm{dual}}(\lambda_n(t))$ for all positive $\lambda_n(t)$.

A number of approaches have been used to maximize the dual function $L_{\mathrm{dual}}(\lambda_n(t))$. The first, and most popular, involves using the subgradient method. This is a generalization of the gradient or steepest descent method, for nondifferentiable functions. In general, the Lagrange function L_{dual} $(\lambda_n(t))$ is nondifferentiable. The subgradient of $L_{\mathrm{dual}}(\lambda_n(t))$ with respect to one of the multipliers $\lambda_n(t)$ is given as

$$\frac{\partial J_{\mathrm{dual}}(\lambda_n(t))}{\partial \lambda_n(t)} = P_{\mathrm{R},n}(t) - \sum_{i=1}^{N} R_{i,n}\left(P_{\mathrm{g}_i}^*(t),\, u_i^*(t)\right), \tag{14.47}$$

where $P_{\mathrm{g}_i}^*(t)$ and $u_i^*(t)$ are the solution to the relaxed problem with multipliers $\lambda_n(t)$. That is, the derivative of the Lagrangian corresponding to a change in $\lambda_n(t)$ is equal to the difference between the requirement and the value of the left-hand side of the constraint evaluated at the solution of the relaxed problem. The subgradient method to update $\lambda_n(t)$ is

$$\lambda_n^{k+1}(t) = \mathrm{Max}\left[0,\, \lambda_n^k(t) + t_k\left\{P_{\mathrm{R},n}(t) - \sum_{i=1}^{N} R_{i,n}\left(P_{\mathrm{g}_i}^*(t),\, u_i^*(t)\right)\right\}\right], \tag{14.48}$$

where
$\lambda_n^k(t)$ is the kth update of $\lambda_n(t)$
t^k is a scalar step length

A number of forms of t^k could be used as long as t^k as $k \to \infty$ and

$$\sum_{k=1}^{\infty} t^k \to \infty. \tag{14.49}$$

Many authors have used the following form $t^k = 1/c + dk$ where c and d are constants. Different constants would be given for the different requirements as long as the conditions on t^k are met. Fisher [3] recommends the following form:

$$t^k = \frac{\mu_n^k\left(L_{\mathrm{dual}}\left(\lambda_n^k\right) - L^*\right)}{\left(P_{\mathrm{R},n}(t) - \sum_{i=1}^{N} R_{i,n}(P_{\mathrm{g}_i}^*(t),\, u_i^*(t))\right)^2} \tag{14.50}$$

Lauer et al. [13] use a different approach to minimizing the dual functions that uses second derivative information.

Different authors use variations of the above Lagrangian methods to ensure that the generation and reserve requirements are met and that the algorithms converge to near optimal solutions. These variations include replacing the spinning and supplemental reserve requirements by constraints that just involve the units' upper limits, and using modifications to the subgradient formulas.

14.4.4　Searching for a Feasible Solution

The problem considered is very sensitive to changes of multipliers. Therefore it is important to start the search for a feasible solution at a point that is fairly close to the dual optimal.

The decoupled subproblems of the dual problem interact through the two Lagrangian multipliers. These multipliers are interpreted as the prices per unit power generation and spinning reserve, respectively, that the system is willing to pay to preserve the power balance and fulfill the spinning-reserve requirement during each hour. Increasing $\lambda_1(t)$ and $\lambda_2(t)$ may lead to the commitment of more generating units and an increase in the total generation and spinning-reserve contribution during hour t. The reverse effect is obtained by decreasing the two multipliers.

The feasible search is based on the above relationship between the unit commitment and the Lagrangian multipliers. The values of $\tilde{\lambda}_1$ and $\tilde{\lambda}_2$ are adjusted repeatedly, based on the amount of violation of the relaxed constraints (the power balance and the spinning-reserve constraints). Therefore the subproblems are solved after each adjustment and the iterations continue until a feasible suboptimal solution is located.

The commitment schedule is very sensitive to the variation of the multipliers. For example, if a system contains several units whose cost characteristics are nearly identical, a minor modification of the Lagrangian multipliers during a particular hour may turn all of these units on or off provided that the original values of the multipliers during that hour are close to the incremental costs of generations and spinning-reserve contributions of these units. Thus the modification of the multipliers should be determined in an appropriate manner; otherwise, the number of committed units during some periods may be more than required. Note that the commitment of a generating unit depends on the values of the multipliers and its commitment states during preceding hours due to the minimum running and shutdown time constraints. This dependence complicates determining the appropriate values of the multipliers.

A simple algorithm is presented in Ref. [5] to find a feasible solution with an additional set of restrictions designed to limit unnecessary commitment of the generating units. An economic dispatch algorithm is then applied to this commitment schedule to find the exact power generation of each generating unit and improve the generating schedule. The flowchart of the searching algorithm, which is implemented to find the feasible solution, is shown in Figure 14.3.

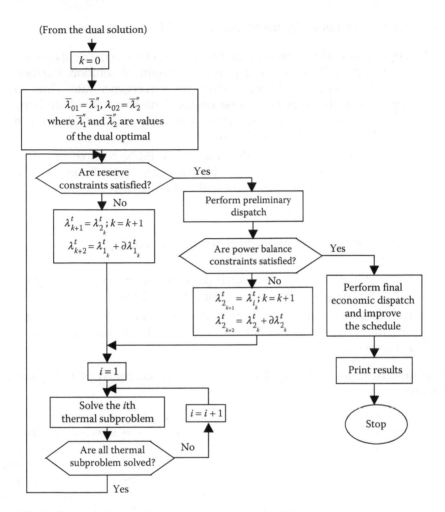

FIGURE 14.3
Flowchart to find feasible solution.

Considering the MW spinning-reserve constraints, the inequality of Equation 14.12 does not provide any upper-bound restrictions. Common-sense for an economic schedule requires that there should not be too much excess MW reserve. As a result, in the searching algorithms, the following constraints are included implicitly to test the validity of the commitment schedule.

$$\sum_{i=1}^{N} u_i(t)P_{g_i}^{max} \leq P_D(t) + P_R(t) + P_E(t). \tag{14.51}$$

Selecting $P_E(t)$ is guided by the following considerations:

1. Upper bounds limit the solution space to be close to the dual optimal point. It may, however, lead to miss the optimal solution. Furthermore, the value of $P_E(t)$ may affect the convergence rate. Thus an appropriate choice of $P_E(t)$ is necessary. Unfortunately, there is no rigorous basis for selecting $P_E(t)$. In the algorithm of Ref. [5], a heuristic procedure is introduced.

2. Criterion that the excess in MW reserve should be reduced to a minimum is not implemented in Ref. [5]. This criterion indicates that the shutdown of any committed unit will violate the reserve constraints.

The weak points of this criterion are

1. Minimum reserve margin does not always correspond to the best dispatch.

2. Computation time for the search process will be increased significantly since the feasible region is tightly restricted.

3. It is possible that none of the solutions can satisfy these tight bounds and the minimum running/shutdown time constraints simultaneously.

The procedure for assigning the values to $E(t)$ suggested in Ref. [5] is as follows:

Step 1: For every hour t, $P_E(t)$ is first set to the sum of the maximum powers of the two most inefficient committed units. This value is selected because the commitment states of these two units are usually subject to the modification of the multipliers during the searching process. It is not a constant and depends on units, which have been committed according to the schedule being examined. In the search process, this bound actually discards those schedules, which allows more than two units to be shutdown during hour t without violating the spinning-reserve constraints.

Step 2: If a feasible solution cannot be found within a reasonable number of iterations, this may be due to the bounds during some hours being too tight. Thus no schedule can satisfy the minimum running/shutdown time and the bounds simultaneously. As a result, for those hours during which no combination of committed units have been found to satisfy Equations 14.12 and 14.32, the values of $P_E(t)$ are increased during these hours successively until a feasible solution is found.

Step 3: The procedure proposed in Ref. [5] is claimed to usually give a satisfactory feasible solution before proceeding to the final refinement of the schedule. Even though it cannot guarantee the true optimal solution, the extent of suboptimality of the solution can be estimated. Since the dual optimal solution is a lower bound of the original commitment problem, the relative difference between the cost of the suboptimal schedule and this bound may determine the quality of the solution.

14.5   Updating $\lambda_n(t)$ in the Unit Commitment Problem

There exist many techniques for updating "lambda" in the searching algorithms of the unit commitment problem. Table 14.2 summarizes two key contributors to this topic and the merits and drawbacks to the proposed method.

TABLE 14.2

Comparison of Selected Techniques for Updating $\lambda_n(t)$ in Solving the UC Problem

Technique	Merits	Limitations
Case A: Tong and Shahidepour [20]	Overcome the approach of heuristic methods	Linear interpolation has no state memory
	Lagrangian relaxation approach is applied to determine a feasible suboptimal schedule	May require additional computations of initial increments for lambda (hence slow convergence)
	Linear interpolation for lambda adjustments can be enhanced with via a bisection method	Discrete controls may cause oscillatory behavior around the feasible points
Case B: Merlin and Sandrin [15]	Overcome the approach of heuristic methods	Direction scaling for improved convergence does not consider relaxation of discretized generation levels and the ramp rate constraints
	Improved or modified subgradient method for lambda iteration	
	Improved convergence via adjustment of tolerance relative to the smallest unit	

14.5.1   Case A: Updating $\lambda_n(t)$

In the updating process of the searching algorithm, values of $\partial\lambda_{1,k}(t)$ and $\partial\lambda_{2,k}(t)$ should be determined. These two values are set to zero if the power balance constraint and the reserve constraint are satisfied, respectively, during hour t. However, if either of these two constraints is violated during hour t, the following two methods are applied to determine the unknowns:

1. Linear interpolation
   Define

$$G_k(t) = \left(\sum_{i=1}^{N} P_{g_i}(t) \right)\Bigg|_{\tilde{\lambda}_1=\tilde{\lambda}_{1,k};\tilde{\lambda}_2=\tilde{\lambda}_{2,k}} \qquad (14.52)$$

then

$$\partial\lambda_{1,k}(t) = \frac{P_D(t) - G_k(t)}{G_k(t) - G_{k-1}(t)} (\lambda_{1,k}(t) - \lambda_{1,k-1}(t)). \qquad (14.53)$$

Define

$$H_k(t) = \left(\sum_{i=1}^{N} u_i(t) \right) \Bigg|_{\tilde{\lambda}_1 = \tilde{\lambda}_{1,k};\ \tilde{\lambda}_2 = \tilde{\lambda}_{2,k}} \tag{14.54}$$

then

$$\partial\lambda_{2,k}(t) = \frac{P_D(t) + P_R(t) - H_k(t)}{H_k(t) - H_{k-1}(t)} (\lambda_{2,k}(t) - \lambda_{2,k-1}(t)). \tag{14.55}$$

2. Bisection method

$$\partial\lambda_{1,k}(t) = \left[\frac{P_D(t) - G^{\min}(t)}{G^{\max}(t) - G^{\min}(t)} (\lambda_1^{\max}(t) - \lambda_1^{\min}(t)) \right],$$
$$- (\lambda_{1,k}(t) - \lambda_1^{\min}(t)) \tag{14.56}$$

where

$$G^{\max}(t) = \left(\sum_{i=1}^{N} P_{g_i}(t) \right) \Bigg|_{\lambda_1(t) = \lambda_1^{\max}(t)} \tag{14.57}$$

with

$$P_D < G^{\max}(t) \tag{14.58}$$

and

$$G^{\min}(t) = \left(\sum_{i=1}^{N} P_{g_i}(t) \right) \Bigg|_{\lambda_1(t) = \lambda_1^{\min}(t)} \tag{14.59}$$

with

$$G^{\min}(t) > P_D. \tag{14.60}$$

The bounds are adjusted after the calculation of the total generation using the updated multipliers using the following rules:

IF $G_{k+1}(t) > P_D(t)$, THEN

$(\lambda_1^{\max}(t), G^{\max}(t))$ be replaced by $(\lambda_{1,k+1}(t), G_{k+1}(t))$;

OTHERWISE

$(\lambda_1^{\min}(t),\ G^{\min}(t))$ be replaced by $(\lambda_{1,k+1}(t),\ G_{k+1}(t))$;

The determination of $\partial\lambda_{2,k}(t)$ is based on a similar approach.

These methods do not work satisfactorily if they are implemented independently. The first method usually branches back and forth around a feasible solution because the relationship between the generation and the multipliers is stepwise. The second method is difficult to apply since the generation during each hour is not merely determined by its corresponding multipliers. The generation during a particular hour (t) is a function of all the multipliers, though the multipliers of that specific hour $(\partial\lambda_1(t)$ and $\partial\lambda_2(t))$ may present the dominant effect. In Ref. [5], these two methods are used together. The linear interpolation provides the first few guesses, then the bisection method is applied. Sometimes the linear interpolation may be recalled if it is found that the feasible solution at a particular hour does not fall within the bounds, based on the change of multipliers during other time intervals. Using this approach, a feasible solution will be obtained within a reasonable computation time.

14.5.2 Case B: Updating $\lambda_n(t)$

Updating the multipliers using Equation 14.24 caused slow convergence [15]. The method adopts the following strategy assuming that the current solution on the kth iteration is $\tilde{P}_i^k(t)$, and $u_i^k(t)$:
Calculating the updates for $\lambda_1(t)$:

Two variations are considered:

1. If the reserve constraints are not met

$$\sum_{i=1}^{N} u_i^k(t)\tilde{P}_i^k(t) = \tilde{P}_D^k(t) \neq P_D(t) \qquad (14.61)$$

   with

$$\lambda_n^{(k+1)}(t) = \text{Max}\left(0,\ \lambda_n^{(k)}(t) + t_k\left(P_D(t) - P_D^{(k)}(t)\right)\right), \qquad (14.62)$$

   where
   $\lambda_n^{(k)}(t)$ is the kth update of $\lambda_n(t)$
   t^k is the scalar step length defined by $t^k = (1/(c + dk))$ with constants c and d

2. If the reserve constraint is met. This means that units designated as online have been loaded to their maximum capacity in an ascending order of incremental cost (priority order). In this case the incremental

cost of the last loaded unit is denoted by $\beta_k(t)$, and is used in the following updating formula:

$$\lambda_{i,k+1}(t) = \lambda_{i,k}(t) + (1 - \alpha_x)\beta_k(t), \tag{14.63}$$

where α_r is a relaxation constant taken as 0.6.

Calculating the updates for $\lambda_1(t)$ updating $\lambda_2(t)$:
  Using Equation 14.32, we obtain

$$\sum_{i=1}^{N} u_i(t)P_{g_i}^{max} \leq P_D(t) + P_R(t) + P_E^{(k)}(t). \tag{14.64}$$

Initially, one set $P_E^{(0)}(t) = 0$, subsequent selection of $P_E^{(k)}(t)$ is done as follows:

1. If the reserve constraints are met within the tolerance of $P_E^{(k)}(t)$

$$P_D(t) + P_R(t) \leq \sum_{i=1}^{N} u_i(t)P_{g_i}^{max} \leq P_D(t) + P_R(t) + P_E^{(k)}(t). \tag{14.65}$$

The multiplier and tolerance are left unchanged.

$$\lambda_{2,k+1}(t) = \lambda_{2,k}(t). \tag{14.66}$$

2. If the reserve constraint is not met within the tolerance of $P_E^{(k)}(t)$.
The modification to the multiplier given by Equation 14.62 as

$$\lambda_n^{(k+1)}(t) = \text{Max}\left[O, \lambda_n^k(t) + t_k\left(P_D(t) - P_D^k(t)\right) \right.$$
$$\left. - \sum_{i=1}^{N} \left(P_{g_i}^{max}u_i^{(k)}(t) + P_E^{(k)}(t)\right) \right], \tag{14.67}$$

where t^k is a scalar step length, with

$$t^k = \frac{1}{c' + d'k}, \tag{14.68}$$

where c' and d' are the constants.

  The updating of the tolerance term is done under two considerations:

1. If $\displaystyle\sum_{i=1}^{N} u_i^{(k)}(t)P_{g_i}^{max} \geq P_D(t) + P_R(t) + P_E^{(k)}(t). \tag{14.69}$

The tolerance is left unchanged and $P_E^{(k+1)}(t) = P_E^{(k)}(t)$.

$$2. \text{ If } \sum_{i=1}^{N} u_i^{(k)}(t) P_{g_i}^{\max} < P_D(t) + P_R(t) + P_E^{(k)}(t). \qquad (14.70)$$

The tolerance is increased and $P_E^{(k+1)}(t) = P_E^{(k)}(t) + \varepsilon$, where ε is a constant whose value is equal to the maximum power of the smallest unit of the system.

14.6 Unit Commitment of Thermal Units Using Dynamic Programming

For small and medium size systems dynamic programming was proposed as a solution technique. Dynamic programming has many advantages; the chief advantage being a reduction in the dimensionality of the problems. Suppose we have four units on a system and any combination of them could serve the single load. These would be a maximum number $(2^4 - 1 = 15)$ of combinations to test.

However, if a strict priority is imposed, there are only four combinations to try as follows:

Priority 1
Unit

Priority 1 unit + priority 2 unit
Priority 1 unit + priority 2 unit + priority 3 unit
Priority 1 unit + priority 2 unit + priority 3 unit + priority 4 unit

The imposition of a priority list arranged in the order of the full-load average cost rate would result in a critical dispatch and commitment only if

1. No load costs are zero.
2. Unit input–output characteristics are linear between zero and full load.
3. There are no other restrictions.
4. Start-up costs are a fixed amount.

Assumptions
The main assumptions for applying dynamic programming to unit commitment problem are

1. A state consists of an array of units with specified units operating and the rest off-line.
2. There are no costs for shifting down a unit.

3. There is a strict priority order, and in each interval a specified minimum amount of capacity must be operating.

4. Start-up cost of a unit is independent of the time it has been off line.

14.6.1 Dynamic Programming Approaches to Unit Commitment Problem

14.6.1.1 Backward Dynamic Programming Approach

The first dynamic programming approach uses a backward (in time) approach in which the solution starts at the last interval and approaches back to the initial points [17].

There are l_{max} intervals in the period to be considered. The dynamic programming equations for the computations of the minimum total fuel cost during a time (or load) period l are given by the recursive equation:

$$F(L,K) = \underset{\{\Phi\}}{\text{Min}} \left(C^{min}(l,K) + S(l, K:\Phi, K+1) + F^{min}(\Phi, K+1) \right), \qquad (14.71)$$

where
    $F^{min}(l,k)$ is the minimum total fuel cost from state k in interval l to the last interval l_{max}
    $C^{min}(l,k)$ is the minimum generation cost in supplying the load interval l given state k
    $S(l, k:\Phi, k+1)$ is the incremental start-up cost going from state k in the lth interval to state Φ in the $(l+1)$th interval
    $\{\Phi\}$ is the set feasible states in the interval $(k+1)$

The production cost $C(l,k)$ is obtained by economically dispatching the units on line in state k. A path is a schedule starting from a state in an interval l to a final interval l_{max}. An optimal path is one for which the total fuel cost is minimum.

14.6.1.2 Forward Dynamic Programming Approach

The backward dynamic programming does not cover many practical situations. For example if the start-up cost of a unit is a function of the time it has been off line then forward dynamic programming approach is more suitable since the previous history of the unit can be computed at each stage.

There are other practical reasons for going forward. The initial conditions are easily specified and the computations can go forward in time as long as required and as long as computer storage is available.

The solution to the forward dynamic programming is done through the recursive equation given by Pang et al. [5], and was improved by Synder and Power [7]. The recursive formula to compute the minimum cost during interval l with combination k is given by

$$F(L,K) = \underset{\{L\}}{\text{Min}} \left(C^{min}(l,K) + S(l-1, L:\Phi, l, K) + F^{min}(l-1, K) \right), \qquad (14.72)$$

where
    $F^{min}(l,k)$ is the minimum total fuel cost to arrive at state (l,k)
    $C^{min}(l,k)$ is the minimum production cost for state (l,k)
    $S(l-1,k:l,k)$ is the transition cost from state $(l-1, 1)$ to state (l,k)
    (l,k) is the kth combination in interval l

For the forward dynamic programming approach a strategy is defined as the transition, or path, from one state at given hour to a state at the next hour.
    Now let X be the number of states to search each period.
    N be the number of strategies, or path, to save at each step.
    Figure 14.4 clarifies this definition and Figure 14.5 shows the application of the recursive formula for forward dynamic programming approach.

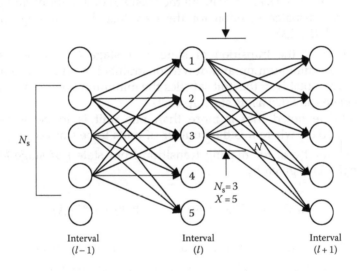

FIGURE 14.4
States and strategies ($N_s = 3$ and $X = 5$).

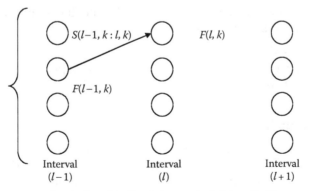

FIGURE 14.5
Application of forward dynamic programming technique.

The forward dynamic programming approach can be summarized in the following steps:

Step 1: Start with the first interval $l = 1$, enumerate all feasible combinations that satisfy

a. Expected load
b. Specified amount of spinning reserve usually 25% of the load at that time interval

In this regard, the economic dispatch problem will be performed to calculate the value of $C^{min}(l,k)$ for each feasible kth combination at stage l.

Step 2: For stage $(l + 1)$, enumerate all the feasible combinations and perform the economic dispatch solution for the new load level at stage $(l + 1)$ to calculate $C^{min}(l + 1,k)$.

Step 3: Check if the transition of state k at stage l to state j at stage $(l + 1)$ satisfies minimum up and down constraints if a unit to be started up or shutdown. Notably, if the unit satisfies the minimum downtime constraint then calculate the $S(l,k)$ as its start-up cost.

It should be noted that if more than one unit is to be started then $S(l,k) = \sum_{i=1}^{n} S_i(t)$, where n is the number of units to be started up.

Step 4: The total cost for making transition from state j at stage l to state j at stage $(l + 1)$ is given by

$$F(l,k) = C^{min}(l,k) + S(l - 1, k : \Phi, k) + F^{min}(l,k). \tag{14.73}$$

Step 5: Calculate all the $F(l,k)$ due to all feasible transition from stage $(l - 1)$.

Step 6: Find the minimum $F^*_{cost}(K,J)$ and save it, also save the path that leads to this optimal one.

Step 7: Proceed in time to the next stage and repeat steps 2–6.

Step 8: When you reach the last stage calculate the minimum total cost and trace back to find the optimal solution.

The flowchart is shown in Figure 14.6.

Figures 14.7 shows a case using forward dynamic programming for three combinations at each stage, for three time intervals.

14.6.2 Case Study

For this case study, two cases will be considered. The first is a priority list schedule; the second is the same case with complete enumeration. Both cases ignore hot-start costs and minimum up and down times.

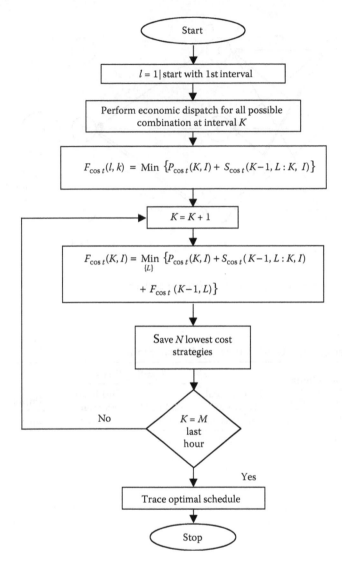

FIGURE 14.6
Flowchart for forward dynamic programming.

In order to make the required computations more efficient, a simplified model of the unit characteristics is used. Four units are to be committed to serve an 8 h pattern for the expected loads. Tables 14.3 through 14.5 show the characteristics, load pattern, and initial status for the case study.

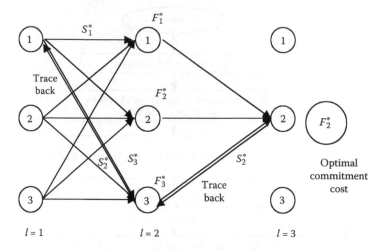

FIGURE 14.7
Flowchart of the forward dynamic programming problem.

TABLE 14.3

Unit Characteristics, Load Pattern, and Initial Status

| | Unit Limits | | | Loading Costs | | | |
| | $P_{g_i}^{max}$ | $P_{g_i}^{min}$ | Incremental Heat Rate | No Load[a] | Full Load Average | Minimum Time (h) | |
Unit	(MW)	(MW)	(Btu/kW h)	($/h)	($/MW h)	Up	Down
1	80	25	10,440	213.00	23.54	4	2
2	250	60	9,000	585.62	20.34	5	3
3	300	75	8,730	684.74	19.74	5	4
4	60	20	11,900	252.00	28.00	1	1

[a] A unit is not allowed to operate at zero output; that is, if a unit is online, it must be loaded between its minimum and maximum. If it is off-line, it must have zero output and its operating cost will be 0 $/h. Fuel costs are 2 $/MBtu.

TABLE 14.4

Unit Characteristics, Load Pattern, and Initial Status

| | Initial Conditions | Start-Up Costs | | |
Unit	(−, Off-line; +, Online)	Hot	Cold	Cold Start
1	−5	150	350	4
2	8	170	400	5
3	8	500	1100	5
4	−6	0	0.02	0

TABLE 14.5

Load Pattern

Hour (h)	Load (MW)
1	450
2	530
3	600
4	540
5	400
6	280
7	290
8	500

Case 1

In case the units are scheduled according to a strict priority order. That is, units are committed in order until the load is satisfied. The total cost for the interval is the sum of the eight dispatch costs plus the transitional costs for starting any units. A maximum of 24 dispatches must be considered as shown in Table 14.6.

TABLE 14.6

Capacity Ordering of the Units

State	Unit Combination: 1 2 3 4 Units	Maximum Capacity (MW)
15	1 1 1 1	690
14	1 1 1 0	630
13	0 1 1 1	610
12	0 1 1 0	550
11	1 0 1 1	440
10	1 1 0 1	390
9	1 0 1 0	380
8	0 0 1 1	360
7	1 1 0 0	330
6	0 1 0 1	310
5	0 0 1 0	300
4	0 1 0 0	250
3	1 0 0 1	140
2	1 0 0 0	80
1	0 0 0 1	60
0	0 0 0 0	0

For case 1 the only states examined at each hour consists of capacity ordering of the units as shown in Table 14.7

TABLE 14.7

Capacity Ordering of the Units for States 5, 12, 14, 15

State	Unit Combination 1 2 3 4 Units	Maximum Capacity (MW)
5	0 0 1 0	300
12	0 1 1 0	550
14	1 1 1 0	630
15	1 1 1 1	690

Note that this is the priority order, that is

State 5 = Unit 3
State 12 = Unit 3 + unit 2, state 14 = unit 3 + unit 2 + unit 1
State 15 = Unit 3 + unit 2 + unit 1 + unit 4

For the first 4 h, only the last three states are of interest. The sample calculations illustrate the technique. All possible commitments start from state 12 since this was given as the initial condition. For hour one the minimum cost is state 12 and so on. The results for the priority-ordered case are as follow in Table 14.8:

TABLE 14.8

Capacity Ordering of the Units for States 1–4

Hour	State with Minimum Total Cost	Pointer for Previous Hour
1	12 (9,208)	12
2	12 (19,857)	12
3	14 (32,472)	12
4	12 (43,300)	14

Note that state 13 is not reachable in this strict priority ordering.

$$F(l, k) = \min_{\{L\}} \left(C^{\min}(l,k) + S(l - 1, k : l, k) + F^{\min}(l - 1, k) \right). \tag{14.74}$$

The allowable states are

{0010, 0110, 1110, 1111} = {5, 12, 14, 15}
in hour 0{l} = {12} initial condition

$l=1$: First hour

k

15 $F(1,15) = P(1,15) + S(0,12; 1,15)$
$= 9861 + 350 = 10211$

14 $F(1,14) = 9493 + 350 = 9843$

12 $F(1,12) = 9208 + 0 = 9208$

$l=2$: Second hour

The feasible states are $\Phi = \{12, 14, 15\}$.

Therefore, $N_k = 3$. Suppose two strategies are saved at each stage, then N_s is 2.

$\{1, N_s\} = \{12, 14\}$

k

15 $F(2,15) = \{P(2,15) + S(1, k; 2, 15) + F(1,k)\}$

$$= 11301 + \min \left\{ \begin{array}{c} 350 + 9{,}208 \\ 0 + 9{,}843 \end{array} \right\}$$

$$= 20860.$$

Also, the process is repeated until the fourth hour where the final cost at $l=4$ will be the minimum commitment cost. The complete schedule is obtained by retracing the steps over the specified time period 4 h.

14.7 Illustrative Problems

PROBLEM 14.7.1

A system has four units, the system data are given in Table 14.9. Solve unit commitment using

    a. Priority order method

    b. Dynamic programming

    c. Lagrange relaxation method

TABLE 14.9

System Data for Four (4) Units Dispatch

Unit	Maximum (MW)	Minimum (MW)	Incremental Heat Rate (Btu/kW h)	No-Load Cost ($/h)	Full-Load Average Cost ($/MW h)	Minimum Times (h) Up	Minimum Times (h) Down
1	80	25	10,440	213.00	23.54	4	2
2	250	60	9,000	585.62	20.34	5	3
3	300	75	8,730	684.74	19.74	5	4
4	60	20	11,900	252.00	28.00	1	1

(continued)

TABLE 14.9 (continued)

Initial Conditions		Start-Up Costs		
Units	Hours Off-Line (−) or Online (+)	Hot ($)	Cold ($)	Cold Start (h)
1	−5	150	350	4
2	8	170	400	5
3	8	500	1,100	5
4	−6	0	0.02	0

Load Pattern	
Hour	Load (MW)
1–6	450
7–12	530
13–18	600
19–24	540

PROBLEM 14.7.2

Given the unit data in Tables 14.10 to 14.13, use forward dynamic programming to find the optimum unit commitment schedules covering the 8 h period.

TABLE 14.10

Unit Limits and Heat Rates for the UC Problem

Unit	Maximum (MW)	Minimum (MW)	Incremental Heat Rate (Btu/kW h)	No-Load Energy Input (M Btu/h)	Start-Up Energy (M Btu)
1	500	70	9,950	300	800
2	250	40	10,200	210	380
3	150	30	11,000	120	110
4	150	30	11,000	120	110

Note: Load data (all time periods = 2 h).

TABLE 14.11

Hourly Load Curve Data

Time Period	Load (MW)
1	600
2	800
3	700
4	950

TABLE 14.12

Start-Up and Shutdown Rules

Units	Minimum Uptime (h)	Minimum Downtime (h)
1	2	2
2	2	2
3	2	4
4	2	4

Note: Fuel cost = 1.00 $/MBtu.

TABLE 14.13

Unit Combinations and Operating Costs for Different Road Levels

Combination	Unit 1	Unit 2	Unit 3	Unit 4	Operating Cost Load 600 MW	Operating Cost Load 700 MW	Operating Cost Load 800 MW	Operating Cost Load 950 MW
A	1	1	0	0	6,505	7,525	X	X
B	1	1	1	0	6,649	7,669	8,705	X
C	1	1	1	1	6,793	7,813	8,833	10,475

Note: 1, up; 0, down.

14.8 Conclusions

This chapter presented unit commitment as an operation scheduling function for management of generation resources for a short-time horizon of 1 day or at most 1 week. Different unit commitment operational constraints were fully addressed and discussed. Different approaches for solving the unit commitment problem were presented starting from the oldest and the most primitive method, priority list method. An illustrative example is presented. A practical approach that is suitable for large-scale power system employs Lagrangian relaxation technique is fully discussed. Different major procedures in problem formulation, search for a feasible solution through the minimization of the duality gap, updating the multiplier, and formation of single unit relaxed problem are shown. Also, several algorithms that employ the same approach were discussed. An approach that is suitable for small and medium power system employing dynamic programming is also discussed. The different assumptions for applying dynamic programming to unit commitment were fully discussed. A comparison between forward and backward dynamic programming approaches was made. The computational procedures involved in applying both approaches to unit commitment were shown. The forward dynamic approach was utilized in solving the problem.

An illustrative example is presented showing the different computational procedures in solving the unit commitment problem for a sample power system.

14.9 Problems

PROBLEM 14.9.1

A system has four units, the system data are given in Table 14.14. Solve unit commitment using

    a. Priority order method

    b. Dynamic programming

    c. Lagrange relaxation method

    d. Genetic algorithm

TABLE 14.14

Unit Limits, Heat Rates, Operating Conditions, and Hourly Loading

Unit	Maximum (MW)	Minimum (MW)	Incremental Heat Rate (Btu/MW h)	No-Load Cost ($/h)	Full-Load Average Cost ($/MW h)	Up	Down
1	100	25	9,800	200	30	5	1
2	200	50	8,000	600	25	5	3
3	250	60	8,500	550	20	6	4
4	80	30	10,000	250	24	3	2

	Initial Conditions	Start-Up Costs		
Units	Hours Off-Line (−) or Online (+)	Hot ($)	Cold ($)	Cold Start (h)
1	−5	200	400	4
2	8	220	450	5
3	8	350	800	5
4	−6	0	0.0	0

Load Pattern	
Hour	Load (MW)
1	500
2	550
3	600
4	560
5	450
6	300
7	350
8	500

PROBLEM 14.9.2

Table 14.15 presents the unit characteristics and load pattern for a five-unit-four time period problem. Each time period is 2 h long as shown in the load pattern data and operating conditions shown in Table 14.16. The input–output characteristics are approximated by a straight line from min to max generation so that the incremental heat rate is constant. Unit no-load and start-up cost are given in terms of heat energy requirements.

TABLE 14.15

Unit Characteristic Data for the UC Problem

Unit	Maximum (MW)	Net Full-Load Heat Rate (Btu/kW h)	Incremental Heat Rate (Btu/kW h)	Minimum (MW)	No-Load Cost ($/h)	Start-Up Cost ($/MBtu)	Minimum Times (h) Up/Down
1	200	11,000	9,900	50	220		8
2	60	11,433	10,100	15	80	30	8
3	50	12,000	10,800	15	60	25	4
4	40	12,900	11,900	5	40	20	4
5	25	13,500	12,140	5	34	24	4

TABLE 14.16

Load Pattern and Operating Conditions

Hours	MW Load	Conditions
	Load Pattern	
		Initially (prior to hour 1) only unit is on and has been on for 4 h
1–2	250	Ignore losses, spinning reserve, etc.
3–4	320	The only requirement is that the generation should be able to supply
5–6	110	the load
7–8	75	Fuel cost for all units may be taken as 1.40 R/MBtu

a. Develop the priority list for these units and solve for the optimum unit commitment. Use a strict priority list with a search range of three ($X=3$) and save no more than three strategies ($N=3$). Ignore min up/min down times for units.

b. Solve the same commitment problem using the strict priority list with $X=3$ and $N=$ as in (a), but obey min up/min down times rules.

c. (Optional) Find the optimum unit commitment without use of strict priority list (i.e., all 32 units on/off combinations are valid). Restrict search range to decrease your effort. Obey the min up/min down times rules.

PROBLEM 14.9.3

Given the unit data in the Tables 14.17 to 14.20 use forward dynamic programming to find the optimum unit commitment schedules covering the 8 h period.

The following tables give the characteristic of all combinations you need as well as the operating cost for each at the loads in the load data. The symbol * indicates that a combination cannot supply the load. The starting conditions are (a) at the beginning of the first period units 1 and 2 are up and (b) units 3 and 4 are down and have been down for 8 h.

TABLE 14.17

Unit Characteristic Data for the UC Problem

	Limits of the Unit				
Unit	$P_{g_i}^{max}$ (MW)	$P_{g_i}^{min}$ (MW)	Incremental Heat Rate Btu/kW h	No Load Energy (MBtu/h)	Start-Up Energy (MBtu/h)
1	500	70	9,950	300	800
2	250	40	10,200	210	380
3	150	30	11,000	120	110
4	150	30	11,000	120	110

Note: Load data (all time periods $=2$ h).

TABLE 14.18

Hourly Load Curve Data

Time, t	Load, $P_D(t)$ (MW)
1	600
1	800
3	700
4	950

TABLE 14.19

Start-Up and Shutdown Rules

Unit	Minimum Uptime (HR)	Minimum Downtime (HR)
1	2	2
2	2	2
3	2	4
4	2	4

Note: Fuel cost $=1.00$ $/MBtu.

TABLE 14.20

Unit Switching Sequences for Different Load Combinations

	Units (1-on, 0-off)				Load, P_D (MW)			
Combinations	1	2	3	4	600	700	800	950
A	1	1	0	0	6505	7,525	*	*
B	1	1	1	0	6649	7,669	8,705	*
C	1	1	1	1	6793	7,813	8,833	10,475

References

1. Ayoub, A. K. and Patton, A. D., Optimal thermal generating unit commitment, *IEEE Transactions on Power Apparatus and Systems*, PAS-90, 1752–1756, 1971.
2. Baldwin, C. J., Dale, K. M., and Dittrich, R. F., A study of economic shut down of generating units in a daily dispatch, *IEEE Transactions on Power Apparatus and Systems*, PAS-78, 1272–1284, 1960.
3. Chowdhury, N. and Billinton, R., Unit commitment in interconnected generating systems using a probabilistic technique, Paper 89 SM 715-4 PWRS, IEEE Summer Power Meeting, Long Beach, CA, July 1989.
4. Cohen, A. I. and Wan, S. H., A method for solving the fuel constrained unit commitment problem, *IEEE Transactions on Power System*, PWRS-2, 608–614, August 1987.
5. Cohen, A. I. and Yoshimura, M., A branch-and-bound algorithm for unit commitment, *IEEE Transactions on Power Apparatus and Systems*, PAS-102, 444–451, February 1983.
6. Dillon, T. S., Edwin, K., Kochs, H. D., and Taud, R. J., Integer programming approach to the problem of optimal unit commitment with probabilistic reserve determination, *IEEE Transactions on Power Apparatus and Systems*, PAS-97/6, 21542166, November–December 1978.
7. Everett, H. E., Generalized Lagrange multiplier method for solving problems of optimum allocation of resources, Operations Research, Vol. 11, pp. 399–417, May–June 1963.
8. Graham, W. D. and McPherson, G., Hydro-thermal unit commitment using a dynamic programming approach, Paper C 73 452-0, IEEE PES Summer Meeting, Vancouver, July 1973.
9. Gruhl, J., Schweppe, F., and Ruane, M., Unit commitment scheduling of electric power systems, in L. H. Fink and K. Carlson, Eds., *Systems Engineering for Power Status and Prospects*, Henniker, NH, 1975.
10. Happ, H. H., Johnson, R. C., and Wright, W. J., Large scale hydro thermal unit commitment method and results, *IEEE Transactions on Power Apparatus and Systems*, PAS-90, 1373–1383, 1971.
11. Hobbs, W. J., Hermon, G., Shebid, S., and Warner, G., An enhanced dynamic programming approach for unit commitment, *IEEE Transactions on Power Systems*, PWRS-3, 1201–1205, 1988.
12. Jain, A. V. and Billinton, R., Unit commitment reliability in a hydro-thermal system, Paper C 73 096-5, IEEE PES Winter Power Meeting, New York, 1973.
13. Lauer, G. S., Bertsekas, D. P., Sandell, N. R. Jr., and Posbergh, T. A., Solution of large-scale optimal unit commitment problems, *IEEE Transactions on Power Apparatus and Systems*, PAS-101, 79–86, January 1982.
14. Lowery, P. G., Generating unit commitment by dynamic programming, *IEEE Transactions Power Apparatus Systems*, PAS-85, 422–426, May 1966.
15. Merlin, A. and Sandrin, P., A new method for unit commitment at electricite de France, *IEEE Transactions on Power Applied System*, PAS-102, 1218–1225, May 1983.
16. Momoh, J. A. and Zhu, J. Z., Application of AHP/ANP to unit commitment in the deregulated power industry, *Proceedings of IEEE SMC'98*, California, 1998, pp. 817–822.

17. Van den Bosch, P. P. J., and Honderd, G., A solution of the unit commitment problem via decomposition and dynamic programming, *IEEE Transactions on Power Apparatus and Systems*, PAS-104, 1684–1690, July 1985.

18. Pang, C. K. and Chen, H. C., Optimal short-term thermal unit commitment, *IEEE Transactions on Power Applied Systems*, PAS-95, 1336–1346, July–August 1976.

19. Pang, C. K., Sheble, G. B., and Albuyeh, F., Evaluation of dynamic programming methods and multiple area representation for thermal unit commitments, *IEEE Transactions Power Apparatus Systems*, PAS-100, 1212–1218, March 1981.

20. Tong, S. K. and Shahidehpour, S. M., A combination of Lagrangian relaxation and linear programming approaches for fuel constrained unit commitment problem, *IEEE Proceedings—Generation, Distribution and Transmission*, 136 (Part C), 162–174, May 1989.

21. Van Meeteren, M. P., Berry, B. M., Farah, J. L., Kamienny, M. G., Enjamio, J. E., and Wynne, W. T., Extensive unit commitment function including fuel allocation and security area constraints, *IEEE Transactions on Power Systems*, PWRS-1(4), 228–232, 1986.

22. Zhuang, F. and Galiana, F. D., Unit commitment by simulated annealing, Paper 1989 SM 659–4 PWRS, IEEE Summer Power Meeting, Long Beach, CA, July 1989.

23. Ozturk, U. A., Mazumdar, M., and Norman, B. A., A solution to the stochastic Unit Commitment problem using chance constrained programming, *IEEE Transactions on Power Systems*, 19(3), 1589–1598, August 2004.

24. Motto, A. L. and Galiana, F. D., Unit commitment with dual variable constraints, *IEEE Transactions on Power Systems*, 19(1), 330–338, February 2004.

25. Ongsakul, W. and Petcharaks, N., Unit commitment by enhanced adaptive Lagrangian relaxation, *IEEE Transactions on Power Systems*, 19(1), 620–628, February 2004.

26. Rajan, C. C. A. and Mohan, M. R., An evolutionary programming-based Tabu search method for solving the unit commitment problem, *IEEE Transactions on Power Systems*, 19(1), 577–585, February 2004.

27. Fu, Y., Shahidehpour, M. and Li, Z., Long-term security-constrained Unit Commitment: hybrid Dantzig-Wolfe decomposition and subgradient approach, *IEEE Transactions on Power Systems*, 20(4), 2093–2106, November 2005.

28. Guan, X., Guo, S., and Zhai, Q., The conditions for obtaining feasible solutions to security-constrained Unit Commitment problems, *IEEE Transactions on Power Systems*, 20(4), 1746–1756, November 2005.

29. Li, Z. and Shahidehpour, M., Security-constrained Unit Commitment for simultaneous clearing of energy and ancillary services markets, *IEEE Transactions on Power Systems*, 20(2), 1079–1088, May 2005.

30. Simopoulos, D. N., Kavatza, S. D., and Vournas, C. D., Reliability constrained unit commitment using simulated annealing, *IEEE Transactions on Power Systems*, 21(4), 1699–1706, November 2006.

31. Carrion, M. and Arroyo, J. M., A computationally efficient mixed-integer linear formulation for the thermal unit commitment problem, *IEEE Transactions on Power Systems*, 21(3), 1371–1378, August 2006.

32. Saber, A. Y., Senjyu, T., Miyagi, T., Urasaki, N., and Funabashi, T., Fuzzy unit commitment scheduling using absolutely stochastic simulated annealing, *IEEE Transactions on Power Systems*, 21(2), 955–964, May 2006.

15

Genetic Algorithms

15.1 Introduction

In many engineering disciplines a large spectrum of optimization problems has grown in size and complexity. In some instances, the solution to complex multidimensional problems by using classical optimization techniques is sometimes difficult or expensive. This realization has led to an increased interest in a special class of searching algorithm, namely, evolutionary algorithms (EAs) [1,3]. In general, these are referred to as stochastic optimization techniques and their foundations lie in the evolutionary patterns observed in living things.

In this area of operational research, there exist several primary branches:

1. Genetic algorithms (GAs)
2. Evolutionary programming (EP)
3. Evolutionary strategies (ES)

To date the GA is the most widely known technology. This optimization technique has been applied to many complex problems in the fields of industrial and operational engineering. In power systems, well-known applications include unit commitment, economic dispatch, load forecasting, reliability studies, and various resource allocation problems.

15.1.1 General Structure of GAs

The typical structure of GAs was described by Goldberg [4]. Essentially, GAs are referred to as stochastic search techniques that are based on the Darwinian thinking of natural selection and natural genetics. In general, GAs start with an initial set of random solutions that lie in the feasible solution space. This random cluster of solution points is called a population. Each solution in the population represents a possible solution to the optimization problem and is therefore called a chromosome. The chromosome is a string of symbols based on the uniqueness of two-state machines; they are commonly binary bit strings.

15.2   Definition and Concepts Used in Genetic Computation

GAs have their foundation both in natural biological genetics and in modern computer science (Table 15.1). As such, nomenclature use in this is inherently a mix of both natural and artificial intelligence.

To understand the roots of GAs, we briefly look at the biological analogy. In biological organisms, a chromosome carries a unique set of information that encodes the data on how the organism is constructed. A collection or complete set of chromosomes is called a phenotype. Also, within each chromosome are various individual structures called genes, which are specific coded features of the organism. With this basic understanding, the following terminologies and concepts are summarized.

15.2.1   Evolutionary Algorithms

EAs represent a broad class of computer-based problem-solving systems. Their key feature is the evolutionary mechanisms that are at the root of formulation and implementation. Of course, EAs by themselves represent a special class of new intelligent system (IS) used in many global optimization algorithms. Figure 15.1 shows the various categories of IS and the position of the GA as one of the more commonly known EP techniques [5–8].

Overall, EAs share the common structure of evolution of individuals in a competitive environment by the processes of selection, mutation, and reproduction. These processes are functions of the simulated performance of each individual as defined by the environment. In EAs, a unique population of structures is maintained based on the search operators. Search operations use probabilistic rules of selection in the evolution process while ensuring that the integrity or fitness of new generation is continuously improved at each stage of the optimization process. Therefore, the reproduction mechanism is primarily focused on the fitness of the individuals in the population, while exploiting the available information. Furthermore, these robust

TABLE 15.1

Terminology in Genetic Algorithms

GA Terms	Corresponding Optimization Description
Chromosome	Solution set
Gene	Part of solution
Alleles	Value of gene
Phenotype	Decoded solution
Genotype	Encoded solution
Locus	Position of gene

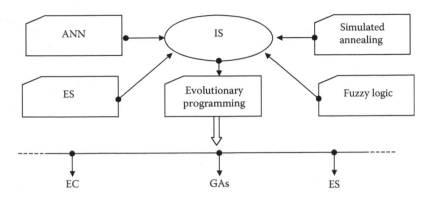

FIGURE 15.1
Common classifications of IS.

and powerful adaptive optimization search mechanisms use recombination and mutations to perturb individuals (parents and offspring), yielding new generations to be evaluated.

Over the past few decades, global optimization algorithms that imitate natural evolutionary principles have proved their importance in many applications. These applications include annealing processes, evolutionary computations (EC), artificial neural networks (ANN), and expert systems (ES) (Figure 15.1).

15.2.2 Genetic Programming

Genetic programming is a useful extension of the genetic model of learning or adaptation into the space of programs. In this special type of programming, the objects that constitute the population are not fixed-length character strings that typically encode feasible solutions to the optimization problem. Rather, the objects that constitute the population are programs that yield candidate solutions to the optimization when executed. In genetic programming, these are expressed as sparse trees rather than lines of code. For example, a simple program to perform the operation $X*Y-(A+B)*C$ would be represented as shown in Figure 15.2.

The programs in the population are composed of elements from the function and terminal sets. In genetic programming, the crossover operation is implemented by taking random selections of the subtree in the individuals. The selection is done according to the fitness of the individuals, and the exchange is done by the crossover operator. Notably, in general genetic programming, mutation is not used as a genetic operator. Genetic programming applications are used by physicists, biologists, engineers, economists, and mathematicians.

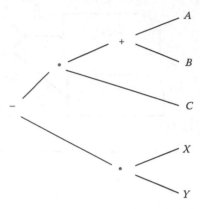

FIGURE 15.2
Simple structure demonstrating the operation
$X*Y - (A + B)*C$.

15.3   GA Approach

GAs are general-purpose search techniques based on principles inspired by the genetic evolutionary mechanism observed in the populations of natural systems and living organisms. Typically, there are several stages in the optimization process:

Stage 1. Creating an initial population

Stage 2. Evaluating the fitness function

Stage 3. Creating new populations

15.3.1   GA Operators

Various operators are used to perform the tasks of the stages in a GA: the production or elitism operator, crossover operator, and the mutation operator. The production operator is responsible for generating copies of any individual that satisfy the goal function. That is, they either pass the fitness test of the goal function or otherwise are eliminated from the solution space.

The crossover operator is used for recombination of individuals within the generation. The operator selects two individuals within the current generation and performs swapping at a random or fixed site in the individual string (Figure 15.3). The objective of the crossover process is to synthesize bits of knowledge from the parent chromosomes that will exhibit improved performance in the offspring. The certainty of producing better performing offspring via the crossover process is one important advantage of GAs.

Finally, the mutation operator is used as an exploratory mechanism that aids the requirements of finding a global extrema to the optimization problem. Basically it is used to randomly explore the solution space by flipping

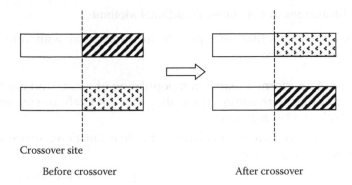

Crossover site

Before crossover   After crossover

FIGURE 15.3
Crossover operation on a pair of strings.

bits of selected chromosomes or candidates from the population. There is an obvious trade-off in the probability assigned to the mutation operator. If the frequency were high, the GA would result in a completely random search with a high loss of data integrity. On the other hand, too low an activation probability assigned to this operator may result in an incomplete scan of the solution space.

15.3.2 Major Advantages

GAs have received considerable attention regarding their potential as a novel optimization technique. There are several major advantages when applying GAs to optimization problems.

1. GAs do not have many mathematical requirements for optimization problems. Due to their evolutionary nature, GAs will search for solutions without regard to the specific inner workings of the problem. They can handle any kind of objective function and any kind of constraint (i.e., linear or nonlinear) defined on discrete, continuous, or mixed search spaces.

2. Ergodicity of evolution operators makes GAs very effective at performing global search (in probability). The traditional approaches perform local search by a convergent stepwise procedure, which compares the values of nearby points and moves to the relative optimal points. Global optima can be found only if the problem possesses certain convexity properties that essentially guarantee that any local optima is a global optima.

3. GAs provide us with a great flexibility to hybridize with domain-dependent heuristics to make an efficient implementation for a specific problem.

15.3.3 Advantages of GAs Over Traditional Methods

The main advantages that GAs present in comparison with conventional methods are as follows:

1. Since GAs perform a search in a population of points and are based on probabilistic transition rules, they are less likely to converge to local minima (or maxima).
2. GAs do not require well-behaved objective functions, hence easily tolerate discontinuities.
3. GAs are well adapted to distributed or parallel implementations.
4. GAs code parameters in a bit string and not in the values of parameters. The meaning of the bits is completely transparent for the GA.
5. GAs search from a population of points and not from a single point.
6. GAs use transition probabilistic rules (represented by the selection, crossover, and mutation operators) instead of deterministic rules.

Nevertheless, the power of conventional methods is recognized. The GA should only be used when it is impossible (or very difficult) to obtain efficient solutions by these traditional approaches.

15.4 Theory of GAs

15.4.1 Continuous and Discrete Variables

Real values can be approximated to the necessary degree by using a fixed-point binary representation. However, when the relative precision of the parameters is more important than the absolute precision, the logarithm of the parameters should be used instead.

Discrete decision variables can be handled directly through binary (or n-ary) encoding. When functions can be expected to be locally monotone, the use of Gray coding is known to better exploit that monotony.

15.4.2 Constraints

Most optimization problems are constrained in someway. GAs can handle constraints in two ways, the most efficient of which is by embedding these in the coding of the chromosomes. When this is not possible, the performance of invalid individuals should be calculated according to a penalty function, which ensures that these individuals are, indeed, poor performers. Appropriate penalty functions for a particular problem are not necessarily easy to design, since they may considerably affect the efficiency of the genetic search.

15.4.3 Multiobjective Decision Problems

Optimization problems very seldom require the optimization of a single objective function. Instead, there are often competing objectives, which should be optimized simultaneously. In opposition to single-objective optimization problems, the solution for a multiobjective optimization problem is not a single solution but a set of nondominated solutions. The task of finding this set of solutions is not always an easy one. GAs have the potential to become a powerful method for multiobjective optimization, keeping a population of solutions, and being able to search for nondominated solutions in parallel.

15.4.4 Other GA Variants

The simple GA has been improved in several ways. Different selection methods have been proposed [4] that reduce the stochastic errors associated with roulette wheel selection. Ranking has been introduced as an alternative to proportional fitness assignment, and has been shown to help avoidance of premature convergence and to speed up the search when the population approaches convergence. Other recombination operators have been proposed, such as the multiple point and reduced-surrogate crossover. The mutation operator has remained more or less unaltered, but the use of real-coded chromosomes requires alternative mutation operators, such as intermediate crossover. Also, several models of parallel GAs have been proposed, improving the performance and allowing the implementation of concepts such as that of genetic isolation. This method works well with bit string representation. The performance of GAs depends on the performance of the crossover operator used.

The crossover rate p_c is defined as the ratio of the number of offspring produced in each generation to the population size (denoted pop size). A higher crossover rate allows exploration of more of the solution space and reduces the chances of settling for a false optimum; but if this rate is too high, a lot of computational time will be wasted.

Mutation is a background operator that produces spontaneous random changes in various chromosomes. A simple way to achieve mutation would be to change one or more genes. In the GA, mutation serves the crucial role of either

1. Replacing the genes lost from the population during the selection process so that they can be tried in a new context, or
2. Providing the genes that were not present in the initial population

The mutation rate p_m is defined as the percentage of the total number of genes in the population and it controls the rate at which new genes are introduced into the population for trial. If it is too low, many genes that would have been useful are never tried. But if it is too high, there will be many random populations, the offspring will start losing their resemblance to the parents, and the algorithm will lose the ability to learn from the history of the search.

GAs differ from conventional optimization and search procedures in several fundamental ways. Goldberg has summarized these as follows:

1. GAs work with a coding of solution sets, not the solutions themselves.
2. GAs search from a population of solutions, not a single solution.
3. GAs use payoff information (fitness function), not derivatives or other auxiliary knowledge.
4. GAs use probabilistic transition rules, not deterministic rules.

15.4.5 Coding

Each chromosome represents a potential solution for the problem and must be expressed in binary form in the integer interval $I = [0,21]$. We could simply code X in binary base, using four bits (such as 1001 or 0101). If we have a set of binary variables, a bit will represent each variable. For a multivariable problem, each variable has to be coded in the chromosome.

15.4.6 Fitness

Each solution must be evaluated by a fitness function to produce a specific value. This objective function is used to model and characterize the problem to be solved. In many instances, the fitness function can be simulated as the objective function used in classical optimization problems. In such cases, these optimization problems may be unconstrained or constrained. For the latter case, a Lagrangian or penalty approach can be used in formulating a suitable fitness function.

Notably, the fitness function does not necessarily have to be in closed mathematical form. It can also be expressed in quantitative form and, in power systems applications, with fuzzy models.

15.4.7 Selection

The selection operator creates new populations or generations by selecting individuals from the old population. The selection is probabilistic but biased toward the best as special deterministic rules are used. In the new generations created by the selection operator, there will be more copies of the best individuals and fewer copies of the worst. Two common techniques for implementing the selection operator are the stochastic tournament and roulette wheel approaches [4].

1. *Stochastic tournament*: This implementation is suited to distributed implementations and is very simple: every time we want to select an individual for reproduction, we choose two, at random, and the best wins with some fixed reliability, typically 0.8. This scheme can be

enhanced by using more individuals in the competition or even considering evolving winning probability.

2. *Roulette wheel*: In this process, the individuals of each generation are selected for survival into the next generation according to a probability value proportional to the ratio of individual fitness over total population fitness; this means that on average the next generation will receive copies of an individual in proportion to the importance of its fitness value.

15.4.8 Crossover

The recombination in the canonical GA is called single-point crossover. Individuals are paired at random with a high probability that crossover will take place. In the affirmative case, a crossover point is selected at random and, say, the rightmost segments of each individual are exchanged to produce two offspring.

Crossover in the canonical GA mutation consists of simply flipping each individual bit with a very low probability (a typical value would be $P_m = 0.001$). This background operator is used to ensure that the probability of searching a particular subspace of the problem space is never zero, thereby tending to inhibit the possibility of ending the search at a local, rather than a global, optimum.

15.4.9 Parameters

Like other optimization methods, GAs have certain parameters such as

1. Population size
2. Genetic operations probabilities
3. Number of individuals involved in the selection procedure, and so on

These parameters must be selected with maximum care, for the performance of the GA depends largely on the values used. Normally, the use of a relatively low population number, high crossover, and low-mutation probabilities are recommended. Goldberg [4] analyzes the effect of these parameters in the algorithms.

15.5 Schemata Theorem

GAs work based on the concept and theory of schema. A schema is a similarity template describing a subset of strings with similarities at certain string positions. If, without loss of generality, we consider only chromosomes represented with binary genes in {0,1}, a schema could be $H = {}^*001^*1$, where the character is $*$ is a "wild card," meaning that the value of 0 or 1 at such a position is undefined. The strings $A = 100101$ and $B = 000111$ both include the schema H because the string alleles match the schema positions 2, 3, 4, and 6.

For binary strings or chromosomes of length L (number of bits or alleles), the number of schemata is 3^L. But the schemata have distinct relevance; a schema 0^{*****} is more vague than 011^*1^* in representing similarities between chromosomes; and a schema $1^{****}0$ spans a larger portion of the string than the schema $1^{**}0^{**}$.

A schema H may be characterized by its order $o(H)$, which is the number of its fixed positions and by its defining length $d(H)$, which is the distance between its first and its last fixed position. For the schema $G = 1^{**}0^{**}$, we have $o(G) = 2$ and $d(G) = 3$.

We now reason about the effect of reproduction on the expected number of different schemata in a population. Suppose that at a given time step t (a given generation) there are m examples of a particular schema H in a population $P(t)$; we have $m = m(H,t)$. Reproduction generates a copy of string A_i with probability $P_i = f_i / \sum f_i$ (assuming a sampling process known as roulette).

After the process of retaining from the population $A(t)$, a nonoverlapping population of size n with replacement, there is an expectation of having in the population $A(t + 1)$, at time $t + 1$, a number $m(H,t + 1)$ of representatives of the schema H given by

$$m(H,t + 1) = m(H,t)\frac{f(H)}{\sum f_i},\tag{15.1}$$

where $f(H)$ is the average fitness of the chromosomes including the schema at H at time t. If we introduce the average fitness of the entire population as

$$f_{av} = \frac{1}{n}\sum f_i,$$

we can write

$$m(H,t + 1) = m(H,t)\frac{f(H)}{f_{av}}.\tag{15.2}$$

This means that a particular schema replicates in the population proportionally to the ratio of the average fitness of the schema by the average fitness of the population. So, schemata that have associated an average fitness above the population average will have more copies in the following generation, while those with average fitness below the population average will have a smaller number of copies. Suppose now that a given schema remains with average fitness above the population average by an amount Cf_{av} with c constant; we could then rewrite Equation 15.2 as Equation 15.3,

$$m(H,t+1) = m(H,t)\frac{f_{av} + Cf_{av}}{f_{av}} = (1+C)m(H,t). \tag{15.3}$$

Assuming a stationary value of c, we obtain

$$m(H,t) = m(H,0)(1+C)^t. \tag{15.4}$$

The effect is clear: an exponential replication in a population of above-average schemata.

A schema may be disrupted by crossover, if the crossover point falls within the defining length spanned by the schemata (we reason with single-point crossover to keep it simple). The survival probability of a schema under a crossover operation performed with probability P_c is

$$P_s \geq 1 - P_c\frac{d(H)}{L-1}. \tag{15.5}$$

Combining reproduction and crossover, we can write the following estimation as shown in Equation 15.6,

$$m(H,t+1) \geq m(H,t)\frac{f(H)}{f_{av}}\left[1 - P_c\frac{d(H)}{L-1}\right]. \tag{15.6}$$

We see now that the survival of a schema under reproduction and crossover depends on whether it is above or below the population average and whether it has a short or long definition length. To add the effect of admitted mutation, randomly affecting a single position with probability P_m we must notice that a schema survives if each of its $o(H)$ fixed positions remains unaffected by mutation. Therefore, the probability of the surviving mutation is $(1-P_m)^{o(H)}$, which can be approximated by Equation 15.7,

$$1 - o(H)P_m \quad \text{for } P_m \ll 1. \tag{15.7}$$

We can conclude that in a process with reproduction, crossover, and mutation we can expect that a particular schema will have a number of copies in generation $t+1$ given approximately by Equation 15.8,

$$m(H,t+1) \geq m(H,t)\frac{f(H)}{f_{av}}\left[1 - P_c\frac{d(H)}{L-1} - o(H)P_m\right]. \tag{15.8}$$

This constitutes the schemata theorem, which is at the root of the building block hypothesis—that highly fit, short (low-order) schemata form partial solutions to a problem, and that a GA will combine these building blocks leading to better performance and the optimum of the problem.

15.6   General Algorithm of GAs

During successive iterations, called generations, the chromosomes are evaluated, using some measures of fitness. To create the next generation, new chromosomes, called offspring, are formed by either

1. Merging two chromosomes from the current generation using a crossover operator, or
2. Modifying a chromosome using a mutation operator

A new generation is formed by

1. Selecting, according to the fitness values, some of the parents and offspring.
2. Rejecting others to keep the population size constant. Fitter chromosomes have higher probabilities of being selected.

After several generations, the algorithms converge to the best chromosome, which, it is hoped, represent the optimum or suboptimal solution to the problem.

Now, let $P(t)$ and $C(t)$ be parents and offspring in the current generation t. Then the general structure of GAs is described in the following procedure:

Begin
    $T \leq 0$:
    initialize $P(t)$;
    evaluate $P(t)$;
    while (not termination condition) **do**
        recombine $P(t)$ to yield $C(t)$;
        evaluate $C(t)$;
        select $P(t)$ to yield $C(t)$
        $t \leftarrow t+1$;
    end
end

Usually, initialization is assumed to be random. Recombination typically involves crossover and mutation to yield offspring. In fact, there are only two kinds of operations in GAs: (1) genetic operations—crossover and mutation and (2) evolution operation—selection. The genetic operations mimic the process of heredity to create new offspring at each generation. The evolution operation mimics the process of Darwinian evolution to create populations from generation to generation. This description differs from the paradigm given by Holland [9] where selection is made to obtain parents for recombination.

Crossover is the main genetic operator. It operates on two chromosomes at a time and generates offspring by combining both chromosomes' features. A simple way to achieve crossover would be to choose a random cutpoint and generate the offspring by combining the segment of the other parent to the right cutpoint.

15.7 Application of GAs

Some of the most successful applications of GAs are listed below.

15.7.1 Control System Engineering

In this field, the problems that are presented to engineers are of high complexity and almost always multiobjective. GAs are powerful tools, especially when used with other existing tools.

15.7.2 Timetabling

A very common manifestation of this kind of problem is the timetabling of exams or classes in universities, and the like. In the exam timetabling case, the fitness function for a genome representing a timetable involves computing degrees of punishment for various problems with the timetable, such as clashes, instances of students having to take consecutive exams, and also instances of students having three or more exams in one day. The modular nature of the fitness function has the key to the main potential strength of using GAs for this sort of thing as opposed to the conventional methods of search and constraint programming. The power of the GA approach is the ease with which it can handle arbitrary kinds of constraints and objectives. Very few other timetabling methods, for example, deal with such objectives at all, which shows how difficult it is (without GAs) to graft the capacity to handle arbitrary objectives onto the basic requirement. The proper way to weight/handle different objectives in the fitness function relation to the general GA dynamics remains, however, an important research problem.

15.7.3 Job-Shop Scheduling

The job-shop scheduling problem is a very difficult one, which so far seems best addressed by sophisticated brand-and-bound search techniques. However, we will see increasingly better results on using GAs on fairly large benchmarks. A crucial aspect of such work (as with any GA application) is the method used to encode schedules. Concerning the point of using GAs at all on half job-shop scheduling problems, the same analysis applies here as suggested above for timetabling. The GA approach enables relatively arbitrary constraints and objectives to be incorporated painlessly into a single optimization method. It is unlikely that GAs will outperform specialized knowledge-based systems and conventional operations research-based approaches to such problems in terms of raw solution quality, however. GAs offer much greater simplicity and flexibility. So, for example, they may be the best method for quick high-quality solutions, rather than finding the best possible solution at any cost.

Similar to job scheduling is the open-shop scheduling problem (OSSP) which shows reliable achievement of results within less than 0.23% of optimal on moderately large OSSPs. A simpler version of the job-shop problem is the flow-shop sequencing problem. In contrast to job-shop scheduling, some maintenance scheduling problems consider which activities to schedule within a planned maintenance period, rather than seeking to minimize the total time taken by the activities. The constraints on which parts may be taken out of service for maintenance at a particular time may be very complex, particularly as they will in general interact.

15.7.4 Management Sciences

GAs have been successfully used in market forecasting with well-known systems such as the prediction and state estimation applications.

15.7.5 Game Playing

GAs can be used to evolve behaviors for playing games. Work in evolutionary game theory typically encompasses the evolution of a population of players who meet randomly to play a game, which they each must adopt one of a limited number of moves. Suppose it is just two moves, X and Y. The players receive a reward analogous to Darwinian fitness, depending on which combination of moves occurs and which move they adopted. In more complicated models there may be several players and several moves.

The players iterate such a game a series of times and then move to a new partner. At the end of all such moves, the players will have a cumulative payoff, their fitness. The real key to using GAs is to come up with an encoding to represent player's strategies, one that is amenable to crossover and to mutation. Possibilities include supposing at each iteration that a player adopts X with some probability (and Y with one minus such). A player can thus be represented as a real number or a bit-string by interpreting the decimal value of the bit-string as the inverse of the probability.

15.8 Application to Power Systems

Here we present a few simplified versions of models applied to power systems, with the purpose of helping readers understand more clearly the potential behind the GA principles (Table 15.2) [12–21].

TABLE 15.2

Genetic Applications in Electric Power Systems

Area	Fields
Expansion or structural planning	Generation-transmission
	Transmission/distribution
	VAr planning, capacitor placement
Operation planning	Unit commit, generator scheduling
	Load dispatch
	React, power, dispatch, volt, control
	Maintenance scheduling
	Security assessment
Generation/transmission and	Loss minimization switching
distribution operation	Alarm processing, fault diagnosis
	Service restoration
	Load management
	Load forecasting
	State estimation
	Facts
Analysis	Power flow
	Harmonics

15.8.1 GAs in the Unit Commitment Problem

Generally, the unit commitment problem (UCP) is one that involves the determination of the optimal set of generating units within the next 1–7 days. The UCP is to minimize operation costs, transition costs, and no-load costs. The operational cost is mainly fuel cost while transitional costs include start-up and shutdown costs of the units. Idle, banking, or standby cost constitute the no-load costs. Genetic-based unit commitment implementation consists of initialization, cost calculations, elitism, reproduction, crossover, standard mutation, economic dispatch calculations, and intelligent mutation of the unit commitment schedules. An explanation of each part of the genetic-based unit commitment algorithm implementation follows.

The initialization is explained for one member of the population (one unit commitment schedule). The number of population members used in this research was 50. A member of the population consists of a matrix with dimension equal to the number of generators by the number of scheduling periods. This matrix represents the on/off status of the generating units. The first initialization step consists of finding the 10 cheapest economic dispatches for each hour that meet system demand and a 10% spinning reserve. A member of the population is then created by randomly choosing 1 of the 10 cheapest economic dispatches for each hour.

15.8.1.1 UCP Statement

In the UCP under consideration, one is interested in a solution that minimizes the total operating cost of the generating units during the scheduling time horizon while several constraints are satisfied.

15.8.1.1.1 Objective Function

The overall objective function of the UCP of N generating units during the scheduling time horizon T (e.g., 24 h) is

$$F = \sum_{t=1}^{T} \sum_{i=1}^{N} [u_i(t) \cdot F_i(E_i(t)) + S_i(t) + f_i(t)],$$

where
  $u_i(t)$ is the status of unit I at hour (on $= 1$, off $= 0$)
  $S_i(t)$ is the start-up/shutdown status of unit i at hour t
  $E_i(t)$ is the energy output from unit i at time t
  $S_i(t)$ is the start-up cost of the ith unit at hour t
  $f_i(t)$ is the ramping cost of the ith unit at hour t

The production cost $F_i(t)$ of a committed unit i is conventionally taken in a quadratic form:

$$F_i(t) = \alpha_i E_i^2(t) + \beta_i E_i(t) + \gamma_i,$$

where α_i, β_i, and γ_i are the cost function parameters of unit i. The start-up cost $S_i(t)$ is a function of the downtime of the unit.

15.8.1.1.2 Constraints

The constraints that have been taken into consideration in this work may be classified into two main groups.

15.8.1.1.2.1 System Constraints

1. Load demand constraints

$$\sum_{i=1}^{N} P_{g_i}(t) \geq \text{net demand},$$

   where net demand is the system peak demand at hour t.
2. *Spinning reserve*: The spinning reserve is the total amount of generation capacity available from all units synchronized (spinning) on the system minus the present load demand:

$$\sum_{i=1}^{N} P_{g_i}(t) \geq (\text{net demand} + \text{spinning reserve}).$$

15.8.1.1.2.2 *Unit Constraints* The constraints on the generating units are as follows:

1. *Generation limits*: In case the units should maintain a given amount of reserve, its upper bounds must be modified accordingly. Therefore, we have

$$P_{g_i}^{\max} \geq P_{g_i} \geq P_{g_i}^{\min},$$

where $P_{g_i}^{\max}$ and $P_{g_i}^{\min}$ are the minimum and maximum generation limits (MW) of unit i, respectively.

2. Minimum up/down times:

$$T_{\text{off}_i} \geq T_{\text{down}_i}$$

$$T_{\text{on}_i} \geq T_{\text{up}_i}$$

for units where

T_{up_i} and T_{down_i} are the unit i minimum up/down times

T_{off_i} and T_{on_i} are the time periods during which unit i is continuously on/off

- Unit initial status
- Crew constraints
- Unit availability (e.g., must run, unavailable, or fixed output [MW])
- Unit derating

15.8.1.2 GA Implementation in the GTS Algorithm

The details of the GA components implementation are described and summarized as follows:

1. *Solution coding*: The solution in the UCP is represented by a binary matrix (U) of dimension $T \times N$ (Figure 15.4a). The proposed method for coding is a mix between binary and decimal numbers. Each column vector in the solution matrix (which represents the operation schedule of one unit) of length T is converted to its equivalent decimal number. The solution matrix is then converted into one row vector (chromosome) of N decimal numbers $(U_1, U_2, \ldots, U_N)$, each representing the schedule of one unit (Figure 15.4b). Typically the numbers $U_1, U_2, \ldots, U_N$ are integers ranging between 0 and $2^N - 1$. Accordingly, a population of size NPOP is stored in a matrix NPOP $\times N$ (Figure 15.4c).

H	Units							
R	1	2	3	4	.	.	.	N
1	1	1	0	0	.	.	.	1
2	1	1	0	0	.	.	.	1
3	1	0	1	0	.	.	.	0
.	.	.	.	.	.	.	.	.
T	0	1	0	1	.	.	.	0

(a)

U_1	U_2	U_3	U_4	.	.	.	U_N

(b)

U_1	U_2	U_3	U_4	.	.	.	U_N
U_1	U_2	U_3	U_4	.	.	.	U_N
.	.	.	.	.	.	.	.
U_1	U_2	U_3	U_4	.	.	.	U_N

(c)

FIGURE 15.4
(a) Binary solution matrix U, (b) equivalent decimal vector (one chromosome), and (c) population of size NPOP chromosomes.

2. *Fitness function*: Unlike the previous solutions of the UCP using GAs the fitness function is taken as the reciprocal of the total operating cost, since we are generating always feasible solutions. The fitness function is then scaled to prevent premature convergence. Linear scaling is used which requires a linear relationship between the original fitness function and the scaled one.

3. *Crossover*: To speed up the calculations, the crossover operation is done between two chromosomes in their decimal form. Two parents are selected according to the roulette wheel rule. Two positions in the two chromosomes are selected at random. The decimal numbers are exchanged between the two parents to produce two children. The two children are then coded into their binary equivalents and checked for constraint violation (load demand and reserve constraints). If the constraints are not satisfied a repair mechanism is applied to restore feasibility to the produced children.

4. *Mutation*: The mutation operation is done by randomly selecting any chromosome with a prespecified probability. The selected chromosome is then coded into its binary equivalent. A unit number and a

time period are randomly selected. Then the proposed rules are applied to reverse the status of this unit keeping the feasibility of the unit constraints related to its minimum up and down times. A checked for the changed time periods, and correction if necessary, for the reserve constraints is then made.

5. *Generating feasible trial solutions*: We have proposed some rules to generate randomly a feasible trial solution as a neighbor to an existing feasible solution. These rules were designed to achieve the minimum up/down constraints satisfaction, which are the most difficult constraints in the UCP, while the reserve constraints are checked and corrected, if necessary, using a repair mechanism. The main idea of these rules could be summarized in two points: the difference between minimum up and down times of a unit is subtracted from the on or off hours of that unit; and the unit status is reversed randomly at certain hours ranging between 0 and this difference.

6. *Repair mechanism*: Due to applying the crossover and mutation operations, the reserve constraints might be violated. A repair mechanism to restore the feasibility of these constraints is applied and described as follows:

   a. Pick at random one of the off units at one of the violated hours.

   b. Apply the rules in Section 15.5 to switch the selected unit from off to on, keeping the feasibility if there are downtime constraints. Check for the reserve constraints at this hour, If satisfied go to another hour. Otherwise, repeat the process at the same hour for another unit.

This procedure has proven faster than algorithms that use penalty functions.

15.8.1.3 Proposed Algorithm

In solving the UCP, two types of variables need to be determined: $U_i(t)$, which are 0–1 (binary) variables, and the units' output energy variables $E_i(t)$, which are continuous variables. The first is a combinatorial optimization. The nonlinear optimization economic dispatch program (EDP) is simultaneously solved via a quadratic programming routine.

The flowchart of the GA for the UCP is given in Figure 15.5. The major steps of the algorithm are summarized as follows:

1. Create an initial population by randomly generating a set of feasible solutions (chromosomes).

2. Evaluate each chromosome by solving EDP.

3. Determine the fitness function of each chromosome in the population.

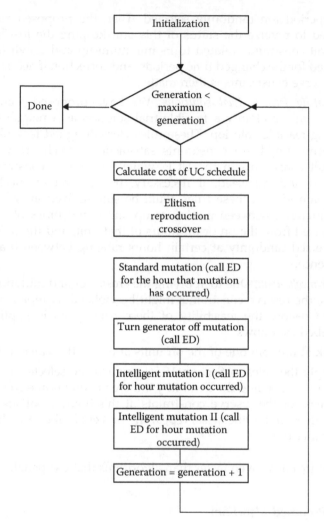

FIGURE 15.5
Flowchart of the GA for unit commitment.

4. Apply GA operators to generate new populations as follows:
   a. Copy the best solution from the current to the new population.
   b. Use the TS algorithm to generate new members in the new population (typically 1%–10%) of the population size as neighbors to randomly selected solutions in the current population.
   c. Apply the crossover operator to complete the members of the new population.
   d. Apply the mutation operator to the new population.
5. Apply the SA algorithm to test the members of the new population.

15.8.1.3.1 Stopping Criteria

There are several possible stopping conditions for the search. In our implementation, we stop the search if one of the following two conditions is satisfied in the given order.

1. Number of iterations performed since the best solution last changed is greater than a prespecified maximum number of iterations.
2. Maximum allowable number of iterations (generations) is reached.

The calculation of cost of the UC schedule consists of the following:

1. If a unit breaks the minimum up-time constraint, the unit is charged as if it were on standby for those hours. A temporary matrix is then created from the original UC schedule, except the standby hours are set to 1 instead of 0.
2. If a unit breaks the minimum downtime constraint in the temporary matrix created in Step 1, the unit is charged as if it were on standby for the additional number of hours needed to satisfy the constraint. The temporary matrix then has those hours set to X instead of 0.
3. Using the temporary matrix created from Steps 1 and 2, the startup, shutdown, and banking costs are calculated for each unit.
4. Fuel cost for each UC schedule is calculated by summing the ED cost for each hour.

Elitism ensures that the best individuals are never lost in moving from one generation to the next. The elitism subroutine combines the two populations and determines the best results from both populations in order of decreasing fitness value. It then saves distinct members that have the highest fitness into the next generation. The amount is determined by the difference in the generator on/off matrices. Reproduction is the mechanism by which the most highly fit members in a population are selected to pass on information to the next population of members. The fitness of each member is determined by taking the inverse of the cost of each member's UC schedule and then ordering the population members by increasing cost. Then each member is assigned fitness according to its rank. The members kept for reproduction are determined by roulette wheel selection. This type of reproduction is called rank-based fitness. Crossover is the primary genetic operator that promotes the exploration of new regions in the search space. Crossover is a structured, yet randomized, mechanism of exchanging information between strings. Crossover begins by selecting at random two members previously placed in the mating pool during reproduction. A crossover point is then selected at random and information from one member, up to the crossover point, is exchanged with the other member, thus creating two new members for the next generation.

The unit commitment procedure searches for global optimal solutions considering the following costs and constraints:

- Unit shutdown and start-up costs
- Operation costs (usually fuel costs)
- Ramping costs
- Load satisfaction
- Ramping constraints
- Spinning reserve criteria
- Maximum wind penetration
- Minimum unit downtimes
- Minimum start-up times (as function of downtime)

15.8.1.3.2   *Coding*

Usually in a period of 48 h, one generator will only change state a few times. Therefore, one would only need a few bits to denote the change of state, instead of 48 bits (one for each hour, if the time step is 1 h). Moreover, during long periods, the same set of generators may be on or off. They can be represented in block. We have therefore divided the scheduling period into a number of blocks, where we admit that the composition of generators on and off is constant (requiring only a number of bits equal to the number of generators). We have also added to each block a number c of bits defining an advance or delay for each one. There is also one extra bit per block allowing a block to be disabled, so that the number of active blocks is variable and adapts to the solution. The number of bits to press a solution is therefore given by $b(n + c + 1)$ and the number b of blocks needed depends on the regularity of the load curve. For instance, in a problem with 25 generators and an hourly scheduling for 48 h (a long-term UCP), the direct encoding of every hour would require a chromosome with 1200 bits, while the same problem divided into 10 blocks (with 2 bits for advance or delay) would require chromosomes of 280 bits in length only. Presently, the number of blocks and the relative size of each block are decisions made by the operator at the modeling stage of the problem (not done automatically).

15.8.1.3.3   *Selection and Deterministic Crowding*

Selection is directed under the deterministic crowding approach [10], which randomly pairs all population elements in each generation. Each pair of parents will undergo mutation and crossover, and yield two children. The members of each set of two parents and two children are then classified and paired according to a similarity measure such as the Hamming distance between chromosomes (the number of differences in the bits). Finally, competition is established in each pair and the two fittest elements will enter the next generation. The deterministic crowding approach can successfully perform niching and preserve the diversity in the GA.

15.8.1.3.4 Neighborhood Digging

At certain generations, a process called neighborhood digging is launched. It tries to displace the start-up and shutdown times of generators in order to improve solutions, selected with a given probability.

The general neighborhood digging tests are

1. Enabling/disabling some blocks
2. Shutting down or starting up generators within some blocks
3. Anticipating/postponing the start of a new block
4. Avoiding unnecessary shutdowns/start-ups of generators

A set of rules has been derived to guide the neighborhood digging process, for instance, for postponement and anticipation of start-up and shutdown of generators, in order to improve the computational efficiency of the process. The decision to adopt changes in the solutions may come either from feasibility verification or from fitness calculation. If better solutions are found, their chromosome coding is added to the genetic pool.

The neighborhood digging procedure has been extensively tested, namely, against a simple GA (where the neighborhood search is disabled) and its superior performance has been confirmed.

15.8.1.3.5 Chromosome Repair

Chromosome repair is applied to infeasible solutions, in an attempt to bring them to feasibility. It consists of a series of very simple and heuristic tests, so that the process remains computationally efficient. If a feasible solution is built, it replaces the original infeasible one in the genetic pool.

15.8.1.3.6 Crossover

The module adopts two-point crossover. In this UCP, with the type of chromosome coding adopted, it proves to lead to superior algorithm performance. Besides costs, the fitness function also includes penalty values for not having all constraints satisfied. The highest penalty is assigned to schedules that do not meet the load. Other penalties apply to schedules that do not respect minimum downtimes or start-up delays. A third set of penalties is included for schedules that do not satisfy a spinning reserve criterion, taken as a fuzzy constraint.

The application allows the user to select two spinning reserve criteria— either a percentage of the load or a probabilistic risk value. The best fitness is therefore assigned to solutions with the lowest value of cost plus penalties.

15.8.2 Load Shedding: A Model with GA

Under-frequency load shedding is used to alleviate load-generation imbalance following a sudden loss of generation or interconnection. A large frequency reduction happens when generators are not able to supply the connected system load or when generator governors are not fast enough to respond to such disturbances. In these latter situations, although the system may remain stable, very large frequency deviations may cause the actuation of frequency protection relays (such as the ones installed in thermal power station auxiliaries) leading the system to collapse. To avoid this situation, utilities usually have shedding schemes that disconnect selected loads throughout the entire power system; besides protecting against frequency collapse, this technique is also used to prevent deep drops in system frequency. However, sometimes the amount of the disconnected load is far more than necessary, which may also lead to large over-frequency oscillations. Load shedding is accomplished by using frequency-sensitive relays that detect the onset of decay in the system frequency. Both frequency and rate-of-change of frequency are measured. Load shedding is usually implemented in stages with each stage triggered at a different frequency level or (df/dt) setting. This staging is performed to prioritize the shedding of the loads, where the least important loads are to be shed first.

Usually, this priority policy is associated with the social impact of the disconnection action, without any other concern related to the load behavior during the emergency conditions. Specifically, frequency-dependent loads present better dynamic characteristics during the power imbalance phenomenon and may be kept connected. The following paragraphs describe a new approach based on GA that aims at determining, for a given loss of generation, the minimum amount of load to be disconnected and the step-actuation factor used for that purpose (minimum frequency relay setting level and df/dt operation). This calculation also takes into consideration the dynamic characteristics of loads (namely, frequency dependence) and their location on the network. The minimization of the load curtailment is subjected to a set of constraints, such as

1. Priority loads should be kept in operation
2. Minimum system frequency deviation
3. Maximum residual system frequency value
4. Maximum system frequency excursion

An optimal load shedding scheme of this type can then be formally considered as a large, nonlinear, discrete optimization problem subject to multiple constraints. To implement the load shedding strategies, some assumptions must be introduced for the modeling of load buses. In the example and results below, without loss of generality, it has been assumed that there are four feeders in each load bus. These feeders are sharing the busload such that, in each feeder, the combination of load types is different.

Load shedding is triggered in the example by two actuation types: frequency level and frequency rate of change. In this work, 49, 48.5, and 48 Hz are, respectively, the first, second, and third levels of frequency settings for load shedding. Simultaneously, the setting of the frequency change rate is fixed at 0.5 Hz/s. This means that whenever system frequency reaches one of the three levels of the frequency setting or if the system frequency declines more than 0.5 Hz/s, a scheduled amount of load will be disconnected. Time delays in the actuation of the frequency relays have also been considered.

15.8.2.1 Coding

The choice of how to encode solutions in a chromosome is of primary importance to the success of the GA approach. In this work, each bit corresponds to one of the feeders of the system (possible location for a disconnecting device operated by a relay). The chromosome is divided into three parts, each part representing a shedding stage. Obviously, those feeders that are assigned to one stage would not be available in others. Thus, each load bus must be coded using 12 bits of the chromosome. This coding example also shows that some bits may become irrelevant—the meaning of the chromosome depends entirely on the fitness function, that is, on the interpretation of the user.

15.8.2.2 Fitness

A fitness function (fit) defined as follows is used to assess the quality of each shedding solution coded in an individual chromosome:

$$\text{fit} = 100 \left(1 - \frac{|f_{mn}| + m}{A}\right)\left(1 - \frac{|f_{mx}|}{B}\right)\left(1 - \frac{|f_{in}|}{C}\right)\left(\frac{D \cdot P_{\text{los}} - P_{\text{shed}}}{E \cdot P_{\text{los}}}\right),$$

where A, B, C, D, and E are weight parameters dependent on the system under analysis and on the importance attributed to each frequency deviation, and $m = 1$ is an offset value. For each individual (solution), P_{los} is the amount of generation lost and P_{shed} is the amount of load to be shed. The values f_{min}, f_{max}, and f_{final} are, respectively, the minimum, maximum, and final frequency deviation values obtained from a step-by-step numerical solution of the differential equation that characterizes the system dynamic behavior. To speed up the procedure several conditions were included during the fitness assessment: solutions were discarded as soon as they were found infeasible and other filtering procedures were included to avoid the need for evaluating the full dynamic behavior of the system.

15.8.2.3 Initial Population

If I_s and O_s in each chromosome are chosen with equal probability, then approximately half the feeders will be set ready for shedding. In other words, half the loads would be ready to be shed regardless of the amount of generation lost. This type of approach for choosing the initial population makes the optimization process faster and more efficient.

15.8.2.4 Genetic Operators

The work described in Ferreira et al. [11] uses the canonical GA using a fixed-size, nonoverlapping population scheme with each new generation created by the selection operator and altered by single-point crossover and mutation, according to fixed operator probabilities.

15.9 Illustrative Examples

Example 15.9.1

Solve the following linear programming problem using the GA.

Maximize $Z = -3x_1 + 5x_2$

Subject to the constraints:

$x_1 + 2x_2 \geq 2.0$

$0 \leq x_1 \leq 3.0$

$0 \leq x_1 \leq 3.0.$

SOLUTION

Using an appropriate GA shell, the optimal value of the objective is $Z^* = 15$ with $x_1^* = 0$ and $x_2^* = 3$.

Example 15.9.2

Use the GA to minimize the following nonlinear problem.

Minimize $Z = \sin x + \sin \dfrac{10x}{3} + \ln -0.84x + 3$

Subject to the boundary conditions:

$2.26 \leq x \leq 11.86.$

SOLUTION

The global optimal solution was found to be $x^* = 10.91397$ with $Z^* = 5.74288$, which is a global minima.

Example 15.9.3

Use the GA to minimize the following nonlinear problem.

Minimize $Z = \displaystyle\sum_{j=1}^{4} [a_j + b_j x(j) + c_j x^2(j)]$

Subject to the boundary conditions:

$$-10.0 \le x_j \le 10.0 \; \forall x_j \in \{1,4\}$$

and the coefficients of the polynomial are

$$
\begin{aligned}
a_1 &= 2.9391848E - 0; & b_1 &= 5.6514290E - 1; & c_1 &= 1.3223214E - 2; \\
a_2 &= 2.6497593E - 0; & b_2 &= -4.0084286E - 2; & c_2 &= 1.0892857E - 3; \\
a_3 &= 2.9166987E - 0; & b_3 &= -1.7450143E - 1; & c_3 &= 6.6666964E - 2; \\
a_4 &= 2.8071963E - 0; & b_4 &= -2.7918714E - 2; & b_4 &= 4.7189286E - 3.
\end{aligned}
$$

SOLUTION

For this GA calculation, the following parameters were used:

Maximum number of chromosomes (bits per individual) $= 75$
Maximum number of generations to run by the GA $= 50$
Maximum number of individuals (maximum population size) $= 500$
Population size of a GA run $= 500$
Mutation probability $= 0.05$
Crossover probability (uniform in this case) $= 0.5$.

After 25060 iterations, the GA results yield:

$$
\begin{aligned}
x_1^* &= -10.000000000000000 & x_2^* &= 10.000000000000000 \\
x_3^* &= 2.273911598073273 & x_4^* &= 2.959407945041670.
\end{aligned}
$$

This yields an objective value of $Z^*(x_1, x_2, x_3, x_4) = 1.649608831366383$, which is a global solution to the constrained optimization problem.

15.10 Conclusions

This chapter discussed one of the EAs, the GA, and its applications in the power system. The definition and concepts used in GAs were presented in Section 15.2. The GA approach and theory were discussed in Sections 15.3 and 15.4, followed by the schemata theorem. In Section 15.6, the general procedure of a GA was discussed. The general approach of GAs was discussed in Section 15.7. Section 15.8 introduced applications of GAs to power systems.

We have considered several various algorithms and heuristic approaches for optimization of large systems. The leading method in evolution programming is the genetic algorithm (GA) for unit commitment. It has been widely applied to other Power System problems such as voltage VAr and state estimation (SE) problems.

15.11   Problem Set

PROBLEM 15.11.1

Minimize $f(x) = (x_1 - 1)^2 + (x_2 - 1)^2$

Subject to

$g_1(x) = x_1 + 2x_2 \leq 10$

$x_1, x_2 \geq 10.$

Solve by classical optimization and GAs.

PROBLEM 15.11.2

Minimize $f(x) = x_1^2 + x_2^2 - 2x_1 - 2x_2$

Subject to

$g_1(x) = x_1 + x_2 - 4 = 0$

$g_2(x) = x_1 - x_2 - 2 = 0$

$x_1, x_2 \geq 0.$

Solve by classical optimization and GAs.

PROBLEM 15.11.3

Using GAs, solve the following multiobjective nonlinear problem.

Maximize $f(x) = f_1(x) + f_2(x) + f_3(x)$

Subject to

$x_1 + x_2 \leq 5$

$x_1 + x_2 - x_3 = 0$

$x_1 > 0, x_2 \geq 0, x_3 > 0,$

where

$$f_1(x_1) = 20x_1 - 2x_1^2$$
$$f_x(x_2) = 16x_2 - x_2^2$$
$$f_3(x_3) = -x_3^2.$$

PROBLEM 15.11.4

Two generators supply the total load of 85 MW. The generator cost functions of the generators are given as follows:

$$f_1(P_{G_1}) = 120 + 10P_{G_1} + 0.02P_{G_1}^2 \quad 10 \leq P_{G_1} \leq 160 \text{ MW}$$

$$f_2(P_{G_2}) = 160 + 12P_{G_2} + 0.015P_{G_2}^2 \quad 30 \leq P_{G_1} \leq 100 \text{ MW}.$$

Use GA to obtain the optimum fuel cost.

References

1. Albrecht, R., Reeves, C., and Steele, N. *Artificial Neural Nets and Genetic Algorithms*, Springer-Verlag, New York, 1993.
2. Bauer, R. *Genetic Algorithms and Investment Strategies*, Wiley, New York, 1994.
3. Gen, M. and Cheng, R. *Genetic Algorithms and Engineering Design*, Wiley, New York, 1997.
4. Goldberg, D. E. *Genetic Algorithms in Search, Optimization and Machine Learning*, Addison-Wesley, Reading, MA, 1989.
5. Koza, J. R. *Genetic Programming*, MIT Press, Cambridge, MA, 1992.
6. Miranda, V. et al. Genetic algorithms in optimal multistage distribution. Network planning, *IEEE Transactions PES*, 9(4), 1994, 1927–1933.
7. Mori, H. A genetic approach to power system topological observability, *Proceedings of the IEEE International SCS*, New York, 1991.
8. Winter, G. et al. *Genetic Algorithms in Engineering and Computer Science*, Wiley, New York, 1996.
9. Holland, J. H. *Adaptation in Natural and Artificial Systems*, University of Michigan Press, Ann Arbor, MI, 1975.
10. Davis, L., Ed. *Handbook of Genetic Algorithms*, van Nostrand Reinhold, New York, 1991.
11. Ferreira, J. R., Peças Lopes, J. A., and Saraiva, J. T. Identification of preventive control procedures to avoid voltage collapse using genetic algorithms, *Proceedings of PSCC'99—Power System Computation Conference*, Trondheim, Norway, 1999.
12. Gil, E., Bustos, J., and Rudnick, H. Short-term hydrothermal generation scheduling model using a genetic algorithm, *IEEE Transactions on Power Systems*, 18(4), Nov. 2003, 1256–1264.
13. Mashhadi, H. R., Shanechi, H. M., and Lucas, C. A new genetic algorithm with Lamarckian individual learning for generation scheduling, *IEEE Transactions on Power Systems*, 18(3), Aug. 2003, 1181–1186.
14. Abdel-Magid, Y. L. and Abido, M. A. Optimal multiobjective design of robust power system stabilizers using genetic algorithms, *IEEE Transactions on Power Systems*, 18(3), Aug. 2003, 1125–1132.
15. Damousis, I. G., Bakirtzis, A. G., and Dokopoulos, P. S. Network-constrained economic dispatch using real-coded genetic algorithm, *IEEE Transactions on Power Systems*, 18(1), Feb. 2003, 198–205.

16. Masoum, M. A. S., Ladjevardi, M., Jafarian, A., and Fuchs, E. F. Optimal placement, replacement and sizing of capacitor Banks in distorted distribution networks by genetic algorithms, *IEEE Transactions on Power Delivery*, 19(4), Oct. 2004, 1794–1801.
17. Zoumas, C. E., Bakirtzis, A. G., Theocharis, J. B., and Petridis, V. A genetic algorithm solution approach to the hydrothermal coordination problem, *IEEE Transactions on Power Systems*, 19(3), Aug. 2004, 1356–1364.
18. Damousis, I. G., Bakirtzis, A. G., and Dokopoulos, P. S. A solution to the unit-commitment problem using integer-coded genetic algorithm, *IEEE Transactions on Power Systems*, 19(2), May 2004, 1165–1172.
19. Chiang, C. Improved genetic algorithm for power economic dispatch of units with valve-point effects and multiple fuels, *IEEE Transactions on Power Systems*, 20(4), Nov. 2005, 1690–1699.
20. Li, F., Pilgrim, J. D., Dabeedin, C., Chebbo, A., and Aggarwal, R. K. Genetic algorithms for optimal reactive power compensation on the national grid system, *IEEE Transactions on Power Systems*, 20(1), Feb. 2005, 493–500.
21. Yan, W., Liu, F., Chung, C. Y., and Wong, K. P. A hybrid genetic algorithm-interior point method for optimal reactive power flow, *IEEE Transactions on Power Systems*, 21(3), Aug. 2006, 1163–1169.

16

Functional Optimization, Optimal Control, and Adaptive Dynamic Programming

16.1 System Performance Evaluation and Optimization of Functionals

Classical control and optimization allow for use of state objectives and constraints with the use of mathematical optimization methods to solve for optimum acceptable or sufficient parameters. When dynamic or finite systems are presented, then the criterions for optimum are different for the system with multiple inputs and multiple outputs. These systems are usually found in control theory as well as in dynamic optimization. The objective of optimal control, dynamic control of the system is based on selecting appropriate methods for minimizing or maximizing some performance criteria [14,15].

In this chapter, we will review the fundamental approach for optimizing functional optimization, followed by systems that require the use of optimal control approaches. Then, we will recast the problem in DP problem followed by the use of approximate dynamic programming (ADP) to improve the challenge of dynamic programming (DP) optimization. The drawback curse of computational burden is overcome by using the new concepts of predictivity and dynamics or stochastic nature of system performance.

Finally, new trends in optimization being used in aerospace and intelligent control are being exploited in power systems. And introduction of its use in power system stability, unit commitment (UC), and optimal dispatch problems is introduced.

16.1.1 Extremization of Functionals

Consider the classical control problem formulated in a state variable form. We shall adopt the state variables of the process as $x_1(t), x_2(t), \ldots, x_n(t)$ with control variables as $u_1(t), u_2(t), \ldots, u_n(t)$.

The first-order differential equation is given by

$$\dot{x}(t) = g(x(t), u(t), t). \tag{16.1}$$

Here, $x(t) = [x_1(t), x_2(t), \ldots, x_n(t)]^T$ is denoted as the state vector and $u(t) = [u_1(t), u_2(t), \ldots, u_n(t)]^T$ as the control vector.

The state vector equation can be written as

$$\dot{x}(t) = g(x(t), u(t), t). \tag{16.2}$$

The time listing defined for time ranges between t_0 and t_f for which the state ranges and the control spread (spacial) the trajectory between t_0 and t_f. For every control function at time t, the state x has a value at $t_1 \to x(t_1)$.

DEFINITION 16.1

A state trajectory which satisfies the state variable constraints during the entire time interval $[t_0, t_f]$ is called an admissible trajectory. The set of admissible state trajectory will be denoted by X and $x \in X$.

DEFINITION 16.2

Similarly, a control limiting which satisfies the control constraints during the entire time interval $[t_0, t_f]$ is called admissible control. The set of admissible control by U and the notation $u \in U$ means the control listing is admissible.

16.1.2 Performance Measure

The functional performance measure as a system to be evaluated qualitatively is as follows:

$$f(y) = \int_{t_0}^{T} g(y, \dot{y}, t) dt, \tag{16.3}$$

where T is free to change but $y(T)$ and $y(t_0)$ are fixed.

In the optimal control measure is represented as

$$J = h(x(t_f), t_f) + \int_{t_0}^{t_f} g(x(t_f), u(t), t) dt, \tag{16.4}$$

where
t_0 and t_f are the initial and final time, respectively
h and g are scalar functions

Stated in optimal control: Find the admissible control u^* which causes the system $\dot{x}(t) = g(x(t), u(t), t)$ to follow an admissible trajectory x^* that minimizes the performance measure

$$J = h(x(t_f), t_f) + \int_{t_0}^{t_f} g(x(t_f), u(t), t)dt,$$ (16.5)

where
u^* is the optimal control
x^* is the optimal trajectory

We develop two theorems to solve problems of functional optimization. Assume continuity requirement at $y(t) = x(t)$ and $g(y, \dot{y}, t)$ is continuous up to the first derivative, a theorem about a can be stated as follows:

THEOREM 16.1

If x with $T = b$ is an extrenum of the functional, then it is necessary that

$g_x - \frac{d}{dt}g_{\dot{x}} = 0$ (Euler–Lagrange equation for some λ and with $a \leq t \leq b$.
$g(b) - g_{\dot{x}}(b)\dot{x}(b) = 0$

If T is fixed, the last equation is disregarded.

THEOREM 16.2

Any solution of Theorem 16.1 is a minimum if there exist P_1 and P_2, $\beta \geq 0$, and $b > a$ to satisfy the following conditions:

1. $A(\beta) \geq 0$ and $B(\beta) \geq 0$
2. $h^T A(\beta)h = 0$, $\dot{h}^T B(\beta)\dot{h} = 0$ and $P_1 h + P_2 \dot{h} = 0$ imply that $h = 0$ almost everywhere
3. $g_t(b) - g_x(b)\dot{x}(b) \geq 0$ with $\dot{x}(b) \neq 0$

If T is fixed, the last equation is disregarded.

16.1.3 Theorems of Optimization of Constrained Functionals

There are two kinds of constraints to be considered in the optimization of functionals. They are

1. $f_i(y, \dot{y}, t) = 0$ $i \in (1, m)$

2. $\int_{t_a}^{T} f_i(y, \dot{y}, t)dt = K$ $i \in (1, p)$,

where all the functions are real, scalar, and single valued. The constants K_i remains unchanged during the process of optimization until the convergence criteria are met. It is required to handle the cases of inequality constraints.

$$\text{Min } f(y) = \int_{t_a}^{T} g(y, \dot{y}, t) dt. \tag{16.6}$$

Subject to

$$f_i(y, \dot{y}, t) = 0 \quad i \in (1, m) \tag{16.7}$$

$$\int_{t_0}^{T} f_i(y, \dot{y}, t) dt = K \quad i \in (1, p) \tag{16.8}$$

$$K^{\min} \leq K \leq K^{\max}. \tag{16.9}$$

The terminal states $y(a)$ and $y(T)$ are fixed but the upper limit is free to change.

Now, by defining vector form of the functionals, we obtain the following representations.

$$F(y, \dot{y}, t) = [f_1(y), f_2(y), \ldots, f_m(y)]^{\mathrm{T}} \tag{16.10}$$

and

$$G(y, \dot{y}, t) = [g_1(y, \dot{y}, t), g_2(y, \dot{y}, t), \ldots, g_p(y, \dot{y}, t)]^{\mathrm{T}}. \tag{16.11}$$

Such that, in matrix form:

$$F(y, \dot{y}, t) = 0 \tag{16.12}$$

$$\int_{t=a}^{t=T} G(y, \dot{y}, t) dt = K, \tag{16.13}$$

where

$$K = [K_1, K_2, \ldots, K_p]^{\mathrm{T}}. \tag{16.14}$$

The Lagrange for the problem is

$$L(y, \dot{y}, t) = g(y, \dot{y}, t) + \lambda(t) F(y, \dot{y}, t) + \lambda G(y, \dot{y}, t), \tag{16.15}$$

where

$$\lambda(t) = [\lambda_1(t), \lambda_2(t), \ldots, \lambda_m(t)]^T \tag{16.16}$$

$$\lambda = [\lambda_1, \lambda_2, \ldots, \lambda_p]^T. \tag{16.17}$$

Note $\lambda(t)$ and λ are the Lagrange vectors of m and p rows, respectively. The former is a function of time and the latter is constant with respect to time. The argument t in $\lambda(t)$ should not be distinguished from the nontime-varying λ.

To write theorem of optimizing the functional with constraints, the following notations are given:

1. $U_{xx} = L_{xx} + \beta F_x^T F_x$ (16.18)

2. $U_{\dot{x}\dot{x}} = L_{\dot{x}\dot{x}} + \beta F_{\dot{x}}^T F_{\dot{x}}$ (16.19)

3. $U_{\dot{x}\dot{x}} = L_{\dot{x}\dot{x}} + \beta F_x^T F_{\dot{x}}$ (16.20)

4. $A(\beta) = U_{xx} - P_1^T P_1$ (16.21)

5. $B(\beta) = U_{\dot{x}\dot{x}} - P_2^T P_2,$ (16.22)

where β is nonnegative number and P_1 and P_2 are conformal matrices which are generally functions of time.

The L- and F-functions without argument denote the value of the trajectory $x(t)$ which is to be determined. The specific values are given as function value at $L(b)$, $L_x(b)$, and $L_{\dot{x}}(b)$ at time $t = b$.

THEOREM 16.3

If x with $T = b$ is an optimum of the constrained functional, then it is necessary that

1. $L_x = (dL_{\dot{x}}/dt) = 0$ for some λ and $\lambda(t)$ in $a \leq t \leq b$
2. $L(b) = L_{\dot{x}}(b)\dot{x}(b)$

If T is fixed, the last equation is disregarded.

THEOREM 16.4

Any solution of Theorem 16.3 is a minimum if there exist P_1 and P_2, $\beta \geq 0$, and $b > a$ to satisfy the following conditions:

1. $A(\beta) \geq 0$ and $B(\beta) \geq 0$
2. $h^T A(\beta)h = 0$, $\dot{h}^T B(\beta)\dot{h} = 0$ and $P_1 h + P_2 \dot{h} = 0$ imply that $h = 0$ almost everywhere
3. $L_t(b) - L_x(b)\dot{x}(b) \geq 0$ with $\dot{x}(b) \neq 0$

If T is fixed, the last equation is disregarded.

16.1.4 Summary of Procedure for Optimizing Constrained Functionals

There are four steps as in the formulation presented in Chapter 4 with static optimization of functions. Constrained functional optimization problems by applying the foregoing theorems give rise to the following steps:

1. Formulation of the Lagrange functions. This is given as

$$F(y, \dot{y}, t) = [f_1(y), f_2(y), \ldots, f_m(y)]^T \qquad (16.23)$$

and

$$G(y, \dot{y}, t) = [g_1(y, \dot{y}, t), g_2(y, \dot{y}, t), \ldots, g_p(y, \dot{y}, t)]^T, \qquad (16.24)$$

where $f_i(y, \dot{y}, t) = 0$ is the differentiable constraints of the F-function. The function of the form $g_i(y, \dot{y}, t)$ is the integrand of the G-Function (integral constraints) such that

$$K_{i1} \leq \int_{t_a}^{T} g_i(y, \dot{y}, t)dt \leq K_{i2} \qquad (16.25)$$

The integral is considered to be a constant K_i with $K_{i1} \leq K_i \leq K_{i2}$ during the optimization process. The Lagrange function is then defined as

$$L(y, \dot{y}, t) = g(y, \dot{y}, t) + \lambda(t)F(y, \dot{y}, t) + \lambda G(y, \dot{y}, t), \qquad (16.26)$$

where

$$f(y) = \int_{t_a}^{T} g_i(y, \dot{y}, t)dt. \qquad (16.27)$$

2. Determination of optimum candidates(s) by applying Theorems 16.1 and 16.2

3. Sufficiency test by applying Theorems 16.3 and 16.4

4. Further optimization: Similar to the case for functions, we now have the case where $f(y)$ is a function of K. In this process, further optimization is to be sought with respect to K_i in the interval $K_{i1} \leq K_i \leq K_{i2}$ for all i. The optimum of K_i may occur at the boundaries of the constraints.

16.2 Solving the Optimal Control Problem

Optimal control problems are defined in terms of performance measures [10]. We begin here with the formulation given as follows from the generalized optimal control problem

$$\dot{x}(t) = a(x(t), u(t), t),$$

where $u^* \in U$ is the optimal control to the spatial trajectory of $x(t)$ where $x^* \in X$ that minimizes a performance function given by

$$J = h(x(t_f), t_f) + \int_{t_0}^{t_f} g(x(t_f), u(t), t)dt. \tag{16.28}$$

Given t_f is free, the control vector is admissible over the time interval, and the function $g(x(t_f), u(t), t)$ is differentiable and continuous over the same trajectory time span.

This performance index (PI) is selected and modified for different class of problems as follows.

Minimum-time problems
To transfer a system from an arbitrary initial state $x(t_0) = x_0$ to a specified target set S in minimum time. The performance measure to be minimized is

$$J = t_f - t_0 \tag{16.29}$$

$$= \int_{t_0}^{t_f} dt. \tag{16.30}$$

Examples of minimum-time control problems include interception of attacking aircraft and missiles, optimization of spacecraft docking machines, operational modes of radar or gun systems.

Terminal control problems
The challenge problem is to minimize the final state of a system from the desired value $r(t_f)$. Candidate performance measure is

$$J = \sum_{i=1}^{n} [x_i(t_f) - r_i(t_f)]^2. \tag{16.31}$$

Since both positive and negative deviations are equality undesirable, the error is squared to enforce this condition in its scale form. Alternatively, absolute measures or l_1-norms could be used to yield:

$$J = \sum_{i=1}^{n} \|x_i(t_f) - r_i(t_f)\|. \tag{16.32}$$

For general purposes, a quadratic form has been shown easier to handle and is presented in matrix form as

$$J = [x(t_f) - r(t_f)]^T [x(t_f) - r(t_f)]. \tag{16.33}$$

The term $\|x_i(t_f) - r_i(t_f)\|$ is called the norm of the vector $[x_i(t_f) - r_i(t_f)]$.

For the general matrix notation, we introduce a real symmetric $n \times n$ weighting matrix H that is positive semidefinite or nonnegative definite by definition. This allows greater generality, such that

$$J = [x(t_f) - r(t_f)]^T H [x(t_f) - r(t_f)] \tag{16.34}$$

or

$$J = \|x(t_f) - r(t_f)\|_H^2. \tag{16.35}$$

Minimum-control-effort problems
The challenge here is to transfer a system from an arbitrary initial state $x(t_0) = x_0$ to a specified target set S, with a minimum expenditure of control effort. The minimum control effort has many forms. These include:

a.
$$J = \int_{t_0}^{t_f} u(t) dt \tag{16.36}$$

(for single control variable system)

b. If there are several controls, and the rate of expenditure of control effort of the ith control is $c_i| u_i |$ for $i \in (1,m)$, then minimizing

$$J = \int_{t_0}^{t_f} \sum_{i=i}^{m} [\beta_i|u_i(t)]dt \tag{16.37}$$

would minimize the control effort expanded. All coefficients are nonnegative weighting factors.

c. Minimum energy dissipated where the goal is to minimize the performance measure given by

$$J = \int_{t_0}^{t_f} u^2(t)dt \tag{16.38}$$

or, for several control inputs the general form of performance, the corresponding performance measure is

$$J = \int_{t_0}^{t_f} [u^T(t)Ru(t)]dt \tag{16.39}$$

$$= \int_{t_0}^{t_f} ||u(t)||_R^2 \, dt, \tag{16.40}$$

where $\mathbf{R}(t)$ is a real symmetric positive definite weighting matrix (i.e., $z^h R z > 0$ for $\forall z \neq 0$). The R-elements may be time-dependent during the control-effort expenditure interval $[t_0, t_f]$.

The choice of the performance measure depends on the expression that minimizes the system error or maximizes the system performance.

We redefine Equation 16.39 as part of optimal control derived from satisfactory requirements for

1. Admissible trajectory, $x \in S$
2. Admissible control, $u \in U$

And admissible control is said to be optimal if it generates and admissible trajectory and also optimizes the PI. That is

$$\underset{v \in P}{\text{Opt}} \, I(y_a, v, a) = I(y_a, v, a),$$

where $u = u(t)$ if an optimal control. The trajectory $x = x(t)$ generates an admissible control $u = u(t)$ within the closed interval $a \leq t \leq T$ is called an optimal trajectory. Note that x intersects S at $x_b = x(b)$.

The ϕ-function
Let $y = y(t)$ in the interval $a \leq t \leq T$ be a point in the admissible trajectory, then the scalar function is defined in the interval by

$$\phi(y, t) = \operatorname*{Opt}_{v \in P} I(y_a, v, a) = \operatorname*{Opt}_{v \in P} \left\{ \int_t^T g[y(\tau), v(\tau), \tau] d\tau + K[t_T, T] \right\}$$

The ϕ-function plays an important role in the statement and proof of the optimum principle.

The costate vector and Hamiltonian
A row n-vector $\lambda = \lambda(t)$ known as the costate vector is defined by the derivative

$$\lambda(t) = \phi_y(y, t)|_{y=x} = \phi_x(x, t),$$

where $x = x(t)$ is a point on the optimal trajectory. A scalar function $\eta = \eta(t)$ is defined as

$$\eta(t) = \phi_y(y, t)|_{y=x} = \phi_x(x, t).$$

Here, both λ and η are functions of time.
   The Hamiltonian is defined by

$$H(y, v, \phi_y, t) = g(y, v, \phi_y, t) + \phi_y(y, v, t),$$

which is a scalar function determinable in the interval $a \leq t \leq T$ whenever the control is given. It should be noted that the vectors $\mathbf{x}$, $\mathbf{u}$, λ, and η without the argument are the values at time t. Sometimes they are specified with argument t for emphasis.

16.2.1   Continuous Optimum Principle

Optimization involving constraints of differential forms is particularly important in modern design of control systems. Calculus of variation is one of the techniques that can be used to solve some of the optimal problems. But, it cannot handle problems such as optimal control that involves piecewise-continuous control variable or dynamic equation that are not differentiable with respect to the control variables.

In 1958, L.S. Pontryagin formulated and proved a maximum principle which relaxes the conditions on both control variables and dynamic equations required by the variational method. Since then, different versions of the maximum principle have been developed for practical uses. Considered in this chapter is a generalized optimum (minimum or maximum) principle that can be used to solve a wide class of optimal problems. The well-known Euler–Lagrange equation can be obtained as a special ease of the optimum principle.

16.2.2 Formulation of the Problem

We now summarize the problem formulation and the discussion on the costate vector and the Hamiltonian as follows:

1. System or dynamic equations

   The dynamic system under consideration is described by the ordinary differential equation

   $$\dot{y}(t) = f(y(t), v(t), t), \tag{16.41}$$

   where $y = y(t)$ is a column n-vector function of t and $v = v(t)$ is a column r-vector function of t. They are referred to as the state and control vectors. The f-function is assumed to be differentiable with y and t but only continuous with v.

2. Performance index (PI)

   The optimality of a solution for the problem is measured by the scalar function

   $$I(y_a, v, T) = \int_{t_0}^{T} g[(y(\tau), v, T)]d\tau + K(y_T, T), \tag{16.42}$$

   where a is fixed, T is free, $y_a = y(a)$ and $y_T = y(T)$. The control vector (r) is admissible and denoted by $v(\tau) \in P$ for all $a \leq \tau \leq T$. The g-function is assumed to be differentiable with y and t but only continuous with v. The K-function is assumed differentiable with both y_T and T.

3. Terminal condition

   The terminal state y_T is restricted to be a point in a manifold S described by the parametric equations $y_T = h(s, T)$ where s is a column q-vector. The h-function is assumed to be differentiable with both s and T. The composite function $K(y_T, T) = K[h(s, T), T] = \overline{K}(s, T)$ is also differentiable with both s and T. A point is a special case of the manifold when $q = 0$.

4. Admissible control

Let P denote all the piecewise continuous functions of t in a subset of R_r which is the r-space. Then, a control is called admissible if $v(t) \in P$. The value of $v(t)$ at a jump point (if any) is defined here to be the limit from the right. This definition requires that $v(t)$ be specified at a jump point $t = d$ by the intervals: $t < d$ and $t \geq d$. Such a specification assures the existence of $v(t)$ for all $a \leq t \leq T$.

5. Admissible trajectory

Any solution of (6-1) with an admissible control started with y_a and terminated in S is called an admissible trajectory.

6. Optimal control

An admissible control is said to be optimal if it generates an admissible trajectory and also optimizes (minimizes or maximizes) the PI. That is

$$\underset{v \in P}{\text{Opt}}\, I(y_a, v, T) = I(y_a, u, a), \tag{16.43}$$

where u is an optimal control. The trajectory $x = x(t)$ generated by the optimal control $u = u(t)$ for $a \leq t \leq b$ according to (6-1) is called an optimal trajectory. Note that x intersects S at $X_b = x(b)$.

7. ϕ-Function

Let $y = y(t)$ in the interval $a \leq t \leq T$ be a point on an admissible trajectory, then the function (scalar) is defined in the interval by $a \leq t \leq T$

$$\phi(y,t) = \underset{v \in P}{\text{Opt}}\, I(y_a, v, T) = \underset{v \in P}{\text{Opt}} \left\{ \int_t^T g[(y(\tau), v(\tau), \tau)]d\tau + K(y_T, T) \right\}. \tag{16.44}$$

The ϕ-function plays an important role in the statement and proof of the optimum principle. It has the following properties: (1) it exists because y is on an admissible trajectory at t, (2) it is a point function depending on y and t only since the control $v(\tau)$ has been chosen, and (3) it is differentiable from the right at y due to the assumption $v(t) = v(t^+)$.

8. Costate vector and Hamiltonian

A row n-vector $\lambda = \lambda(t)$ known as the costate vector is defined by the derivative

$$\lambda = \phi_y(y,t)|_{y=x} = \phi_x(x,t), \tag{16.45}$$

where $x = x(t)$ is a point on an optimal trajectory. A scalar function $\eta = \eta(t)$ is defined by

$$\eta = \phi_t(y,t)|_{y=x} = \phi_t(x,t). \tag{16.46}$$

Both λ and η are functions of x and t. The Hamiltonian is defined by

$$H(y,v,\phi_y,t) = g(y,v,t) + \phi_y(y,t)f(y,v,t), \qquad (16.47)$$

which is a scalar function determinable in the interval a $a \leq t \leq T$ whenever the control is given. It should be noted that the vectors x, v, u, λ, and η without argument are the values at t. Sometimes, they are specified with argument t for emphasis.

16.2.3   Theorems for the Pontryagin Maximum Principle (PMP)

The continuous optimum principle is aimed at solving the problem formulated in Section 16.2.1. It specifies a set of necessary conditions that can be utilized to determine the optimal control and trajectory. The principle may be regarded as a generalization of the well-known Pontryagin maximum principle (PMP). It can be stated as the following form of theorem.

THEOREM 16.5

If u is an optimal control and x is the associated optimal trajectory which intersects S at T=b, then there exist a row vector λ and η and a scalar in $a \leq t \leq T$ such that

(a) $\underset{v \in P}{\mathrm{Opt}}\, H(x, v, \lambda, t) + \eta = H(x, u, \lambda, t) + \eta = 0$ $\qquad$ (16.48)

(b) $\dot{\lambda} + \lambda f_x(x, u, t) + g_x(x, u, t) = 0$ $\qquad$ (16.49)

(c) $\dot{\eta} + \lambda f_t(x, u, t) + g_t(x, u, t) = 0$ $\qquad$ (16.50)

(d) $\lambda(b)h_s(s, b) = \overline{K}_s(s, b)$ $\qquad$ (16.51)

(e) $\lambda(b)h_b(s, b) + \eta(b) = \overline{K}_b(s, b)$ $\qquad$ (16.52)

Condition (a) implies that x, λ, and η are kept constant during the optimization with respect to v.

The optimal control is sought by optimizing the Hamiltonian at each t in the interval $a \leq t \leq b$ and hence the equality of condition (a) holds for each t in the interval.

In conditions (d) and (e), $h_s(s, b)$ and $\overline{K}_b(s, b)$ denote the partial differentiations of the functions with respect to b.

16.2.4 Sufficiency Test and Some Special Cases for the Optimum Principle

Consider y to be the trajectory generated by an admissible control v from (y_a, a) to (y_T, T) where y_T is the state that y intersects S at $t = T$.
Recall the ϕ-function

$$\phi(y, t) = \operatorname*{Opt}_{v \in P} I(y_a, v, T) = \operatorname*{Opt}_{v \in P} \left\{ \int_t^T g[(y(\tau), v(\tau), \tau)] d\tau + K(y_T, T) \right\}. \qquad (16.53)$$

It follows that

$$\phi(y, t) = \operatorname*{Opt}_{v(\tau) \in P} I(y, v, T). \qquad (16.54)$$

For $t \leq \tau \leq T$. At $t = T$, the ϕ-function reduces to

$$\phi(y_T, T) = K(y_T, T) \qquad (16.55)$$

For convenience, we will use ϕ_y and ϕ_t to denote the partial derivatives of the ϕ-function with respect to y and t, respectively. In terms of the derivatives, a test of sufficiency for the solution obtained from the optimum principle can be stated as follows.

Sufficiency test: The control u obtained from the optimum principle is a global minimum if $H(y, v, \phi_y, t) + \phi_t \geq 0$ for all $a \leq \tau \leq T$. (In the same breath, reversal of the inequality ensures a global maximum if $H(y, v, \phi_y, t) + \phi_t \leq 0$).
To verify the test, we write condition (a) of the principle as

$$H(x, u, \lambda, t) + \eta = g(x, u, t) + \lambda f(x, u, t) + \eta \qquad (16.56)$$

$$= g(x, u, t) + \lambda \dot{x} + \eta \qquad (16.57)$$

$$= g(x, u, t) + \dot{\phi}(x, t) + \eta \qquad (16.58)$$

$$= 0 \qquad (16.59)$$

for all $a \leq t \leq b$. Integration of both sides from a to b along the optimal trajectory gives

$$\int_a^b g(x, u, t) dt + \int_a^b d\phi(x, t) = 0. \qquad (16.60)$$

Since the integrand of the second integral is a total differential, we obtain from the above equation

$$\int_a^b g(x, u, t)dt + \phi(x_b, b) - \phi(x_a, a) = I(x_a, u, a) - \phi(x_a, a) = 0. \qquad (16.61)$$

On the other hand, the test criterion reveals that

$$H(y, v, \phi_y, t) + \phi_t = g(y, v, t) + \phi_y \dot{y} + \phi_y \qquad (16.62)$$

$$= g(y, v, t) + \dot{\phi}(y, t) \geq 0 \qquad (16.63)$$

for all $a \leq t \leq T$. Integration of both sides from a to T along y gives:

$$\int_a^T g(y, v, t)dt + \int_a^T d\phi(x, t) \geq 0. \qquad (16.64)$$

Carrying out the second integral, we have

$$\int_a^b g(y, v, t)dt + \phi(x_T, T) - \phi(x_a, a) = I(y_a, v, a) - \phi(y_a, a) \geq 0. \qquad (16.65)$$

By the assumption of fixed initial state, we have $x_a = y_a$, and hence $\phi(x_a, a) = \phi(y_a, a)$. It follows from Equations 16.61 and 16.65 and the above equation that

$$I(x_a, u, a) = \phi(x_a, a) = \phi(y_a, a) \leq I(y_a, v, a). \qquad (16.66)$$

This concludes that u is a minimal control and x is the associated minimal trajectory. (Maximal control of u can be similarly verified when the inequality sign of the criterion is reversed.)

Note, in order to apply the sufficiency test, one has to find the scalar function $\phi(y, t)$ and then ϕ_y and ϕ_t. Suggested in Equation 16.61 is the fact that $\phi(y, t) = I(y, u, t)$. In other words, one can determine $\phi(y, t)$ from $I(x_a, u, a)$ by taking x_a and a as parameters and then replace x_a and a by y and t, respectively. Nonnumerical value restrictions on x_a or a in order to apply the test introduces complications.

In general, we conclude that from the Hamiltonian, we apply the necessary conditions of optimality. Then, for optimization, determine the controls that satisfy the sufficiency conditions.

16.2.5   Use of the Optimum Principle for Special Control Problems

1. *Terminal state problems*

   a. *T is fixed.*

   When $T = b$ is fixed, $\phi(x, t)$ is only a function of x at $t = T$. Therefore, $\eta(b)$ cannot be defined by $\phi_t(x_b, t)$. In such a case, condition
   (e) of the continuous principle theorem must be disregarded and
   leave $\eta(b)$ to be determined by the known $t = T$.

   b. *S is a point $(q = 0)$.*

   When x_b is fixed, $\phi(x, t)$ is not a function of x at $t = b$ and hence $\lambda(b)$
   cannot be defined by $\phi_x(x_b, b)$. In such a case, condition (c) must be
   disregarded and leave $\lambda(b)$ to be determined by the known x_b.

2. *Unconstrained Euler–Lagrange equation*

   Consider the special case in which $\dot{y}(t) = f(y, v, t) = v$. Then, it follows
   from the optimum principle that

   a. $g(x, \dot{x}, t) + \lambda u + \eta = 0$

   where u is unconstrained and hence

   $$\frac{dg}{du} + \lambda = 0 \quad \text{or} \quad g_{\dot{x}} + \lambda = 0 \tag{16.67}$$

   b. $\qquad \dot{\lambda} + g_{\dot{x}} = 0 \quad \text{or} \quad g_{\dot{x}} + \dfrac{d}{dt}[-g_x] = g_{\dot{x}} + \dfrac{d}{dt}[-g_x] = 0 \tag{16.68}$

3. *Autonomous systems*

   Consider the special case in which the functions f, g, and h are not
   explicit functions of t, and also K is not an explicit function of T.
   Condition (c) of the optimum principle shows that $\dot{\lambda} = 0$ or $\eta = c = 0$
   for all $a \le t \le b$ since condition (d) requires that $\eta(b) = 0$. This
   concludes that conditions (c) and (e) are to be deleted and $\eta = 0$ for
   free T but $\eta = c$ for fixed T.

4. *Linear and time-invariant systems*

   A wide class of optimization problems in modern controls is of the
   linear and time-invariant form. The constraint equations and also the
   PI are linear and time-invariant. That is

   $$\dot{y}(t) = Ay + Bv \quad \text{and} \quad I = \int_0^b (Cy + Dv + e)dt + K(y_b), \tag{16.69}$$

   where A is n-square, B is nxr, C is Ixn, D is Ixr, e is a scalar constant, and
   K is a scalar function. The manifold is the same as described before.
   Application of the optimum principle yields the following result:

   a. $\qquad (\lambda A + C)x + (\lambda B + D)uu + e = 0 \quad$ are all $0 \le t \le b$. $\tag{16.70}$

   The optimal u does not exist unless it is constrained.

b. $$\dot{\lambda} + \lambda A + C = 0 \qquad (16.71)$$

c. $$\lambda(b)h_s(s) = \overline{K}_s(s) \quad \text{where } b \text{ is free.} \qquad (16.72)$$

The solutions of λ from (b) and $x(t)$ from the constraint can be expressed by

$$\lambda(t) = \lambda(0)e^{-At} - \int_0^t Ce^{-A(t-\tau)} \, d\tau \qquad (16.73)$$

$$x(t) = e^{At}x(0) + \int_0^t e^{A(t-\tau)}Bu(\tau) \, d\tau. \qquad (16.74)$$

Now, define a matrix function $E(t)$ in such a way that $E(t) = \int_0^t e^{A\tau} \, d\tau$ or $dE(t) = e^{At} \, dt$. Then, in terms of $E(t)$, the above vectors can be written as

$$\lambda(t) = [\lambda(0) = CE(t)]e^{-At} \qquad (16.75)$$

$$e^{-At}x(t) = x(0) - \int_0^t dE(-\tau)Bu(\tau)d\tau. \qquad (16.76)$$

Let us consider the bang–bang control as generated by bounded u_i for all i. Integration by parts reveals that

$$\int_0^t dE(\tau-)Bu(\tau) \, d\tau = E(-t)Bu(t) - \int_0^t E(-\tau)B \, du(\tau). \qquad (16.77)$$

Then, by multiplying e^{At} we obtain

$$x(t) = e^{At}[x(0) - E(-t)Bu(t) + M(t)], \qquad (16.78)$$

where

$$M(t) = \int_0^t E(-\tau)B \, du(\tau) = \sum E(-t_s)B[u(t_s^+) - u(t_s^-)]. \qquad (16.79)$$

The summation includes all the jumps that are caused by the switching time t_s, which is less than t. The tedious part of solving the problem is the evaluation of e^{At} and $E(t)$. Similarity transformation can be employed but it requires eigenvalues and eigenvectors. Series expansion can be used to evaluate them on computer but t must be specified numerically.

16.2.6   Regulator Problem and Riccati Equation

As a wide class of control problem, it is required to find an optimal control that keeps a state vector close to a prescribed trajectory. The control effort is to be reduced as much as possible and the state vector constrained by a set of linear differential equations.

In term of the optimum principle, we can formulate the problem as to minimize a PI with

$$g(y, v, t) = 0.5[q(t) - z(t)]^T D(t)[q(t) - z(t)] + 0.5v^T Q(t)v(t) \qquad (16.80)$$

and

$$K(y_T, T) = 0.5[q(t) - d]^T G[q(t) - d]. \qquad (16.81)$$

Subject to

$$f(y, v, t) = A(t)y(t) + B(t)v(t) + w(t). \qquad (16.82)$$

and

$$q(t) = C(t)y(t).$$

The terminal T is assumed to be fixed but $y(T)$ is free or S is the whole space. The matrices that involve in the above equations are

1. $v(t)$ is an unconstrained control vector and $y(t)$ is the associated state vector, $q(t)$ is an output vector, $z(t)$ is a reference vector, and d is a constant vector. They have respectively r, n, m, m, and m components.
2. G is symmetric, positive semidefinite, and constant.
3. $D(t)$ is symmetric, positive semidefinite, and function of t.
4. $Q(t)$ is symmetric, positive-definite, and function of t.
5. $C(t)$ is mxn and function of t.

Some special cases of the problem are given names as below:

The linear servomechanism problem: $d = z(T)$.
The tracking problem: $w(t) = 0$ and $d = z(T)$.
The output regulator problem: $w(t) = 0$, $z(r) = 0$, and $d = 0$.
The state regulator problem: $w(t) = 0$, $z(t) = 0$, $d = 0$, and $C(t) = I_n$.

To solve the problem by using the optimum principle, we will omit the argument unless there is an ambiguity. The principle gives the following conditions:

(a)
$$H_{min} = 0.5[Cx - z]^T D(t)[Cx - z] + 0.5u^T Qu$$
$$+ \lambda(Ax + Bu + w) + \eta, \qquad (16.83)$$

where $u = -Q^{-1}B^T\lambda^T$. This holds true because H is a quadratic function of u and Q is positive semidefinite.

(b)
$$\dot{\lambda} + \lambda A + (Cx - z)^T DC = 0 \qquad (16.84)$$

(c)
$$\lambda = (Cx - d)^T GC \quad \text{at } t = T. \qquad (16.85)$$

The problem is extremely hard, if not impossible, to solve because the boundary condition of x is given at $t = a$ but that of λ is given at $t = T$. It is for this reason that the problem is to be solved indirectly for an n-square matrix function $F(t)$ and an n-vector $h(t)$ related by

$$\lambda^T = Fx + h \qquad (16.86)$$

We have dropped the argument of time here for simplicity. By differentiating this equation with respect to time and then substituting $\dot{x}$ yield

$$\dot{\lambda}^T = \dot{F}x + F\dot{x} + \dot{h} \qquad (16.87)$$
$$= (\dot{F} + FA - FBQ^{-1}B^TF)x + (\dot{h} = FBQ^{-1}B^Th + Fw). \qquad (16.88)$$

On the other hand, we have from condition (b) that

$$\dot{\lambda}^T + (A^TF + C^TDC)x + A^Th - C^TDz = 0. \qquad (16.89)$$

Cancellation/elimination of $\dot{\lambda}$ gives

$$(\dot{F} + FA - FBQ^{-1}B^TF + A^TF + C^TDC)x + (A^TF + C^TDC)x \qquad (16.90)$$
$$+ \dot{h} + (A^T - FBQ^{-1}B^T)h + Fw - C^TDz = 0. \qquad (16.91)$$

The functions F and h will be determined in such a way that

$$\dot{F} + FA - FBQ^{-1}B^TF + A^TF + C^TDC = 0 \qquad (16.92)$$

and

$$\dot{h} + (A^T - FBQ^{-1}B^T)h + Fw - C^TDz = 0. \qquad (16.93)$$

The boundary conditions of F and h satisfy at $t = T$

$$F = C^TGC \qquad (16.94)$$

and

$$h = -C^TGD. \qquad (16.95)$$

Equation 16.92 is known as the Riccati equation which is nonlinear but can be solved by numerical methods. After $F(t)$ and $h(t)$ have been solved from Equations 16.92 and 16.93 subject to the specified boundary conditions, the state vector is to be solved from the following equation:

$$\dot{x} = Ax + Bu + w = Ax - BQ^{-1}B^{T}\lambda^{T} + w \tag{16.96}$$

$$= Ax - BQ^{-1}B^{T}(Fx + h) + w \tag{16.97}$$

$$= (A - BQ^{-1}B^{T}F)x + (w - BQ^{-1}B^{T}h) \tag{16.98}$$

subject to the given initial condition $x(a)$.

It is not hard to verify that F and h are thus solved will make x and λ satisfy all the conditions (a), (b), and (d) as obtained from the optimum principle earlier. The optimal control is then determined by

$$u = -Q^{-1}B^{T}\lambda^{T} = -Q^{-1}B^{T}(Fx + h). \tag{16.99}$$

which may be constructed by F, h, and x in practical applications.

The matrix F^{T} as being solved from Equation 16.92 is a symmetrical matrix. It can be shown by transposing Riccati's equation to have

$$\dot{F}^{T} + F^{T}A^{T} - F^{T}BQ^{-1}B^{T}F^{T} + F^{T}A + C^{T}DC = 0. \tag{16.100}$$

The solution of F^{T} from this equation is the same as that from Riccati. This holds true as they are subject to the same terminal conditions that $F = F^{T} = C^{T}GC$ at $t = T$.

Furthermore, the Riccati equation can be converted to linear differential equation but the computational burden of the solution increases as the number of variables is doubled. Such linear differential system of equation can be solved analytically in principle but the state transition matrix is often times difficult to determine for many practical applications and especially when the system dimension is large. The Riccati equations, however, become more useful when solving time-invariant systems where the $A, B, C, D,$ and Q matrices are constants with respect to time.

16.3 Selected Methods of Determining the Control Functions for Convergence of Optimum Principle

Two methods that are commonly used to minimize the control function that minimizes the performance measure in Sections 16.1 and 16.2 are not discussed. We have discussed Pontryagin's principle and the PMP. The principles give rise to necessary conditional statement that must hold on an optimal trajectory. We will now begin our discussion by introducing the DP method.

16.3.1 Dynamic Programming Method

Consider the classical time-dependent example given by

$$\dot{x}(t) = a(x(t), u(t), t), \tag{16.101}$$

where we seek to find $u^*(t) = f(x(t), t)$ from the admissible vector space $u \in U$ in the within the limits of the time interval with maximum value t_f.

The function f is the optimal control law with feedback or optimal policy which provides information on how to generate $u(t)$ at time t for the corresponding state value $x(t)$.

16.3.2 Principle of Optimality Is Used to Find $u^*(t)$ (Richard Bellman's Method)

From Chapter 15, we define Bellman's optimality principle as: An optimal policy has the property that whatsoever the initial state and the initial decisions are, the remaining decisions must constitute and optimal policy with regard to the state resulting from the first decision.

Dynamic programming is a computational technique which extends the decision-making concept to sequence of decisions which together define an optimal policy and trajectory.

In a generalized form of dynamic systems represented as

$$\dot{x}(t) = Ax(t) + Bu(t). \tag{16.102}$$

The performance measure or cost to be minimized is

$$J(x(t), u(t), \lambda, t) = x^2(t) + \lambda \int_{t_0}^{t_f} u^2(t) \mathrm{d}t, \tag{16.103}$$

where t_f is the specified final time and λ is the weighting factor to permit adjustment or the relative importance of the two terms in J. And, $x(t)$ and $u(t)$ are squared vectors since the sign values of these quantities are of importance.

The performance measure reflects the desire to derive the final state of $x(t)$ as close to zero as possible without excessive and/or expensive control effort.

For generality in DP, we write

$$C_{02}(x(0), u(0)) = J_{01}(x(0), u(0)) + J^*(x(1)). \tag{16.104}$$

The cost or the optimal trajectory is given by

$$J_{02}^*(x(0)) = \underset{u(0)}{\text{Min}} \left[J_{01}(x(0), u(0)) + J_{12}^*(x(1)) \right].$$ (16.105)

$C_{02}(x(0), u(0))$ is the minimum cost of operation over the last two stated for one quantized value of $x(0)$ given a particular trial quantized value of $u(0)$.

$J_{01}(x(0), u(0))$ is the cost of operation in the interval $k=0$ to $k=1$ for specified quantized values of $x(0)$ and $u(0)$.

$J_{12}^*(x(1))$ is the cost of the optimal last-stage trajectory, which is a function of the state $x(1)$.

$J_{02}^*(x(0))$ is the minimum cost of operation over the last two pages for one quantized value of $x(0)$ given a particular trial quantized value of $u(0)$.

The law that minimizes

$$J = h(x(t_f)) + \int_{t_0}^{t_f} g(x(t), u(t), t) dt,$$ (16.106)

where

$$u \in U \quad \text{and} \quad 0 \le t_0 \le t \le t_f.$$ (16.107)

The principle of optimality in discrete form converges to

$$J_{N-K,N}^*(x(N-K)) = \underset{u(N-K)}{\text{Min}} \left\{ g_D(x(N-K), u(N-K)) \right.$$

$$\left. + J_{N-K+1,N}^*(a_D(x(N-K), u(N-K))) \right\}.$$ (16.108)

This is the desired recursive equation of DP problems. The derivation of the recurrence equation has also revealed another important concept of the imbedding principle. That is, the optimum policy and the minimum cost possible for a K-stage process are contained in the results for an N-stage process, provided $N \ge K$.

Now, DP using discrete form is ideally for digital programming. An alternate method for solving the DP is by nonlinear partial differential equation defined from the Hamilton–Jacobi–Bellman (H–J–B) equation, which is presented in this chapter. Again, we are trying to define the control problem in the form

$$\dot{x}(t) = a(x(t), u(t), t).$$ (16.109)

This is to be controlled to minimize its performance measure given by the Hamilton–Jacobi index denoted by

$$J = h(x(t_f), t_f) + \int_{t_0}^{t_f} g(x(\tau), u(\tau), \tau) d\tau.$$ (16.110)

The *h*- and *g*-functions are specified; t_0 and t_f are fixed; and τ is the dummy variable of integration.

Now let us use the imbedding principle to include this problem in a more general sense by considering the performance measure given by

$$J\left(\mathbf{x}(t), t, \underset{t_0 \leq \tau \leq t_f}{\mathbf{u}(\tau)}\right) = h(\mathbf{x}(t_f), t_f) + \int_{t_0}^{t_f} g(\mathbf{x}(\tau), \mathbf{u}(\tau), \tau) d\tau, \tag{16.111}$$

where $t \leq t_f$, and $\mathbf{x}(t)$ can be any admissible state value. $\mathbf{u}^*(t)$ will be the optimal control listing in specified interval with $t_0 \leq t \leq t_f$ of the trajectory.

Let us now attempt to determine the controls that minimize Equation 16.111 for all admissible $\mathbf{x}(t)$, and for all $t \leq t_f$. The minimum cost-to-go function is therefore

$$J^*(\mathbf{x}(t), t) = \underset{\substack{\mathbf{u}(\tau) \\ t_0 \leq \tau \leq t_f}}{\text{Min}} \left\{ \int_{t_0}^{t_f} g(\mathbf{x}(\tau), \mathbf{u}(\tau), \tau) d\tau + h(\mathbf{x}(t_f), t_f) \right\}. \tag{16.112}$$

Subdivision of the interval yields

$$J^*(\mathbf{x}(t), t) = \underset{\substack{\mathbf{u}(\tau) \\ t_0 \leq \tau \leq t_f}}{\text{Min}} \left\{ \int_{t_0}^{t+\Delta t} g d\tau + \int_{t+\Delta t}^{t_f} g d\tau + h(\mathbf{x}(t_f), t_f) \right\}. \tag{16.113}$$

The principle of optimality now requires

$$J^*(\mathbf{x}(t), t) = \underset{\substack{\mathbf{u}(\tau) \\ t_0 \leq \tau \leq t+\Delta t}}{\text{Min}} \left\{ \int_{t_0}^{t+\Delta t} g d\tau + J^*(\mathbf{x}(t + \Delta t), t + \Delta t) \right\}, \tag{16.114}$$

where $J^*(\mathbf{x}(t + \Delta t), t + \Delta t)$ is the minimum cost of the process for the time interval $t + \Delta t \leq \tau \leq t_f$ with the initial state $\mathbf{x}(t + \Delta t)$. Equation 16.114 yields the trace toward the final optimum for $\mathbf{u}^*(t)$.

By assuming existence of continuity and existence of the second partial derivatives of $J^*(\mathbf{x}(t + \Delta t), t + \Delta t)$ that are bounded, the Taylor series expansion of J^* in the neighborhood of $(\mathbf{x}(t), t)$ gives

$$J^*(\mathbf{x}(t), t) = \underset{\substack{\mathbf{u}(\tau) \\ t_0 \leq \tau \leq t+\Delta t}}{\text{Min}} \left\{ \int_{t_0}^{t+\Delta t} g d\tau + J^*((\mathbf{x}), t) + \left[\frac{\partial}{\partial t} J^*((\mathbf{x}), t) \right] \Delta t \right.$$

$$+ \left[\frac{\partial}{\partial t} J^*((\mathbf{x}), t) \right]^T [(\mathbf{x}(t + \Delta t), t + \Delta t)]$$

$$\left. + O^2(\Delta t) \text{ higher order terms} \right\}. \tag{16.115}$$

Note that $O^2(\Delta t)$ denotes the terms containing $[\Delta t]^2$ and higher orders of Δt that are present in the approximation of the integral and its associated Taylor series truncation. For small perturbations in Δt, we develop

$$J^*(\mathbf{x}(t), t) = \underset{\mathbf{u}(\tau)}{\text{Min}} \{g(\mathbf{x}(t), \mathbf{u}(t), t)\Delta t + J^*((\mathbf{x}), t)$$

$$+ J_t^*((\mathbf{x}), t)\Delta t + J_x^{*\text{T}}(\mathbf{x}(t), t)[\mathbf{a}(\mathbf{x}(t), \mathbf{u}(t), t]\Delta t$$

$$+ O^2(\Delta t) \text{ terms.} \tag{16.116}$$

By eliminating the J^* terms that are independent on $\mathbf{u}(t)$ in the optimization process, and taking the derivative with respect to time in the limit as $\Delta t \to 0$, we obtain

$$0 = \underset{\mathbf{u}(\tau)}{\text{Min}} \{g(x(t), u(t), t) + J_x^{*\text{T}}(x(t), t)[a(x(t), u(t), t)]\}. \tag{16.117}$$

To set the boundary value for the Equation 16.117, we set $t = t_f$. Also, from Equation 16.112, we deduce that

$$J^*(\mathbf{x}(t_f), t_f) = h(\mathbf{x}(t_f), t_f). \tag{16.118}$$

We now define the Hamiltonian H as

$$\underline{H}(\mathbf{x}(t), \mathbf{u}(t), J_x^*(t)) \triangleq g(\mathbf{x}(t), \mathbf{u}(t), t) + J_x^{*\text{T}}(\mathbf{x}(t), t)[\mathbf{a}(\mathbf{x}(t), \mathbf{u}(t), t)] \tag{16.119}$$

and

$$\underline{H}(\mathbf{x}(t), \mathbf{u}^*(\mathbf{x}(t), J_x^*, t), J_x^*, t) = \underset{\mathbf{u}(\tau)}{\text{Min}} \underline{H}(\mathbf{x}(t), \mathbf{u}(t)J_x^*, t). \tag{16.120}$$

Here, the minimizing control is a function of $\mathbf{x}(t), J_x^*$, and t. These definitions converge to the Hamilton–Jacobi equation, which is stated as

$$0 = J_x^*(\mathbf{x}(t), t) + \underline{H}(\mathbf{x}(t), \mathbf{u}^*(\mathbf{x}(t), J_x^*, t), J_x^*, t) \tag{16.121}$$

This Hamilton–Jacobi equation is similar to the Bellman's recurrence equation shown earlier. As such, it is formally referred to as the H–J–B equation.

16.3.3 Relationship between Dynamic Programming and the Minimum Principle

Given the control problem in the form

$$\dot{x}(t) = a(\mathbf{x}(t), \mathbf{u}(t), t), \tag{16.122}$$

we have found the performance measure denoted by

$$J = h(\mathbf{x}(t_f), t_f) + \int\limits_{t_0}^{t_f} g(\mathbf{x}(\tau), \mathbf{u}(\tau), \tau) d\tau, \qquad (16.123)$$

which when minimized provides us information on the optimum control and state vectors, $\mathbf{u}^*(t)$ and $\mathbf{x}^*(t)$ from their respective admissible spaces. In other words, the goal of finding $\mathbf{u}^* \in U$ is to force the system to behave in an ideal or almost ideal manner such that the performance measure is minimized.

In DP, we show that the H–J–B equation derivation resulted in

$$0 = J_x^*(\mathbf{x}(t), t) + \underset{\mathbf{u}(\tau)}{\text{Min}}\, H(\mathbf{x}(t), \mathbf{u}(t), J_x^*, t). \qquad (16.124)$$

That is

$$0 = J_x^*(\mathbf{x}(t), t) + \underset{\mathbf{u}(\tau)}{\text{Min}}\, \left\{ g(\mathbf{x}(t), \mathbf{u}(t), t) + J_x^{*\mathrm{T}}(\mathbf{x}(t), t)[\mathbf{a}(\mathbf{x}(t), \mathbf{u}(t), t)] \right\}. \qquad (16.125)$$

Also, Equation 16.125 must satisfy the trajectory of $J^*(\mathbf{x}(t), t)$ with Taylor series assumptions and approximation on J_x^* and J_t^*.

If $(\mathbf{x}^*(t), t)$ is a particular point in the state-time space, the H–J–B corresponding to this point satisfying the relationship is

$$\begin{aligned} g(\mathbf{x}^*(t), \mathbf{u}^*(t), t) &+ J_x^{*\mathrm{T}}(\mathbf{x}^*(t), t)[\mathbf{a}(\mathbf{x}^*(t), \mathbf{u}^*(t), t)] \\ &= \underset{\mathbf{u}(\tau)}{\text{Min}}\, \left\{ g(\mathbf{x}^*(t), \mathbf{u}(t), t) + J_x^{*\mathrm{T}}(\mathbf{x}^*(t), t)[\mathbf{a}(\mathbf{x}^*(t), \mathbf{u}(t), t)] \right\} \end{aligned} \qquad (16.126)$$

for all $t_0 \le t \le t_f$, this gives the point $(\mathbf{x}^*(t), t)$ such that

$$0 = J_x^{*\mathrm{T}}(\mathbf{x}^*(t), t) + g(\mathbf{x}^*(t), \mathbf{u}^*(t), t) + J_x^{*\mathrm{T}}(\mathbf{x}^*(t), t)[\mathbf{a}(\mathbf{x}^*(t), \mathbf{u}^*(t), t)]. \qquad (16.127)$$

For this first-order equation, given that if t_f is fixed and $x(t_f)$ is free, the boundary conditions that holds true is

$$J^*(\mathbf{x}^*(t_f), t_f) = h(\mathbf{x}^*(t_f), t_f). \qquad (16.128)$$

Also, in the Pontryagin minimum principle applied to the optimal control problem stated by Equation 16.122, we find that

$$\frac{\partial}{\partial t}(\mathbf{x}^*(t)) = \dot{\mathbf{x}}^*(t) = \frac{\partial}{\partial \mathbf{p}} H(\mathbf{x}^*(t), \mathbf{u}^*(t), \mathbf{p}^*(t), t) \qquad (16.129)$$

$$\frac{\partial}{\partial t}\mathbf{p}^*(t) = \dot{\mathbf{p}}^*(t) = -\frac{\partial}{\partial \mathbf{p}} H(\mathbf{x}^*(t), \mathbf{u}^*(t), \mathbf{p}^*(t), t) \qquad (16.130)$$

$$H(\mathbf{x}^*(t), \mathbf{u}^*(t), \mathbf{p}^*(t), t) \le H(\mathbf{x}^*(t), \mathbf{u}(t), \mathbf{p}^*(t), t) \qquad (16.131)$$

for the set of admissible controls, i.e., $\forall \mathbf{u}^* \in U$ and for $t_0 \le t \le t_f$, are the necessary optimality conditions for $\mathbf{u}^*$ to be an optimal control and $\mathbf{x}^*$ to be the corresponding optimal trajectory. The resulting boundary conditions for the first-order state-costate differential equations of the Pontryagin minimum principle are therefore

$$\mathbf{x}^*(t_0) = \mathbf{x}_0 \qquad (16.132)$$

and

$$\mathbf{p}^*(t_f) = \frac{\partial}{\partial \mathbf{x}} \underline{h}(\mathbf{x}^*(t_f), t_f). \qquad (16.133)$$

By applying the definition of the Hamiltonian, we get

$$\underline{H}(\mathbf{x}(t), \mathbf{u}(t), \mathbf{p}(t), (t)) \underline{\triangleq} g(\mathbf{x}(t), \mathbf{u}(t), t) + \mathbf{p}^{\mathrm{T}}(t)[\mathbf{a}(\mathbf{x}(t), \mathbf{u}(t), t)]. \qquad (16.134)$$

By applying the Pontryagin minimum principle of Equations 16.129 and 16.130, we get

$$\dot{\mathbf{x}}(t) = \mathbf{a}(\mathbf{x}^*(t), \mathbf{u}^*(t), t) \qquad (16.135)$$

$$\dot{\mathbf{p}}^*(t) = -\left[\frac{\partial}{\partial \mathbf{x}} \mathbf{a}(\mathbf{x}^*(t), \mathbf{u}^*(t), t)\right]^{\mathrm{T}} \mathbf{p}^*(t) - \left[\frac{\partial}{\partial \mathbf{x}}(g(\mathbf{x}^*(t), \mathbf{u}^*(t), t))\right]. \qquad (16.136)$$

Also, Equation 16.131 implies that

$$\underline{H}(\mathbf{x}^*(t), \mathbf{u}^*(t), t) = \operatorname*{Min}_{\mathbf{u}(\tau)} \underline{H}(\mathbf{x}^*(t), \mathbf{u}(t), \mathbf{p}^*(t), t) \qquad (16.137)$$

or

$$g(\mathbf{x}^*(t), \mathbf{u}^*(t), \mathbf{p}^*(t), t) + \mathbf{p}^{*\mathrm{T}}(t)[\mathbf{a}(\mathbf{x}^*(t), \mathbf{u}^*(t), t)]$$
$$= \operatorname*{Min}_{\mathbf{u}(\tau)} \left\{ g(\mathbf{x}^*(t), \mathbf{u}(t), t) + \mathbf{p}^{*\mathrm{T}}(t)[\mathbf{a}(\mathbf{x}^*(t), \mathbf{u}(t), t)] \right\}. \qquad (16.138)$$

Comparing Equations 16.136 and 16.138, we note that both equations are identical and in fact,

$$J_{\mathbf{x}}^*(\mathbf{x}^*(t), t) = \mathbf{x}^*(t). \qquad (16.139)$$

Furthermore, this conclusion is consistent with the fact that the equations of the Pontryagin minimum principle can be derived from the H–J–B's functional equations. The derivation is left to the reader.

Additionally, it is worthy to note that, since $J_t^*(\mathbf{x}(t), t)$ does not depend on $\mathbf{u}(t)$, Equation 16.126 can be written as

$$0 = \underset{\mathbf{u}(\tau)}{\text{Min}} \left\{ J_t^*(\mathbf{x}^*(t), t) + g(\mathbf{x}^*(t), \mathbf{u}(t), t) + J_x^{*\mathrm{T}}(\mathbf{x}^*(t), t)[\mathbf{a}(\mathbf{x}^*(t), \mathbf{u}(t), t)] \right\}. \quad (16.140)$$

Overall, given the values of the optimal state $\mathbf{x}^*(t)$, the control $\mathbf{u}^*(t)$ minimizes the RHS of Equation 16.140, and the minimum value is zero.

16.3.4 Section Summary

In summary, we have considered a typical nonlinear system that is characterized by a minimizing cost function. The solution requires numerical approach to solve its performance measures, several of which have been described in this chapter. The application of both functional optimization and the H–J–B equation requires setting up the necessary conditions for optimality and the boundary conditions [14].

The Euler–Lagrange equation has also been applied and has applications for off-line computations in iterative or recursive computations.

In optimal control, at the onset of starting the trajectory that seeks to filter admissible points that do not improve the performance measure, the best control is approximated. Successive iteration improves the solution until the search is exhausted.

Notably, the class of problem that has been addressed so far has dependency on time. Least-square estimation or minimization of the PI is possible provided that there are no stochastic behavior or predictivity requirements within the system. For real-time applications and practical systems, this need is important. The next sections of this chapter extend on the principles of DP and optimal control to include adaptive dynamic critics (ADC). Here, the optimization method has the goal to minimize the expected value of the cost function with respect to the control vectors, state variables, system performance, and the penalties in the presence of probabilistic disturbances and uncertainties.

The foundation of ADC is in the H–J–B equation, which forms a basis for real-time approximate optimal control in dynamic environments [1–4].

16.4 Adaptive Critics Design (ACD) and ADP

16.4.1 Background to Complex Intelligent Networks

ADC are neural network (NN) designs capable of optimization over time under conditions of noise and uncertainties. This family of adaptive critics lends itself to other exiting NNs-based techniques for control and optimization [8,9,11]. This NN adapt to the various functions of complex, nonlinear systems.

As family of ACDs was proposed by Werbos as a new optimized technique combining the concepts of reinforcement learning and approximate DP [12,13].

For a series of control actions in time sequences requiring an optimal control law, the entire method determines the adaptive control law adequately based on performance measures that adapts to the system state in time. The adaptive critic method delivers the optimal control laws for a system by successively adapting two artificial neural networks (ANNs). These are

1. Action NN that dispenses the control signals.
2. Critic NN that learns the desired PI for some functions associated with the PI. It gives or provides refinement of the action network.

Both networks have no knowledge of the optimization trajectory but has the ability to learn and tune the control decisions toward the known desired goal functions of the overall system. Also, these two NNs approximate H–J–B's equation associated with optimal control theory.

In distinguishing the features among ADP and optimal control, optimal policies, laws, and value functions are designed and modeled as preceptor structures (NN computations) whose steps are improved over time by solving recursive relationships of DP where the final time, system dynamics, and trajectory leading to ADP converge to the optimal values [9].

The adaptation process starts with a nonoptimal, arbitrarily chosen value of the control set in time by the action network. Whereas the critic network guides the action network toward the optimal solution, at each successive adaptation, it is desired that neither of the networks needs to know or has any information of an optimal trajectory. Rather, only the desired costs need to be known.

Furthermore, these integrated networks determine the optimal control policy for the entire range of initial conditions without external training (as discussed in the chapter and sections on DP). We claim that DP provides a search which tracks backward from the initial step to the final step relating the suboptimal paths. This curse of dimensionality hurts DP because of the excessive computation of burden.

In contrast, in supervised learning, an ANN training algorithm uses a desired output which is compared to the actual output and determines the error term to be used to allow the network, typically back propagation (BP) algorithms, to learn. This error term is determined by using it sensitivity with respect to various parameter changes or input to the network. The use of BP algorithms is linked to reinforcement learning via the critic network to attain desirable attributes.

The use of the critic network removed the task of learning from the action network by simplifying a new criteria call the "cost-to-go" or "strategic utility" function. This corresponds to the function J of Bellman's equations in DP. The critic network therefore allows us to develop one-to-one relationship with DP. Hence, we will also review this mapping with the essential characteristic of DP, namely:

1. Epochs.
2. States that may be finite states or partially observed states as in partially observable Markov decision processes (POMDP) with continuous state variables [2,4] or nondeterministic state variables.
3. Actions may have probabilistic rules associated with them.
4. Rewards may have probabilistic roles as well.
5. Disturbances may be nondeterministic.

With the following definition of nomenclature $x(t)$ where $x(t) \in X$ state valued variables that are often times probabilistic at the discretized time instance t, comparable to s modeled in a Markov decision process frequently used in reinforcement learning.

The control action is given by $u(t)$, comparable to a in a Markov decision process environment; and $r(t)$ is the binary reinforcement signal provided by the external environment.

The reward-to-go with discount factor γ is now given as

$$R(t) = r(t+1) + \gamma r(t+2) + \gamma^2 r(t+3) + \cdots = \sum_{k=1}^{\infty} \gamma^{k-1} r(t+k). \qquad (16.141)$$

This is the reward-to-go, cost-to-go, or value function. In reinforcement leaning, $R(t)$ or an approximation of $R(t)$ is usually represented as $V(t)$, $Q'(t)$, or $J(t)$. In this chapter, we shall adopt that $J(t)$ is the approximate value of $R(t)$, a natural extension of optimal control notations.

16.4.2 From DP to Adaptive or "Approximate" Dynamic Programming (ADP)

ADP is a computational intelligence technique that incorporates time dependency of deterministic or stochastic data and under conditions of noise and uncertainty for future optimal control decision-making. Also called "reinforcement learning," "adaptive critics," "neural-dynamic programming," and "approximate DP". ADP considers the optimization over time by using learning approximation to handle problems that severally challenge conventional methods due to their very large scale and lack of sufficient prior knowledge. ADP overcomes the "curse" of dimensionality in DP. Traditionally, there is only one exact and efficient way to solve problems in optimization over time, in general case where noise and nonlinearity are present: DP.

ADP determines optimal control laws for a system by successively adapting two NNs: one is action network (which dispenses the control signals) and the other is critic network (which learns the desired PI for some function associated with the PI). ADP is designed to maximize the expected value of the sum of future utility over all future time periods:

$$\text{Maximize} \sum_{k=0}^{\infty} (1 + r)^{-k} U(t + k).\tag{16.142}$$

Figure 16.1 shows the general associations of the critic network, action network, and the plant under control. Here, $x(t)$ is the system state, $u(t)$ is the action, $J(t)$ is the secondary or strategic utility function.

In DP, the user supplies both a utility function—the function to be maximized—and a stochastic model of the external plant or environment. However, the ADP design attempts to approximate DP in the general case. The cost of running true DP is proportional to the number of possible states in the plant or environment; that number, in turn, grows exponentially with the number of variables in the environment. Therefore, approximate methods are needed even with many small-scale problems. ADP is defined more precisely as designs that include a critic network—a network whose output is an approximation of the J-function, or to its derivatives, or to something very closely related to these two, the action network in an adaptive critic system is adapted so as to maximize J in the near-term future. To maximize future utility subject to constraints, you can simply train the action network to obey those constraints when maximizing J; the validity of DP itself is not affected by such constrains.

DP is used to solve for another function, J, which serves as a secondary or strategic utility function. The key theorem is that any strategy of action that maximizes J in the short term will also maximize the sum of U over all future times. J is a function of $R(t)$, where $R(t)$ is complete state description of the plant to be controlled at time t and $u(t)$ are the vector of actions. DP converts

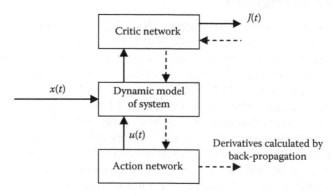

FIGURE 16.1
Action and critic networks in ADP.

a problem in optimization over time into a "simple" problem in maximizing J just one step ahead in time.

$$J(R(t)) = \max_{u(t)} \left(U(R(t), u(t)) \right) + \frac{\langle J(R(t+1)) \rangle}{1 + r} - U_0, \tag{16.143}$$

where r and U_0 are constants that are needed only in infinite-time-horizon problems (and then only sometimes), and where the angle brackets refer to expectation value.

Adaptive critic designs may be defined as design that attempts to approximate DP in the general case. The cost of running true DP is proportional to the number of possible states in the plant or environment; that number, in turn, grows exponentially with the number of variables in the environment. Therefore, approximate methods are needed even with many small-scale problems. Adaptive critic designs are defined more precisely as designs that include a critic network—a network whose output is an approximation of the J-function, or to its derivatives, or to something very closely related to these two, the action network in an adaptive critic system is adapted so as to maximize J in the near-term future.

To maximize future utility subject to constraints, you can simply train the action network to obey those constraints when maximizing J; the validity of DP itself is not affected by such constraints.

16.4.3 Critic Network Variants

Different types of critic networks were developed over the years. The primary work of Werbos provides the variants of the family of critics, which leads to the approximation of the J-function. They are briefly explained here with their respective design architectures and these critic designs are based on foundations of DP.

There are three ADP families—heuristic dynamic programming (HDP), dual heuristic programming (DHP), and globalized dual heuristic programming (GDHP). Each kind of DP method has its own advantages and disadvantages described in [9,12].

Briefly, the variants include:

1. HDP, which adapts a critic network whose output is an approximation of $J(R(t))$.

2. DHP, which adapts a critic network whose outputs represent the derivatives of $J(R(t))$.

3. GDHP, which adapts a critic network whose output is an approximation of $J(R(t))$, but adapts it so as to minimize errors in the implied derivatives of J. GDHP tries to combine the best of HDP and DHP.

In general, DHP is advantageous over HDP as it builds derivative terms over time directly and reduces the probability of error introduced by BP. However, DHP is more difficult to implement than HDP because of the additional computations needed to solve the derivatives of *J*-function and its complexity increases as the size of the problem grows bigger. The third method, GDHP, combines HDP and DHP. It is more accurate but involves increased complexity than both HDP and DHP.

Also, HDP intends to break down, through very slow learning, as the size of a problem grows larger. However, DHP is more difficult to implement. The three methods listed above all yield action-independent critics, there are also ways to adapt a critic network that inputs $R(t)$ and $u(t)$ to produce action-dependent heuristic dynamic programming (ADHDP) and action-dependent dual heuristic programming (ADDHP).

DHP
DHP is based on differentiating the Bellman equation. Before performing the differentiation, we have to decide how to handle $u(t)$. One way is simply to define the function $u(R(t))$ as that function of R which, for every R, maximizes the right-hand side of the Bellman equation. With that definition (for the case $r = 0$), the Bellman equation becomes:

$$J(R(t)) = U(R(t), u(t)) + \langle J(R(t+1)) \rangle - U_0, \tag{16.144}$$

where we must also consider how $R(t+1)$ depends on $R(t)$ and $u(R(t))$. Differentiating, and applying the chain rule, we get

$$\lambda_i(R(t)) = \frac{\partial J(R(t))}{\partial R_i(t)} = \frac{\partial}{\partial R_i(t)} U(R(t), u(R(t))) + \left\langle \frac{\partial J(R(t+1))}{\partial R_i(t)} \right\rangle$$

$$= \frac{\partial J(R(t), u(t))}{\partial R_i(t)} + \sum_j \frac{\partial U(R, u)}{\partial u_j} \cdot \frac{\partial u_j R(t)}{\partial R_i(t)} + \sum_j \left\langle \frac{\partial J(R(t+1))}{\partial R_i(t+1)} \cdot \frac{\partial R_j(t+1)}{\partial R_i(t)} \right\rangle$$

$$+ \sum_{j,k} \left\langle \frac{\partial J(R(t+1)}{\partial R_j(t+1)} \cdot \frac{\partial R_j(t+1)}{\partial u_k(t)} \cdot \frac{\partial u_k(t)}{\partial R_j(t)} \right\rangle \tag{16.145}$$

For DHP, the RHS of the function is evaluated by including the full model of the plant dynamics. It used models for both critic and controller training, for example, $(\partial X_j(t+1)/\partial X_i(t))$ and $(\partial X_j(t+1)/\partial U_i(t))$.

HDP
HDP is a procedure for adapting a network or function, $J(R(t), W)$, which attempts to approximate the function, $J(R(t))$. From the ADP network using HDP, a variation of the J-function is given by

$$J(R(t)) = \underset{u(t)}{\text{Max}}\, (U(R(t), u(t))) + \frac{\langle J(R(t+1)) \rangle}{1+r} - U_0(R(t_0), u(t_0)). \tag{16.146}$$

For simplicity, we will assume problems such that we can assume $U_0 = 0$. HDP is a procedure for adapting a network or function.
Furthermore,

$$\frac{\partial J(R(t))}{\partial R_i(t)} = \frac{\partial}{\partial R_i(t)} U(R(t), u(R(t))) + \left\langle \frac{\partial J(R(t+1))}{\partial R_i(t)} \right\rangle. \tag{16.147}$$

Here, the derivative of optimize J-functions with respect to input R is obtained via BP through the critic network and this gives rise to the computation of value functions. This critic estimates $J(R(t))$ based directly on the plant states $X(t)$. Since these data are available directly from the plant, critic training does not need a plant model for its calculations.

GDHP
In GDHP, this critic approximates both the J-function and its derivatives. The state variables are input and the output are both $J(t)$ and $J_x(t)$. The GDHP therefore models both critic and controller training.

ADHDP or "Q-learning"
The critic training is the same as for HDP. This is however easier since the control variables are input to the critic network. The derivatives of the J-function with respect to the control are terms $\partial X(t)/\partial U(t)$ obtained directly from BP through the critic network. The ADHP critic does not use the plant models in the training process.
Furthermore, if we defined a new quantity:

$$J'(R(t), u(t)) = U(R(t), u(t)) + \frac{\langle J(R(t+1)) \rangle}{1+r}. \tag{16.148}$$

By substituting Equations 6.143 and 6.147, we may derive a recurrence rule for J'

$$J'(R(t), u(t)) = U(R(t), u(t)) + \underset{u(t+1)}{\text{Max}} \frac{\langle J'(R(t+1), u(t+1)) \rangle}{1+r}. \tag{16.149}$$

ADHDP adapts a critic network, $J'(R(t), u(t), W)$, which attempts to approximate J' as defined in Equation 6.148.

ADDHP
The ADDHP uses both states and control variables as inputs and output of the gradient of the J-function with respect to the state and control variables.
That is

$$\frac{\partial J(t)}{\partial X_i(t)} \quad \text{and} \quad \frac{\partial X_j(t)}{\partial U_i(t)}.$$

It uses a plant model for critic training but not for controller training. The method uses DHP critic training process, but gets its derivative model from the controller training (obtained directly from the critic output).

ADGDHP

This critic has used both states and control variables as inputs, and they output both the values of the J-function and its gradient with respect to state and control variables. As with the GDHP, critic training utilizes both the HDP and DHP recursions, and controller training as in ADDHP. Therefore, ADGDHP features a model from critic training but not for controller training.

Table 16.1 shows the summary of the critic/controller training needs of the six adaptive critics and Table 16.2 shows a summary of the different features of the three main variants of ACD discussed.

TABLE 16.1

Summary of Critic and Controller Training for ACDs

ACD Variants or Structures	Model Needed for Training	
	Critic	Controller
HDP		×
ADHDP		
DHP	×	×
ADDHP	×	
GHDP	×	×
ADGHDP	×	

TABLE 16.2

Summary of the Salient Features of ADP Variants

ADP Method	J-Function	Advantage	Disadvantage
HDP	Critic network whose output is an approximation of J-function: $$J(R(t)) = \underset{u(t)}{\text{Max}} \{U(R(t), u(t))\} + \frac{\langle J(R(t+1)) \rangle}{1+r}$$	Easy to formulate	The size of a problem grows bigger
DHP	Adapts a critic network whose outputs represent the derivatives of J-function: $$J(R(t)) = U(R(t), u(t)) + \langle J(R(t+1)) \rangle - U_0$$	Since DHP builds derivative terms over time directly, it reduces the probability of error introduced by BP	More difficult to implement because of derivatives of J
ADHDP	$$J'(R(t), u(t)) = U(R(t), u(t)) + \underset{u(t+1)}{\text{Max}} \frac{\langle J'(R(t+1), u(t+1)) \rangle}{1+r}$$	Combine HDP and DHP, and add new input to the system	Difficult to form the model

16.5 Architecture of ACDs

Implementation of ADP for online purposes is of best value to the industry [12]. The goal of direct ADP or neural dynamic programming is to optimize a given performance measure by learning to choose appropriate control actions through interactions with the environment. Learning is performed without an explicit model of the system or plant under control. Instead, information relating to the system is captured directly by both the action and critic networks through learning.

Figure 16.2 shows the typical representation of direct neural dynamic programming. The solid lines denote system information flow, while the dashed lines represent error back-propagation paths to reduce the squared Bellman error.

The structure includes two networks, action and critic, as building blocks. The critic network is trained toward optimizing a total reward-to-go objective, namely to balance the Bellman equation. The action network is trained such that the critic output approaches an ultimate objective of success, $R^*(t)$. A fractional convergence sequence as a sufficiency condition is generally required of the learning structure such that the ultimate performance of R can be derived.

Also, during the learning process, the action network is constrained by the critic to generate controls that optimize the future reward-to-go instead of only temporarily optimal solutions. In contrast to usual supervised NN learning applications, there are no readily available training sets of input–output pairs used for approximating the overall objective function $R(t)$ in terms of a least squares fit. Instead, both the control action U and the critic output J are updated according to an error function that changes from one time step to the next. In the online learning control implementation, the controller is naïve when it starts to control. This is because, initially, both the action and critic networks possess random weights/parameters. Once a system state is observed, an action will be subsequently produced based on the parameters in the action network. A better control under the specific system state should

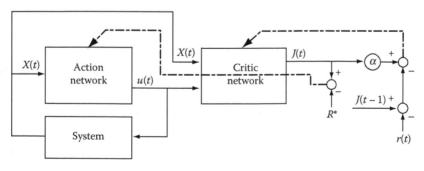

FIGURE 16.2
A typical representation of direct neural dynamic programming.

result in a reduced Bellman's error. This set of system operations will be reinforced through memory or association between states and control output in the action network. Otherwise, the control will be adjusted through tuning the weights in the action network to minimize the Bellman error.

16.5.1 Critic Networks

The overall goal is to minimize Bellman's error by adaptation of weights. The critic network is used to provide an output, which is an approximation of $R(t)$, the total future rewards-to-go or "cost-to-go."

The cost (or reward) function $R(t)$ at time t is given by

$$R(t) = r(t+1) + \gamma r(t+2) + \gamma^2 r(t+3) + \cdots \tag{16.150}$$

$$= \sum_{k=1}^{\infty} \gamma^{k+1} r(t+k), \tag{16.151}$$

where γ is the discount factor.

The reward-to-go is

$$R(t) \simeq J(t) = V(t) \text{ or } Q(t). \tag{16.152}$$

So, from the prediction of errors from Bellman's error we get:

$$e_C(t) = \gamma J(t) - [J(t-1) - r(t)] = [r(t) + \gamma J(t)] - J(t-1) \tag{16.153}$$

and hence the objective function to minimize the critic network using a minimum square error function is given by

$$\text{Min } E_C(t) = \frac{1}{2} e_C^2(t) = \frac{1}{2} e_C^T R e_C. \tag{16.154}$$

Learning in critic requires selection of weights that are updated recursively. The standard formulation to obtain $w_C(t)$ involves

$$w_C(t+1) = w_C(t) + \Delta w_C(t) \tag{16.155}$$

and

$$\Delta w_C(t) = l_C e_C \left[-\frac{\partial e_C(t)}{\partial w_C(t)} \right], \tag{16.156}$$

where l_C is the learning rate of the critic network and

$$\frac{\partial E_C(t)}{\partial w_C(t)} = \frac{\partial E_C(t)}{\partial J(t)} \cdot \frac{\partial J(t)}{\partial w_C(t)}. \tag{16.157}$$

The algorithm for the critics network is given below.

16.5.2 Action Networks

The principle in adapting the action network is to propagate error between the ultimate performance objective denoted by R^* or its approximate J-function from the critic network. Since r_s is the successful reinforcement signal to obtain R^* using $r_s(1-\alpha)^{-1}$ in ADP, an action network of linear or nonlinear systems, $x(t)$ is used as input to create controls $u(t)$.

The weight update in the action network (epochs) can be formulated using a minimum error calculation that is updated via successive training of the associated weighting coefficients.

The error of the action network is computed from the approximate J-function and the output of the critic network as

$$e_A(t) = J(t) - R^*(t). \tag{16.158}$$

The goal of the action network is to optimize the a minimum square error function given by

$$\text{Min } E_A(t) = \frac{1}{2}e_A^2(t) = \frac{1}{2}e_C^T R e_C. \tag{16.159}$$

By training via dynamic action weights, the action network may be reported by the function $u(t) = f(x, w_A)$ at the minimum error performance where we introduce training weights $w_A(t)$ for the action network. The weights are updated weights by

$$w_A(t+1) = w_A(t) + \Delta w_A(t), \tag{16.160}$$

where

$$\Delta w_A(t) = l_A e_A \left[-\frac{\partial e_A(t)}{\partial w_A(t)} \right] \tag{16.161}$$

with

$$\frac{\partial E_A(t)}{\partial w_A(t)} = \frac{\partial E_A(t)}{\partial J(t)} \cdot \frac{\partial J(t)}{\partial w_A(t)}. \tag{16.162}$$

Here, l_A is the nonzero learning rate of the action network that is decreased over time to very small values.

Application: A selected algorithm for using HDP
We now summarize the procedure for estimating lambda using ADP concept of the DHP method.

Step 1: Obtain training/target data for the system and pretrain the action and the critic networks. Obtain $R(t)$, $u(t)$, and $R(t+1)$.

Step 2: Compute the error action and the critic networks

$$e_A(t) = u(t) - \langle u(t) \rangle \qquad\qquad 16.163(a)$$

$$e_C(t) = J(t) - \gamma J(t+1) - U(t). \qquad\qquad 16.163(b)$$

Step 3: After the error is minimized, obtain action network output $u(t)$ as a function of $R(t)$ and $R(t+1)$ and compute $u(t+1)$ using $R(t+1)$.

Step 4: Input $u(t)$, $R(t)$, $u(t+1)$, $R(t+1)$ to the critic network and solve for $J(t+1)$.

Step 5: Compute the utility function given by

$$J(R(t)) = U(R(t), u(t)) + \langle J(R(t+1)) \rangle / (1+r) - U_0.$$

Step 6: Use the critic network to compute the derivatives of the *J*-function given by

$$J(R(t)) = \underset{u(t)}{\text{Max}}\, (U(R(t), u(t))) + \frac{\langle J(R(t+1)) \rangle}{1+r}. \qquad (16.164)$$

Step 7: Backpropagate and update the weights of the critic and action network.

Update the weights of the critic and action networks, check the errors, and terminate the algorithm as optimal results are found.

16.5.3 ACDs Comparative Studies

The distinguishing characteristics of each of the following main ACD are described below.

HDP used a parametric structure call "Actor" to approximate the control law. It predicts the trajectory toward an approximate value function. Notably, the critic serves the role of reevaluation of the utility function and provides the actor with a performance measure.

In DHP, the actor also approximates the control law. The critic mechanism approximates the derivatives of the value function with respect to the input training set. It is known to be faster than HDP in computational steps.

GHDP combines the advantages of both HDP and DHP characteristics. The critic approximates both the value of the *J*-function and its derivatives.

The action-dependent versions of the approaches again referred to as approximate value function that depends explicitly on the controls.

16.5.4 Summary

An ADC construct of parametric structures, namely (i) the critic that approximates the value or cost-to-go-function, (ii) the actor that approximates the control law and (iii) the function to be approximated are chosen.

An algorithm that updates them at every cycle or successive iteration can be obtained from the policy improvement over time and the value determination operation so that we produce two sequences of functions that converge to the optimal solution.

Utility functions and state variables: Users of ADC are concerned with two main issues:

1. Deciding what to include in the $X(t)$ state vector that represents the input to the critic and to the controller. (Hint: Use your engineering intuition and whatever rigorous knowledge is available to satisfy yourself that the controller and the critic have sufficient data to assume that at every point in the corresponding state space, there will exist a unique action for the controller to take).

2. Clarifying the utility function is defined as $U(t) = \|X(t) - X^*(t)\|$. Different relative weights are used to decide which component of $X(t)$ is included in the definition of $U(t)$ to accomplish the targets. For instance, a design objective that monitors the utility function may be to reduce the distance to zero or to reduce the velocity error to zero. The weighting factors are tuned to arrive optimally at such goals without compromising the system performance by too aggressively adjusting the participation of $X_i(t)$ in $U(t)$. The definition of $U(t)$ generates some interesting linkages between ADP and the HJB theory.

16.6 Typical Architectures of Variants or ADP (Critics Illustrations)

HDP critic adaptation
HDP is a procedure for adapting a network or function, $J(R(t), W)$. We have utilized an approximate function, $J(R(t))$, which is a small perturbation of the Bellman equation,

$$J(R(t)) = \underset{u(t)}{\text{Max}} \left(U(R(t), u(t)) + \frac{\langle J(R(t+1)) \rangle}{1+r} \right). \tag{16.165}$$

This is the case where model-based ANN is used with supervised learning. It is simply model-dependent design where the cost-to-go J-function after $\Delta\hat{x}(t)$ has been estimated with three times delays (TDL) as

$$J(R(t)) = \sum_{k=0}^{\infty} \gamma^k U(R(t+k)), \tag{16.166a}$$

where γ is the discount factor for finite horizon problem in the domain $0 < \gamma < 1$.

$U(t)$ is the utility function or local cost (to be discussed later) and $x(t)$ is the input vector to the critic network. The inputs to the two critic networks are outputs from the model NNs, which are model based with appropriate time-delay values. The critic network is trained forward in time for real-time analysis/operation relevant to optimization over time for practical applications.

Figure 16.3 shows the HDP critic NN design for adaptation and learning. As before, U is the utility function of the local cost and $R(t)$ is an input vector to the critic network. The critic network NN is trained forward in time (multitime steps ahead), which is an important asset for real-time operation. In this architecture, critic networks have the same inputs but delayed differently. This allows us to compute their corresponding outputs $J(t)$ and $\hat{J}(t+1)$. The second critic estimates the $\hat{J}(t)$-function or cost-to-go function at time $t+1$ by using a model NN to get inputs one step ahead of the time sequence. Thus, this further allows for the computation of $\hat{J}(t+1)$, the output of the critic NN at time t.

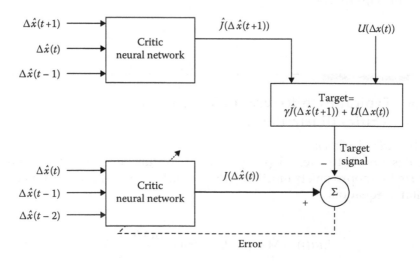

FIGURE 16.3
HDP critic NN adaptation and learning.

The critic network is then used to minimize the error over time by the following performance error measure

$$\| E_1(t) \| = \frac{1}{2} \sum_t E_{C1}^2(t), \qquad (16.166b)$$

where

$$E_{C1}(t) = J(\Delta \hat{x}(t)) - \gamma \hat{J}(\hat{x}(t+1)) - U(\Delta x(t)), \qquad (16.167)$$

where $\Delta x(t)$ is the change in $x(t)$, a vector of the observables of the plant of any available states present.

For critic adaptation, $J(\Delta \hat{x}(t+1))$ gives the trajectory and $\gamma \hat{J}(\hat{x}(t+1)) - U(\Delta x(t))$ assures the target or desired value $\hat{J}(\Delta \hat{x}(t))$ is achieved.

For the action network, the goal is to optimize the a minimum square error function given by

$$\|E_2(t)\| = \frac{1}{2} \sum_t E_{A1}^2(t), \qquad (16.168)$$

where

$$E_{A1}(t) = \frac{\partial J(t)}{\partial A(t)} \qquad (16.169)$$

and

$$\frac{\partial J(t)}{\partial A(t)} = \frac{\partial J(t)}{\partial \Delta \hat{x}(t)} \cdot \frac{\partial \Delta \hat{x}(t)}{\partial A(t)}. \qquad (16.170)$$

The weight change in the action network $\Delta w_{A1}(t)$ can therefore be written as

$$\Delta w_{A1}(t) \propto \frac{\partial E_2(t)}{\partial w_A(t)}. \qquad (16.171)$$

This can be further expanded to get:

$$\Delta w_{A1}(t) = -\alpha E_{A1}(t) \frac{\partial E_{A1}(t)}{\partial w_{A1}(t)} \qquad (16.172)$$

$$\Delta w_{A1}(t) = -\alpha \frac{\partial J(t)}{\partial A(t)} \frac{\partial}{\partial w_{A1}} \left(\frac{\partial J(t)}{\partial A(t)} \right). \qquad (16.173)$$

Here, the learning rate alpha is positive and is required to decrease over time to very small values.

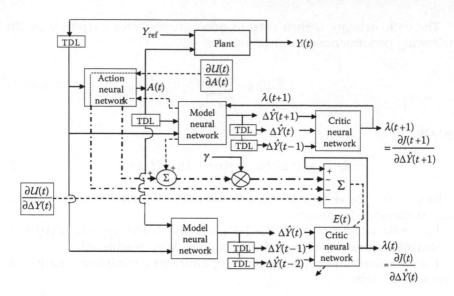

FIGURE 16.4
DHP critic NN adaptation.

DHP critic NN adaptation
Figure 16.4 shows the overall HDP critic NN design for adaptation and learning.

In the DHP scheme, application of the chain rule for derivatives yields

$$\frac{\partial J[\Delta Y(t+1)]}{\partial \Delta Y_i(t)} = \sum_{i=1}^{n} \lambda_i(t+1)\frac{\partial Y_i(t+1)}{\partial \Delta Y_i(t)} + \sum_{k=1}^{m}\sum_{i=1}^{n} \lambda_i(t+1)\frac{\partial \Delta Y_i(t+1)}{\partial A_k(t)}\frac{\partial A_k(t)}{\partial \Delta Y_i(t)},$$

(16.174)

where

$$\lambda_i(t+1) = \partial J[\Delta Y(t+1)]/\partial \Delta Y_i(t+1).$$ (16.175)

Also n and m are the numbers of outputs of the model and the action NNs, respectively. The n components of the vector $E(t)$ are subsequently determined by

$$E(t) = \frac{\partial J[\Delta Y(t)]}{\partial \Delta Y_i(t)} - \gamma \frac{\partial J[\Delta Y(t+1)]}{\partial \Delta Y_i(t)} - \frac{\partial U(t)}{\partial \Delta Y_i(t)} - \sum_{k=1}^{m} \frac{\partial U(t)}{\partial A_k(t)}\frac{\partial A_k(t)}{\partial \Delta Y_i(t)}. \quad (16.176)$$

DHP action NN adaptation
Figure 16.5 shows the HDP action NN design for adaptation and learning. It is used to propagate $\lambda(t+1)$ back through the model network to the action network.

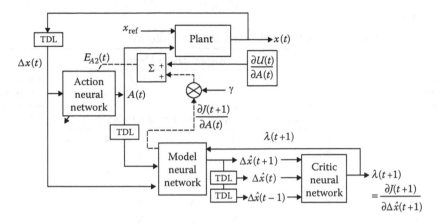

FIGURE 16.5
DHP action NN adaptation.

The goal of the adaptation process can be expressed as

$$\frac{\partial U(\Delta x(t))}{\partial A(t)} + \gamma \frac{\partial J(\Delta \hat{x}(t))}{\partial A(t)} = 0. \tag{16.177}$$

The error signal for the action network adaptation is given by

$$E_{A2}(t) = \frac{\partial U(\Delta x(t))}{\partial A(t)} + \gamma \frac{\partial \hat{J}(\Delta \hat{x}(t))}{\partial A(t)}. \tag{16.178}$$

Also, the incremental changes for updating the weights are computed while applying BP using

$$\Delta w_{A2} = -\alpha \left[\frac{\partial U(\Delta x(t))}{\partial A(t)} + \gamma \frac{\partial \hat{J}(\Delta \hat{x}(t))}{\partial A(t)} \right]^{\mathrm{T}} \frac{\partial A(t)}{\partial W_{A2}}. \tag{16.179}$$

Again, the alpha learning is positive and decreasing and w_{A2} are the weights associated with the DHP action NN.

16.7 Applications of DSOPF to Power Systems Problems

In this section of the chapter, we show two solved examples of applications of new optimization techniques to power systems research work listed in Table 16.3 [3,5,6,7]. Two cases of ADP applications are now considered. These are (i) unit commitment and (ii) optimal network reconfiguration.

TABLE 16.3

Optimization Methods for Selected Power System Problems

Selected Power System Challenges	Optimization Methods								Classical Methods (LP, NLP, IP, etc.)
	DA	Optimal Control	Risk Assessment	IS	DP	ADP	AHP	Game Theory	
Reliability	■		■				■	■	
Fault analysis/3R's	■			■		■			
Unit commitment	■	■		■	■	■	■	■	
DSOPF		■		■		■			■
Control coordination	■			■	■	■		■	■
Stability and DSA		■	■	■		■			
State estimation		■	■						■

3R's, reconfiguration, restoration, and remedial control; DA, decision analysis; IS, intelligent systems; DP, dynamic programming; ADP, adaptive dynamic programming; AHP, analytical hierarchical processes; LP, linear programming; NLP, nonlinear programming; IP, interior point; DSOPF, dynamic stochastic optimal power flow; DSA, dynamic security assessment.

Case 1: *ADP for solving the power system unit commitment (UC) problem*
The objective function of the UC problem can be formulated as the sum of costs of all the units over time, and presented mathematically as [5,7]:

$$F = \sum_{t=1}^{T} \sum_{i=1}^{N} \left[u_i(t) F_i(E_i(t)) + S_i(t) \right]. \tag{16.180}$$

The constraint models for the UC optimization problem are as follows.
System energy balance

$$0.5 \sum_{i=1}^{N} \left[u_i(t) P_{g_i}(t) + u_i(t-1) P_{g_i}(t-1) \right] = P_D(t). \tag{16.181}$$

Energy and power exchange

$$E_i(t) = 0.5 \left[P_{g_i}(t) + P_{g_i}(t-1) \right]. \tag{16.182}$$

System spinning reserve requirements

$$\sum_{i=1}^{N} u_i(t) P_{g_i}(t) \geq P_D(t) + P_R(t). \tag{16.183}$$

Unit generation limits

$$P_{g_i}^{\min} \leq P_{g_i}(t) \leq P_{g_i}^{\max} \tag{16.184}$$

with $t \in \{1,T\}$ and $i \in \{1,N\}$ in all cases where

F is the total operation cost on the power system

$E_i(t)$ is the energy output of the ith unit at hour t

$F_i(E_i(t))$ is the fuel cost of the ith unit at hour t

$u_i(t)$ is the ratio of generation output and capability

N is the total number of units in the power system

T is the total time under which UC is performed

$P_{g_i}(t)$ is the power output of the ith unit at hour t

$P_{g_i}^{max}$ is the maximum power output of the ith unit

$P_{g_i}^{min}$ is the minimum power output of the ith unit

$S_i(t)$ is the start-up cost of the ith unit at hour t

In the reserve constraints, there are various classifications for reserve and these include units on spinning reserve and units on cold reserve under the conditions of banked boiler or cold start.

Lagrange relaxation is being used regularly to solve UC problems. It is much more beneficial for utilities with a large number of units since the degree of suboptimality goes to zero as the number of units increases. It has also the advantage of being easily modified to model characteristics of specific utilities. It is relatively easy to add unit constraints. The main disadvantage of Lagrangian relaxation is its inherent suboptimality.

$$
L(\lambda,\mu,\nu) = \sum_{t=1}^{T} \sum_{i=1}^{N} [C_i(P_{g_i}(t)) + S_i(x_i(t))]
$$
$$
+ \lambda(t)\left(P_d(t) + P_R(t) - \sum P_{g_i}\right) + \mu(t)\left(P_{g_i}^{max} - P_{g_i}\right), \quad (16.185)
$$

where $\lambda(t)$ and $\mu(t)$ are the multipliers associated with the requirement for time t.

Solution approach using ADP variant for the UC problem
ADP is able to optimize the system over time under conditions of noise and uncertainty. If optimal operation samples are used to train the networks of the ADP, the NN can learn how to commit or adapt the generators and follow the operators' patterns. When load is changed, it can change the operation according to the load changing. Figure 16.6 shows the schematic diagram for implementations of HDP.

The input of the action network is the states of generators and the action is how to adjust the output of generators. The output J presents the cost-to-go function and the task is to minimize the J-function.

In this diagram, the input is the state variable of the network, and it is the cost of generation vector. It can be presented as $X = [C(P_{g_i})]$. And the output is control variables of units, and it is the adjustment of unit

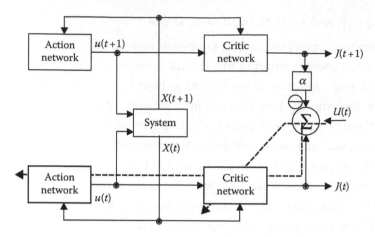

FIGURE 16.6
The scheme of implementation of HDP.

generation, presented as: $u = [\Delta P_g]$. The utility function is local cost, so it is a cost function about unit generation within any time interval. It can be presented as $U = f(P,t)$.

After transposing the power system variables using the guidelines above, the schema of implementation of HDP include the following computations.

The error of the critic network is

$$e_C(t) = \gamma J(t) - J(t+1) - U(t) \qquad (16.186)$$

and the updating weight using:

$$w_C(t+1) = w_C(t) + \Delta w_C(t) \qquad (16.187)$$

and

$$\Delta w_C(t) = \eta_C \left[-\frac{\partial e_C(t)}{\partial w_C(t)} \right], \qquad (16.188)$$

where

$$\frac{\partial E_C}{\partial w_{Cij}^{(1)}} = \frac{\partial E_C}{\partial e_C} \cdot \frac{\partial e_C}{\partial y_{Ck}} \cdot \frac{\partial h_{Ck}}{\partial h'_{Ck}} \cdot \frac{\partial h'_{Cj}}{\partial w_{Cij}^{(1)}} \qquad (16.189)$$

$$= \gamma e_C \cdot \left[\frac{1}{2} \left(1 - h_{Cj}^2 \right) \right] \cdot w_{Cj}^{(2)} x_i \qquad (16.190)$$

$$\frac{\partial E_C}{\partial w_{Cjk}^{(2)}} = \frac{\partial E_C}{\partial e_C} \cdot \frac{\partial e_C}{\partial y_{Ck}} \cdot \frac{\partial y_{Ck}}{\partial w_{Cjk}^{(2)}} = \gamma e_C y_{Ck} \tag{16.191}$$

I is the number of elements in R vector

J is the number of hidden layer node

K is the number of output layer node

M is the number of elements in u (action) vector

h'_C is the hidden layer input nodes

h_C is the hidden layer output nodes

y'_C is the output layer input nodes

y_C is the output layer output nodes

$w^{(1)}{}_C$ is the weights between input and hidden layers

$w^{(2)}{}_C$ is the weights between hidden and output layers

x is the input layer nodes

The error of the action network is computed as

$$e_A(t) = J(t) - U(t) \tag{16.192}$$

and the updating weight is

$$w_A(t+1) = w_A(t) + \Delta w_A(t) \tag{16.193}$$

and

$$\Delta w_A(t) = \eta e_A \left[-\frac{\partial e_A(t)}{\partial w_A(t)} \right], \tag{16.194}$$

where

$$\frac{\partial E_A}{\partial w_{Ajk}^{(2)}} = \frac{\partial E_A}{\partial e_A} \cdot \frac{\partial e_A}{\partial J_k} \cdot \frac{\partial J_k}{\partial y_{Ak}} \cdot \frac{\partial y_{Ak}}{\partial y'_{Ak}} \cdot \frac{\partial y'_{Ak}}{\partial w_{Ajk}^{(2)}} \tag{16.195}$$

$$= \gamma e_A h_{Aj} \cdot \left[\frac{1}{2}\left(1 - h_{Aj}^2\right) \right] \cdot \left[\sum_{j=1}^{J} w_{Cj}^{(2)} \frac{1}{2}\left(1 - h_{Cj}^2\right) w_{Cij}^{(1)} \right] \tag{16.196}$$

$$\frac{\partial E_A}{\partial w_{Aij}^{(1)}} = \frac{\partial E_A}{\partial e_A} \cdot \frac{\partial e_A}{\partial J_k} \cdot \frac{\partial J_k}{\partial y_{Ak}} \cdot \frac{\partial y_{Ak}}{\partial y'_{Ak}} \cdot \frac{\partial y'_{Ak}}{\partial w_{Aij}^{(1)}} \tag{16.197}$$

$$= \gamma e_A w_{Ajk}^{(2)} x_i \cdot \left[\frac{1}{2} \left(1 - h_{Aj}^2 \right) \right] \cdot \left[\frac{1}{2} \left(1 - y_{Ak}^2 \right) \right]$$

$$\cdot \left[\sum_{j=1}^{J} w_{Cj}^{(2)} \frac{1}{2} \left(1 - h_{Cj}^2 \right) w_{Cij}^{(1)} \right]. \tag{16.198}$$

The structure of the NN in HDP is shown in Figure 16.7.

Also, the corresponding calculation steps are as follows:

Step 1: Use the sample data to pretrain the action network. The error is the difference between the output and the real value.

Step 2: Use the sample data to train the critic network with the pretrained and unchanged action network. Use Equations 16.171 through 16.176 to update the weights. Then begin to apply the mature ADP network in the real work.

Step 3: Input the current state data $X(t)$ to the action network.

Step 4: Get the output $u(t)$ of the action network. Input $u(t)$ to the system function to get the state of next time $X(t+1)$.

Step 5: Use the state of next time $X(t+1)$ to get the action of next time $u(t+1)$.

Step 6: Input the action and state of different time $u(t)$, $X(t)$ and $u(t+1)$, $X(t+1)$ to different critic network, respectively, and J-function for different time $J(t)$, $J(t+1)$ are obtained.

Step 7: Backpropagate and update the weights of the critic and action network using Equations 16.171 through 16.183. Then time $t = t+1$. Go to step 3.

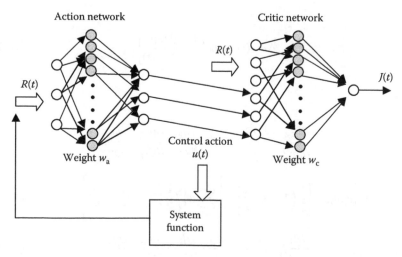

FIGURE 16.7
The structure of the NN in HDP.

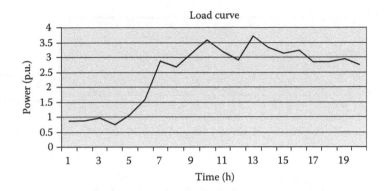

FIGURE 16.8
Load curve of a 3-generator, 6-node system.

Results of ADP computation for the UC problem
Figure 16.8 shows the load duration curve used for this small 5-bus test system. There are three generators in the system and the network parameters and cost function for this simple parameter in this example is given in Ref. [5].

Figure 16.9 shows the control action impact on the *J*-function of output vs. expected *J*-function [*J*]. The closeness of the line graphs indicates that the ADP method generates correct results.

After training, the HDP can give the generation plan, which is very close to the optimal plan. The HDP method can deal with the dynamic process of UC, and easily to get a global optimal solution, which is difficult for classical

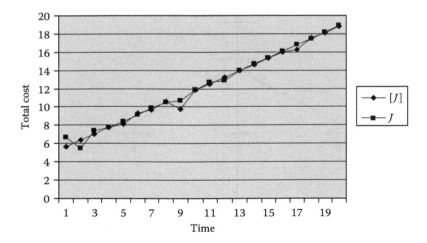

FIGURE 16.9
Comparison of expected [*J*] vs. actual *J*.

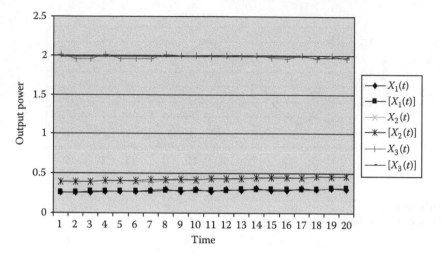

FIGURE 16.10
Generation schedule for the UC problem solved using ADP.

optimization methods. Figure 16.10 shows that generation schedule of three generators system.

In Figure 16.11, X_1, X_2, and X_3 present the output of the three generators respectively, and $[X_1]$, $[X_2]$, and $[X_3]$ present the expected (or say, optimal) output of the three generators, respectively.

UC problem is a large-scale, mixed-integer, and dynamic optimization problem. The ADP method is employed for solving the UC problem over

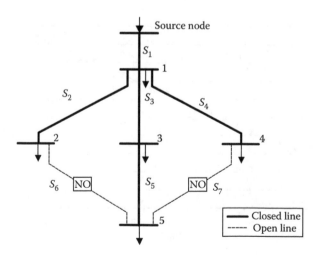

FIGURE 16.11
Small-scale power system for reconfiguration problem.

time and obtains the global optimization solution with the constraints in load dynamics and topology changes.

Case 2: *ADP for optimal network reconfiguration*
Distribution networks are generally configured radially for effective and noncomplicated protection schemes. Under normal operation conditions, distribution feeders may be reconfigured to satisfy objectives of minimum distribution line losses, optimum voltage profile and relieve the overloads in the network. Power system reconfiguration problem has the objectives:

- Minimum distribution line losses
- Optimum voltage profile
- Relieve the overloads in the network

The minimum distribution line loss optimization problem of the reconfigured distribution systems is formulated as follows:

$$\text{Minimize } \sum |z_b i_b| \tag{16.199}$$

$$\text{s.t. } [A]\, i = I \tag{16.200}$$

where
z_b is the impedance of the branch
I_b is the complex current flow in the branch b
i is the m-vector of complex branch currents

A is the $n \times m$ network incidence matrix, whose entries are

$= +1$ if branch b starts from the node p
$= -1$ if the branch b starts from the node b
$= 0$ if the branch is not connected to the node p

m is the total number of the branches
n is the total number of network nodes
I is the n-vector of complex nodal injection currents

The illustrative example problem is solved by using integer interior point method presented in Ref. [5], here the ADP method for the 5-bus system shown below in Figure 16.11 is utilized.

It involves the development of a framework of ADP which involves (a) action network, (b) critic network, and (c) the plant model, as shown in Figure 16.12 for network distribution reconfiguration.

The algorithm to solve this problem using ADP is presented in Figure 16.13.

In order to solve the optimal distribution reconfiguration problem by using the ADP algorithm, we need to model and specify each part of the system

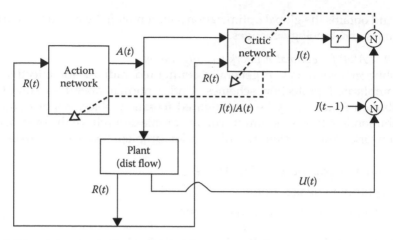

FIGURE 16.12
ADP structure for the network reconfiguration problem.

structure shown in Figure 16.13. There are four major parts in the system structure: action vectors, state vectors, immediate rewards, and the plant. The system is tested with a 5-bus and a 32-bus system. We discuss the different parts of the ADP implementation structure as follows.

Rewards (*utility function*) ~ Optimal reconfiguration involves selection of the best set of branches to be opened, one from each loop, such that the resulting radial distribution systems has the desired performance. Amongst the several performance criteria considered for optimal network reconfiguration, the one selected is the minimization of real power losses. Application of the ADP to optimal reconfiguration of radial distribution systems is linked to the choice of an immediate reward U, such that the iterative value of J is minimized, while the minimization of total power losses is satisfied over the whole planning period. Thus, we compute the immediate reward as

$$U = -Total\ losses \tag{16.201}$$

Action vectors ~ If each control variable A_i is discretized in d_{u_i} levels (e.g., branches to be opened one at each loop of radial distribution systems), the total number of action vectors affecting the load flow is

$$A = \prod_{i=1}^{m} d_{u_i} \tag{16.202}$$

Here, m expresses the total number of control variables (e.g., total number of branches to be switched out).

The control variables comprise the sets of branches to be opened, one from each loop. From the network above, we can easily deduce from the simple

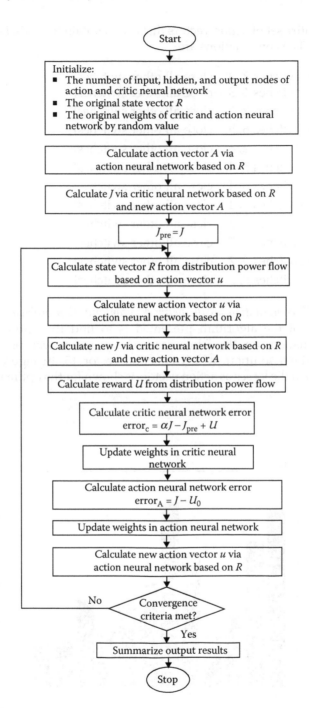

FIGURE 16.13
Flowchart for ADP-based optimal reconfiguration strategy.

system the entire set of action vectors that can maintain the radial structure of the network. The combinations are

A_1: {open switches 2, 3   close all other switches}
A_2: {open switches 6, 3   close all other switches}
A_3: {open switches 2, 5   close all other switches}
A_4: {open switches 6, 5   close all other switches}
A_5: {open switches 2, 4   close all other switches}
A_6: {open switches 3, 4   close all other switches}
A_7: {open switches 6, 4   close all other switches}
A_8: {open switches 5, 4   close all other switches}
A_9: {open switches 2, 7   close all other switches}
A_{10}: {open switches 3, 7   close all other switches}
A_{11}: {open switches 6, 7   close all other switches}
A_{12}: {open switches 5, 7   close all other switches}

Results of ADP computation for the network reconfiguration problem
The purpose of the algorithm presented is to find the optimal switches status combination, for the 5-bus case. The program was used to determine the optimal solution, which is action vector 15. In Figure 16.14, the minimization of the losses as action vectors is shown for the optimal switching sequence.

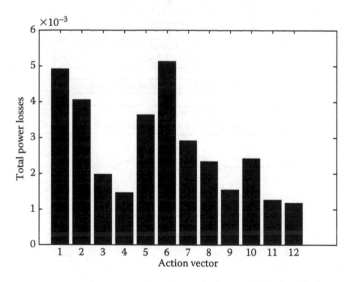

FIGURE 16.14
Action vector performance during system training.

After the initialization, the action network generates the first action vector by random number, the action vector then inputs into the critic vector with state variables. With the output of critic network J and immediate cost U, the new error for action and critic network could be obtained. The weights in those two networks then can be updated based on BP rules. After sufficient iterations, the system will output the result. In our case, it is the optimal action vector, which is the best switches status combination with the minimum losses.

Optimal training of the weights of ADP action vectors was obtained and used to minimize the losses in the reconfigured network. We recommend extending this study to large-scale aerospace power system while addressing the multiobjective challenges of restoration, reconfiguration, and remedial control.

References

1. D. V. Prokhorov and D. C. Wunsch, Adaptive critic designs, *IEEE Transaction on Neural Network*, 8, 997–1007, Sep. 1997.
2. J. Si and Y. Wang, On-line learning control by association and reinforcement, *IEEE Transaction on Neural Networks*, 12, 264–276, Mar. 2001.
3. J. Si, A. G. Barto, W. B. Powell, and D. C. Wunsch, *Handbook of Learning and Approximate Dynamic Programming*, pp. 561–596. IEEE Press Series on Computational Intelligence, 2004, August 2004, Wiley-IEEE Press.
4. G. K. Venayagamoorthy, R. G. Harley, and D. C. Wunsch, Dual heuristic programming excitation neuro-control for generators in a multi-machine power system, *IEEE Transaction on Industry Applications*, 39, 382–394, Mar./Apr. 2003.
5. J. A. Momoh and Y. Zhang, Unit commitment using adaptive dynamic programming (ADP), *13th International Conference on Intelligent Systems Application to Power Systems (ISAP)*, Arlington, Virginia, Nov. 6–10, 2005.
6. J. M. Lee and J. H. Lee, Approximate dynamic programming strategies and their applicability for process control: A review and future directions, *International Journal of Control, Automation, and Systems*, 2(3), 263–278, Sep. 2004.
7. A. L. Motto and F. D. Galiana, Unit commitment with dual variable constraints, *IEEE Transactions on Power Systems*, 19(1), 330–338, Feb. 2004.
8. P. Werbos, Optimization methods for brain-like intelligent control, *Proceedings of the 34th IEEE Conference on Decision and Control*, New Orleans, LA, 1995
9. J. A. Momoh, Towards dynamic stochastic optimal power flow, in J. Si, A. G. Barto, W. B. Powell, and D. Wunsch (Eds.), *Handbook of Learning and Approximate Dynamic Programming*, pp. 561–598. Wiley, New York, 2004.
10. R. F., Stengel, *Optimal Control and Estimation*, Dover Publications, Inc., Mineola, New York, 1986.
11. R. L. Welch and G. K. Venayagamoorthy, HDP based optimal control of a grid independent PV System, *IEEE Power Engineering Society (PES) General Meeting*, Montreal, Canada, June 18–22, 2006.

12. P. J. Werbos, A brain-like design to learn optimal decision strategies in complex environments, in M. Karny, K. Warwick, and V. Kurkova (Eds.), *Dealing with Complexity: A Neural Networks Approach.* Springer, London, 1998.
13. P. J. Werbos, *Roots of Backpropagation.* Wiley, New York, 1994.
14. H. T. Chieh, *Applied Optimization Theory and Optimal Control.* Feng Chia University, Taiwan Press, Taiwan, 1990.
15. D. E. Krik, *Optimal Control Theory.* Prentice Hall Inc., Englewood Cliffs, NJ, 1970.

Index

A

Action-dependent dual heuristic programming, 570–572
Action-dependent heuristic dynamic programming, 570–572
Active power, definition of, 19
Active power generations, 401
   cost minimization, 389
Adaptation process, goal of, 581
Adaptive critics design
   architecture of
      action network, 575–576
      critic networks, 574
      direct neural dynamic programming, 573
   characteristics of, 576
   complex intelligent networks, 565–567
   definition of, 569
   neural network (NN) designs, 565
   stochastic model, 568
   users of, 577
Adaptive dynamic critics (ADC), 565
ADDHP, see Action-dependent dual heuristic programming
ADHDP, see Action-dependent heuristic dynamic programming
ADP, see Approximate dynamic programming
Affine-scaling methods, 197, 200
AGC, see Automatic generation control
Analog-to-digital conversion, 369
ANNs, see Artificial neural networks
Apparent power
   equivalent definition of, 20
   formula for, 19
   sinusoidal waveforms, 18
   three-phase systems, 26
Approximate dynamic programming, 539
   action and critic networks in, 568
   action vector performance during system training, 592

architectures of
   DHP action NN adaptation, 580–581
   DHP critic NN adaptation, 580
   error minimization, 579
   HDP critic adaptation, 577–578
   HDP critic NN adaptation and learning, 578
computational intelligence technique, 567
critic network variants, 569–572
from DP, 567–569
dual heuristic programming (DHP), 569
globalized dual heuristic programming (GDHP), 569
heuristic dynamic programming (HDP), 569
network reconfiguration problem, computation for, 590, 592
and optimal control, features of, 566
optimal distribution reconfiguration problem, 589
optimal reconfiguration strategy, flowchart for, 591
types of
   dual heuristic programming, 569–570
   globalized dual heuristic programming (GDHP), 569, 571
   heuristic dynamic programming, 569–571
UC problem solving, generation schedule for, 588
variants, features of, 572
Armature heating, 31
Artificial neural networks, 264, 511, 566
Augmented Lagrangian function, 417, 420
Automatic control systems, 274
Automatic differentiation software, 416
Automatic generation control